Statistical Thermodynamics and Microscale Thermophysics

Many of the exciting new developments in microscale engineering are based on the application of traditional principles of statistical thermodynamics. This book offers a modern view of thermodynamics, interweaving classical and statistical thermodynamic principles and applying them to current engineering systems. It begins with coverage of microscale energy storage mechanisms from a quantum mechanics perspective and then develops the fundamental elements of classical and statistical thermodynamics. Next, applications of equilibrium statistical thermodynamics to solid, liquid, and gas phase systems are discussed. The remainder of the book is devoted to nonequilibrium thermodynamics of transport phenomena and an introduction to nonequilibrium effects and noncontinuum behavior at the microscale.

Although the text emphasizes mathematical development, it includes many examples and exercises that illustrate how the theoretical concepts are applied to systems of scientific and engineering interest. It offers a fresh view of statistical thermodynamics for advanced undergraduate and graduate students, as well as practitioners, in mechanical, chemical, and materials engineering.

Van P. Carey is a Professor in the Mechanical Engineering Department at the University of California, Berkeley. The main focus of his research is development of advanced computational models of microscale thermophysics and transport in multiphase systems.

Statistical Thermodynamics and Microscale Thermophysics

VAN P. CAREY

University of California, Berkeley

CAMBRIDGE
UNIVERSITY PRESS

PUBLISHED BY THE PRESS SYNDICATE OF THE UNIVERSITY OF CAMBRIDGE
The Pitt Building, Trumpington Street, Cambridge, United Kingdom

CAMBRIDGE UNIVERSITY PRESS
The Edinburgh Building, Cambridge CB2 2RU, UK   http://www.cup.cam.ac.uk
40 West 20th Street, New York, NY 10011-4211, USA   http://www.cup.org
10 Stamford Road, Oakleigh, Melbourne 3166, Australia

First published 1999

Typset in Times Roman 10/12 pt. in LATEX 2_ε [TB]

*A catalog record for this book is available from
the British Library.*

Library of Congress Cataloging-in-Publication Data

Carey, V. P. (Van P.)
    Statistical thermodynamics and microscale thermophysics / Van P.
Carey.
        p.   cm.
      ISBN 0-521-65277-4 (hb). – ISBN 0-521-65420-3 (pb)
      1. Statistical thermodynamics.   2. Thermodynamics.   I. Title.
    QC311.5.C36   1999
    621.402′1 – dc21   98-45449
                                                                            CIP

ISBN 0 521 65277 4 hardback
ISBN 0 521 65420 3 paperback

Transferred to digital printing 2004

To Lee F. Carey, Bart Conta, and Dennis G. Shepherd,
three of the best applied thermodynamicists I ever met

Contents

Nomenclature

a	normalized bulk velocity, $= w_0/(2N_A k_B/\bar{M})^{1/2}$
a_R	Redlich–Kwong constant
a_v	van der Waals constant
$A^{\Sigma*}$	interface area
b_R	Redlich–Kwong constant
b_v	van der Waals constant
B	second virial coefficient
c	particle speed
$\hat{c}_P$	molar specific heat at constant pressure
$\hat{c}_V$	molar specific heat at constant volume
C	third virial coefficient
e_e	electron charge
E	energy
E_x	electric field in the x direction
f	fugacity
$f(\varepsilon)$	average occupancy of a microstate with energy ε
$f(\mathbf{w}, \mathbf{z}, t)$	fractional particle velocity distribution
f_{ep}	distribution function for an ensemble of particles
f_{es}	ensemble number density
f_{pv}	particle velocity distribution
$f_{>c}$	fraction of molecules with speeds greater than c
F	Helmholtz free energy
g	degeneracy
$g(r)$	radial distribution function
$g(\nu)$	frequency distribution function
$g^*(\varepsilon)$	distribution function for electron quantum states
g_e	gravitational acceleration
$\hat{g}$	molar Gibbs function
g_i	degeneracy of energy level i
G	Gibbs function
h	Planck's constant
$\hat{h}$	molar specific enthalpy
$\hat{h}_l$	molar specific enthalpy of saturated liquid
$\hat{h}_v$	molar specific enthalpy of saturated vapor
$\hbar$	$= h/2\pi$
H	enthalpy
$\hat{H}$	Hamiltonian
I	moment of inertia
j	number flux of molecules
j_m	mass flux of molecules

$J_{a,x}$	flux of species a in the x direction
$J_{S,x}$	flux of entropy in the x direction
$J_{U,x}$	flux of energy in the x direction
$\mathbf{J}_a$	flux of species a
$\mathbf{J}_S$	flux of entropy
$\mathbf{J}_U$	flux of energy
k_B	Boltzmann constant
k_t	thermal conductivity
k_x	force constant for linear restoring force
K	wavenumber, $= 2\pi/\lambda$
K_c	equilibrium constant
K_p	equilibrium constant
L	characteristic system dimension
L_{ij}	kinetic coefficients
L_{ms}	mean spacing between particles
$L_{\nabla T}$	temperature gradient length scale, $= T/\nabla T$
Lz	Lorentz number, $= k_t/\sigma_e T$
m	mass per particle
M	mass per molecule for polyatomic molecules
M	molecular mass
n_j	number of ensemble members in energy level ε_j
$\langle n_{l'} \rangle$	mean number of particles in microstate l'
N	number of particles
N_a	number of species a particles
$N_{a,i}$	number of species a particles in energy level i
N_A	Avogadro's number
$\tilde{N}_a$	number of species a particles per unit volume
N_a'	instantaneous number of species a particles
p	momentum
p_i	generalized momentum
p_x	momentum in the x direction
$\tilde{p}$	$= P_r - 1$
$\tilde{p}$	fluctuation probability density function
$\hat{p}_j$	generalized momentum
P	pressure
P_a	partial pressure for species a
P_c	critical pressure
P_r	reduced pressure, $= P/P_c$
P_{sat}	equilibrium saturation pressure
$\tilde{P}$	probability
q	molecular partition function
q_a	molecular partition function for particle species a
q_e	partition function for electronic energy storage
q_i	generalized coordinate
q_{int}	partition function for internal energy storage
q_{nucl}	partition function for nuclear energy storage
q_{rot}	partition function for rotational energy storage

$q_{\text{rot, nucl}}$	partition function for rotational and nuclear energy storage
q_{tr}	partition function for translational energy storage
q_{vib}	partition function for vibrational energy storage
$\hat{q}_j$	generalized coordinate
Q	canonical partition function
r_{e}	equilibrium bubble or droplet radius
R	universal gas constant, $= N_A k_B$
R_{H}	Rydberg constant
$\hat{s}$	molar entropy
$\hat{s}_{\text{f}}$	molar internal energy for saturated liquid
$\hat{s}_{\text{g}}$	molar entropy for saturated vapor
S	system entropy
$\tilde{S}$	entropy per unit volume
S'	instantaneous entropy
$S_{\text{e}}^{\Sigma*}$	surface excess entropy
$\tilde{t}$	$= T_{\text{r}} - 1$
T	temperature
T_{c}	critical temperature
T_{m}	equilibrium melting temperature
T_m	temperature in phase m
T_{r}	reduced temperature $= T/T_{\text{c}}$
T_{sat}	equilibrium saturation temperature
u	velocity in the x direction
$\hat{u}$	molar specific internal energy
$\hat{u}_{\text{f}}$	molar specific internal energy for saturated liquid
$\hat{u}_{\text{g}}$	molar specific internal energy for saturated vapor
U	system internal energy
$\tilde{U}$	internal energy per unit volume
U'	instantaneous internal energy
$\hat{U}$	potential function in quantum system
$U_{\text{e}}^{\Sigma*}$	surface excess internal energy
v	velocity in the y direction
v	mass specific volume
v_{g}	group velocity
$\hat{v}$	molar specific volume
$\hat{v}_{\text{c}}$	critical molar specific volume
$\hat{v}_{\text{r}}$	reduced specific volume, $= \hat{v}/\hat{v}_{\text{c}}$
$\tilde{v}$	$= \hat{v}_{\text{r}} - 1$
V	system volume
w	velocity in the z direction
w_0	bulk velocity normal to interface
W	number of microstates corresponding to a particular macrostate
x	Cartesian spatial coordinate
X_i	mole fraction of species i
y	Cartesian spatial coordinate
z	Cartesian spatial coordinate
Z	compressibility factor, $= P\hat{v}/RT$

Z	partition function for the microcanonical ensemble
Z_a	partition function for the microcanonical ensemble for species a
Z_N	configuration integral for single-species system
Z_{N_a,N_b}	configuration integral for a binary mixture

Greek

β	$= 1/k_B T$
β_T	coefficient of thermal expansion
γ	analysis parameter
γ	ratio of specific heats, $= \hat{c}_p / \hat{c}_v$
γ_a	activity coefficient for species a
Γ_a	bulk motion correction factor for $a > 0$
Γ_{-a}	bulk motion correction factor for $a < 0$
$\Gamma_e^{\Sigma^*}$	surface excess mass
δ	liquid film thickness
δQ	differential heat interaction (positive into system)
δW	differential work interaction (positive out of system)
$\Delta \hat{h}_{lv}$	molar latent heat of vaporization, $= \hat{h}_v - \hat{h}_l$
ε	energy
ε_i	energy level i
$\varepsilon_{i'}$	energy microstate i'
ε_F	Fermi energy
ε_0	characteristic energy in the Lennard–Jones 6–12 pair potential
$\hat{\varepsilon}_A$	absolute Seebeck coefficient of material A
$\hat{\varepsilon}_{AB}$	thermoelectric power for a thermocouple of materials A and B
θ	angular coordinate in spherical polar coordinates
θ_D	Debye temperature, $= h\nu_D / k_B$
θ_E	Einstein temperature, $= h\nu_E / k_B$
θ_F	Fermi temperature, $= \mu_0 / k_B$
θ_{rot}	rotational temperature, $= h^2 / (8\pi^2 I k_B)$
$\theta_{rot,m}$	$= \theta_{rot}$ for linear molecules, $= (\theta_{rot,A}\theta_{rot,B}\theta_{rot,C})^{1/3}$ for nonlinear polyatomic molecules
θ_{vib}	vibrational temperature, $= h\nu / k_B$
κ_T	isothermal compressibility
λ	wavelength
λ_c	mean free path between collisions
Λ	thermal de Broglie wavelength
Λ	lattice wave amplitude
μ_a	chemical potential (per molecule) for species a
μ_0	Fermi energy
$\hat{\mu}_a$	molar chemical potential for species a, $= \mu_a N_A$
μ_m	absolute viscosity
ν	frequency
ν_D	Debye cutoff frequency
ν_E	Einstein characteristic frequency
$\bar{\nu}_c$	mean collision frequency

ξ	total number of translational and rotational energy storage modes
Ξ	grand canonical partition function
$\hat{\pi}_{AB}$	Peltier coefficient for a junction of material A and material B
ρ_N	number density of particles
σ_c	collision cross section
σ_e	electrical conductivity
σ_i	interfacial tension
σ_{LJ}	characteristic distance in the Lennard–Jones 6–12 pair potential
σ_{lg}	interfacial tension at a liquid–gas interface
σ_s	symmetry number
σ_x	standard deviation in variable x
σ_x^2	variance of variable x
$\hat{\sigma}$	accommodation coefficient
τ	time interval
τ_r	relaxation time
$\hat{\tau}_A$	Thomson coefficient for material A
$\langle \tau_c \rangle$	mean time between collisions
ϕ	intermolecular pair potential
ϕ	electric potential
ϕ	angular coorinate in spherical polar coordinates
ϕ_a	fugacity coefficient for species a
Φ_N	potential energy in system containing N particles
ψ	time-independent wave function
Ψ	wave function
ω	angular frequency, $= 2\pi\nu$
ω	acentric factor
Ω	total number of system microstates

Preface

The structure of this book is designed to facilitate coherent development of classical and statistical thermodynamic principles. The book begins with coverage of microscale energy storage mechanisms from a modern quantum mechanics perspective. This information is then incorporated into a statistical thermodynamics analysis of many-particle systems with fixed internal energy, volume, and number of particles. From this analysis emerges the definitions of entropy and temperature, the extremum principle form of the second law, and the fundamental relation for the system properties. The third chapter takes the concepts derived from the statistical treatment and uses mathematical techniques to expand the macroscopic thermodynamics framework. By the end of the third chapter, the full framework of classical thermodynamics is established, including definitions of all commonly used thermodynamic properties, relations among properties, different forms of the second law, and the Maxwell relations.

In the fourth chapter, statistical ensemble theory is covered, building on the initial statistical treatment in Chapter 2 and the expanded macroscopic framework developed in Chapter 3. The canonical ensemble and grand canonical ensemble formalisms are developed, and the relations developed from these formalisms are used to explore the significance of fluctuations in thermodynamics systems. By the end of the fourth chapter all the fundamental elements of classical and statistical thermodynamics have been established. Chapters 5–7 deal with applications of equilibrium statistical thermodynamics to solid, liquid, and gas phase systems.

The final three chapters of the text cover thermal phenomena that involve nonequilibrium and/or noncontinuum effects. Throughout these chapters, implications of the statistical thermodynamics theory based on microscale characteristics of the system are considered in tandem with predictions of classical thermodynamics. Chapter 8 presents an integrated treatment of multiphase systems, which includes classical aspects of phase transitions and phase equilibria as well as nonequilibrium features associated with metastable states and intrinsic stability limits. It draws upon the van der Waals model to illustrate how the fundamental equation together with necessary conditions for equilibrium define the saturation conditions and the limits of intrinsic stability for a pure fluid or a binary mixture. In this chapter, thermodynamic similitude and near critical behavior are also examined from statistical and classical thermodynamic viewpoints.

Chapter 9 covers nonequilibrium thermodynamics of transport phenomena. Here again, classical as well as microscale statistical perspectives are considered. The relations among fluxes and potentials are obtained from classical theory, while the key Onsager reciprocity relations are obtained by considering the statistics of microscale fluctuations. Applications involving thermoelectric effects are discussed in detail. Chapter 10 uses the tools of statistical and classical thermodynamics developed earlier in the text to analyze nonequilibrium effects and noncontinuum behavior at the microscale. There are many important examples in which these effects are important. The intent in Chapter 10 is to introduce these effects by exploring a limited number of important examples. Kinetic theory and the Boltzmann transport equation are developed because they are commonly used as the basis for analysis of nonequilibrium or noncontinuum systems. Nonequilibrium and noncontinuum aspects of multiphase systems and electron transport in solids are examined in detail. The final section of Chapter 10 examines length scales and time scales at which classical and continuum theories become suspect. This defines the range of conditions for which classical and continuum theories are expected to be accurate. Although limited in its coverage, Chapter 10 is intended to provide an introduction to microscale aspects of nonequilibrium and noncontinuum phenomena and to illustrate how they relate to the theoretical framework developed in the preceding chapters.

Many aspects of microscale thermophysics can be fully understood only after a basic understanding of statistical thermodynamics, nonequilibrium thermodynamics, and kinetic theory is attained. Chapter 10 is therefore a fitting capstone to this text. It brings the presentation full circle in the sense that the treatment begins with microscale energy storage, develops equilibrium classical and statistical thermodynamics theory for macroscopic systems, and then returns to microscale systems to explore the limitations of macroscopic theory and to examine how tools from statistical thermodynamics and nonequilibrium thermodynamics can be used to analyze microscale thermophysics and transport phenomena.

Virtually all other statistical thermodynamics texts develop statistical thermodynamics for a system containing one species of particle. This makes it difficult to expand the analytical framework to handle systems that contain a mixture of different particle or molecular species. In this text, a binary mixture of two species of particles or molecules is considered throughout the analytical development. By so doing, the pure system behavior is captured by setting the number of particles of one species to zero, and the interactions between different particles in a binary mixture are indicated directly. Furthermore, the symmetry of the terms in the resulting relations for the two species generally makes it clear how the equations should be extended to handle a mixture of three or more molecular species. This presentation should better prepare engineers for the increasing number of important applications involving multicomponent mixtures.

Although the text emphasizes mathematical development, it includes examples in each chapter and exercises at the end of each chapter that illustrate how the theoretical concepts are applied to systems of scientific and engineering interest. An effort has been made to include examples and problems that span a wide variety of modern applications. Some of the exercises ask students to develop and run computer programs to tackle more challenging problems. Inclusion of these problems and examples is motivated by the author's belief that a full understanding of these concepts is best developed by working with them directly. Extensive use of the van der Waals model is made in the examples throughout the text. This is done for two reasons. First, the van der Waals model can be related back to molecular characteristics through the statistical thermodynamics framework. This connection illustrates how relations among properties are connected to properties of the molecules themselves. Second, although very idealized, the van der Waals model exhibits most of the features of real fluids, including the existence of a critical point and first-order phase transitions. The examples demonstrate how a model of this type can be used with the results of statistical and classical thermodynamics theory for quantitative engineering analysis.

The topic sequence here is somewhat unconventional for a statistical thermodynamics text. I have therefore included a brief introductory remarks section at the beginning of each chapter of the text to provide the reader with an overview of the contents of the chapter and a sense of how that chapter fits into the flow of ideas in the text.

The structure of the book makes it suitable as a primary text for a graduate-level course in advanced engineering thermodynamics or as a source of foundation material in an advanced specialty course on microscale transport phenomena. For a graduate course on advanced engineering thermodynamics, the first seven chapters would be the primary source material for the course and some special topics from the remainder of the text could be included. For an advanced graduate course on microscale transport phenomena, the material in Chapters 1–7 could be reviewed briefly and the material in Chapters 8–10 could provide source material for an introduction to the thermodynamics of microscale systems and microscale transport phenomena. The foundation thus provided could then be followed by an in-depth exploration of microscale transport phenomena in different system types.

Thanks are due to reviewers of the early manuscript versions of this text: Richard Buckius of the University of Illinois, Edward McAssey, Jr. of Villanova University, and Donald Fenton of Kansas State University. Their many valuable criticisms and suggestions are greatly appreciated.

Quantum Mechanics and Energy Storage in Particles

This text differs from most statistical thermodynamics textbooks in that it does not deal exclusively with statistical aspects of thermodynamics. Instead, it attempts to weave together statistical and classical elements to develop the full theoretical framework of thermodynamics. Chapter 1 begins this development by establishing the basic features of energy storage at the atomic and molecular levels. It contains a very short introduction to basic aspects of quantum mechanics. The quantum models discussed in this chapter are models of energy storage modes found in common molecules. Conclusions regarding energy levels and their degeneracy for these modes of energy storage are cornerstones of the statistical thermodynamic theory developed in later chapters.

1.1   Microscale Energy Storage

Since this text is designed for graduate-level engineering instruction, it is likely that the reader has already encountered some elements of thermodynamics in previous courses and very likely that he or she has some idea of the usefulness of thermodynamic analysis for systems of scientific and technological interest. Mechanical or chemical engineers who are thoroughly versed in classical equilibrium thermodynamics may wonder what a statistical development of thermodynamics has to offer beyond the tools provided by classical thermodynamics.

In response to such an inquiry, we can identify two main benefits of developing a statistical thermodynamic theory. First, by design, statistical thermodynamics theory provides a link between macroscopic "classical" thermodynamic analysis of system behavior and the microscopic characteristics of the atoms, molecules, or subatomic particles that make up the system. To better understand the nature of establishing such a link, consider the length-scale and time-scale ranges shown in Figure 1.1. The coordinate axes in this diagram span just about the entire range of length and times scales associated with systems and processes in the known universe. The shortest length scales are associated with the interior structure of atoms and the dimensions of subatomic particles, while the largest length scales are associated with intergalactic distances.

In a similar manner, the shortest time scales are associated with microscale interactions of atoms and/or fundamental particles, and the longest is the age of the universe. Relativity theory requires that matter and energy cannot move faster than the speed of light. Since natural processes generally are the result of the motion of matter or photons, processes having very large length scales and very small times scales cannot be attained. Hence the upper left region of the diagram is in some sense a forbidden zone. Statistical thermodynamics is conceptually important because it provides the theoretical link between microscale interactions and systems undergoing processes at much larger length and time scales.

The second benefit of developing statistical thermodynamic theory is that it provides analytical tools for engineering and scientific analysis. With increasing frequency, engineers and scientists are seeking to work with systems whose behavior lies outside the bounds of

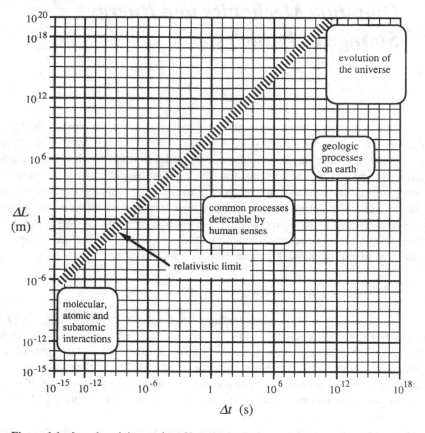

Figure 1.1 Length and time scales of known processes.

classical thermodynamic analysis. The areas of microelectro-mechanical systems (MEMS), fusion reactor development, and ultrafast laser processing of materials are just three examples of engineering applications in which important phenomena are beyond the bounds of classical thermodynamics theory. Engineers specializing in the thermosciences will need tools from statistical thermodynamics and kinetic theory to deal with these and other new applications areas.

In developing the theoretical foundation of thermodynamics, we first seek to define how energy is stored in systems composed of large numbers of particles. In this text, the term *particles* is meant to include subatomic particles, atoms, or molecules. As a first step in dealing with such systems, we must first establish how individual particles themselves store energy. Some insight into this issue can be obtained by considering the simple models of atomic and molecular structures shown in Figure 1.2. In general, such particles may store energy as translational energy or by modifying their internal state. Internal energy storage is directly linked to the internal structure of the particle. In a single atom, internal energy storage or release is accompanied by changes in nuclear or electronic energy levels. Molecules composed of two or more atoms may also store energy internally by altering the nuclear state or electron energy levels of one or more of the atoms. In addition, molecules may store energy in kinetic energy of rotation about the molecule's center of mass and in

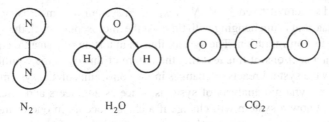

Figure 1.2   Models of typical molecular structures.

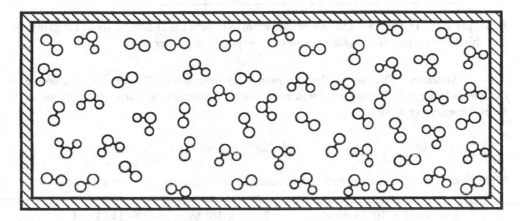

Figure 1.3   A typical system of molecules.

vibration of the molecule. The latter mode of energy storage results when atoms oscillate about their equilibrium positions in the molecular structure.

Our ultimate goal is to develop analytical tools that can be used to analyze a system of particles, like that shown schematically in Figure 1.3. The particles shown in the figure are molecules. The system contains a mixture of diatomic molecules and molecules containing three atoms. Note that translation of the molecules leads to collisions that cause energy to be exchanged among the different modes of energy storage. At the microscopic level, the distribution of energy within each molecule and within the system is constantly changing with time.

In analyzing systems like that shown in Figure 1.3, we will assume that the mass of each species of particle and energy are conserved at both the microscopic and macroscopic level. Conservation of energy is a key foundation element in both classical and statistical thermodynamics. For that reason, conservation of energy is generally referred to as the *first law of thermodynamics*. The volume V and the number of each species of particle in the system $N_a, N_b, \ldots$ are taken to be intrinsic properties of any system. We will assume that they can be measured if necessary and, if desired, we can specify their values.

The *internal energy* of the system, U, is defined as the total energy of the particles in the system, including the translational and internal energy, and any energy stored in force interactions between the particles. If we can isolate the system so no energy crosses its boundary, the first law requires that the internal energy is constant, independent of time, even though particle collisions may alter the way in which the energy is stored within the system. For an isolated system at fixed $U, V, N_a, N_b, \ldots$, a system will macroscopically attain an equilibrium state that is unchanging with time. At the microscopic level, the system is constantly changing its configuration. The system thus is in a state of dynamic equilibrium. We will see in later sections of this text that the nature of this dynamic equilibrium strongly influences how the system reacts to changes in the constraints on the system. This issue is central to thermodynamic analysis of systems, since as engineers and scientists, we often want to predict how a system will change if a change occurs in one or more of the macroscopic constraints on the system (such as the specified values of U, V, or particle numbers).

Example 1.1   The mean speed of a helium atom in helium gas at 27°C is about 1,020 m/s. Use this information to estimate the internal energy of the gas per kmol.

Solution   The mass per helium atom m is known to be 6.65×10^{-27} kg. Assuming the internal energy is due only to kinetic energy of the molecules, we can estimate the mean kinetic energy per atom as

$$e_k = \tfrac{1}{2}mc^2 = \left(\tfrac{1}{2}\right) 6.65 \times 10^{-27}(1{,}020)^2 = 3.46 \times 10^{-21} \text{ J}.$$

Multiplying by Avogadro's number, we obtain the internal energy per kmol:

$$\hat{u} = (6.02 \times 10^{26})3.46 \times 10^{-21} = 2.082 \times 10^6 \text{ J/kmol} = 2{,}082 \text{ kJ/kmol}$$

Pioneering efforts to construct statistical theories of the behavior of large collections of particles were made by Maxwell, Boltzmann, and Gibbs in the last half of the nineteenth century. Because these efforts preceded the development of quantum theory, simple classical mechanics treatments of molecule energy storage mechanisms were incorporated into these initial statistical mechanics models. In fact, the statistical mechanics models of Maxwell and Boltzmann were highly controversial when they were first proposed because the existence of atoms and molecules was not widely accepted by scientists at that time. The work of Planck, Einstein, and others in the early part of the twentieth century established the existence of atoms, molecules, and eventually the subatomic structure of matter. Quantum theory provided the means for predicting the manner in which energy is stored in atoms, molecules, and subatomic particles. This made it possible to extend the statistical mechanics concepts developed by Boltzmann and Gibbs to a wide variety of system types.

In this text, the current knowledge of the atomic structure of matter is taken as the starting point for development of macroscopic thermodynamics. Macroscopic equilibrium thermodynamics is a consequence of the nature of microscale energy storage and the statistical behavior of very large numbers of particles. We will therefore first establish the basic elements of quantum theory necessary to determine the rules of energy storage and energy exchange in atoms, molecules, and subatomic particles. We will then construct a statistical analysis of a system composed of a large number of particles that obey such rules. Once these two components are blended together to form a statistical thermodynamic framework, we will explore the application of the framework to a variety of systems.

1.2   A Review of Classical Mechanics

Classical mechanics analysis of the motion and mechanical energy storage of a particle can be cast in at least three ways. The Newtonian formulation of classical mechanics can be stated as

$$\frac{d\mathbf{p}}{dt} = \mathbf{F}, \tag{1.1}$$

where $\mathbf{p}$ is the particle momentum and $\mathbf{F}$ is the force exerted on the particle.

The Lagrangian formulation of classical mechanics is stated in terms of the *Lagrangian* L defined as

$$L(x, y, z, u, v, w) \equiv K(u, v, w) - \hat{U}(x, y, z), \tag{1.2}$$

where $\hat{U}(x, y, z)$ is the potential energy of the particle and K is the kinetic energy of the particle given by

$$K(u, v, w) = \frac{m}{2}[u^2 + v^2 + w^2]. \tag{1.3}$$

In the above relation, m is the mass of the particle. Noting, for the x direction, that

$$\frac{\partial L}{\partial u} = \frac{\partial K}{\partial u} = mu = p_x, \qquad \frac{\partial L}{\partial x} = -\frac{\partial \hat{U}}{\partial x} = F_x, \tag{1.4}$$

Newton's equation can be written as

$$\frac{d}{dt}\left(\frac{\partial L}{\partial u}\right) = \frac{\partial L}{\partial x}. \tag{1.5}$$

If we generalize the notation as

$$x = q_1, \quad y = q_2, \quad z = q_3, \qquad u = \dot{q}_1, \quad v = \dot{q}_2, \quad w = \dot{q}_3, \tag{1.6}$$

the equations of motion can then be written as

$$\frac{d}{dt}\left(\frac{\partial L}{\partial \dot{q}_j}\right) = \frac{\partial L}{\partial q_j}, \qquad j = 1, 2, 3. \tag{1.7}$$

An extraordinary and very useful property of Lagrange's equations of motion is that they have the same form in any coordinate system (useful since it is sometimes easier mathematically to define $\hat{U}$ in a particular coordinate system). A set of equations like (1.7) can be written for each particle in a multiparticle system.

A third formulation of classical mechanics is the Hamiltonian formulation. To construct the Hamiltonian formulation of classical mechanics, we begin by defining generalized momenta as

$$p_j = \frac{\partial L}{\partial \dot{q}_j}, \qquad j = 1, 2, \ldots, 3N \quad \text{(for a system of } N \text{ particles)}. \tag{1.8}$$

Each generalized momentum p_j is said to be conjugate to coordinate q_j. Since each momentum is linearly proportional to the velocity $\dot{q}_j$, it is a simple task to replace any velocity terms in the formulation with the corresponding generalized momentum p_j. We do so throughout the definition of the Lagrangian and define the *Hamiltonian H* for a system of N particles to be

$$\hat{H}(p_1, p_2, \ldots, p_{3N}, q_1, q_2, \ldots, q_{3N}) = \sum_{j=1}^{3N} p_j \dot{q}_j - L(p_1, p_2, \ldots, p_{3N}, q_1, q_2, \ldots, q_{3N}).$$

$$(1.9)$$

In general, for systems of particles we expect that the total kinetic energy of the particles is given by

$$K = \sum_{j=1}^{3N} a_j(q_1, q_2, q_3, \ldots, q_{3N}) \dot{q}_j^2,$$

$$(1.10)$$

and by definition

$$p_j = \frac{\partial L}{\partial \dot{q}_j} = \frac{\partial K}{\partial \dot{q}_j} = 2a_j \dot{q}_j.$$

$$(1.11)$$

The coefficient a_j varies depending on whether the generalized velocity is linear or angular. Substituting this result and the definition of L into the relation defining $\hat{H}$ yields

$$\hat{H} = K + \hat{U}.$$

$$(1.12)$$

Note that the Hamiltonian equals the total system energy. If the Lagrangian is not an explicit function of time, we can differentiate the relation defining $\hat{H}$ to obtain

$$d\hat{H} = \sum_j^{3N} (\dot{q}_j dp_j + p_j d\dot{q}_j) - \sum_j^{3N} \frac{\partial L}{\partial \dot{q}_j} d\dot{q}_j - \sum_j^{3N} \frac{\partial L}{\partial q_j} dq_j.$$

$$(1.13)$$

Using the following relations obtained above:

$$\frac{\partial L}{\partial q_j} = \dot{p}_j, \qquad \frac{\partial L}{\partial \dot{q}_j} = p_j$$

$$(1.14)$$

the relation for $d\hat{H}$ becomes

$$d\hat{H} = \sum_j^{3N} \dot{q}_j \, dp_j - \sum_j^{3N} \dot{p}_j \, dq_j.$$

$$(1.15)$$

Since $\hat{H} = \hat{H}(p_1, p_2, \ldots, p_{3N}, q_1, q_2, \ldots, q_{3N})$ it also can be mathematically stated that

$$d\hat{H} = \sum_j^{3N} \left(\frac{\partial \hat{H}}{\partial p_j} \right) dp_j + \sum_j^{3N} \left(\frac{\partial \hat{H}}{\partial q_j} \right) dq_j.$$

$$(1.16)$$

By equating coefficients of the dp_j and dq_j terms in Eqs. (1.15) and (1.16), we obtain Hamilton's equations of motion

$$\left(\frac{\partial \hat{H}}{\partial p_j} \right) = \dot{q}_j, \qquad \left(\frac{\partial \hat{H}}{\partial q_j} \right) = -\dot{p}_j, \qquad j = 1, 2, \ldots, 3N.$$

$$(1.17)$$

These $6N$ first-order equations completely describe the dynamics of a system of N particles. Given the initial conditions, we could try to solve these equations to predict the

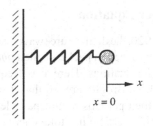

Figure 1.4

macroscopic behavior of a system of molecules. However, to do so is undesirable for two reasons. First, solving so many equations is very difficult, and second, at the microscopic level quantum effects are not included. Hamilton's equations have been presented here because they demonstrate that the energy of the system is a function of appropriately chosen coordinates and conjugate momenta. This concept will be central to our development of a statistical view of the thermodynamics of the system in subsequent chapters.

Example 1.2 The mass in the mass–spring system in Figure 1.4 oscillates about the position $x = 0$. The spring exerts a restoring force F that is linearly dependent on x, obeying the relation

$$F = -kx,$$

where k is a constant that characterizes the spring. Find the Lagrangian and Hamiltonian for the system and show that Eq. (1.5) reduces to Newton's equation of motion.

    Solution At any instant, the kinetic energy K is $(1/2)mu^2$, where $u = dx/dt$. The potential energy is

$$\hat{U} = -\int_0^x F\, dx = -\int_0^x -kx\, dx = \frac{1}{2}kx.$$

The Lagrangian and Hamiltonian are therefore given by

$$L = \tfrac{1}{2}mu^2 - \tfrac{1}{2}kx^2$$

and

$$\hat{H} = \tfrac{1}{2}mu^2 + \tfrac{1}{2}kx^2.$$

Substituting the relation for L into Eq. (1.5), we obtain

$$\frac{d}{dt}\left(\frac{\partial L}{\partial u}\right) = \frac{\partial L}{\partial x},$$

$$\frac{d}{dt}(mu) = -kx = F.$$

Thus, Eq. (1.5) reduces to the time derivative of momentum being equal to the applied force, which is essentially Newton's law of motion.

1.3   Quantum Analysis Using the Schrödinger Equation

As a result of the work of Planck and others, by 1920, the dual nature (wave/particle) of radiant energy was widely accepted. Radiant energy is carried in packets called *photons*, each with energy $h\nu$, which have particle and wavelike qualities. Here ν is frequency and $h = 6.63 \times 10^{-34}$ Js is Planck's constant. In 1924, de Broglie argued that if radiant energy could have both wave and particle properties, then perhaps matter particles also have wavelike qualities. By analogy, he postulated that if a particle has total energy ε and momentum p, then it has a wavelength λ and frequency ν associated with it such that

$$\varepsilon = h\nu \tag{1.18a}$$

and

$$p = h/\lambda. \tag{1.18b}$$

For a particle traveling with momentum mv this implies that

$$\lambda = h/mv. \tag{1.19}$$

Example 1.3   In 1927 Davisson and Germer demonstrated the wavelike scattering of an electron beam reflected from a nickel surface. The wavelength of the electron beam determined from the interference pattern and the known spacing of the nickel atoms agreed well with that predicted by de Broglie's formula. The mass of an electron is 9.109×10^{-31} kg. If they travel at 0.05 times the speed of light, it follows that

$$\lambda = \frac{6.63 \times 10^{-34}}{9.109 \times 10^{-31}(.05 \times 3 \times 10^8)} = 4.85 \times 10^{-11} \text{m} = 0.485 \,\text{Å}.$$

For visible light $\lambda = 4{,}000\text{--}8{,}000$ Å. Thus an electron beam can provide higher image resolution since its wavelength is much shorter than visible light. This is the motivation behind the development of the electron microscope.

The coordinates $q_1, q_2, \ldots$ and momenta $p_1, p_2, \ldots$ that quantify the dynamics of an N-particle system define a multidimensional space referred to as *phase space*. The wavelike nature of particles implies that they are really smeared in phase space with no definite position and momentum. It can be argued, based on the de Broglie relations, that for any measurement, the uncertainty in a particle's position, Δx, and in its corresponding momentum, Δp_x, are constrained by the relation

$$|\Delta x \, \Delta p_x| \geq \frac{h}{4\pi} = \frac{\hbar}{2}, \tag{1.20a}$$

where

$$\hbar = \frac{h}{2\pi}. \tag{1.20b}$$

Equation (1.20a) is a quantitative statement of the *Heisenberg uncertainty principle*. We will consider this issue further in Section 1.6.

Because matter has some wave characteristics, it is expected that some wave equation governing its behavior should exist. If we consider the following mathematical representation of a one-dimensional (1-D) wave:

$$\Psi - C \exp\{i[2\pi x/\lambda) - 2\pi \nu t]\},$$

(1.21)

and differentiate, it is easy to show that

$$-i\hbar \frac{\partial \Psi}{\partial x} = \frac{h}{\lambda}\Psi = p_x\Psi$$

(1.22)

and

$$i\hbar \frac{\partial \Psi}{\partial t} = h\nu\Psi = \varepsilon\Psi.$$

(1.23)

This suggests the identification of $-i\hbar(\partial/\partial x)$ with p_x and $i\hbar(\partial/\partial t)$ with ε. In a nonrelativistic, classical (3-D) system, the total Hamiltonian (total energy) for a particle is given by

$$\hat{H} = K + \hat{U} = \frac{p^2}{2m} + \hat{U}(\mathbf{r}),$$

(1.24)

where

$$p^2 - p_x^2 + p_y^2 + p_z^2,$$

(1.25a)

$$\mathbf{r} = x\mathbf{i} + y\mathbf{j} + z\mathbf{k}.$$

(1.25b)

The above arguments imply that

$$p_x^2 = -i\hbar\frac{\partial}{\partial x}\left(-i\hbar\frac{\partial}{\partial x}\right) = -\hbar^2\frac{\partial^2}{\partial x^2},$$

(1.26a)

$$p_y^2 = -i\hbar\frac{\partial}{\partial y}\left(-i\hbar\frac{\partial}{\partial y}\right) = -\hbar^2\frac{\partial^2}{\partial y^2},$$

(1.26b)

$$p_z^2 = -i\hbar\frac{\partial}{\partial z}\left(-i\hbar\frac{\partial}{\partial z}\right) = -\hbar^2\frac{\partial^2}{\partial z^2}.$$

(1.26c)

Substituting Eqs. (1.26) and (1.25a) into (1.24), the right-hand side becomes the Hamiltonian operator

$$\hat{H} = \frac{-\hbar^2}{2m}\left[\frac{\partial^2}{\partial x^2} + \frac{\partial^2}{\partial y^2} + \frac{\partial^2}{\partial z^2}\right] + \hat{U}(\mathbf{r}).$$

(1.27)

The Hamiltonian $\hat{H}$ is interpreted as equaling the total energy ε, which, in turn, is identified with $i\hbar(\partial/\partial t)$. Replacing $\hat{H}$ with $i\hbar(\partial/\partial t)$, and applying the operators on both sides to the wave function Ψ, yields

$$i\hbar\frac{\partial \Psi}{\partial t} = \frac{-\hbar^2}{2m}\nabla^2\Psi + \hat{U}(\mathbf{r})\Psi.$$

(1.28)

With a little rearranging this can be written as

$$\nabla^2\Psi - \frac{2m}{\hbar^2}\hat{U}(\mathbf{r})\Psi = \left(\frac{2m}{i\hbar}\right)\frac{\partial \Psi}{\partial t}.$$

(1.29)

This is the *Schrödinger equation* for the system wave function Ψ.

The wave version of quantum mechanics theory is built upon the following postulate:

♦   *The state of a quantum-mechanical system is completely specified by a function* $\Psi(\mathbf{r}, t)$ *that depends on the coordinates of the particles and on time. This function, called the* wave function *or* state function, *has the property that* $\Psi^*(\mathbf{r}, t)\,\Psi(\mathbf{r}, t)$ $dx\,dy\,dz$ *is the probability that the particle lies in the volume element* $dx\,dy\,dz$ *located at location* $\mathbf{r}$ *at time* t. *(Note that* Ψ^* *is the complex conjugate of* Ψ.)

The wave function Ψ is determined by solving the Schrödinger equation.

For Ψ to have the property indicated in the postulate, Ψ must satisfy

$$\int_V \Psi\Psi^*\,dV = 1. \tag{1.30}$$

If solution of the Schrödinger equation is postulated to be of the form

$$\Psi(\mathbf{r}, t) = f(t)\psi(\mathbf{r}) \tag{1.31}$$

then substituting into Eq. (1.29) and rearranging leads to

$$\left(\frac{i\hbar}{f}\right)\frac{df}{dt} = \frac{1}{\psi}\left[\hat{U}(\mathbf{r})\psi - \frac{\hbar^2}{2m}\nabla^2\psi\right]. \tag{1.32}$$

Note that the left side of the above equation is a function only of time whereas the right side is a function only of position. Hence, each side must be a constant, which we will designate as C_0. Setting the left side equal to C_0 yields

$$f = f_0\,\exp\{-iC_0t/\hbar\}, \tag{1.33}$$

where f_0 is the initial value of f at $t = 0$. Substituting this result into Eq. (1.31) yields

$$\Psi(\mathbf{r}, t) = f_0\psi(\mathbf{r})\exp\{-iC_0t/\hbar\}. \tag{1.34}$$

Differentiating gives

$$i\hbar\frac{\partial\Psi}{\partial t} = C_0\Psi = \varepsilon\Psi, \tag{1.35}$$

from which we conclude that the separation constant C_0 is the energy,

$$C_0 = \varepsilon. \tag{1.36}$$

The relation for f is therefore

$$f = f_0\,\exp\{-i\varepsilon t/\hbar\}. \tag{1.37}$$

Thus, the time variation is oscillatory with fixed amplitude. Boundary conditions and potential energy effects are manifested in the spatial variation. Setting the right side of the separated Schrödinger equation equal to ε yields

$$\hat{U}(\mathbf{r})\psi - \frac{\hbar^2}{2m}\nabla^2\psi = \varepsilon\psi. \tag{1.38}$$

The above equation is the *time-independent Schrödinger equation.* The time-independent Schrödinger equation with appropriate boundary conditions generally forms a Sturm–Liouville system for which ε is an eigenvalue. For such a system it is known that solutions exist only for discrete values of ε. Solution of the time-independent Schrödinger equation will specify the allowable quantum energy levels for the system.

1.4   Model Solutions of the Time-Independent Schrödinger Equation

Several simple model solutions of the time-independent Schrödinger equation can be used as models of energy storage in atoms and molecules. For this reason we will examine these simple model solutions in this section.

A Free Particle in a Box – 1-D Case

For the one-dimensional particle-in-a-box system, the potential function that confines the particle to a finite x domain is defined as

$$\hat{U} = 0 \quad \text{for } 0 < x < L, \tag{1.39a}$$

$$\hat{U} \to \infty \quad \text{at } x = 0 \quad \text{and} \quad x = L, \tag{1.39b}$$

(see Figure 1.5).

Note that $\hat{U}$ becomes infinite at the walls due to the instantaneous reversal of momentum at each wall. Since a particle with finite energy cannot exist where $\hat{U} \to \infty$, the probability of finding the particle must approach zero at $x = 0$ and $x = L$. Assuming ψ is continuous, we adopt the boundary conditions

$$\psi = 0 \quad \text{at } x = 0+, x = L- \tag{1.40}$$

and take $\hat{U} = 0$ over the entire interval. For the 1-D case, the time-independent Schrödinger equation becomes an ordinary differential equation:

$$\left(-\frac{\hbar^2}{2m}\right)\frac{d^2\psi}{dx^2} = \varepsilon_x \psi. \tag{1.41}$$

By direct substitution, it is easily shown that the solution must have the form

$$\psi = A \sin\left(\sqrt{\frac{2m\varepsilon_x}{\hbar^2}}x\right). \tag{1.42}$$

To satisfy the boundary conditions it is necessary that $L\sqrt{2m\varepsilon_x/\hbar^2}$ be integer multiples of π:

$$L\sqrt{2m\varepsilon_x/\hbar^2} = n_x\pi, \qquad n_x = 1, 2, 3, \ldots,$$

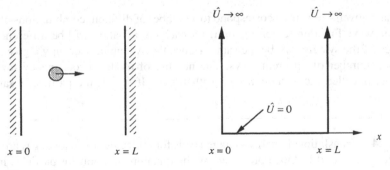

Figure 1.5

or

$$\varepsilon_x = \frac{\hbar^2 \pi^2}{2mL^2} n_x^2,$$ (1.43)

where n_x is the quantum number. Substituting and requiring that

$$\int_V \Psi \Psi^* \, dV = 1$$

yields $A = \sqrt{2/L}$. The full wave function solutions are therefore

$$\psi = \sqrt{\frac{2}{L}} \sin\left(\frac{\pi n_x x}{L}\right), \qquad n_x = 1, 2, 3, \ldots.$$ (1.44)

A Free Particle in a 3-D Box

For the 3-D case, the time-independent equation becomes

$$\left(-\frac{\hbar}{2m}\right) \left[\frac{\partial^2 \psi}{\partial x^2} + \frac{\partial^2 \psi}{\partial y^2} + \frac{\partial^2 \psi}{\partial z^2}\right] = \varepsilon \psi.$$ (1.45)

Extending the reasoning used for the 1-D case, we have taken $\hat{U} = 0$ inside the box and ψ is taken to be zero at all walls:

$$\psi = 0 \quad \text{at } x = 0, x = L, y = 0, y = L, z = 0, z = L.$$ (1.46)

Using separation of variables, it can be easily shown that solutions satisfying the boundary conditions and the requirement that

$$\int_V \Psi \Psi^* \, dV = \int_0^L \int_0^L \int_0^L \Psi \Psi^* \, dx \, dy \, dz = 1$$ (1.47)

are of the form

$$\psi = \left(\frac{2}{L}\right)^{3/2} \sin\left(\frac{\pi n_x x}{L}\right) \sin\left(\frac{\pi n_y y}{L}\right) \sin\left(\frac{\pi n_z z}{L}\right),$$ (1.48)

where n_x, n_y, and n_z must be chosen from the set $\{1, 2, 3, \ldots\}$. It follows directly from this solution form that

$$\varepsilon = \left(\frac{h^2}{8mV^{2/3}}\right) \left(n_x^2 + n_y^2 + n_z^2\right).$$ (1.49)

Note that a given $\varepsilon = \varepsilon_i$ may correspond to a number of different combinations of n_x, n_y, and n_z. The wave function (eigenfunction) for each combination will be different but the total energy of the system will be the same. Hence the system may occupy a given energy level ε_i in a number of different ways. The number of modes of occupancy for the ith energy level is called the *degeneracy* g_i of that level. If $g_i = 1$, the particle is said to be *nondegenerate*.

Example 1.4   Translational energy storage levels for molecules in an ideal gas are usually assumed to correspond to those predicted by the quantum solution for particles in a 3-D box. Use the particle-in-a-box solution to determine the separation between the first two translational energy levels for neon atoms in a system with a volume of 1,000 cm^3.

Solution   The energy difference between the lowest two nonzero energy levels specified by Eq. (1.49) is given by

$$\Delta\varepsilon = \left(\frac{h^2}{8mV^{2/3}}\right)[(2^2 + 1^2 + 1^2) - (1^2 + 1^2 + 1^2)] = 3\left(\frac{h^2}{8mV^{2/3}}\right).$$

Substituting $h = 6.63 \times 10^{-34}$ Js, $V = 0.001$ m^3, and $m = 3.35 \times 10^{-26}$ kg for neon yields

$$\Delta\varepsilon = 3\left(\frac{(6.63 \times 10^{-34})^2}{8(3.35 \times 10^{-26})(0.001)^{2/3}}\right) = 4.92 \times 10^{-40} \text{ J.}$$

This illustrates that the typical separation between translational quantum states is so small for macroscopic systems as to be unmeasureable. We will show later that the mean translational energy of an atom in an ideal gas is $(3/2)k_B T$, where $k_B = 1.38 \times 10^{-23}$ J/K and T is the absolute temperature. (We will define absolute temperature more precisely later. For now we will accept it as a suitably defined temperature scale in the units kelvin, abbrevated as K.) At 290 K (near room temperature), the mean translational energy of a molecule is 6.0×10^{-21} J, which is much larger than the separation between levels computed above.

The Harmonic Oscillator

Consider a particle moving about an equilibrium position at $x = 0$ with a restoring force linearly dependent on x:

$$F = -kx. \tag{1.50}$$

The simplest example of such a system is the mass–spring system shown schematically in Figure 1.4 in which the mass can move in one dimension. The potential energy stored in the spring is given by

$$\hat{U}(x) = -\int_0^x F\,dx = \int_0^x kx\,dx = \frac{kx^2}{2}. \tag{1.51}$$

The standing-wave Schrödinger equation then becomes

$$\left(-\frac{\hbar^2}{2m}\right)\frac{d^2\psi}{dx^2} + \left(\frac{kx^2}{2} - \varepsilon\right)\psi = 0. \tag{1.52}$$

The probability of finding the particle at $x \to \pm\infty$ must be zero since $\hat{U} \to \infty$ at those locations. This implies the boundary conditions

$$\psi(+\infty) = \psi(-\infty) = 0. \tag{1.53}$$

Defining

$$\beta = 2m\varepsilon/\hbar^2 \tag{1.54a}$$

and

$$\alpha^2 = km/\hbar^2 \tag{1.54b}$$

we can write Eq. (1.52) as

$$\frac{d^2\psi}{dx^2} + [\beta - (\alpha x)^2]\psi = 0.$$

(1.55)

Executing the transform

$$\xi = \alpha^{1/2}x,$$

(1.56a)

$$\psi = Q(\xi)\exp\{-\xi^2/2\}$$

(1.56b)

Eq. (1.55) becomes

$$\frac{d^2Q}{d\xi^2} - 2\xi\frac{dQ}{d\xi} + \left[\frac{\beta}{\alpha} - 1\right]Q = 0.$$

(1.57)

Note that the $\exp\{-\xi^2/2\}$ factor automatically forces the solution to satisfy the boundary conditions at $x \to \pm\infty$. Equation (1.57) is the *Hermite equation*, which has solutions only when

$$\left[\frac{\beta}{\alpha} - 1\right] = 2n, \qquad n = 0, 1, 2, 3, \ldots.$$

(1.58)

The resulting solutions are Hermite polynomials, defined as

$$H_n(\xi) = (-1)^n e^{\xi^2}\frac{d^n}{d\xi^n}(e^{-\xi^2}).$$

(1.59)

Substituting into the transform relation for ψ and invoking the normalization condition (1.30) yields wave solutions of the form

$$\psi_n(x) = \left[\sqrt{\frac{\alpha}{\pi}}\frac{1}{2^n n!}\right]^{1/2} H_n(\alpha^{1/2}x)\exp\{-\alpha x^2/2\}.$$

(1.60)

Expressing Eq. (1.58) in terms of the original variables, we obtain the following relations for the quantum energy levels:

$$\varepsilon = \left(n + \frac{1}{2}\right)h\left[\frac{1}{2\pi}\sqrt{\frac{k}{m}}\right], \qquad n = 1, 2, 3, \ldots.$$

(1.61)

From a classical analysis of a harmonic oscillator of this type, $(1/2\pi)\sqrt{k/m}$ can be identified as the frequency of the oscillator, which we denote as v:

$$\varepsilon = \left(n + \tfrac{1}{2}\right)hv, \qquad n = 1, 2, 3, \ldots,$$

(1.62)

$$v = (1/2\pi)\sqrt{k/m}.$$

(1.63)

For the harmonic oscillator, the levels of a quantum-mechanical harmonic oscillator are often represented graphically in a diagram like that shown in Figure 1.6.

Note that the ground-state energy (for $n = 0$) is not zero. The energy of the oscillator can be written

$$E_{\text{HO}} = \left(p_x^2/2m\right) + (kx^2/2).$$

(1.64)

A zero value of E_{HO} would require both p_x and x to be exactly zero, violating the uncertainty principle.

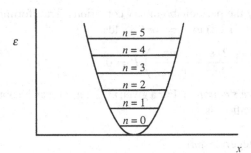

Figure 1.6

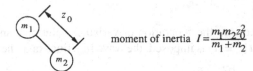

moment of inertia $I = \dfrac{m_1 m_2 z_0^2}{m_1 + m_2}$

Figure 1.7

The Rigid Rotor

For the rigid rotor shown in Figure 1.7, the potential energy $\hat{U}$ is 0 for the free rotation in two angular directions. Expressing the Laplacian in spherical coordinates, the standing-wave Schrödinger equation then becomes

$$\frac{1}{\sin\theta}\frac{\partial}{\partial\theta}\left(\sin\theta\frac{\partial\psi}{\partial\theta}\right) + \frac{1}{\sin^2\theta}\frac{\partial^2\psi}{\partial\phi^2} = -\frac{2I}{\hbar^2}\varepsilon\psi. \tag{1.65}$$

Postulating a separation of variables solution,

$$\psi = P(\theta)\Phi(\phi), \tag{1.66}$$

we substitute, taking m^2 to be the separation constant, to obtain

$$\frac{d^2\Phi}{d\phi^2} = -m^2\Phi, \tag{1.67}$$

$$\frac{1}{\sin\theta}\frac{d}{d\theta}\left(\sin\theta\frac{dP}{d\theta}\right) + \left(\frac{2I\varepsilon}{\hbar^2} - \frac{m^2}{\sin^2\theta}\right)P = 0. \tag{1.68}$$

The 2π periodicity of the solutions requires that

$$\Phi(0) = \Phi(2\pi), \quad \left(\frac{d\Phi}{d\phi}\right)_0 = \left(\frac{d\Phi}{d\phi}\right)_{2\pi},$$

$$P(0) = P(2\pi), \quad \left(\frac{dP}{d\theta}\right)_0 = \left(\frac{dP}{d\theta}\right)_{2\pi}. \tag{1.69}$$

Solutions to Eq. (1.67) are of the form

$$\Phi = A \sin m\phi + B \cos m\phi, \tag{1.70}$$

where $m = \pm 1, \pm 2, \pm 3, \ldots$ to satisfy the periodic boundary conditions. Transforming the independent variable in the second Eq. (1.68) to $\xi = \cos\theta$ yields

$$(1 - \xi^2)\frac{d^2 P}{d\xi^2} - 2\xi\left(\frac{dP}{d\xi}\right) + \left(\frac{2I\varepsilon}{\hbar^2} - \frac{m^2}{1 - \xi^2}\right)P = 0. \tag{1.71}$$

For integer values of m, this is *Legendre's equation*. For a given m, solutions can be obtained for an infinite number of ε values specified as

$$\varepsilon = \frac{\hbar^2}{2I}l(l + 1), \qquad l = \text{integer} \geq |m|. \tag{1.72}$$

Solutions are the *associated Legendre polynomials*

$$P_l^m(\xi) = \frac{(1 - \xi^2)^{m/2}}{2^l l!}\frac{d^{l+m}}{d\xi^{l+m}}(\xi^2 - 1)^l. \tag{1.73}$$

If these results are back-substituted into the separation of variables solutions, and the normalization condition ($\int\int\int \psi^*\psi \, dV = 1$) is imposed, the wave functions take the form

$$\psi = \frac{1}{\sqrt{2\pi}}\left[\frac{(2l + 1)(l - m)!}{2(l + m)!}\right]^{1/2} P_l^m(\cos\theta)\, e^{im\phi}. \tag{1.74}$$

Note that each l corresponds to one ε (energy level) and l is therefore the principal (rotational) quantum number. For each l there can be any m in the range specified by $l \geq |m|$. There are $2l + 1$ values of m permissible for each l and hence for each ε. Thus, the degeneracy g_l of energy level ε_l is

$$g_l = 2l + 1. \tag{1.75}$$

Increased degeneracy corresponds to increased degrees of freedom for energy storage.

Example 1.5   The rigid-rotor solution is commonly used to model rotational energy storage in molecules in an ideal gas. Use the rigid-rotor solution to determine the separation between the first two energy levels for nitrogen molecules in an ideal gas.

   Solution   The energy difference between the lowest two nonzero energy levels specified by Eq. (1.72) is given by

$$\Delta\varepsilon = \left(\frac{\hbar^2}{2I}\right)[2(2 + 1) - 1(1 + 1)] = 4\left(\frac{\hbar^2}{2I}\right).$$

The nuclei of the atoms of nitrogen with mass 2.33×10^{-26} kg are separated by a mean distance of about 1.6×10^{-10} m. The moment of intertia of the nitrogen molecule is given by

$$I = \tfrac{1}{2}mz_0^2 = \tfrac{1}{2}(2.33 \times 10^{-26})(1.6 \times 10^{-10})^2 = 3.0 \times 10^{-46} \text{ kg m}^2.$$

Substituting the result for I and $\hbar = 6.63 \times 10^{-34}/2\pi$ Js in the relation for $\Delta\varepsilon$ yields

$$\Delta\varepsilon = 4\left(\frac{(6.63 \times 10^{-34}/2\pi)^2}{2(3.0 \times 10^{-46})}\right) = 7.42 \times 10^{-23} \text{ J}.$$

We will show later that the mean rotational energy of a nitrogen molecule in an ideal gas is $k_B T$, where T is the absolute temperature and $k_B = 1.38 \times 10^{-23}$ J/K. By comparison, at

290 K (near room temperature) the mean rotational energy of a nitrogen molecule is 4.0×10^{-21} J, which is about 50 times larger than the separation between levels computed above.

The simple model systems for which we have developed quantum mechanical solutions to the Schrödinger equation will be used in later sections to model specific energy storage mechanisms within atoms and molecules. Before proceeding further it is illuminating to examine how an entire atom can be treated quantum mechanically. The simplest treatment of this type is the model of the hydrogen atom described in the next section.

1.5 A Quantum Mechanics Model of the Hydrogen Atom

The Schrödinger equation described in previous sections can be used to describe the hydrogen atom as a whole. For the hydrogen atom the nucleus consists of one proton and one electron is in orbit around the nucleus (see Figure 1.8).

The attractive force between electron and nucleus is given by Coulomb's law:

$$F = C_C e_e^2 / r^2,$$

where $e_e = 1.602 \times 10^{-19}$ coulombs is the charge of one electron. We therefore define the potential energy for the system as

$$\hat{U} = -\int_{\infty}^{r} F \, dr = -\frac{C_1}{r}, \qquad C_1 = C_C e_e^2. \tag{1.76}$$

The negative sign is applied because stored energy increases as $r \to \infty$. Note that the mass of an electron $= 9.1 \times 10^{-31}$ kg and the mass of a proton $= 1.67 \times 10^{-27}$ kg. Since the proton mass m_p is much larger than the electron mass m_e, we idealize the nucleus as being stationary, with the electron orbiting around it. The Schrödinger equation then takes the form

$$\frac{\hbar^2}{2m_e} \nabla^2 \psi + \left(\varepsilon + \frac{C_1}{r} \right) \psi = 0. \tag{1.77}$$

Postulating a solution of the form $\psi = Y(\theta, \phi) R(r)$, substituting, and executing the usual separation of variable technique results in the following equations for Y and R:

$$\left(\frac{1}{\sin \theta} \right) \frac{d}{d\theta} \left(\sin \theta \frac{dY}{d\theta} \right) + \left(\frac{1}{\sin^2 \theta} \right) \frac{d^2 Y}{d\phi^2} = - \left(\frac{2I}{\hbar^2} \right) \varepsilon Y, \tag{1.78}$$

$$\left(\frac{1}{r^2} \right) \frac{d}{dr} \left(r^2 \frac{dR}{dr} \right) + \left(\frac{2m_e}{\hbar^2} \right) \left[\varepsilon + \frac{2C_1}{r} - \frac{l(l+1)\hbar^2}{2I} \right] R = 0, \tag{1.79}$$

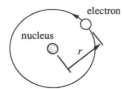

Figure 1.8

where

$$I = m_e r^2.$$ (1.80)

The equation for Y is the rigid-rotor equation (with the same periodic boundary conditions) solved above. It follows directly that solutions for Y are

$$Y = \frac{1}{\sqrt{2\pi}} \left[\frac{(2l+1)(l-m)!}{2(l+m)!} \right]^{1/2} P_l^m(\cos\theta) e^{im\phi}.$$ (1.81)

To solve the radial Eq. (1.79), we define

$$a = \sqrt{-\frac{\hbar^2}{8m_e \varepsilon}}, \qquad \eta = \frac{r}{a}, \qquad L(\eta) = a\eta^{-l} R(a\eta) e^{\eta/2}.$$ (1.82)

For this system, the energy zero point is taken to correspond to infinite separation of the electron and nucleus. The definition of a above includes the minus sign inside the square root because the energy levels for finite separation are negative. Invoking these definitions, the equation for R becomes

$$\eta \frac{d^2 L}{d\eta^2} + (k+1-\eta)\frac{dL}{d\eta} + n_L L = 0,$$ (1.83)

where

$$k = 2l+1, \qquad n_L = \frac{2am_e C_1}{\hbar} - l - 1.$$ (1.84)

This equation, the *associated Laguerre equation*, has solutions known as the *associated Laguerre polynomials*:

$$L_{n_L}^k(\eta) = -\sum_{j=0}^{n_L-k} \frac{(-1)^j (n_L!)^2 \eta^j}{(n_L - k - j)!(k+j)!j!}.$$ (1.85)

Rearranging the relation for n_L and substituting for a yields

$$\varepsilon = \frac{-m_e C_1^2}{2\hbar^2 (n_L + l + 1)^2}.$$ (1.86)

Setting $n = n_L + l + 1$ yields

$$\varepsilon = \frac{-m_e C_1^2}{2\hbar^2 n^2}.$$ (1.87)

The negative sign on the right side of the relation for ε is consistent with the requirement that increasing n correspond to increasing energy. Solving for $R(r)$ yields

$$R(r) = \left[\frac{(n-l-1)!}{2na[(n+l)!]^3} \right]^{1/2} (r/a)^l e^{-r/2a} L_{n+l}^{2l+1}(r/a).$$ (1.88)

Note that a is a function of n and

$$L_{n+l}^{2l+1}(r/a) = \sum_{j=0}^{n-l-1} \frac{(-1)^{j+1}[(n+l)!]^2 (r/a)^j}{(n-l-1-j)!(2l+1+j)!j!}.$$ (1.89)

For a given n, l can vary from zero to $n - 1$ and give a valid radial solution. As shown for the rigid-rotor solution, for each l, m can have any of $2l + 1$ values. Hence, for this solution:

$n =$ principal quantum number (specifying energy level),
$l =$ angular quantum number,
$m =$ magnetic quantum number,

and for a given n, the limits on l and m are:

$$0 \leq l \leq n - 1, \qquad -l \leq m \leq l.$$

It follows directly that the total degeneracy for a given n value is given by

$$g_n = \sum_{l=0}^{n-1} (2l + 1) = n^2. \tag{1.90}$$

Radiation emission or absorption results in a change of energy level for the hydrogen atom. For emission, the photon energy $h\nu$ is given by

$$h\nu = \varepsilon_1 - \varepsilon_2 = \frac{2\pi^2 m_e C_1^2}{h^2} \left[\frac{1}{n_2^2} - \frac{1}{n_1^2} \right]. \tag{1.91}$$

This relation can be written in terms of the wavenumber k_{12} as

$$k_{12} = \frac{\nu_{12}}{c_l} = R_H \left[\frac{1}{n_2^2} - \frac{1}{n_1^2} \right], \tag{1.92a}$$

where R_H is the Rydberg constant given by

$$R_H = \frac{2\pi^2 m_e C_1^2}{c_l h^3}. \tag{1.92b}$$

Based on experimental determinations of the (Balmer) series of emission spectral lines for hydrogen, Johannes Rydberg, in a study done in the 1890s, concluded that $R_H = 109,720 \text{ cm}^{-1}$. The quantum-theory relation indicated above predicts $R_H = 109,680 \text{ cm}^{-1}$. The agreement of the theoretical prediction with the measurements is indeed impressive.

Example 1.6 The Balmer series of emission spectral lines for hydrogen corresponds to electrons falling from levels above $n = 2$ down to the $n = 2$ level. Show that the Balmer series corresponds to wavelengths between 3,640 and 6,563 Å.

Solution Since the wavenumber $k_{12} = 1/\lambda_{12}$, Eq. (1.92a) can be written

$$\frac{1}{\lambda_{12}} = R_H \left[\frac{1}{n_2^2} - \frac{1}{n_1^2} \right].$$

For the Balmer series, $n_2 = 2$, and so

$$\lambda_{12} = \frac{1}{R_H} \left[\frac{1}{4} - \frac{1}{n_1^2} \right]^{-1}.$$

The largest wavelength in the series corresponds to $n_1 = 3$. Substituting and using $R_H = 109,720 \text{ cm}^{-1}$, we find

$$\lambda_{12} = 6.562 \times 10^{-5} \text{ cm} = 6,562 \,\text{Å}.$$

The smallest wavelength in the series corresponds to $n_1 \to \infty$. In this case, the relation for λ_{12} yields

$$\lambda_{12} = 3.646 \times 10^{-5} \text{ cm} = 3,646 \,\text{Å}.$$

Thus, the Balmer series spans the wavelength range from 3,646 Å to 6,562 Å.

1.6   The Uncertainty Principle

To further explore the nature of the uncertainty principle, we now wish to reconsider the one-dimensional particle in a box discussed earlier. For that system, we found that

$$\psi = \sqrt{\frac{2}{L}} \sin\left(\frac{\pi n_x x}{L}\right), \qquad n_x = 1, 2, 3, \ldots, \tag{1.93}$$

$$\varepsilon_x = \frac{\hbar^2 \pi^2}{2mL^2} n_x^2. \tag{1.94}$$

Because $\psi^* \psi \, dx$ is interpreted as a probability, it can be used to calculate averages and variances. For the particle in a 1-D box, ψ has no imaginary part and is therefore its own complex conjugate:

$$\psi^* = \sqrt{\frac{2}{L}} \sin\left(\frac{\pi n_x x}{L}\right). \tag{1.95}$$

It follows that

$$\psi^* \psi \, dx = \left(\frac{2}{L}\right) \sin^2\left(\frac{\pi n_x x}{L}\right) dx \tag{1.96}$$

is the probability that a particle is found between x and $x + dx$. The average value of x, the mean position of the particle, is obtained by integrating x, weighted by its probability, over the box dimensions:

$$\langle x \rangle = \int_0^L x \left(\frac{2}{L}\right) \sin^2\left(\frac{\pi n_x x}{L}\right) dx. \tag{1.97}$$

Evaluation of the integral yields

$$\langle x \rangle = \frac{L}{2} \quad \text{for all } n_x. \tag{1.98}$$

We can also evaluate the spread of the variation of x by computing the variance of x (from the mean) defined as

$$\sigma_x^2 = \int_0^L (x - \langle x \rangle)^2 \psi^* \psi \, dx. \tag{1.99}$$

Expanding, we obtain

$$\sigma_x^2 = \int_0^L x^2 \psi^* \psi \, dx - 2\langle x \rangle \int_0^L x \psi^* \psi \, dx + \langle x \rangle^2 \int_0^L \psi^* \psi \, dx$$

$$= \langle x^2 \rangle - 2\langle x \rangle^2 + \langle x \rangle^2$$

$$= \langle x^2 \rangle - \langle x \rangle^2. \tag{1.100}$$

$\langle x^2 \rangle$ is computed as

$$\langle x^2 \rangle = \int_0^L x^2 \left(\frac{2}{L} \right) \sin^2 \left(\frac{\pi n_x x}{L} \right) dx$$

$$= \left(\frac{L}{2\pi n_x} \right)^2 \left(\frac{4\pi^2 n_x^2}{3} - 2 \right). \tag{1.101}$$

Substituting (1.98) and (1.101) into (1.100), we get

$$\sigma_x^2 = \langle x^2 \rangle - \langle x \rangle^2 = \left(\frac{L}{2\pi n_x} \right)^2 \left(\frac{\pi^2 n_x^2}{3} - 2 \right). \tag{1.102}$$

The standard deviation in x is therefore given by

$$\sigma_x = \left(\frac{L}{2\pi n_x} \right) \left(\frac{\pi^2 n_x^2}{3} - 2 \right)^{1/2}. \tag{1.103}$$

Similarly computing the average energy or momentum is a bit trickier because these quantities are represented by operators:

$$\hat{H} = -\frac{\hbar^2}{2m} \frac{d^2}{dx^2} \quad \text{(energy)}, \tag{1.104}$$

$$p_x = -i\hbar \frac{d}{dx} \quad \text{(momentum)}. \tag{1.105}$$

We now consider the question: In integrating to get the average, does the operator work on $\psi^*(x)\psi(x)$ or on ψ or ψ^* alone? To address this question we consider the following equation, which results from the definition of the Hamiltonian,

$$\hat{H}\psi = \varepsilon\psi. \tag{1.106}$$

Multiplying both sides by ψ^* and integrating, we obtain

$$\int_0^L \psi^* \hat{H} \psi \, dx = \varepsilon \int_0^L \psi^* \psi \, dx = \varepsilon. \tag{1.107}$$

This indicates that to get ε for $\langle \varepsilon \rangle$ we must sandwich the operator between ψ^* and ψ:

$$\langle \varepsilon \rangle = \int_0^L \psi^* \hat{H} \psi \, dx. \tag{1.108}$$

Based on this same reasoning, we define

$$\langle p_x \rangle = \int_0^L \psi^* p_x \psi \, dx = \int_0^L \psi^* \left(-i\hbar \frac{d}{dx} \right) \psi \, dx. \tag{1.109}$$

Substituting for the wave function yields

$$\langle p_x \rangle = \int_0^L \sqrt{\frac{2}{L}} \sin\left(\frac{\pi n_x x}{L}\right)\left(-i\hbar\frac{d}{dx}\right)\sqrt{\frac{2}{L}}\sin\left(\frac{\pi n_x x}{L}\right)dx, \tag{1.110}$$

which leads to

$$\langle p_x \rangle = 0 \quad \text{(for all } n_x\text{)}. \tag{1.111}$$

Similarly, for $\langle p_x^2 \rangle$:

$$\langle p_x^2 \rangle = \int_0^L \psi^* p_x^2 \psi \, dx = \int_0^L \psi^* \left(-i\hbar\frac{d}{dx}\right)^2 \psi \, dx \tag{1.112}$$

$$= \int_0^L \left[\sqrt{\frac{2}{L}}\sin\left(\frac{\pi n_x x}{L}\right)\right](-\hbar^2)\frac{d^2}{dx^2}\left[\sqrt{\frac{2}{L}}\sin\left(\frac{\pi n_x x}{L}\right)\right]dx,$$

which reduces to

$$\langle p_x^2 \rangle = \frac{\pi^2 n_x^2 \hbar^2}{L^2}. \tag{1.113}$$

Combining these results, we obtain

$$\sigma_p^2 = \langle p_x^2 \rangle - \langle p_x \rangle^2 = \frac{\pi^2 n_x^2 \hbar^2}{L^2}$$

or equivalently

$$\sigma_p = \frac{\pi n_x \hbar}{L}. \tag{1.114}$$

If we substitute this result into the relation for σ_x to eliminate L, we obtain

$$\sigma_x = \left(\frac{\hbar}{2\sigma_p}\right)\left[\frac{\pi^2 n_x^2}{3} - 2\right]^{1/2}, \tag{1.115}$$

which implies that

$$\sigma_x \sigma_p = \left(\frac{\hbar}{2}\right)\left[\frac{\pi^2 n_x^2}{3} - 2\right]^{1/2}. \tag{1.116}$$

For any nontrivial wave function solution $n_x \geq 1$, which implies that the term in square brackets is greater than one. It follows directly that

$$\sigma_x \sigma_p > \left(\frac{\hbar}{2}\right), \tag{1.117}$$

which is the mathematical statement of the Heisenberg uncertainty principle. We can interpret this result as follows. Since $\sigma_p = \pi n_x \hbar/L$, for a free particle ($L \to \infty$) σ_p is zero. A free particle thus has a definite momentum. However, if we localize the position of the particle within a dimension L, it no longer has a definite momentum, and the uncertainty in its momentum is given by $\sigma_p = \pi n_x \hbar/L$. In the limit of exactly specifying the particle's location, the uncertainty in momentum is infinite. The uncertainty principle states that the product of the two uncertainties is of the order of Planck's constant.

Although we have derived the uncertainty principle for the specific case of a particle in a 1-D box, it can be shown to be generally valid. This result implies that a particle does not have an exact point representation in phase space. For the 1-D particle in a box, the

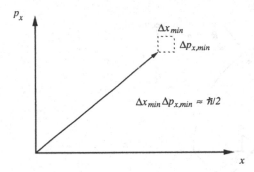

Figure 1.9

most we can say is that it is somewhere inside a region having an area of order $\hbar/2$ (see Figure 1.9). For large systems (large L) with large momentum, phase space is virtually a continuum since $\hbar$ is a very small number.

1.7   Quantum Energy Levels and Degeneracy

In the previous sections of this chapter, we have used basic quantum mechanics theory to explore the energy storage characteristics of several model systems. In later chapters the particle-in-a-box solution will provide the basis for evaluating translational energy storage in atoms, molecules, and other particles. Similarly, the rigid-rotor and harmonic-oscillator solutions will be used as models for rotational energy storage and vibrational energy storage in molecules. The hydrogen atom model considered in Section 1.5 is an example of yet another energy storage mechanism in atoms and molecules: electronic energy storage.

It is important to recognize that in all the models considered in this chapter, the energy levels accessible to the system are quantized – they can take on only specific values corresponding to integer values of appropriate quantum numbers. We have also found that the harmonic-oscillator energy levels can be occupied in only one way, whereas the energy levels of the rigid rotor and the hydrogen atom exhibit varying amounts of degeneracy. We now will examine the issue of degeneracy for the particle-in-a-box model. For the 3-D particle in a box, the quantum energy levels are given by

$$\varepsilon = \frac{h^2}{8mV^{2/3}}\left(n_x^2 + n_y^2 + n_z^2\right),\tag{1.118}$$

where $V = L^3$ and n_x, n_y, n_z are chosen from $\{1, 2, 3, \ldots\}$. In this case, the degeneracy $g(\varepsilon)$ is dictated by the number of ways that $R^2 = 8mV^{2/3}\varepsilon/h^2$ can be written as the sum of three squared positive integers. Thus, g is an erratic function of $R^2(\varepsilon)$ for small R values but becomes smooth at large R. Consider a three-dimensional space spanned by orthogonal coordinates n_x, n_y, n_z. There is a one-to-one correspondence between energy states for the particle and points located at integer values of n_x, n_y, n_z (see Figure 1.10). The equation

$$R^2 = \frac{8mV^{2/3}\varepsilon}{h^2} = \left(n_x^2 + n_y^2 + n_z^2\right)\tag{1.119}$$

represents the surface of a sphere in this space. For large R, the states are so close together that the total number of states with $\varepsilon \leq R^2h^2/8mV^{2/3}$ is essentially equal to the volume of

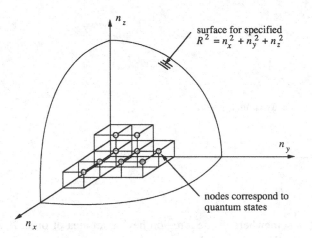

Figure 1.10

one octant of a sphere of radius R. If the number of such states is denoted as n_ε, we have

$$n_\varepsilon = \frac{1}{8}\left(\frac{4\pi R^3}{3}\right) = \frac{\pi}{6}\left(\frac{8mV^{2/3}\varepsilon}{h^2}\right)^{3/2}. \tag{1.120}$$

The number of states in the interval ε to $\varepsilon + \delta\varepsilon$ is interpreted as the degeneracy of energy level ε:

$$g(\varepsilon) = \frac{dn_\varepsilon}{d\varepsilon}\delta\varepsilon = \frac{\pi}{4}\left(\frac{8mV^{2/3}}{h^2}\right)^{3/2}\varepsilon^{1/2}\delta\varepsilon. \tag{1.121}$$

This relation allows us to evaluate the degeneracy for a particle in a box. For common circumstances, this relation predicts an extremely large degeneracy (see example below).

Example 1.7   The mass of electrons is 9.109×10^{-31} kg. Suppose an electron has an energy equivalent to a velocity of 0.05 times the speed of light and is confined in a volume of one cubic millimeter. The energy is

$$\varepsilon = mv^2/2 = 9.1 \times 10^{-31}(0.05 \times 3 \times 10^8)^2/2 = 1.0 \times 10^{-16} \text{ J}.$$

If we take $\delta\varepsilon = 0.01\varepsilon$, the degeneracy is

$$g = \frac{\pi}{4}\left(\frac{8mV^{2/3}}{h^2}\right)^{3/2}\varepsilon^{1/2}\delta\varepsilon$$

$$= \frac{\pi}{4}\left(\frac{8 \times 9.1 \times 10^{-31} \times (.001)^{2/3}}{(6.63 \times 10^{-34})^2}\right)^{3/2}(1.0 \times 10^{-16})^{1/2}(.01 \times 1.0 \times 10^{-16})$$

$$= 5.3 \times 10^{26}.$$

Clearly, the degeneracy of this energy state is very large. For a system having a large number of such particles, the system degeneracy would be enormous!

Table 1.1 *Summary of Relations for Energy Quantum States, Degeneracy, and Selection Rules*

	Energy Quantum States	Degeneracy and Selection Rule
Particle in a 3-D box	$\varepsilon = \dfrac{h^2}{8mV^{2/3}}\left(n_x^2 + n_y^2 + n_z^2\right)$ n_x, n_y, n_z contained in $\{1, 2, 3, \ldots\}$	$g(\varepsilon) = \dfrac{\pi}{4}\left(\dfrac{8mV^{2/3}}{h^2}\right)^{3/2}\varepsilon^{1/2}\delta\varepsilon$ Selection rule: none
Harmonic oscillator	$\varepsilon = \left(n + \tfrac{1}{2}\right)h\nu$ $n = 0, 1, 2, 3, \ldots$	$g(\varepsilon_n) = 1$ Selection rule: Dipole moment must vary with oscillation and $\Delta n = \pm 1$
Rigid rotor	$\varepsilon = \dfrac{\hbar^2}{2I}l(l+1)$ $l = 1, 2, 3, \ldots$	$g(\varepsilon_l) = 2l + 1$ Selection rule: Rotor must have a dipole moment and $\Delta l = \pm 1$
Hydrogen atom	$\varepsilon = \dfrac{-m_e C_1^2}{2\hbar^2 n^2}$ $n = 1, 2, 3, \ldots$	$g(\varepsilon_n) = n^2$ Selection rule: none

As we construct a theoretical framework for thermodynamics in later chapters we will have to account both for the quantized energy levels of particles and the degeneracy of the levels. The quantum energy levels and degeneracy for the systems considered in this chapter are summarized in Table 1.1.

1.8   Other Important Results of Quantum Theory

In addition to the quantum theory specification of energy levels and degeneracies, there are other predictions of quantum theory that are relevant to developments in later chapters. First, quantum theory not only predicts the discrete energy level accessible to the system but also predicts the permissible transitions between energy levels. The allowed transitions from one energy state to another are dictated by selections rules, which generally vary from one system to another.

For the particle in a 3-D box, there are no restrictions on transitions from one state to another. The particle in any state, with any combination of n_x, n_y, and n_z, can absorb or emit a photon and undergo a transition to any other state accessible to the particle. Thus, there is no selection rule for this system.

In contrast, the rigid-rotor and harmonic-oscillator systems do obey selection rules. For the rigid rotor, the selection rule requires that the rotor (which is usually a model for a molecule) must have a permanent dipole moment and only transitions associated with emission or absorption of a photon corresponding to $\Delta l = \pm 1$ are permitted.

For the harmonic oscillator, the selection rule requires that emission or absorption of a photon can occur only if the dipole moment of the oscillator varies during vibration and only for $\Delta n = \pm 1$. As we will discuss further in later chapters, molecules may simultaneously be rotating and vibrating. As a first approximation, the energy storage modes are usually modeled as separate independent energy storage modes. The rotational and vibrational energy of a diatomic molecule may then be given by

$$\varepsilon_{\text{vib,rot}} = \left(n + \frac{1}{2}\right) h\nu_0 + \frac{\hbar^2}{2I}l(l+1). \tag{1.122}$$

The energy levels associated with vibration are generally more widely spaced than those for rotation. In this case the respective selection rules for each mechanism described above still apply ($\Delta n = \pm 1$, $\Delta l = \pm 1$). We will discuss the effect of these selection rules on emission spectra in a later chapter.

For the simple hydrogen atom system discussed in Section 1.5, there are no selection rule requirements on transitions between electronic energy states. This is not generally true for other atoms and molecules, however. When additional electrons are present and when both vibrational and electronic transitions may occur, selection rules specific to each circumstance must be obeyed.

Derivation of selection rules for different system types requires an analysis of photon interaction with the molecule using the full time-dependent Schrödinger equation. Although analysis of this sort is beyond the scope of this introductory presentation, knowledge of the selection rules is important to the understanding of absorption and emission spectra in gases. For our purposes, we will incorporate the selection rules into our analysis as accepted results of quantum theory. Selection rules for the simple systems considered in this chapter are summarized in Table 1.1. The reader interested in more detailed treatment of this issue may consult the references by Liboff [1] and McQuarrie [2] cited at the end of this chapter.

Another noteworthy result of quantum theory is that all known particles generally fall into one of two categories: fermions and bosons. This categorization is associated with the quantum property of spin. Fundamental particles that have half-odd-integral spin are *fermions*. Particles that have integral spin are *bosons*. Electrons and protons are fermions. In addition, a compound particle containing an odd number of fermions will itself be a fermion if the binding energy of the compound particle is high compared with other transitions. An example of such a particle is helium 3 (He^3). Photons and deuterons have integral spin and are therefore bosons. The full significance of the spin property can only be appreciated by examining the quantum mechanics theory in more detail. We will not pursue such an examination here. It is sufficient for us to note that quantum theory requires fermions to obey the Pauli exclusion principle, which states that in a system of N fermion particles, no two fermions may occupy the same quantum state. In a system of N boson particles, there are no restrictions on how many particles may occupy a given microstate. Atoms and molecules for which there are no restrictions on microstate occupancy are also generally categorized as bosons.

In developing a statistical treatment of energy storage in a system of N particles, we therefore must allow for two particle types: fermions and bosons. In the next chapter we will begin to explore the impact of particle type on macroscopic thermodynamic behavior of multiparticle systems.

Exercises

1.1 Using laboratory instrumentation, if we can determine the position of a helium atom to within 1 nm (10^{-9} m), what is the quantum limitation on the uncertainty of any measurements of its velocity?

1.2 We will show later that the mean speed of a molecule or atom in an ideal gas is given by $(8k_B T/\pi m)^{1/2}$, where T is the absolute temperature and $k_B = 1.38 \times 10^{-23}$ J/K. Using this relation, determine and plot the variation of the mean quantum wavelength for a helium atom ($m = 6.64 \times 10^{-27}$ kg) in an ideal gas between absolute temperatures of 5 K and 300 K. What does this suggest about the effect of temperature on quantum effects in an ideal gas?

1.3 As mentioned in Example 1.4, the mean translational energy for an atom in an ideal gas is $(3/2)k_B T$, where T is the absolute temperature and $k_B = 1.38 \times 10^{-23}$ J/K. Assuming the 3-D particle-in-a-box solution applies to the translational quantum levels for neon molecules in an ideal gas, determine the order of magnitude of the largest quantum number among n_x, n_y, and n_z for atoms in a system at $T = 290$ K. Is this value large or small compared to one?

1.4 CO_2 is a linear molecule with a carbon atom ($m = 1.99 \times 10^{-26}$ kg) at the center and an oxygen atom ($m = 2.66 \times 10^{-26}$ kg) at each end. The separation between the carbon and oxygen atoms is about 2.77×10^{-10} m. Use the rigid-rotor quantum solution to determine the separation between the first two energy levels for CO_2 molecules in an ideal gas.

1.5 Use the rigid-rotor solution to determine the first three rotational energy levels of N_2O. This molecule is linear and unsymmetric (N—N—O) with a moment of inertia of 6.69×10^{-46} kg m^2.

1.6 We will show in a later chapter that the vibrational energy storage in a nitrogen molecule can be modeled using the harmonic-oscillator quantum solution if the mass of the oscillator is taken to be the reduced mass $m_r = m/2$ (m being the mass of a nitrogen atom). A suitable value of the spring constant k must also be specified. A value of 2,250 N/m has been suggested for the effective spring constant. Use this value and the harmonic-oscillator solution to predict the separation of the first two vibrational energy levels for the nitrogen atom. We will also show later that in an ideal gas at absolute temperature T, the mean translational energy for a nitrogen atom is $(3/2)k_B T$ where $k_B = 1.38 \times 10^{-23}$ J/K. Compare the separation between the two lowest vibrational levels to the mean translational energy of a nitrogen molecule in an ideal gas at an absolute temperature of 290 K. (Note that for nitrogen, $m = 5.65 \times 10^{-26}$ kg.)

1.7 For the harmonic oscillator discussed in the text, the wave function for the lowest (ground state) energy level is given by Eq. (1.60) with $n = 0$. (Note that $H_0(\alpha^{1/2}x)$ is identically one for all arguments.) An oscillator in this state has an energy $\varepsilon = \hbar\sqrt{k/m}/2$.

   (a) For a classical harmonic oscillator with specified k, m and total energy ε, derive a relation for the maximum amplitude of the oscillation $x_{max,cl}$.

   (b) For the quantum-mechanical oscillator, determine the probability that the amplitude of the oscillation will exceed $x_{max,cl}$. This behavior is characteristic of quantum-mechanical systems and is responsible for the phenomenon called the *tunnel effect* in which the system is said to have tunneled into a classically forbidden zone.

1.8 The three-dimensional harmonic oscillator shown in Figure 1.11 is a possible model for an atom in a solid crystal lattice. The restoring force in each of the directions x, y, z is linear and represented by a "spring" constant for each direction. You may assume that the oscillations are small enough that the energy stored in any one spring is independent of that stored in any other.

   (a) Write down the appropriate form of the Schrödinger equation for this problem and list the boundary conditions.

   (b) Develop a solution for the wave function and determine the quantum energy levels for the 3-D case. (Hint: Where possible, make use of the results obtained for the 1-D harmonic oscillator.)

   (c) Do the energy levels of this system have a degeneracy greater than 1? Briefly explain your answer.

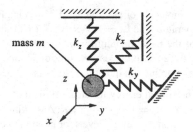

Figure 1.11

1.9 Using the results of the model analysis in the text, calculate the ionization energy of the hydrogen atom. Note that the ionization energy is the energy required to take the electron from the ground state to a totally unbound condition (far from the nucleus). Express your answer in joules.

1.10 We will later show that in an ideal gas at absolute temperature T, the mean translational energy for a helium atom is $(3/2)k_B T$. Use the results of the particle-in-a-box quantum model to predict the degeneracy at ε values within 1% of $(3/2)k_B T$ for a system with a volume of $0.1 \ \text{m}^3$ and an absolute temperature at 280 K. (Note that for helium, $m = 6.64 \times 10^{-27}$ kg.)

1.11 As mentioned in Example 1.5, the mean rotational energy for a nitrogen molecule in an ideal gas is $k_B T$, where T is the absolute temperature and $k_B = 1.38 \times 10^{-23}$ J/K. Assuming the rigid-rotor solution applies to the rotational quantum levels for nitrogen molecules in an ideal gas, determine the quantum number l and the degeneracy g at the mean energy $k_B T$ for nitrogen molecules in a system at $T = 290$ K. Are these values large or small compared to one?

1.12 Carbon dioxide is a linear molecule with a carbon atom ($m = 1.99 \times 10^{-26}$ kg) at the center and an oxygen atom ($m = 2.66 \times 10^{-26}$ kg) at each end. The separation between the carbon and oxygen atoms is about 2.77×10^{-10} m. We will show in a later chapter that the mean rotational energy for a CO_2 molecule in an ideal gas is $k_B T$, where T is the absolute temperature and $k_B = 1.38 \times 10^{-23}$ J/K. Assuming the rigid-rotor solution applies to the rotational quantum levels for CO_2 molecules in an ideal gas, determine the quantum number l and the degeneracy g at the mean energy $k_B T$ for CO_2 molecules in a system at $T = 290$ K. Are these values large or small compared to one?

References

[1] Liboff, R. L., *Introductory Quantum Mechanics*, Addison-Wesley Publishing Company, Reading, MA, 1980.
[2] McQuarrie, D. A., *Quantum Chemistry*, University Science Books, Mill Valley, CA, 1983.

Statistical Treatment of Multiparticle Systems

Using the basic features of microscale energy storage discussed in Chapter 1, in Chapter 2 we develop the foundations of statistical thermodynamics. In doing so, we introduce the concepts of microstates and macrostates and properly account for the fact that particles in fluid systems are generally indistinguishable. The development of the theoretical framework in this and subsequent chapters considers a binary mixture of two particle types. The statistical machinery is applied first to a microcanonical ensemble of systems, each having a specified volume, number of particles, and total internal energy. Definitions of entropy and temperature emerge from this development. Application of the results to a monatomic gas is discussed.

2.1 Microstates and Macrostates

In this chapter we will construct a general statistical mechanics foundation on which we will develop a full equilibrium thermodynamic theory for systems composed of a large number of particles. In doing so we will make use of the information about energy storage derived from quantum theory in the previous chapter.

In analyzing systems of particles, we can deal with the state of a system at two levels: the microstate of the system and the macrostate of the system. The system *microstate* is the detailed configuration of the system at a microscopic level. To specify the microstate we would have to specify the quantum state (including the position) of each particle in the system. If we observe a system at a macroscopic level, we can, at best, distinguish some of the gross characteristics of the system. Here, the term macroscopic refers to observations that are averaged in some sense over time and length scales that are much larger than those characterizing molecular motion. A nontechnical analogy of this circumstance is an impressionist painting viewed from across the room. At that distance, the eye averages together the microscale brush strokes. In a macroscopic sense, the image is well defined, even though we cannot discern every individual brush stroke.

An additional point can be extracted from the analogy with impressionist paintings. Because the eye averages the contributions of individual brush strokes, different microscopic arrangements of brush strokes may average to produce essentially the same macroscopic image when seen from across the room. Thus, one macrostate can correspond to many microstates. The same is true of thermodynamic systems composed of many particles. The *macrostate* is a macroscopically distinguishable system characteristic that corresponds to a collection of system microstates that macroscopically "look" the same.

Equilibrium in classical thermodynamics has an intrinsically macroscopic viewpoint. It implies that the macrostate of the system is unchanging with time. However, we expect that the microstate will continuously change with time owing to the motion of the particles and collisions among particles. If the macrostate changes with time, the associated phenomena lie in the domain of nonequilibrium thermodynamics – a topic we will explore in later sections of this text. From a microscopic perspective, macroscopic equilibrium results when a system is constrained in such a way that it spends the overwhelming majority of the time in microstates that correspond to one macrostate.

The statistical treatment of systems of many particles is based on the following postulate:

◆   *All microstates consistent with the constraints on the system are equally probable.*

Since, as discussed above, each macrostate corresponds to many microstates, it follows directly that:

◆   *The macrostate having the largest number of (equally probable) microstates is most probable (i.e., most likely to be observed).*

These concepts are central to the development of a statistical treatment of multiparticle systems. We will expand them in more quantitative ways later in this chapter. We now must bring in appropriate computational machinery to make the treatment functional. This is done in the next section. Some of the computational machinery is based on results of combinatorial analysis and basic probability theory, which are reviewed in Appendix I. Readers unfamiliar with these areas are urged to read Appendix I before proceeding to the next section.

2.2   The Microcanonical Ensemble and Boltzmann Statistics

The use of ensemble concepts was developed to a refined methodology by J. Willard Gibbs in the early 1900s. We could pursue a statistical analysis that focuses on the energy states of individual particles within a system, as Boltzmann did in his pioneering work on statistical mechanics. One drawback of such an approach is that it becomes more complicated to use when the particles in the system are interdependent in such a way that the total system energy cannot be cleanly divided among the particles. The ensemble formalism has the advantage that it can be extended in a straightforward way to treat systems of interdependent particles.

For our purposes, an ensemble is defined as a collection of a very large number of identical systems that are all subject to the same macroscopic constraints. The analysis we will develop is referred to as *ensemble theory*. Each system in the ensemble is identical in its imposed macroscopic constraints, but each system in the ensemble is in a different microstate. In a single real system with the imposed macroscopic constraints, the microstate will vary with time owing to random fluctuations at the microscopic level resulting from particle motion and particle collisions. The ensemble of systems can be viewed as a list of the microstates from which the real system is randomly selecting its instantaneous microstate. This interpretation is the basis of the *ergodic hypothesis*, which may be stated as follows:

◆   *The ensemble exhibits the same average properties in space (over the ensemble) as a single system exhibits in time.*

It follows from this hypothesis that the ensemble average of a property is the same as the time average of that property for the single system. Each system in the ensemble must be in one of the microstates consistent with the imposed constraints. As a result, different constraints define different types of ensembles.

In this chapter we will consider a specific type of ensemble in which all systems consist of N_a particles of species a and N_b particles of species b confined within a specified volume V. Each system in the ensemble has the same total internal energy U. We denote this energy

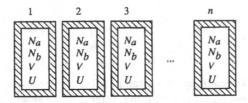

Figure 2.1 The microcanonical ensemble.

content as *internal energy* to differentiate it from kinetic energy associated with bulk motion of the entire system or potential energy associated with the bulk system. Specifically, the internal energy is the sum total of the energy associated with translation of the particles and the energy stored within the particles. The motivation for picking a system containing a binary mixture of two components is that it will allow us to examine how the statistical framework is affected by the presence of two species. Note that by setting N_a or N_b to zero we should be able to obtain results that are applicable to a single-component system.

Each ensemble member system is isolated from all other systems so that N_a, N_b, V, and U values are fixed. As indicated schematically in Figure 2.1, there are n systems in the ensemble, where n is very large. In particular, n must be large enough so that the ensemble includes every microstate accessible to the system within the imposed constraints.

For this type of ensemble, there are no work interactions, no heat transfer, and no mass exchange among the systems. In later chapters we will consider other types of ensembles where one or more of these interactions among systems will be possible. For now, however, we will specifically focus on the microcanonical ensemble.

Our objective in this section is to use the microcanonical ensemble as a framework to explore Boltzmann statistics. In the systems in the ensemble, particles of each species all have access to the same set of quantum energy states for that particle type under the specified system constraints. Each system in the ensemble is in one of a large number of microstates that are possible for the system with the specified N_a, N_b, V, and U values. We define

$\Omega =$ the total number of system microstates consistent with the values of N_a,

N_b, V, and U specified for the ensemble member systems.

If we had the ability to examine one of the systems at a microscopic level, we could, in principle, determine the energy level in which each particle resides and we could count the number of particles that reside in each energy level for that specific system in the ensemble. At the end of this process, we would have two sets of occupation numbers $\{N_{a,i}\}$ and $\{N_{b,j}\}$, where each $N_{a,i}$ is the number of particles of type a in energy level i having energy ε_i, and each $N_{b,j}$ is the number of particles of type b in energy level j with energy ε_j. Note that in general, the energy levels ε_i and ε_j will have degeneracies g_i and g_j, respectively. Each energy level may therefore correspond to many microstates.

The sets of occupation numbers $\{N_{a,i}\}$ and $\{N_{b,j}\}$ must satisfy the following constraints on the system:

$$\sum_{i=0}^{\infty} N_{a,i} = N_a, \tag{2.1}$$

$$\sum_{j=0}^{\infty} N_{b,j} = N_b, \tag{2.2}$$

and

$$\sum_{i=0}^{\infty} \varepsilon_i N_{a,i} + \sum_{j=0}^{\infty} \varepsilon_j N_{b,j} = U, \tag{2.3}$$

where the summation limit ∞ denotes a summation over all possible energy levels. If we could similarly determine the occupation numbers for all the systems in the ensemble, we would find that, in many instances, systems will have the same pair of occupation number sets. This is a consequence of the fact that a given set of occupation numbers can correspond to many different microstates.

The sets of $\{N_{a,i}\}$ and $\{N_{b,j}\}$ values dictate how energy is distributed in the system, since $N_{a,i}/N_a$ is the fraction of type a particles with energy ε_i and $N_{b,j}/N_b$ is the fraction of the b particles with energy ε_j. Because the total system energy is limited, there is a finite limit to the energy level attainable by a single particle and therefore the occupation number values must approach zero as $i \to \infty$ or $j \to \infty$. A specific pair of $\{N_{a,i}\}$ and $\{N_{b,j}\}$ distributions thus indicates, in a gross or macroscopic way, the state of the system. We therefore designate each pair of occupation number sets $\{N_{a,i}\}$ and $\{N_{b,j}\}$ satisfying the above global constraints (2.1)–(2.3) as being a macrostate of the system.

We assume that we can distinguish individual particles of both species from one another (i.e., they are *distinguishable*). In a later section we will reconsider the validity of this assumption, but for now we will accept it as correct for the system we are considering.

Based on the arguments described above, a macrostate is presumed to correspond to a pair of occupation number sets $\{N_{a,i}\}$ and $\{N_{b,j}\}$ satisfying the global constraints (2.1)–(2.3). For species a, the number of ways of dividing the N_a distinguishable particles among the energy levels in conformance with the occupation number set $\{N_{a,i}\}$ is equal to the number of ways of dividing N_a distinguishable objects into groups of $N_{a,0}, N_{a,1}, \ldots, N_{a,i}, \ldots$. From basic combinatorial analysis (see Appendix I) this is given by

$$\left\{ \begin{array}{l} \textit{number of ways of dividing } N_a \textit{ particles} \\ \textit{among energy levels as dictated by } \{N_{a,i}\} \end{array} \right\} = N_a! \prod_{i=0}^{\infty} \frac{1}{N_{a,i}!}. \tag{2.4}$$

However, for species a, each energy level ε_i has degeneracy g_i, which means that the $N_{a,i}$ particles in that energy level may be in any of g_i microstates. As discussed in Section 1.8, we also know that particles may be bosons or fermions. Here, we will assume that our particles are bosons, and hence they are free to occupy any microstate without restriction. (We will return to consider fermions in a later section.)

To get a relation for the number of microstates corresponding to an occupancy number set, we must multiply the contribution of each energy level in the above equation by the number of ways that the $N_{a,i}$ particles can be distributed among the g_i microstates for that energy level. As described in Appendix I, the number of ways of distributing $N_{a,i}$ distinguishable particles in g_i microstates is equal to $g_i^{N_{a,i}}$. Multiplying this by the corresponding term in Eq. (2.4) for each energy level, we obtain the following relation for the number of ways of arranging the species a particles among the microstates for a specified occupancy number set:

$$\left\{ \begin{array}{l} \textit{number of ways of arranging} \\ \textit{species a particles among} \\ \textit{microstates dictated by } g_i \textit{ and } \{N_{a,i}\} \end{array} \right\} = N_a! \prod_{i=0}^{\infty} \frac{g_i^{N_{a,i}}}{N_{a,i}!}. \tag{2.5}$$

Based on identical arguments, we can write a similar relation for species b:

$$\left\{ \begin{array}{l} \textit{number of ways of arranging} \\ \textit{species b particles among} \\ \textit{microstates dictated by } g_j \textit{ and } \{N_{b,j}\} \end{array} \right\} = N_b! \prod_{j=0}^{\infty} \frac{g_j^{N_{b,j}}}{N_{b,j}!}. \tag{2.6}$$

For each arrangement of the a particles, the number of possible arrangements of b particles is given by Eq. (2.6). The total number of combined arrangements of the N_a and N_b particles is therefore equal to the product of the expressions on the right-hand sides of Eqs. (2.5) and (2.6). This total number of arrangements is the total number of microstates for the specified pair of $\{N_{a,i}\}$ and $\{N_{b,j}\}$ distributions. Designating this number of microstates as W, we can therefore write

$$W(\{N_{a,i}\}, \{N_{b,j}\}) = N_a! \left(\prod_{i=0}^{\infty} \frac{g_i^{N_{a,i}}}{N_{a,i}!} \right) N_b! \left(\prod_{j=0}^{\infty} \frac{g_j^{N_{b,j}}}{N_{b,j}!} \right). \tag{2.7}$$

Note that for real systems $W \approx O(N_A!) \approx O(10^{26}!)$. It follows that the total number of possible microstates consistent with the system constraints is equal to the sum of the W values for all possible sets of occupation numbers:

$$\Omega = \sum_{\{N_{a,i}\}} \sum_{\{N_{b,j}\}} W(\{N_{a,i}\}, \{N_{b,j}\})$$

$$= \sum_{\{N_{a,i}\}} \sum_{\{N_{b,j}\}} \left[N_a! \left(\prod_{i=0}^{\infty} \frac{g_i^{N_{a,i}}}{N_{a,i}!} \right) N_b! \left(\prod_{j=0}^{\infty} \frac{g_j^{N_{b,j}}}{N_{b,j}!} \right) \right], \tag{2.8}$$

where the notation

$$\sum_{\{N_{a,i}\}} \sum_{\{N_{b,j}\}}$$

indicates a sum over all possible sets of occupation numbers consistent with constraints (2.1)–(2.3).

We will denote the probability of an observed state as $\tilde{P}$(observed state). From the above observations it follows that

$$\tilde{P}(\text{macrostate } \{N_{a,i}\}, \{N_{b,j}\}) = \frac{W(\{N_{a,i}\}, \{N_{b,j}\})}{\Omega},$$

$$\tilde{P}(\text{any given microstate}) = \frac{1}{\Omega}.$$

From these results we can also interpret W as

$$W = \frac{\tilde{P}(\text{the specified macrostates})}{\tilde{P}(\text{any given microstate})}.$$

W is sometimes interpreted as a disorder number or a randomness number. In general, W is very large because $\tilde{P}$ (any given microstate) is small.

Since all microstates are equally probable, the macrostate with the largest number of corresponding microstates is most probable and hence is most likely to be observed in a real system constrained to specified N_a, N_b, V, and U values. Consider as a common example

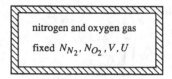

Figure 2.2   An example system.

of such a thermodynamic system a mixture of oxygen and nitrogen (similar to air) in a
closed, rigid, perfectly insulated container (Figure 2.2).

Macroscopic measurements indicate that at equilibrium, the macrostate observed for the
system does not change detectably over long periods of time. It is clear that during such long
time intervals, the system microstate changes many times due to motion of the molecules
and collisions. If all available microstates are equally probable, a single macrostate is likely
to be observed for long time intervals only if the number of microstates for that macrostate
is much larger than the number for any other macrostate.

In terms of our ensemble, this implies that the overwhelming majority of the member
systems will be in microstates that correspond to one macrostate (or equivalently, one pair
of occupation number sets). For the system considered here, we therefore seek to find the
pair of occupation number sets $\{N_{a,i}\}$, $\{N_{b,j}\}$ that maximizes W subject to the constraints
(2.1)–(2.3). We expect that this set of occupation numbers are those that characterize the real
system at equilibrium. Once determined, this distribution will provide the means to relate microscopic characteristics of the particles to macroscopic properties of the system at equilibrium. These results will form the foundation of our development of thermodynamics theory.

The paired sets of occupation numbers that maximize $\ln W$ also maximize W. Since
attempting to maximize $\ln W$ is mathematically simpler than attempting to maximize W,
we will adopt the objective of finding the sets of occupation numbers $\{N_{a,i}\}$, $\{N_{b,j}\}$ that
maximize $\ln W$ subject to the constraints (2.1)–(2.3).

Taking the natural log of both sides of Eq. (2.7) yields

$$\ln W = \ln N_a! + \sum_{i=0}^{\infty} N_{a,i} \ln g_i - \sum_{i=0}^{\infty} \ln N_{a,i}! + \ln N_b! + \sum_{j=0}^{\infty} N_{b,j} \ln g_j - \sum_{j=0}^{\infty} \ln N_{b,j}!.$$

$$(2.9)$$

Since N_a and N_b are typically very large, we make use of *Stirling's approximation* (see
Appendix I),

$$\ln N! \cong N \ln N - N \quad \text{(for large } N\text{)}. \tag{2.10}$$

Using (2.10) to evaluate the $\ln N_a$ and $\ln N_b$ terms, the relation (2.9) for $\ln W$ becomes

$$\ln W = N_a \ln N_a - N_a + \sum_{i=0}^{\infty} N_{a,i} \ln g_i - \sum_{i=0}^{\infty} (N_{a,i} \ln N_{a,i} - N_{a,i}) + N_b \ln N_b$$

$$- N_b + \sum_{j=0}^{\infty} N_{b,j} \ln g_j - \sum_{j=0}^{\infty} (N_{b,j} \ln N_{b,j} - N_{b,j}). \tag{2.11}$$

Using constraints (2.1) and (2.2), this reduces to

$$\ln W = N_a \ln N_a + \sum_{i=0}^{\infty} N_{a,i} \ln g_i - \sum_{i=0}^{\infty} N_{a,i} \ln N_{a,i} + N_b \ln N_b$$

$$+ \sum_{j=0}^{\infty} N_{b,j} \ln g_j - \sum_{j=0}^{\infty} N_{b,j} \ln N_{b,j}. \tag{2.12}$$

Differentiating both sides of the above equation yields

$$d(\ln W) = dN_a \ln N_a + dN_a + \sum_{i=0}^{\infty} dN_{a,i} \ln g_i - \sum_{i=0}^{\infty} (dN_{a,i} \ln N_{a,i} + dN_{a,i})$$

$$+ dN_b \ln N_b + dN_b + \sum_{j=0}^{\infty} dN_{b,j} \ln g_j - \sum_{j=0}^{\infty} (dN_{b,j} \ln N_{b,j} + dN_{b,j}), \tag{2.13}$$

and since constraints (2.1) and (2.2) imply that

$$\sum_{i=0}^{\infty} dN_{a,i} = dN_a = 0, \qquad \sum_{j=0}^{\infty} dN_{b,j} = dN_b = 0,$$

Eq. (2.13) reduces to

$$d(\ln W) = \sum_{i=0}^{\infty} dN_{a,i} \ln g_i - \sum_{i=0}^{\infty} dN_{a,i} \ln N_{a,i} + \sum_{j=0}^{\infty} dN_{b,j} \ln g_j - \sum_{j=0}^{\infty} dN_{b,j} \ln N_{b,j}. \tag{2.14}$$

At the maximum that we seek, $d(\ln W) = 0$. Imposing this requirement on Eq. (2.14) gives

$$\sum_{i=0}^{\infty} dN_{a,i} \ln g_i - \sum_{i=0}^{\infty} dN_{a,i} \ln N_{a,i} + \sum_{j=0}^{\infty} dN_{b,j} \ln g_j - \sum_{j=0}^{\infty} dN_{b,j} \ln N_{b,j} = 0. \tag{2.15}$$

Differentiating the constraint relations yields

$$\sum_{i=0}^{\infty} dN_{a,i} = dN_a = 0, \tag{2.16a}$$

$$\sum_{j=0}^{\infty} dN_{b,j} = dN_b = 0, \tag{2.16b}$$

and

$$\sum_{i=0}^{\infty} \varepsilon_i dN_{a,i} + \sum_{j=0}^{\infty} \varepsilon_j dN_{b,j} = dU = 0. \tag{2.16c}$$

We now make use of the method of Lagrange multipliers. (See Appendix I for a description of this technique.) We multiply the constraint equations by Lagrange multipliers λ_1, λ_2, and λ_3 respectively:

$$\lambda_1 \sum_{i=0}^{\infty} dN_{a,i} = 0, \tag{2.17a}$$

$$\lambda_2 \sum_{j=0}^{\infty} dN_{b,j} = 0, \tag{2.17b}$$

$$\lambda_3 \sum_{i=0}^{\infty} \varepsilon_i dN_{a,i} + \lambda_3 \sum_{j=0}^{\infty} \varepsilon_j dN_{b,j} = 0. \tag{2.17c}$$

We then add these equations and subtract Eq. (2.15) to get

$$\lambda_1 \sum_{i=0}^{\infty} dN_{a,i} + \lambda_3 \sum_{i=0}^{\infty} \varepsilon_i dN_{a,i} - \sum_{i=0}^{\infty} dN_{a,i} \ln g_i + \sum_{i=0}^{\infty} dN_{a,i} \ln N_{a,i}$$

$$+ \lambda_2 \sum_{j=0}^{\infty} dN_{b,j} + \lambda_3 \sum_{j=0}^{\infty} \varepsilon_j dN_{b,j} - \sum_{j=0}^{\infty} dN_{b,j} \ln g_j + \sum_{j=0}^{\infty} dN_{b,j} \ln N_{b,j} = 0,$$

which reduces to

$$\sum_{i=0}^{\infty} (\lambda_1 + \lambda_3 \varepsilon_i - \ln g_i + \ln N_{a,i}) \, dN_{a,i} + \sum_{j=0}^{\infty} (\lambda_2 + \lambda_3 \varepsilon_j - \ln g_j + \ln N_{b,j}) \, dN_{b,j} = 0.$$

$$\tag{2.18}$$

At the maximum, this relation must hold for arbitrary choices of $dN_{a,i}$ and $dN_{b,j}$, which implies that the coefficients of the differential terms must all be zero:

$$\lambda_1 + \lambda_3 \varepsilon_i - \ln g_i + \ln N_{a,i} = 0, \tag{2.19a}$$

$$\lambda_2 + \lambda_3 \varepsilon_j - \ln g_j + \ln N_{b,i} = 0. \tag{2.19b}$$

Solving these relations for $N_{a,i}$ and $N_{b,j}$, we obtain

$$N_{a,i} = g_i e^{-\lambda_1} e^{-\lambda_3 \varepsilon_i}, \tag{2.20a}$$

$$N_{b,j} = g_j e^{-\lambda_2} e^{-\lambda_3 \varepsilon_j}. \tag{2.20b}$$

Using the constraint relations (2.1) and (2.2) together with Eq. (2.20) yields

$$N_a = \sum_{i=0}^{\infty} N_{a,i} = \sum_{i=0}^{\infty} g_i e^{-\lambda_1} e^{-\lambda_3 \varepsilon_i} = e^{-\lambda_1} \sum_{i=0}^{\infty} g_i e^{-\lambda_3 \varepsilon_i}, \tag{2.21a}$$

$$N_b = \sum_{j=0}^{\infty} N_{b,j} = \sum_{j=0}^{\infty} g_j e^{-\lambda_2} e^{-\lambda_3 \varepsilon_j} = e^{-\lambda_2} \sum_{j=0}^{\infty} g_j e^{-\lambda_3 \varepsilon_j}. \tag{2.21b}$$

Combining Eqs. (2.20) and (2.21), we find that

$$\frac{N_{a,i}}{N_a} = \frac{g_i e^{-\lambda_1} e^{-\lambda_3 \varepsilon_i}}{e^{-\lambda_1} \sum_{i=0}^{\infty} g_i e^{-\lambda_3 \varepsilon_i}} = \frac{g_i e^{-\lambda_3 \varepsilon_i}}{\sum_{i=0}^{\infty} g_i e^{-\lambda_3 \varepsilon_i}}, \tag{2.22a}$$

$$\frac{N_{b,j}}{N_b} = \frac{g_j e^{-\lambda_2} e^{-\lambda_3 \varepsilon_j}}{e^{-\lambda_2} \sum_{j=0}^{\infty} g_j e^{-\lambda_3 \varepsilon_j}} = \frac{g_j e^{-\lambda_3 \varepsilon_j}}{\sum_{j=0}^{\infty} g_j e^{-\lambda_3 \varepsilon_j}}. \tag{2.22b}$$

The distributions above are generally written in the forms

$$\frac{N_{a,i}}{N_a} = \frac{g_i e^{-\lambda_3 \varepsilon_i}}{Z_a} \tag{2.23a}$$

and

$$\frac{N_{b,j}}{N_b} = \frac{g_j e^{-\lambda_3 \varepsilon_j}}{Z_b}, \tag{2.23b}$$

where

$$Z_a = \sum_{i=0}^{\infty} g_i e^{-\lambda_3 \varepsilon_i}, \tag{2.24a}$$

$$Z_b = \sum_{j=0}^{\infty} g_j e^{-\lambda_3 \varepsilon_j}. \tag{2.24b}$$

Equations (2.23a) and (2.23b) above are the distributions that maximize $\ln W$ subject to the imposed constraints. They are generalized *Boltzmann distributions* in terms of the as yet undefined constant λ_3. The parameters Z_a and Z_b are termed *partition functions*. In a single-component system only one partition function would be defined, usually just designated as Z. We will see that a partition function is itself a thermodynamic property because it is a unique function of the macrostate of the system.

2.3 Entropy and Temperature

In the analysis in the previous section, note that once sets of $N_{a,i}$ and $N_{b,j}$ values are specified, W can be calculated using Eq. (2.7). Thus, W is a macroscopic property of this system, which indicates the macrostate of this system. For the same reasons, $\ln W$ may be considered to be a macroscopic property of the system. Either W or $\ln W$ could be used as an indicator of the macrostate for a system with specified N_a, N_b, U, and V. Although they both contain the same information, using $\ln W$ as a property is advantageous. To understand why, we first adopt the following definition for a property we call the *entropy* of the system:

$$S = k_B \ln W. \tag{2.25}$$

In the above definition, k_B is a numerical constant that we set to define the units for this property. In accordance with convention, k_B is referred to as the *Boltzmann constant* in recognition of Boltzmann's pioneering work in statistical mechanics on which the derivation of Eq. (2.25) is based.

. Now we will explore how this property behaves when we consider an isolated system composed of two subsystems (see Figure 2.3).

When the entire composite system is at equilibrium, the overall W and S values for each subsystem are uniquely determined:

$$S_1 = k_B \ln W_1, \tag{2.26}$$

$$S_2 = k_B \ln W_2. \tag{2.27}$$

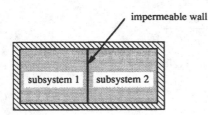

Figure 2.3

By definition, W_1 is equal to the number of ways we can divide N_1 distinguishable particles among the energy microstates of system 1, and for each way of dividing up the N_1 particles in subsystem 1, there are W_2 ways of dividing up the N_2 particles of subsystem 2 among its energy microstates. It follows that the total number of ways in which the two subsystems can divide $N_1 + N_2$ particles among their respective energy microstates is $W_1 W_2$:

$$W_{1\&2} = W_1 W_2. \tag{2.28}$$

Hence, for the overall system

$$S_{1\&2} = k_B \ln\{W_1 W_2\} = k_B \ln W_1 + k_B \ln W_2 = S_1 + S_2. \tag{2.29}$$

Clearly the entropy of the composite system is equal to the sum of the entropies of the two subsystems. Since this analysis can easily be extended to any number of subsystems by considering them two at a time, we conclude that the entropy of a composite system with an arbitrary number of subsystems is additive over the constituent subsystems. This *additivity property* is intuitively attractive and practically useful. When conducting a thermodynamic analysis of a system composed of several subsystems, the definition of this entropy property ensures that the entropy of the overall system is just the sum of the entropies of the component systems. We therefore adopt the definition (2.25) for a new macroscopic property of systems called entropy and include it in our development of the theoretical framework of thermodynamics.

It is important to note that the definition of entropy given in Eq. (2.25) is applicable to any macrostate (with any sets of occupation numbers). If we designate the occupation number sets that maximize W for the binary system considered in the previous sections as $\{N_{a,i}\}^*$ and $\{N_{b,j}\}^*$, then the value of entropy for the system at equilibrium is given by

$$S = k_B \ln W_{\max} \text{ (at equilibrium)}, \tag{2.30a}$$

where

$$W_{\max} = W(\{N_{a,i}\}^*, \{N_{b,j}\}^*). \tag{2.30b}$$

We now wish to return to the issue of evaluating the constant λ_3 in Eqs. (2.23a,b). We begin by substituting Eq. (2.12) into Eq. (2.25) to obtain

$$\frac{S}{k_B} = N_a \ln N_a + \sum_{i=0}^{\infty} N_{a,i} \ln g_i - \sum_{i=0}^{\infty} N_{a,i} \ln N_{a,i} + N_b \ln N_b$$

$$+ \sum_{j=0}^{\infty} N_{b,j} \ln g_j - \sum_{j=0}^{\infty} N_{b,j} \ln N_{b,j}. \tag{2.31}$$

In (2.31) we next evaluate $N_{a,i}$ and $N_{b,j}$ terms in the summation using Eqs. (2.23a,b),

$$\frac{S}{k_{\mathrm{B}}} = N_a \ln N_a + \sum_{i=0}^{\infty} \left(\frac{N_a g_i e^{-\lambda_3 \varepsilon_i}}{Z_a} \right) \ln g_i - \sum_{i=0}^{\infty} \left(\frac{N_a g_i e^{-\lambda_3 \varepsilon_i}}{Z_a} \right)$$

$$\times \ln \left\{ \frac{N_a g_i e^{-\lambda_3 \varepsilon_i}}{Z_a} \right\} + N_b \ln N_b + \sum_{j=0}^{\infty} \left(\frac{N_b g_j e^{-\lambda_3 \varepsilon_j}}{Z_b} \right) \ln g_j$$

$$- \sum_{j=0}^{\infty} \left(\frac{N_b g_j e^{-\lambda_3 \varepsilon_j}}{Z_b} \right) \ln \left\{ \frac{N_b g_j e^{-\lambda_3 \varepsilon_j}}{Z_b} \right\}, \tag{2.32}$$

and reorganize the summations to obtain

$$\frac{S}{k_{\mathrm{B}}} = N_a \ln N_a + \frac{N_a}{Z_a} \sum_{i=0}^{\infty} g_i e^{-\lambda_3 \varepsilon_i} (\ln g_i + \lambda_3 \varepsilon_i - \ln N_a - \ln g_i + \ln Z_a)$$

$$+ N_b \ln N_b + \frac{N_b}{Z_b} \sum_{j=0}^{\infty} g_j e^{-\lambda_3 \varepsilon_j} (\ln g_j + \lambda_3 \varepsilon_j - \ln N_b - \ln g_j + \ln Z_b)$$

$$= N_a \ln N_a + \frac{N_a}{Z_a} \sum_{i=0}^{\infty} \lambda_3 \varepsilon_i g_i e^{-\lambda_3 \varepsilon_i} - \frac{N_a \ln N_a}{Z_a} \sum_{i=0}^{\infty} g_i e^{-\lambda_3 \varepsilon_i}$$

$$+ \frac{N_a \ln Z_a}{Z_a} \sum_{i=0}^{\infty} g_i e^{-\lambda_3 \varepsilon_i} + N_b \ln N_b + \frac{N_b}{Z_b} \sum_{j=0}^{\infty} \lambda_3 \varepsilon_j g_j e^{-\lambda_3 \varepsilon_j}$$

$$- \frac{N_b \ln N_b}{Z_b} \sum_{j=0}^{\infty} g_j e^{-\lambda_3 \varepsilon_j} + \frac{N_b \ln Z_b}{Z_b} \sum_{j=0}^{\infty} g_j e^{-\lambda_3 \varepsilon_j} \tag{2.33}$$

Using Eqs. (2.24a,b) to evalate the last two summations in Eq. (2.33) yields

$$\frac{S}{k_{\mathrm{B}}} = \lambda_3 \sum_{i=0}^{\infty} \varepsilon_i \left(\frac{N_a g_i e^{-\lambda_3 \varepsilon_i}}{Z_a} \right) + \lambda_3 \sum_{j=0}^{\infty} \varepsilon_j \left(\frac{N_b g_j e^{-\lambda_3 \varepsilon_j}}{Z_b} \right) + N_a \ln Z_a + N_b \ln Z_b. \tag{2.34}$$

Equations (2.23a,b) indicate that the expressions in brackets in the i and j summations above are just $N_{a,i}$ and $N_{b,j}$, respectively. We can therefore write Eq. (2.34) as

$$\frac{S}{k_{\mathrm{B}}} = \lambda_3 \left(\sum_{i=0}^{\infty} \varepsilon_i N_{a,i} + \sum_{j=0}^{\infty} \varepsilon_j N_{b,j} \right) + N_a \ln Z_a + N_b \ln Z_b. \tag{2.35}$$

Equation (2.3) implies that the term in brackets is equal to U. Substituting U for this term in (2.35), we obtain

$$\frac{S}{k_{\mathrm{B}}} = \lambda_3 U + N_a \ln Z_a + N_b \ln Z_b. \tag{2.36}$$

We next differentiate (2.36) with respect to λ_3 to get

$$\frac{1}{k_B}\left(\frac{\partial S}{\partial \lambda_3}\right)_{N_a,N_b,V} = U + \lambda_3 \left(\frac{\partial U}{\partial \lambda_3}\right)_{N_a,N_b,V} + \frac{N_a}{Z_a}\left(\frac{\partial Z_a}{\partial \lambda_3}\right)_{N_a,N_b,V} + \frac{N_b}{Z_b}\left(\frac{\partial Z_b}{\partial \lambda_3}\right)_{N_a,N_b,V}.$$

(2.37)

The last two partial derivatives on the right side can be evaluated using the definitions of Z_a and Z_b (Eqs. (2.24a,b)):

$$\frac{N_a}{Z_a}\left(\frac{\partial Z_a}{\partial \lambda_3}\right)_{N_a,N_b,V} = \frac{N_a}{Z_a}\left(\frac{\partial}{\partial \lambda_3}\sum_{i=0}^{\infty} g_i e^{-\lambda_3 \varepsilon_i}\right)_{N_a,N_b,V},$$

(2.38)

$$\frac{N_b}{Z_b}\left(\frac{\partial Z_b}{\partial \lambda_3}\right)_{N_a,N_b,V} = \frac{N_b}{Z_b}\left(\frac{\partial}{\partial \lambda_3}\sum_{j=0}^{\infty} g_j e^{-\lambda_3 \varepsilon_j}\right)_{N_a,N_b,V}.$$

(2.39)

In general, the energy levels and the degeneracies for the particles in the system are expected to be functions of the volume of the system and the number of particles of each species. The particle-in-a-3-D-box quantum solution, which is a model of the translational energy storage in the particles, indicates that the quantum energy levels associated with translation depend directly on the system volume (see Table 1.1). Dependence of the energy levels on system volume and number of particles may also occur if the particles are electrically charged or if longer-range molecular attractive forces are important. In general, it is therefore reasonable to expect that the energy levels and degeneracies are functions only of N_a, N_b, and V:

$$\varepsilon_i = \varepsilon_i(N_a, N_b, V),$$

(2.40a)

$$g_i = g_i(N_a, N_b, V),$$

(2.40b)

$$\varepsilon_j = \varepsilon_j(N_a, N_b, V),$$

(2.41a)

$$g_j = g_j(N_a, N_b, V).$$

(2.41b)

Furthermore, since N_a, N_b, and V are fixed for the system under consideration, the ε_i, g_i, ε_j, and g_j are fixed and can be considered as constants in taking the partial derivative on the right side of (2.35). Evaluating of the derivatives in Eqs. (2.38) and (2.39) and adding the equations together then yields

$$\frac{N_a}{Z_a}\left(\frac{\partial Z_a}{\partial \lambda_3}\right)_{N_a,N_b,V} + \frac{N_b}{Z_b}\left(\frac{\partial Z_b}{\partial \lambda_3}\right)_{N_a,N_b,V} = -\frac{N_a}{Z_a}\sum_{i=0}^{\infty}\varepsilon_i g_i e^{-\lambda_3 \varepsilon_i} - \frac{N_b}{Z_b}\sum_{j=0}^{\infty}\varepsilon_j g_j e^{-\lambda_3 \varepsilon_j}.$$

(2.42)

Using Eqs. (2.3) and (2.23a,b) it can be easily shown that the right side of the above equation is just $-U$:

$$\frac{N_a}{Z_a}\left(\frac{\partial Z_a}{\partial \lambda_3}\right)_{N_a,N_b,V} + \frac{N_b}{Z_b}\left(\frac{\partial Z_b}{\partial \lambda_3}\right)_{N_a,N_b,V} = -U.$$

(2.43)

Substituting Eq. (2.43) into Eq. (2.37) and solving for λ_3 yields

$$\lambda_3 = \frac{(\partial S/\partial \lambda_3)_{N_a,N_b,V}}{k_B(\partial U/\partial \lambda_3)_{N_a,N_b,V}}. \tag{2.44}$$

Since from basic multivariable calculus

$$\frac{(\partial S/\partial \lambda_3)_{N_a,N_b,V}}{(\partial U/\partial \lambda_3)_{N_a,N_b,V}} = \frac{1}{(\partial U/\partial S)_{N_a,N_b,V}}, \tag{2.45}$$

Eq. (2.44) can be converted to

$$\lambda_3 = \frac{1}{k_B(\partial U/\partial S)_{N_a,N_b,V}}. \tag{2.46}$$

Equation (2.46) is the sought-after relation for λ_3. Obviously the derivative $(\partial U/\partial S)_{N_a,N_b,V}$ plays an important role in the statistical thermodynamics framework we are developing. Because of its importance, we define a special symbol T to denote this derivative:

$$T \equiv (\partial U/\partial S)_{N_a,N_b,V}. \tag{2.47}$$

Since T is usually the symbol that designates temperature in thermodynamics texts, the reader may wonder whether this is, in fact, the temperature. Although we have not demonstrated it here, this is an appropriate definition of temperature. In a later section, we will show that this definition is consistent with the usual intuitive understanding of temperature. With this definition, the relation for λ_3 becomes

$$\lambda_3 = \frac{1}{k_B T}. \tag{2.48}$$

The equilibrium distributions of the particles among the available energy states is then given by

$$\frac{N_{a,i}}{N_a} = \frac{g_i e^{-\varepsilon_i/k_B T}}{Z_a}, \tag{2.49a}$$

$$\frac{N_{b,j}}{N_b} = \frac{g_j e^{-\varepsilon_j/k_B T}}{Z_b}, \tag{2.49b}$$

where

$$Z_a = \sum_{i=0}^{\infty} g_i e^{-\varepsilon_i/k_B T}, \tag{2.50a}$$

$$Z_b = \sum_{j=0}^{\infty} g_j e^{-\varepsilon_j/k_B T}. \tag{2.50b}$$

Replacing λ_3 with $1/k_B T$ in Eqs. (2.36) and (2.43) then gives the following relations among thermodynamic properties of the system:

$$\frac{S}{k_B} = \frac{U}{k_B T} + N_a \ln Z_a + N_b \ln Z_b, \tag{2.51}$$

$$U = \frac{N_a k_B T^2}{Z_a} \left(\frac{\partial Z_a}{\partial T}\right)_{N_a,N_b,V} + \frac{N_b k_B T^2}{Z_b} \left(\frac{\partial Z_b}{\partial T}\right)_{N_a,N_b,V}. \tag{2.52}$$

Now that we have developed a definition of temperature, a word about units is in order. Four temperature scales are commonly used. The Kelvin scale uses units defined so that 100 units on the Kelvin scale (K) correspond to the difference in temperature between the boiling point and freezing point of water at one atmosphere pressure. On this absolute scale, the temperature at the triple point of water is 273.16 K. The other common scale is the Rankine scale (°R), which uses units defined such that 180 units on the Rankine scale correspond to the difference in temperature between the boiling and freezing points of water at atmospheric pressure. The triple point of water corresponds to 491.688°R on this scale. Temperatures in the Celsius (°C) and Fahrenheit (°F) scales are related to Kelvin and Rankine scale values as:

$$°C = K - 273.15,$$

$$°F = °R - 459.67.$$

To this point we have not discussed the magnitude or units for the Boltzmann constant. Equations (2.49a,b) imply that the units should be energy divided by temperature. For consistency with the SI system, the appropriate value for the Boltzmann constant is

$$k_B = 1.3805 \times 10^{-23} \text{ J/K.}$$

Example 2.1   A system contains distinguishable species a and species b boson particles at temperature $T = 300$ K. The species a particles can occupy only the energy levels $k_B\theta_a$, $2k_B\theta_a$, or $3k_B\theta_a$ whereas species b particles can occupy only the levels $k_B\theta_b$, $2k_B\theta_b$, and $3k_B\theta_b$. $\theta_a = 60$ K, $\theta_b = 30$ K and each energy level has a degeneracy of one. The system contains $N_A = 6.02 \times 10^{26}$ particles of each species. Determine the distribution of energy levels occupied at equilibrium and the mean energy of a particle in the system.

Solution   The partition function for the species a particles is given by Eq. (2.50a):

$$Z_a = \sum_{i=0}^{\infty} g_i e^{-\varepsilon_i/k_B T} = e^{-k_B\theta_a/k_B T} + e^{-2k_B\theta_a/k_B T} + e^{-3k_B\theta_a/k_B T}$$

$$= e^{-0.2} + e^{-0.4} + e^{-0.6} = 2.038.$$

A similar calculation using Eq. (2.50b) yields $Z_b = 2.464$. The distribution of particles among available energy states for species a is given by Eq. (2.49a). Substituting for the first energy level, we obtain

$$\frac{N_{a,1}}{N_a} = \frac{e^{-\varepsilon_1/k_B T}}{Z_a} = \frac{e^{-k_B\theta_a/k_B T}}{Z_a} = \frac{e^{-0.2}}{2.038} = 0.402.$$

Similar substitutions for the second and third energy levels yield

$$\frac{N_{a,2}}{N_a} = \frac{e^{-0.4}}{2.038} = 0.329, \qquad \frac{N_{a,3}}{N_a} = \frac{e^{-0.6}}{2.038} = 0.269.$$

The values of $N_{a,i}/N_a$ thus determined indicate the fraction of particles that, on the average, occupy the ith energy level for species a. For species b, using Eq. (2.49b) in a similar manner yields

$$\frac{N_{b,1}}{N_b} = \frac{e^{-0.1}}{2.464} = 0.367, \qquad \frac{N_{b,2}}{N_b} = \frac{e^{-0.2}}{2.464} = 0.332, \qquad \frac{N_{b,3}}{N_b} = \frac{e^{-0.3}}{2.464} = 0.301.$$

For species a, the mean particle energy $\langle \varepsilon_a \rangle$ is obtained by taking a weighted average over all accessible levels:

$$\langle \varepsilon_a \rangle = \sum_{i=1}^{3} \frac{N_{a,i}}{N_a} \varepsilon_i = 0.402 k_B \theta_a + 0.329(2) k_B \theta_a + 0.260(3) k_B \theta_a$$

$$= 1.867 k_B \theta_a = 1.867(1.3805 \times 10^{-23})60 = 1.546 \times 10^{-21} \text{ J}.$$

The same method is used for species b:

$$\langle \varepsilon_b \rangle = \sum_{j=1}^{3} \frac{N_{b,j}}{N_b} \varepsilon_j = 0.367 k_B \theta_b + 0.332(2) k_B \theta_b + 0.301(3) k_B \theta_b$$

$$= 1.934 k_B \theta_b = 1.934(1.3805 \times 10^{-23})30 = 8.010 \times 10^{-22} \text{ J}.$$

Thus for this system the mean energy values for the two species differ by about a factor of two.

2.4   The Role of Distinguishability

The analysis in the previous section is based on the premise that the particles are distinguishable. This is appropriate for solid crystals where each atom is localized near a specific point in the crystal lattice. Unfortunately, it is not appropriate for fluids in which the particles (atoms, molecules, or fundamental particles) are free to move about within the system volume. In systems with freely moving particles, the particles of a given species are *indistinguishable*. Our task in this section is to examine how our analysis from the previous section must be changed to account for the indistiguishability of the particles in fluid systems.

In considering modifications to the analysis to account for indistinguishability, we will also account for restrictions on microstate occupancy. As discussed in Section 1.8, virtually all particles fall into two categories: bosons and fermions. There are no restrictions on the number of bosons that may occupy any given microstate. Most atoms and molecules behave as bosons. One exception is ^{3}He, which is a fermion. For a system of fermions, no more than one fermion may occupy a given microstate. Electrons and protons are perhaps the most important examples of particles that behave as fermions. The restriction on the occupancy of microstates for electrons is a consequence of the Pauli exclusion principle.

We first consider systems of bosons. As in the last section we are considering member systems in a microcanonical ensemble having specified N_a, N_b, V, and U. If we consider one energy level occupied by $N_{a,i}$ particles of species a, the number of ways of selecting the

$N_{a,i}$ particles from the total of N_a particles in the system is just 1, because the particles are indistinguishable. No matter which $N_{a,i}$ we choose, the particles occupying the energy level would "look" the same. However, we can permute the particles among the g_i microstates for the energy level in a number of ways. Because the particles are bosons, there are no restrictions on the way that the particles may occupy the microstates. From basic combinatorial analysis (see Appendix I) we can determine the number of these permutations as

$$\left\{ \begin{array}{l} \textit{number of ways of arranging } N_{a,i} \\ \textit{indistinguishable particles among } g_i \textit{ microstates} \end{array} \right\} = \frac{(g_i + N_{a,i} - 1)!}{(g_i - 1)!N_{a,i}!}. \tag{2.53}$$

Because the arrangements within energy levels are independent of one another, the total number of ways of arranging the a species particles among the microstates is the product of such terms for all energy levels:

$$\left\{ \begin{array}{l} \textit{number of ways of arranging} \\ \textit{species a particles among} \\ \textit{microstates dictated by } g_i \textit{ and } \{N_{a,i}\} \end{array} \right\} = \prod_{i=0}^{\infty} \frac{(g_i + N_{a,i} - 1)!}{(g_i - 1)!N_{a,i}!}. \tag{2.54}$$

A similar relation must apply for species b:

$$\left\{ \begin{array}{l} \textit{number of ways of arranging} \\ \textit{species b particles among} \\ \textit{microstates dictated by } g_j \textit{ and } \{N_{b,j}\} \end{array} \right\} = \prod_{j=0}^{\infty} \frac{(g_j + N_{b,j} - 1)!}{(g_j - 1)!N_{b,j}!}. \tag{2.55}$$

For each arrangement of the a particles, the number of possible arrangements of b particles is given by Eq. (2.55). The total number of ways of combining the arrangements of the two species is therefore equal to the product of the right sides of Eqs. (2.54) and (2.55). We designate this total number of microstates for the specified pair of distributions $\{N_{a,i}\}$ and $\{N_{b,j}\}$, as

$$W_{\text{bos}} = \prod_{i=0}^{\infty} \frac{(g_i + N_{a,i} - 1)!}{(g_i - 1)!N_{a,i}!} \prod_{j=0}^{\infty} \frac{(g_j + N_{b,j} - 1)!}{(g_j - 1)!N_{b,j}!}. \tag{2.56}$$

Alternatively, the particles may be indistinguishable fermions. Again, because the particles are indistinguishable, selection of any $N_{a,i}$ particles of species a for energy level i amounts to the same occupancy. The fermions may be permuted among the g_i microstates, but no more than one may occupy a given microstate. Thus a microstate is either occupied by one particle or unoccupied. Combinatorial analysis again provides the means to calculate the number of such arrangements (see Appendix I):

$$\left\{ \begin{array}{l} \textit{number of ways of selecting } N_{a,i} \textit{ microstates} \\ \textit{to be occupied from } g_i \textit{ total microstates} \end{array} \right\} = \frac{g_i!}{N_{a,i}!(g_i - N_{a,i})!}. \tag{2.57}$$

Because the arrangements within the energy levels are independent, the total number of system microstates is the product of terms like the right side of Eq. (2.57) for all energy levels:

$$\left\{\begin{array}{l} \textit{number of ways of arranging} \\ \textit{species a particles among} \\ \textit{microstates dictated by } g_i \textit{ and } \{N_{a,i}\} \end{array}\right\} = \prod_{i=0}^{\infty} \frac{g_i!}{N_{a,i}!(g_i - N_{a,i})!}. \tag{2.58}$$

The corresponding equation for species b is

$$\left\{\begin{array}{l} \textit{number of ways of arranging} \\ \textit{species b particles among} \\ \textit{microstates dictated by } g_j \textit{ and } \{N_{b,j}\} \end{array}\right\} = \prod_{j=0}^{\infty} \frac{g_j!}{N_{b,j}!(g_j - N_{b,j})!}. \tag{2.59}$$

As in the case for bosons, the total number of microstates for the two-species system is the product of the terms on the right sides of Eqs. (2.58) and (2.59). To indicate that this result is for fermions, we denote the total number of microstates as

$$W_{\text{fer}} = \prod_{i=0}^{\infty} \frac{g_i!}{N_{a,i}!(g_i - N_{a,i})!} \prod_{j=0}^{\infty} \frac{g_j!}{N_{b,j}!(g_j - N_{b,j})!}. \tag{2.60}$$

The total number of microstates for either bosons or fermions can be represented by the relation

$$W_{b/f}(\{N_{a,i}\}, \{N_{b,j}\}) = \prod_{i=0}^{\infty} \frac{[g_i + \eta(N_{a,i} - 1)]!}{(g_i - \eta - \xi N_{a,i})! N_{a,i}!} \prod_{j=0}^{\infty} \frac{[g_j + \eta(N_{b,j} - 1)]!}{(g_j - \eta - \xi N_{b,j})! N_{b,j}!}, \tag{2.61}$$

where W for indistinguishable bosons corresponds to

$$\eta = 1, \quad \xi = 0 \quad \text{(indistinguishable bosons),} \tag{2.62}$$

and for indistinguishable fermions we set

$$\eta = 0, \quad \xi = 1 \quad \text{(indistinguishable fermions).} \tag{2.63}$$

For systems containing indistinguishable bosons or indistinguishable fermions, we again seek to find the occupation number sets $\{N_{a,i}\}$, $\{N_{b,j}\}$ that maximize $\ln W$ subject to the constraints (2.1)–(2.3). As before, we interpret the pair of occupation number sets that maximizes $\ln W$ as being the equilibrium distribution for the system. Taking the natural log of both sides of Eq. (2.61) and rearranging yields

$$\ln W_{b/f} = \sum_{i=0}^{\infty} \ln\{[g_i + \eta(N_{a,i} - 1)]!\} - \sum_{i=0}^{\infty} \ln\{[g_i - \eta - \xi N_{a,i}]!\}$$

$$- \sum_{i=0}^{\infty} \ln\{N_{a,i}!\} + \sum_{j=0}^{\infty} \ln\{[g_j + \eta(N_{b,j} - 1)]!\}$$

$$- \sum_{j=0}^{\infty} \ln\{[g_j - \eta - \xi N_{b,j}]!\} - \sum_{j=0}^{\infty} \ln\{N_{b,j}!\}. \tag{2.64}$$

Since all the factorials on the right side of Eq. (2.64) are expected to be large, we apply Stirling's approximation to each:

$$
\ln W_{\mathrm{b/f}} = \sum_{i=0}^{\infty} [[g_i + \eta(N_{a,i} - 1)] \ln\{g_i + \eta(N_{a,i} - 1)\} - [g_i + \eta(N_{a,i} - 1)]]
$$

$$
- \sum_{i=0}^{\infty} [[g_i - \eta - \xi N_{a,i}] \ln\{g_i - \eta - \xi N_{a,i}\} - [g_i - \eta - \xi N_{a,i}]]
$$

$$
- \sum_{i=0}^{\infty} [N_{a,i} \ln N_{a,i} - N_{a,i}] + \sum_{j=0}^{\infty} [[g_j + \eta(N_{b,j} - 1)]
$$

$$
\times \ln\{g_j + \eta(N_{b,j} - 1)\} - [g_j + \eta(N_{b,j} - 1)]] - \sum_{j=0}^{\infty} [[g_j - \eta - \xi N_{b,j}]
$$

$$
\times \ln\{g_j - \eta - \xi N_{b,j}\} - [g_j - \eta - \xi N_{b,j}]] - \sum_{j=0}^{\infty} [N_{b,j} \ln N_{b,j} - N_{b,j}].
$$

$$(2.65)$$

Using the constraint relations (2.1) and (2.2), the above relation simplifies to

$$
\ln W_{\mathrm{b/f}} = \sum_{i=0}^{\infty} [[g_i + \eta(N_{a,i} - 1)] \ln\{g_i + \eta(N_{a,i} - 1)\} - [g_i - \eta - \xi N_{a,i}]
$$

$$
\times \ln\{g_i - \eta - \xi N_{a,i}\} - N_{a,i} \ln N_{a,i} + (1 - \eta - \xi)N_{a,i}]
$$

$$
+ \sum_{j=0}^{\infty} [[g_j + \eta(N_{b,j} - 1)] \ln\{g_j + \eta(N_{b,j} - 1)\} - [g_j - \eta - \xi N_{b,j}]
$$

$$
\times \ln\{g_j - \eta - \xi N_{b,j}\} - N_{b,j} \ln N_{b,j} + (1 - \eta - \xi)N_{b,j}]].
\qquad (2.66)
$$

Differentiating, we obtain

$$
d(\ln W_{\mathrm{b/f}}) = \sum_{i=0}^{\infty} [\eta dN_{a,i} \ln\{g_i + \eta(N_{a,i} - 1)\} + \eta dN_{a,i}
$$

$$
+ \xi dN_{a,i} \ln\{g_i - \eta - \xi N_{a,i}\} + \xi dN_{a,i} - dN_{a,i} \ln N_{a,i} - dN_{a,i}
$$

$$
+ (1 - \eta - \xi) dN_{a,i}] + \sum_{j=0}^{\infty} [\eta dN_{b,j} \ln\{g_j + \eta(N_{b,j} - 1)\}
$$

$$
+ \eta dN_{b,j} + \xi dN_{b,j} \ln\{g_j - \eta - \xi N_{b,j}\} + \xi dN_{b,j}
$$

$$
- dN_{b,j} \ln N_{b,j} - dN_{b,j} + (1 - \eta - \xi) dN_{b,j}].
\qquad (2.67)
$$

Setting $d(\ln W_{b/f}) = 0$ and simplifying yields

$$\sum_{i=0}^{\infty} [\eta \ln\{g_i + \eta(N_{a,i} - 1)\} + \xi \ln\{g_i - \eta - \xi N_{a,i}\} - \ln N_{a,i}] dN_{a,i}$$

$$+ \sum_{j=0}^{\infty} [\eta \ln\{g_j + \eta(N_{b,j} - 1)\} + \xi \ln\{g_j - \eta - \xi N_{b,j}\} - \ln N_{b,j}] dN_{b,j} = 0.$$

(2.68)

Using the method of Lagrange multipliers, we multiply the differential constraint relations (2.16a), (2.16b), and (2.16c) by λ_1, λ_2, and λ_3, respectively, add the resulting equations, and subtract Eq. (2.68). The final result is

$$\sum_{i=0}^{\infty} [\lambda_1 dN_{a,i} + \lambda_3 \varepsilon_i dN_{a,i} - [\eta \ln\{g_i + \eta(N_{a,i} - 1)\}$$

$$+ \xi \ln\{g_i - \eta - \xi N_{a,i}\} - \ln N_{a,i}] dN_{a,i}] + \sum_{j=0}^{\infty} [\lambda_2 dN_{b,j} + \lambda_3 \varepsilon_j dN_{b,j}$$

$$- [\eta \ln\{g_j + \eta(N_{b,j} - 1)\} + \xi \ln\{g_j - \eta - \xi N_{b,j}\} - \ln N_{b,j}] dN_{b,j}] = 0,$$

(2.69)

which simplifies to

$$\sum_{i=0}^{\infty} [\lambda_1 + \lambda_3 \varepsilon_i - \eta \ln\{g_i + \eta(N_{a,i} - 1)\} - \xi \ln\{g_i - \eta - \xi N_{a,i}\} + \ln N_{a,i}] dN_{a,i}$$

$$+ \sum_{j=0}^{\infty} [\lambda_2 + \lambda_3 \varepsilon_j - \eta \ln\{g_j + \eta(N_{b,j} - 1)\} - \xi \ln\{g_j - \eta - \xi N_{b,j}\}$$

$$+ \ln N_{b,j}] dN_{b,j} = 0.$$

(2.70)

At the maximum, the coefficients of the $dN_{a,i}$ and $dN_{b,j}$ terms must all be zero, which implies that

$$\lambda_1 + \lambda_3 \varepsilon_i - \eta \ln\{g_i + \eta(N_{a,i} - 1)\} - \xi \ln\{g_i - \eta - \xi N_{a,i}\} + \ln N_{a,i} = 0,$$

(2.71a)

$$\lambda_2 + \lambda_3 \varepsilon_j - \eta \ln\{g_j + \eta(N_{b,j} - 1)\} - \xi \ln\{g_j - \eta - \xi N_{b,j}\} + \ln N_{b,j} = 0.$$

(2.71b)

For bosons, we set $\eta = 1$ and $\xi = 0$ in Eqs. (2.71a,b) and rearrange to obtain

$$N_{a,i} = g_i e^{-\lambda_1} e^{-\lambda_3 \varepsilon_i} \left(1 + \frac{N_{a,i}}{g_i} - \frac{1}{g_i}\right),$$

(2.72a)

$$N_{b,j} = g_j e^{-\lambda_2} e^{-\lambda_3 \varepsilon_j} \left(1 + \frac{N_{b,j}}{g_j} - \frac{1}{g_j}\right) \quad \text{(for bosons)}.$$

(2.72b)

For fermions, we set $\eta = 0$ and $\xi = 1$ in Eqs. (2.71a,b) and obtain

$$N_{a,i} = g_i e^{-\lambda_1} e^{-\lambda_3 \varepsilon_i} \left(1 - \frac{N_{a,i}}{g_i} \right), \qquad (2.73a)$$

$$N_{b,j} = g_j e^{-\lambda_2} e^{-\lambda_3 \varepsilon_j} \left(1 - \frac{N_{b,j}}{g_j} \right) \quad \text{(for fermions)}. \qquad (2.73b)$$

Equations (2.72a,b) and (2.73a,b) can be solved explicitly for $N_{a,i}$ and $N_{b,j}$, but the resulting relations are complicated and it is difficult to carry the analysis further. Instead, we note that for systems of particles with moderate to high energies, the degeneracy of the most populated energy levels is enormous (see Example 1.4). For such circumstances, the number of microstates for each energy level is much larger than the number of particles in the system occupying that level and very few of the available microstates are occupied. Systems in which this is true are said to exhibit *dilute occupancy*. For conditions that result in dilute occupancy, $N_{a,i}/g_i$, $N_{b,j}/g_j$, $1/g_i$, and $1/g_j$ are negligible compared to one. If we neglect these ratios compared to one in Eqs. (2.72a,b) and (2.73a,b), we find that both sets of equations reduce to

$$N_{a,i} = g_i e^{-\lambda_1} e^{-\lambda_3 \varepsilon_i}, \qquad (2.74a)$$

$$N_{b,j} = g_j e^{-\lambda_2} e^{-\lambda_3 \varepsilon_j}. \qquad (2.74b)$$

Using the constraints (2.1) and (2.2) on $N_{a,i}$ and $N_{b,j}$, we can eliminate the multipliers λ_1 and λ_2 and obtain

$$\frac{N_{a,i}}{N_a} = \frac{g_i e^{-\lambda_3 \varepsilon_i}}{\sum_{i=0}^{\infty} g_i e^{-\lambda_3 \varepsilon_i}}, \qquad (2.75a)$$

$\left(\begin{array}{l} \textit{for indistinguishable bosons or fermions} \\ \textit{in the limit of dilute occupancy} \end{array} \right)$

$$\frac{N_{b,j}}{N_b} = \frac{g_j e^{-\lambda_3 \varepsilon_j}}{\sum_{j=0}^{\infty} g_j e^{-\lambda_3 \varepsilon_j}}, \qquad (2.75b)$$

which is identical to the Boltzmann statistics result obtained for distinguishable particles with no restrictions on microstate occupancy.

Although the form of the distributions (2.75a,b) is the same as Boltzmann statistics for distinguishable particles, the relations for W for indistinguishable bosons and fermions are different from the relation for W that applied to distinguishable bosons. We must therefore reexamine how the change in the relation for W affects the thermodynamic properties for the system. Since by definition $S = k_B \ln W$, we set S/k_B equal to the right side of Eq. (2.66), and use the distributions (2.75a,b) to evaluate $N_{a,i}$ and $N_{b,j}$. In addition, because we have dilute occupancy, we neglect $N_{a,i}/g_i$, $N_{b,j}/g_j$, $1/g_i$, and $1/g_j$ compared to one, where appropriate. For both the boson and fermion particle types, the resulting equation for S/k_B reduces to

$$\frac{S}{k_B} = \lambda_3 U + N_a \ln Z_a - N_a \ln N_a + N_b \ln Z_b - N_b \ln N_b + N_a + N_b. \qquad (2.76)$$

To evaluate λ_3 we use the same approach as was used in Section 2.3 for distinguishable particles. We differentiate with respect to λ_3, evaluate the derivatives and solve for λ_3. The result of this manipulation is

$$\lambda_3 = \frac{1}{k_B(\partial U/\partial S)_{N_a,N_b,V}} = \frac{1}{k_B T}. \tag{2.77}$$

It follows directly that for a system that exhibits dilute occupancy, the distributions and the partition function definitions for indistinguishable bosons or fermions are identical to the Boltzmann statistics case:

$$\frac{N_{a,i}}{N_a} = \frac{g_i e^{-\varepsilon_i/k_B T}}{Z_a}, \tag{2.49a}$$

$$\frac{N_{b,j}}{N_b} = \frac{g_j e^{-\varepsilon_j/k_B T}}{Z_b}, \tag{2.49b}$$

where

$$Z_a = \sum_{i=0}^{\infty} g_i e^{-\varepsilon_i/k_B T}, \tag{2.50a}$$

$$Z_b = \sum_{j=0}^{\infty} g_j e^{-\varepsilon_j/k_B T}. \tag{2.50b}$$

It follows from these relations and the energy constraint (2.3) that the relation for U is identical to that for Boltzmann statistics of distinguishable particles:

$$U = \frac{N_a k_B T^2}{Z_a} \left(\frac{\partial Z_a}{\partial T}\right)_{N_a,N_b,V} + \frac{N_b k_B T^2}{Z_b} \left(\frac{\partial Z_b}{\partial T}\right)_{N_a,N_b,V}. \tag{2.52}$$

Substituting $\lambda_3 = 1/k_B T$ into Eq. (2.76) yields the following relation for the entropy:

$$\frac{S}{k_B} = \frac{U}{k_B T} + N_a \ln\left\{\frac{Z_a}{N_a}\right\} + N_b \ln\left\{\frac{Z_b}{N_b}\right\} + N_a + N_b. \tag{2.78}$$

Note that this relation differs from the entropy relation obtained for distinguishable bosons. The net effect of particle indistinguishability is to reduce the entropy of the system. Equations (2.49), (2.50), (2.52), and (2.78) thus provide the linkage between microscale energy storage and the macroscopic thermodynamic properties for a system of indistinguishable bosons or fermions in the moderate to high energy limit where dilute occupancy occurs.

What about low-energy systems? It turns out that in very cold systems we cannot invoke the dilute occupancy approximation and we must allow for low degeneracy and quantum effects in the analysis of the statistical behavior of such systems. We will return to examine such systems in more detail in Section 6.4. Fortunately, in engineering applications, the overwhelming majority of system types and particle energy levels encountered do lie in ranges where dilute occupancy occurs. The results summarized above therefore provide a useful foundation for thermodynamic analysis of this broad range of system types.

Example 2.2   Initially a system contains one kmol ($N_A = 6.02 \times 10^{26}$ molecules) of water molecules in a liquid phase at atmospheric pressure and 373.2 K. Energy is added to the system at constant pressure until it contains only a vapor phase at 373.2 K. Evelute Z_g/Z_f, where Z_f is the partition function for the initial liquid state and Z_g is the partition function for the final gaseous state.

Solution   For a single-component system, Eq. (2.78) reduces to

$$\frac{S}{k_B} = \frac{U}{k_B T} + N_a \ln \left\{ \frac{Z_a}{N_a} \right\} + N_a.$$

Since $N_a = N_A$, the resulting S and U values are those for one kmole of water, $\hat{s}$ and $\hat{u}$, respectively:

$$\frac{\hat{s}}{k_B} = \frac{\hat{u}}{k_B T} + N_A \ln \left\{ \frac{Z_a}{N_A} \right\} + N_A.$$

Evaluating this relation at the initial liquid state and the final gaseous state and subtracting the two equations yields

$$\frac{\hat{s}_g - \hat{s}_f}{k_B} = \frac{\hat{u}_g - \hat{u}_f}{k_B T} + N_A \ln \left\{ \frac{Z_g}{N_A} \right\} - N_A \ln \left\{ \frac{Z_f}{N_A} \right\}.$$

Rearranging, we get

$$\frac{\hat{s}_g - \hat{s}_f}{k_B} = \frac{\hat{u}_g - \hat{u}_f}{k_B T} + N_A \ln \left\{ \frac{Z_g}{Z_f} \right\},$$

which can be solved for Z_g/Z_f to give

$$\frac{Z_g}{Z_f} = \exp \left\{ \frac{T \hat{s}_{fg} - \hat{u}_{fg}}{N_A k_B T} \right\},$$

where $\hat{s}_{fg} = \hat{s}_g - \hat{s}_f$ and $\hat{u}_{fg} = \hat{u}_g - \hat{u}_f$. Values of $\hat{s}_{fg}$ and $\hat{u}_{fg}$ are tabulated in standard steam tables:

at $T = 373.2$ K: $\hat{s}_{fg} = 108.9$ kJ/kmol K, $\hat{u}_{fg} = 3.758 \times 10^4$ kJ/kmol.

Substituting these values into the above relation for Z_g/Z_f, we obtain

$$\frac{Z_g}{Z_f} = \exp \left\{ \frac{373.2(108.9) - 3.758 \times 10^4}{(6.02 \times 10^{26})1.3805 \times 10^{-23}(373.2)} \right\} = 6.31 \times 10^{158}.$$

Conversion from a liquid to a vapor phase has resulted in an enormous increase in the partition function. From its definition, Eq. (2.50a), it is clear that an increase in Z results from an increase in degeneracy or an increase in the number of accessible energy levels that are comparable to $k_B T$. Conversion of liquid to vapor increases both the degeneracy and the number of accessible energy levels by a large amount.

2.5   More on Entropy and Equilibrium

Entropy

Entropy is such an important property in the thermodynamics theory we are developing that it is useful to clearly document the characteristics of entropy that have emerged from our analysis so far. There are four important characteristics of the entropy property that we have identified in our development. First, we have defined entropy so that the entropy of a composite system is *additive* over the constituent subsystems.

Two additional important properties of the entropy relate to its functional dependence on other properties for a system at equilibrium. We have shown that if we specify the energy U, volume V, and the number of each species of particles in a system, at equilibrium, the value of the system entropy is uniquely defined as the maximum of S for all the macrostates consistent with the imposed constraints. Thus equilibrium can be viewed as a condition dictated by an *extremum principle* whereby the system seeks a state of maximum entropy. This is sometimes referred to as the *entropy maximum principle*. In addition, since the resulting maximized S value depends on the specified U, V, N_a, and N_b, for equilibrium states, S must be a function of U, V, N_a, and N_b:

$$S = S(U, V, N_a, N_b). \tag{2.79}$$

A relation among U, V, N_a, N_b, and S is termed a *fundamental relation* for a thermodynamic system. We will show in the next chapter that all conceivable thermodynamic information about a system can be obtained from the fundamental thermodynamic relation.

The fourth characteristic of entropy relates to its dependence on internal energy U. As the energy of a system increases, it will increasingly occupy higher quantum energy levels. For the 3-D particle-in-a-box quantum solution considered in the first chapter, it was shown that the degeneracy of the energy levels increases monotonically as the energy level increased. Because the particle-in-a-box solution is an appropriate model of translation energy storage in particles, we expect that in any system with translating particles, the system will increasingly occupy energy levels with higher degeneracies as the system energy increases. Furthermore, Eq. (2.31) clearly indicates that entropy will increase monotonically with increasing energy level degeneracy. Taken together, these observations imply that entropy will be a monotonically increasing function of internal energy.

In reaching this conclusion, we have considered only translational energy storage. In systems of molecules that have rotational and vibrational energy storage in addition to translation, these additional mechanisms can be modeled with the rigid-rotor and harmonic-oscillator solutions discussed in Chapter 1. We will see in Chapter 5 that the vibrational energy storage in polyatomic molecules can be modeled as a combination of independent normal modes of vibration, each of which can be considered to be a simple harmonic oscillator. Depending on the structure of the molecule, the vibrational energy storage levels for the molecule may be nondegenerate or its vibrational energy levels may have a degeneracy greater than one. The rigid-rotor model solution implies that the degeneracy of rotational energy levels becomes increasingly large as the energy level increases. The overall effect of combining rotational energy storage and/or vibrational energy storage with translation is essentially the same as translation alone: System entropy will increase monotonically with system energy.

What if there is no translation, as is the case in a crystalline solid? In that case, the atoms are bound to surrounding molecules at a specific mean location. We will examine such a system in detail in a later chapter. Although the simple one-dimensional harmonic oscillator considered in Chapter 1 exhibits energy levels that are nondegenerate, in a three-dimensional lattice of atoms, the number of normal oscillatory modes accessible to the system generally increases as the energy of the system increases. This trend produces a rapid increase in number of possible system microstates as the system energy rises. Equation (2.25) then implies that entropy will steadily increase with system internal energy.

The overall conclusion is that, for solid or fluid systems, the entropy of the system will increase monotonically with increasing system internal energy. This characteristic is a direct result of the fact that, at higher energies, the system generally can span a wider range of energy levels and the degeneracy of the energy levels increases with increasing energy. The important characteristics of the entropy function are summarized below:

(i)   Equilibrium corresponds to a maximum entropy for a system at specified U, V, N_a, and N_b. This is the *entropy maximum principle*.
(ii) The entropy of a composite system with an arbitrary number of subsystems is additive over the constituent subsystems. This is the *additivity property*.
(iii) At equilibrium, the value of entropy is a function of U, V, N_a, and N_b. This function is referred to as the fundamental relation for the system.
(iv) Entropy is a monotonically increasing function of energy.

One final aspect of the definition of entropy is worth noting. In the previous section it was shown that for a system at equilibrium the entropy is given by

$$S = k_B \ln W_{max} \quad \text{(at equilibrium)},\tag{2.30a}$$

where W_{max} is given by Eq. (2.30b). It was also argued that for the system to achieve a macroscopically stable equilibrium, W_{max} must be very much larger than the W value for any other macrostate. As indicated by Eq. (2.8), the total number of accessible microstates Ω must be the sum of the W values for all macrostates. If we designate W_i as the W value for macrostate i, then Ω is computed as

$$\Omega = \sum_{i=1}^{M} W_i = W_1 + W_2 + W_3 + \cdots + W_M,\tag{2.80}$$

where M is the total number of macrostates consistent with the system constraints. In the above expression for Ω there are M positive terms, the largest of which is W_{max}. The sum must be at least as large as W_{max} and is certainly less than M times W_{max}:

$$W_{max} \leq \Omega \leq M W_{max}.$$

Taking the natural log of all terms, we obtain

$$\ln W_{max} \leq \ln \Omega \leq \ln M + \ln W_{max}.$$

With real systems in which the number of particles is on the order of Avogadro's number, the number of microstates W is far greater than the number of macrostates M. In the above relation, $\ln M$ is therefore negligible compared to $\ln W$. It follows that to a high degree of accuracy

$$\ln W_{max} = \ln \Omega\tag{2.81}$$

and the entropy of the system at equilibrium is given by

$$S = k_B \ln \Omega.\tag{2.82}$$

The significance of this result is that it links the entropy of the system at equilibrium to the total number of microstates accessible to the system. It follows from the fact that at equilibrium Ω is virtually equal to W_{max} because W_{max} will be so much larger than any other contribution to Ω. Qualitatively this result implies that, to a high degree of accuracy, we can say that if the number of accessible microstates is doubled, the entropy will increase by a factor of ln 2. Conversely, the magnitude of the entropy is a direct indicator of the number of microstates accessible to the system.

Example 2.3   As in Example 2.2, we here will consider a system that contains one kmol of water molecules ($N_A = 6.02 \times 10^{26}$ molecules) initially in a liquid phase at atmospheric pressure and 373.2 K. Energy is added to the system at constant pressure until it contains only a vapor phase at 373.2 K. Evalute Ω_g/Ω_f, where Ω_f is the total number of accessible microstates for the initial liquid state and Ω_g is the total number of accessible microstates for the final gaseous state.

    Solution   For any system at equilibrium, Eq. (2.82) requires that

$$S = k_B \ln \Omega.$$

Since we are considering a system containing one mole the entropy values in the inital liquid and final vapor states are the molar values, $\hat{s}_f$ and $\hat{s}_g$, repsectively. It follows from the above equation that

$$\hat{s}_f = k_B \ln \Omega_f \quad \text{and} \quad \hat{s}_g = k_B \ln \Omega_g,$$

where Ω_f is the number of microstates accessible to the system in the initial liquid state and Ω_g is the number of microstates accessible to the system in the final vapor state. Combining these equations yields

$$\hat{s}_g - \hat{s}_f = k_B \ln \left(\frac{\Omega_g}{\Omega_f} \right),$$

which can be solved for Ω_g/Ω_f to give

$$\frac{\Omega_g}{\Omega_f} = \exp \left\{ \frac{\hat{s}_{fg}}{k_B} \right\},$$

where $\hat{s}_{fg} = \hat{s}_g - \hat{s}_f$. Values of $\hat{s}_{fg}$ are tabulated in standard steam tables:

    at $T = 373.2$ K: $\hat{s}_{fg} = 108.9$ kJ/kmol K.

Substituting these values into the above relation for Ω_g/Ω_f, we obtain

$$\frac{\Omega_g}{\Omega_f} = \exp \left\{ \frac{108.9 \times 1000}{1.38 \times 10^{-23}} \right\} = 10^{3.4 \times 10^{27}}.$$

Conversion from a liquid to a vapor phase has clearly resulted in an enormous increase in the number of microstates accessible to the system.

Conditions at Equilibrium

    With the knowledge of entropy obtained thus far, we now will more fully explore the macroscopic requirements for equilibrium. Consider the isolated system shown in Figure 2.4 consisting of two subsystems, I and II. Each contains particles of species *a* and species *b*,

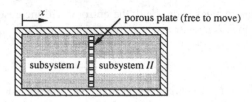

Figure 2.4 A composite system.

and the wall separating the two subsystems is a thin porous metal plate that allows free exchange of heat and species exchange between the subsystems. The plate is free to move so that its x position is unrestricted. The outer boundary of the overall system is rigid and insulated so that no energy or mass exchange with the outside can occur. The values of U, V, N_a, and N_b for the total system are therefore fixed.

For each subsystem, S at equilibrium is a function of U, V, N_a, and N_b in the subsystem. Each subsystem could equilibrate individually at various U, V, N_a, and N_b values that would not violate the specifications of U, V, N_a, and N_b for the overall system. The extremum principle requires that, at equilibrium, the subsystems must reach equilibrium conditions that maximize S for the overall system subject to the overall system constraints on U, V, N_a, and N_b. Additivity of S requires that

$$S = S^{\mathrm{I}} + S^{\mathrm{II}}. \tag{2.83}$$

Differentiating this relation yields

$$dS = dS^{\mathrm{I}} + dS^{\mathrm{II}}. \tag{2.84}$$

The interpretation of the above equation is as follows. Once the overall system has reached equilibrium, molecular-level fluctuations may briefly produce differential exchanges of energy or mass or volume between the subsystems, which will produce changes in the entropy of the subsystems. However, since S for the overall system is at a maximum, the resulting differential change in overall S is zero. Equation (2.84) therefore implies that

$$dS^{\mathrm{I}} + dS^{\mathrm{II}} = 0. \tag{2.85}$$

At equilibrium for each subsystem

$$S^{\mathrm{I}} = S^{\mathrm{I}}\left(U^{\mathrm{I}}, V^{\mathrm{I}}, N_a^{\mathrm{I}}, N_b^{\mathrm{I}}\right), \tag{2.86a}$$

$$S^{\mathrm{II}} = S^{\mathrm{II}}\left(U^{\mathrm{II}}, V^{\mathrm{II}}, N_a^{\mathrm{II}}, N_b^{\mathrm{II}}\right). \tag{2.86b}$$

For subsystem I, we can therefore write the following first-order expansion:

$$dS^{\mathrm{I}} = \left(\frac{\partial S}{\partial U}\right)^{\mathrm{I}}_{N_a,N_b,V} dU^{\mathrm{I}} + \left(\frac{\partial S}{\partial V}\right)^{\mathrm{I}}_{U,N_a,N_b} dV^{\mathrm{I}}$$

$$+ \left(\frac{\partial S}{\partial N_a}\right)^{\mathrm{I}}_{N_b,U,V} dN_a^{\mathrm{I}} + \left(\frac{\partial S}{\partial N_b}\right)^{\mathrm{I}}_{N_a,U,V} dN_b^{\mathrm{I}}. \tag{2.87}$$

We can similarly write for subsystem II

$$dS^{II} = \left(\frac{\partial S}{\partial U}\right)^{II}_{N_a,N_b,V} dU^{II} + \left(\frac{\partial S}{\partial V}\right)^{II}_{U,N_a,N_b} dV^{II}$$

$$+ \left(\frac{\partial S}{\partial N_a}\right)^{II}_{N_b,U,V} dN_a^{II} + \left(\frac{\partial S}{\partial N_b}\right)^{II}_{N_a,U,V} dN_b^{II}. \tag{2.88}$$

We further note that since

$$U = U^I + U^{II}, \qquad V = V^I + V^{II}, \qquad N_a = N_a^I + N_a^{II}, \qquad N_b = N_b^I + N_b^{II}$$

and U, V, N_a, and N_b are fixed constants, differentiating these relations yields

$$dU^I = -dU^{II}, \tag{2.89}$$

$$dV^I = -dV^{II}, \tag{2.90}$$

$$dN_a^I = -dN_a^{II}, \tag{2.91}$$

$$dN_b^I = -dN_b^{II}. \tag{2.92}$$

Substituting Eqs. (2.87) and (2.88) into (2.85) and using Eqs. (2.89)–(2.92), we obtain

$$\left[\left(\frac{\partial S}{\partial U}\right)^I_{N_a,N_b,V} - \left(\frac{\partial S}{\partial U}\right)^{II}_{N_a,N_b,V}\right] dU^I + \left[\left(\frac{\partial S}{\partial V}\right)^I_{U,N_a,N_b} - \left(\frac{\partial S}{\partial V}\right)^{II}_{U,N_a,N_b}\right] dV^I$$

$$+ \left[\left(\frac{\partial S}{\partial N_a}\right)^I_{N_b,U,V} - \left(\frac{\partial S}{\partial N_a}\right)^{II}_{N_b,U,V}\right] dN_a^I$$

$$+ \left[\left(\frac{\partial S}{\partial N_b}\right)^I_{N_a,U,V} - \left(\frac{\partial S}{\partial N_b}\right)^{II}_{N_a,U,V}\right] dN_b^I = 0. \tag{2.93}$$

If $dS = 0$ for all possible exchanges, the perturbation coefficients in square brackets must each be equal to zero. It follows directly that at equilibrium we must have

$$\left(\frac{\partial S}{\partial U}\right)^I_{N_a,N_b,V} = \left(\frac{\partial S}{\partial U}\right)^{II}_{N_a,N_b,V}, \tag{2.94}$$

$$\left(\frac{\partial S}{\partial V}\right)^I_{U,N_a,N_b} = \left(\frac{\partial S}{\partial V}\right)^{II}_{U,N_a,N_b}, \tag{2.95}$$

$$\left(\frac{\partial S}{\partial N_a}\right)^I_{N_b,U,V} = \left(\frac{\partial S}{\partial N_a}\right)^{II}_{N_b,U,V}, \tag{2.96}$$

$$\left(\frac{\partial S}{\partial N_b}\right)^I_{N_a,U,V} = \left(\frac{\partial S}{\partial N_b}\right)^{II}_{N_a,U,V}. \tag{2.97}$$

Using Eq. (2.47),

$$\left(\frac{\partial U}{\partial S}\right)_{N_a,N_b,V} = T, \tag{2.47}$$

Eq. (2.94) can be written

$$\frac{1}{T^{\mathrm{I}}} = \frac{1}{T^{\mathrm{II}}}$$

or equivalently

$$T^{\mathrm{I}} = T^{\mathrm{II}}. \tag{2.98}$$

Thus to achieve equilibrium with respect to internal energy exchange with volume held constant, we require that the T property defined by Eq. (2.47) must be the same in both subsystems. This is the characteristic that we intuitively associate with the concept of temperature:

♦   *When two closed systems of fixed volume are brought into contact, no net exchange of internal energy will occur if the two systems are at the same temperature.*

We therefore conclude that Eq. (2.47) does in fact provide an appropriate definition of temperature T. The temperature so defined is termed the *absolute temperature*.

To analyze the meaning of the other relations (2.95)–(2.97), we first note the following basic fact from multivariable calculus. If y is a function of x and z, then

$$\left(\frac{\partial x}{\partial z}\right)_y = -\left(\frac{\partial x}{\partial y}\right)_z \left(\frac{\partial y}{\partial z}\right)_x. \tag{2.99}$$

For constant N_a and N_b, S is a function of U and V and Eq. (2.99) implies that

$$\left(\frac{\partial S}{\partial V}\right)_{U,N_a,N_b} = -\left(\frac{\partial S}{\partial U}\right)_{N_a,N_b,V} \left(\frac{\partial U}{\partial V}\right)_{N_a,N_b,S} = -\frac{1}{T}\left(\frac{\partial U}{\partial V}\right)_{N_a,N_b,S}. \tag{2.100}$$

We define a property P as

$$P \equiv -\left(\frac{\partial U}{\partial V}\right)_{N_a,N_b,S}. \tag{2.101}$$

Using (2.100) and this new property definition, Eq. (2.95) can be written as

$$\frac{P^{\mathrm{I}}}{T^{\mathrm{I}}} = \frac{P^{\mathrm{II}}}{T^{\mathrm{II}}}. \tag{2.102}$$

Since we have already shown that the temperatures must be equal at equilibrium, this requirement reduces to

$$P^{\mathrm{I}} = P^{\mathrm{II}}. \tag{2.103}$$

Note that this requirement is associated with exchange of energy between the subsystems due to motion of the separating boundary, which causes one system to contract while the other expands. This requirement implies that if two systems are separated by a boundary surface that offers no resistance to applied forces, the two systems are in mechanical equilibrium if this P property is the same in both systems. This notion is consistent with our intuitive understanding of pressure or normal stress exerted by a fluid or solid on a system boundary. We therefore adopt Eq. (2.101) as a definition of *pressure*.

For constant V and N_b, S is a function of U and N_a and Eq. (2.99) implies

$$\left(\frac{\partial S}{\partial N_a}\right)_{V,N_b,U} = -\left(\frac{\partial S}{\partial U}\right)_{V,N_b,N_a}\left(\frac{\partial U}{\partial N_a}\right)_{V,N_b,S} = -\frac{1}{T}\left(\frac{\partial U}{\partial N_a}\right)_{V,N_b,S}. \quad (2.104)$$

For constant V and N_a, S is a function of U and N_b and we similarly find that

$$\left(\frac{\partial S}{\partial N_b}\right)_{V,N_a,U} = -\left(\frac{\partial S}{\partial U}\right)_{V,N_a,N_b}\left(\frac{\partial U}{\partial N_b}\right)_{V,N_a,S} = -\frac{1}{T}\left(\frac{\partial U}{\partial N_b}\right)_{V,N_a,S}. \quad (2.105)$$

We define properties μ_a and μ_b as

$$\mu_a = \left(\frac{\partial U}{\partial N_a}\right)_{V,S,N_b}, \quad (2.106)$$

$$\mu_b = \left(\frac{\partial U}{\partial N_b}\right)_{V,S,N_a}. \quad (2.107)$$

Using Eqs. (2.104) and (2.105) and these new property definitions, Eqs. (2.96) and (2.97) can be written as

$$\frac{\mu_a^{I}}{T^{I}} = \frac{\mu_a^{II}}{T^{II}}, \quad (2.108)$$

$$\frac{\mu_b^{I}}{T^{I}} = \frac{\mu_b^{II}}{T^{II}}. \quad (2.109)$$

Since the temperatures must be equal at equilibrium, these requirements reduce to

$$\mu_a^{I} = \mu_a^{II}, \quad (2.110)$$

$$\mu_b^{I} = \mu_b^{II}. \quad (2.111)$$

These requirements are associated with species exchange between the subsystems. Equations (2.110) and (2.111) imply that if two systems are separated by a permeable boundary surface, the two systems are in mass transfer equilibrium for each species if this μ property is the same in both systems. The properties defined by Eqs. (2.106) and (2.107) are termed the *chemical potential for species a* and the *chemical potential for species b*, respectively. In a mixture there is a corresponding chemical potential for each species present.

In summary, in terms of these newly defined properties, for the system in Figure 2.4 we can state the necessary conditions for equilibrium as being:

$$T^{I} = T^{II}, \qquad P^{I} = P^{II}, \qquad \mu_a^{I} = \mu_a^{II}, \qquad \mu_b^{I} = \mu_b^{II}.$$

It should be noted that these are necessary conditions for equilibrium in the specific system shown in Figure 2.4. If the characteristics of the boundary between the subsystems were changed, the necessary conditions for equilibrum would generally be different. We will explore this issue in more detail in Chapter 3. The analysis in this section serves to illustrate that the definitions of temperature, pressure, and chemical potential make it possible to concisely state the necessary conditions for equilibrium in the system shown in Figure 2.4. We will see in Chapter 3 that for a wide variety of systems, it is often most convenient to specify the necessary conditions for equilibrium in terms of these properties.

We have already discussed units for temperature. Pressure must have units of energy per unit volume, or equivalently, force per unit area. The latter is consistent with the common interpretation of pressure as the force per unit of surface area on the walls or boundary surfaces that enclose a system. In the SI system, the units would be N/m^2. As defined here, chemical potential has the units of energy per molecule. In the SI system, the units would be J. The chemical potential per kmol $\hat{\mu}_a$ and the chemical potential per unit mass $\tilde{\mu}_a$ are also sometimes used in thermodynamic analysis. These are related to the chemical potential per molecule as

$$\hat{\mu}_a = \mu_a N_A,$$

$$\tilde{\mu}_a = \mu_a N_A / \bar{M}_a,$$

where N_A is Avogadro's number and $\bar{M}_a$ is the molecular mass of species a. In the SI system the units of $\hat{\mu}_a$ and $\tilde{\mu}_a$ are J/kmol and J/kg, respectively.

Negative Absolute Temperatures

In Section 2.5, we concluded that for common thermodynamic systems containing large numbers of atoms, molecules, or other particles, the entropy is a monotonically increasing function of internal energy. Most common thermodynamic systems containing atoms and molecules in a fluid or solid state have an infinite number of energy levels. As the temperature of the system is increased, more particles are raised to higher energy levels. This is accompanied by increasing disorder as the particles are distributed over more and more states. As the energy increases, entropy increases and since, by definition, temperature is given by

$$T \equiv (\partial U / \partial S)_{N_a, N_b, V} \tag{2.47}$$

it follows that the (absolute) temperature is positive ($T > 0$) for such systems.

From the definition (2.47) it is clear that the absolute temperature could only be negative if the entropy of the system decreased as its internal energy increased. The discussion above implies that this cannot occur in a system that has an infinite number of energy levels. In such a system, an increase of temperature produces increasing occupancy of higher energy levels. However, Eq. (2.49a) indicates that the ratio of the mean number of particles (of type a) in two adjacent energy levels i and $i + 1$ is given by

$$\frac{N_{a,i+1}}{N_{a,i}} = \frac{g_{i+1}}{g_i} e^{-(\varepsilon_{i+1} - \varepsilon_i)/k_B T}. \tag{2.112}$$

In a system with infinitely many energy levels, if the system energy is finite, the higher energy levels of the system must be less populated than lower energy levels. Since g_{i+1}/g_i is slightly larger than one and $\varepsilon_{i+1} > \varepsilon_i$, $N_{a,i+1}/N_{a,i}$ can be less than one only if the temperature T is positive. As T approaches infinity, $N_{a,i+1}/N_{a,i}$ approaches one. This ratio could equal one only if the system had infinite energy. Equation (2.112) implies that negative T would require that $N_{a,i+1}/N_{a,i}$ be greater than one, which would require more energy than in the limit of infinite temperature. We therefore conclude that negative temperatures are impossible in a system that has an infinite number of energy levels. Since most common systems are of this type, negative temperatures are regarded as an impossibility for virtually all systems encountered in nature and in technological applications.

However, suppose we found a system that had only a finite number of energy levels. It may then be possible to cause a population inversion in a system with finite energy, resulting in $N_{a,i+1}/N_{a,i} > 1$, which would imply that the system has a negative temperature. A system having only two possible energy states is analyzed in Example 2.4. This type of system is shown to have positive temperatures at low energy levels and negative temperatures at high system energy levels in which the majority of the particles are in the higher energy state. Thus, in a system with only two energy states, the population inversion that exists at higher energies results in negative absolute temperatures.

Example 2.4 Particles in a system are distinguishable and can exist in one of only two possible energy levels, each having a degeneracy of one. The energy of the lower ground state is taken to be zero. The upper state has energy ε_1. The system contains N_a particles. Determine how the entropy and temperature vary with energy for this system.

Solution The internal energy of the system is zero if all the particles are in the ground state. If $N_{a,1}$ is the mean number of particles in state 1, then $N_a - N_{a,1}$ must be the mean number in the ground state. It follows that the internal energy of the system must be given by

$$U = N_{a,1}\varepsilon_1.$$

With $N_{a,1}$ particles in state 1, the number of microsates W is the number of ways of putting N_a distinguishable particles into two bins such that there are $N_{a,1}$ in one and $N_a - N_{a,1}$ in the other. It follows from Eq. (2.4) that

$$W = N_a! \left(\frac{1}{(N_a - N_{a,1})!} \right) \left(\frac{1}{N_{a,1}!} \right).$$

Taking the log of both sides and using the fact that $S = k_B \ln W$ yields

$$S/k_B = \ln[N_a!] - \ln[(N_a - N_{a,1})!] - \ln[N_{a,1}!].$$

Using the Stirling approximation, this relation becomes

$$S/k_B = N_a \ln N_a - N_a - (N_a - N_{a,1})\ln(N_a - N_{a,1}) - (N_a - N_{a,1}) - N_{a,1} \ln N_{a,1} - N_{a,1}.$$

Since $U = N_{a,1}\varepsilon_1$, we replace $N_{a,1}$ with U/ε_1. Doing so and rearranging somewhat, we obtain

$$\frac{S}{N_a k_B} = -\left(1 - \frac{U}{N_a \varepsilon_1}\right) \ln\left(1 - \frac{U}{N_a \varepsilon_1}\right) - \left(\frac{U}{N_a \varepsilon_1}\right) \ln\left(\frac{U}{N_a \varepsilon_1}\right).$$

The variation of entropy with internal energy for this system predicted by the above relation is shown in Figure 2.5.

It can be seen that at low energies, the entropy increases with increasing energy. At higher energies, the trend reverses, however. At zero energy, the system is perfectly ordered with all particles in the ground state. At the highest possible energy for the system, $U = N_a \varepsilon_1$, the system is also perfectly ordered with all particles in the higher energy level. The entropy is zero for both of these perfectly ordered states. The entropy is maximum when the particles are evenly divided between the two accessible energy levels. The temperature is the inverse of the slope of the curve in Figure 2.5. At low energies the slope is positive and therefore the temperature is positive. T increases as U increases at low energies, approaching $+\infty$

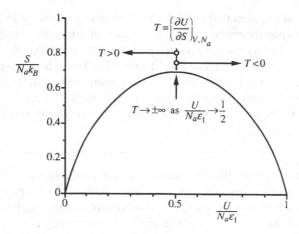

Figure 2.5

as $U \to (1/2)N_a\varepsilon_1$. At high energies, the slope and hence the temperature are negative. T increases as U increases. As U decreases toward $(1/2)N_a\varepsilon_1$, T approaches $-\infty$. Thus, the system has positive temperatures at low energies and negative temperatures at high energies, and $U = (1/2)N_a\varepsilon_1$ is a singular point that corresponds to $T = \pm\infty$.

As noted above, most common systems contain particles with an unlimited number of increasing energy levels, and absolute temperatures for such systems are positive. If a system with unbounded energy levels at a positive temperature is brought into contact with a system at a negative temperature, the equilibrium temperature reached when the systems are brought into contact must be positive because the composite system will have unbounded energy levels.

For the system considered in Example 2.4, negative temperatures correspond to higher system energies than positive temperatures. When a system of the type considered in the example at a positive temperature is brought into contact with a similar system at a negative temperature, energy will transfer from the system having a negative temperature to the system having a positive temperature. This transfer increases the entropy of the composite system, which must be maximized at equilibrium. Thus, negative absolute temperatures are hotter than positive absolute temperatures. The scale of absolute temperature therefore runs from $+0$ K to $+\infty$ K/$-\infty$ K to -0 K.

The two-energy-state system considered in Example 2.4 can be used as a model of the nuclear spin energy storage in a crystal exposed to a magnetic field. In the presence of a magnetic field, the lowest nuclear state may be split into a finite number of nuclear magnetic states. For this model to apply, the energy storage in the nuclear spin states must be decoupled from energy storage elsewhere in the crystal lattice. The nuclear spin storage must reach equilibrium quickly and transfer of energy between nuclear spin storage and other storage modes must be weak. When these conditions are met, the nuclear spin subsystem behaves like as a finite-state system like that considered in Example 2.4 and may exhibit negative temperatures. Abragam and Proctor [1] conducted experiments with LiF crystals that meet these conditions. Their investigation indicated that the lithium and fluorine nuclei may act as two separate subsystems that exhibit different negative temperatures. Further information on the thermodynamics of systems that exhibit negative absolute temperatures can be found in the references by Ramsey [2], Klein [3], Abragam and Proctor [1], Proctor [4], and Muschik [5].

2.6 Maxwell Statistics and Thermodynamic Properties for a Monatomic Gas

Up to this point in this chapter we have talked in general terms about the energy storage in a system of particles. In this section we want to consider a specific system type. We will specifically consider a system containing two species of boson-type particles that store energy only by translational kinetic energy. We will limit the model to conditions where the system exhibits dilute occupancy. This is a good model of a monatomic ideal gas for moderate temperatures where electronic and nuclear energy storage effects are negligible. We begin with the Boltzmann distribution relations generated in Section 2.3:

$$\frac{N_{a,i}}{N_a} = \frac{g_i e^{-\varepsilon_i/k_B T}}{Z_a}, \tag{2.49a}$$

$$\frac{N_{b,j}}{N_b} = \frac{g_j e^{-\varepsilon_j/k_B T}}{Z_b}, \tag{2.49b}$$

where

$$Z_a = \sum_{i=0}^{\infty} g_i e^{-\varepsilon_i/k_B T}, \tag{2.50a}$$

$$Z_b = \sum_{j=0}^{\infty} g_j e^{-\varepsilon_j/k_B T}. \tag{2.50b}$$

For translational storage only, the degeneracy is given by Eq. (1.121) from the particle-in-a-box quantum solution in Section 1.7:

$$g(\varepsilon) = \frac{dn_\varepsilon}{d\varepsilon}\delta\varepsilon = \frac{\pi}{4}\left(\frac{8mV^{2/3}}{h^2}\right)^{3/2} \varepsilon^{1/2}\delta\varepsilon. \tag{1.121}$$

The energy levels for translational storage become more closely spaced as the energy increases. For translation of atom-sized particles at room temperature or above, the spacing of the quantum energy levels is extremely close. For practical purposes, we can consider energy to be a continuous variable. In this continuum limit, the summations in the relation for the partition functions become integrals:

$$Z_a = \sum_{i=0}^{\infty} g_i e^{-\varepsilon_i/k_B T} = \sum_{\text{all }\varepsilon} g(\varepsilon)e^{-\varepsilon/k_B T} = \int_0^\infty \frac{\pi}{4}\left(\frac{8m_a V^{2/3}}{h^2}\right)^{3/2} \varepsilon^{1/2}e^{-\varepsilon/k_B T}d\varepsilon, \tag{2.113a}$$

$$Z_b = \sum_{j=0}^{\infty} g_j e^{-\varepsilon_j/k_B T} = \sum_{\text{all }\varepsilon} g(\varepsilon)e^{-\varepsilon/k_B T} = \int_0^\infty \frac{\pi}{4}\left(\frac{8m_b V^{2/3}}{h^2}\right)^{3/2} \varepsilon^{1/2}e^{-\varepsilon/k_B T}d\varepsilon. \tag{2.113b}$$

Evaluating the integrals on the right side of the above relations yields

$$Z_a = V \left(\frac{2\pi m_a k_B T}{h^2} \right)^{3/2}, \tag{2.114a}$$

$$Z_b = V \left(\frac{2\pi m_b k_B T}{h^2} \right)^{3/2}. \tag{2.114b}$$

In the continuum limit, $N_{a,i}$, the number of a particles in energy level i, is replaced with dN_a, the differential number of particles in the interval from ε to $\varepsilon + d\varepsilon$, and $\delta\varepsilon$ in the degeneracy relations becomes a differential $d\varepsilon$. A similar transformation applies to the distribution for b particles. The discrete distributions (2.49a,b) are thereby transformed into the following continuous distributions:

$$\frac{dN_a}{N_a} = \left(\frac{\pi V}{4Z_a} \right) \left(\frac{8m_a}{h^2} \right)^{3/2} \varepsilon^{1/2} e^{-\varepsilon/k_B T} d\varepsilon, \tag{2.115a}$$

$$\frac{dN_b}{N_b} = \left(\frac{\pi V}{4Z_b} \right) \left(\frac{8m_b}{h^2} \right)^{3/2} \varepsilon^{1/2} e^{-\varepsilon/k_B T} d\varepsilon. \tag{2.115b}$$

Substituting the partition function relations (2.114a,b) into (2.115a,b), we obtain

$$\frac{dN_a}{N_a} = \left(\frac{2}{\sqrt{\pi}} \right) \frac{\varepsilon^{1/2}}{(k_B T)^{3/2}} e^{-\varepsilon/k_B T} d\varepsilon, \tag{2.116a}$$

$$\frac{dN_b}{N_b} = \left(\frac{2}{\sqrt{\pi}} \right) \frac{\varepsilon^{1/2}}{(k_B T)^{3/2}} e^{-\varepsilon/k_B T} d\varepsilon. \tag{2.116b}$$

Thus both species obey the same energy distribution at equilibrium. Note that the right side of the above equations are the probabilities that a particle has energy between ε and $\varepsilon + d\varepsilon$. We can determine the mean energy of a particle as

$$\langle \varepsilon \rangle = \int_0^\infty \varepsilon \left(\frac{2}{\sqrt{\pi}} \right) \frac{\varepsilon^{1/2}}{(k_B T)^{3/2}} e^{-\varepsilon/k_B T} d\varepsilon.$$

Evaluating the integral yields

$$\langle \varepsilon \rangle = \tfrac{3}{2} k_B T. \tag{2.117}$$

Thus the mean energy per particle is the same for species a and species b, regardless of differences in particle mass.

If we denote the speed of the type a and type b particles as c, the energy of the species a particles is related to their speed as

$$\varepsilon = \tfrac{1}{2} m_a c^2, \tag{2.118a}$$

while for species b, the energy is given by

$$\varepsilon = \tfrac{1}{2} m_b c^2. \tag{2.118b}$$

The energy distributions can be converted to speed distributions by substituting the appropriate energy relation (2.118a) or (2.118b) into Eqs. (2.116a,b). The resulting speed distributions for the two species are

$$\frac{dN_a}{N_a} = 4\pi \left(\frac{m_a}{2\pi k_B T}\right)^{3/2} c^2 e^{-m_a c^2/2k_B T} dc, \tag{2.119a}$$

$$\frac{dN_b}{N_b} = 4\pi \left(\frac{m_b}{2\pi k_B T}\right)^{3/2} c^2 e^{-m_b c^2/2k_B T} dc. \tag{2.119b}$$

Note that if the masses of the two particle types are different, they will have distinctly different speed distributions, as dictated by the relations above. The above equations are two embodiments of the *Maxwell speed distribution*, which was first derived as part of the kinetic theory of gases developed by James Clerk Maxwell in the 1860s. This relation emerges as a special case of the Boltzmann distribution.

Example 2.5 Derive a relation for the most probable speed of an atom in a monatomic gas. Use the results to determine the most probable speed for helium atoms at 290 K.

Solution Equation (2.115a) can be interpreted as the probability that a randomly selected atom in a monatomic gas has a speed between c and $c + dc$:

$$\tilde{P}(c) = \frac{dN_a}{N_a} = 4\pi \left(\frac{m_a}{2\pi k_B T}\right)^{3/2} c^2 e^{-m_a c^2/2k_B T} dc.$$

The right side of the above relation exhibits a maximum at a particular c value. This most probable value $c_{mp,a}$ is that which maximizes dN_a/dc. Solving the above relation for dN_a/dc, differentiating, and setting the result equal to zero yields

$$c_{mp,a} = \left(\frac{2k_B T}{m_a}\right)^{1/2}.$$

This relation indicates that the location of the peak in the distribution curve shifts to higher speeds as the temperature increases. This shift is indicated graphically in Figure 2.6.

For helium ($m = 6.64 \times 10^{-27}$ kg) at 290 K, we therefore have

$$c_{mp,a} = \left(\frac{2k_B T}{m}\right)^{1/2} = \left(\frac{2(1.38 \times 10^{-23})(290)}{6.67 \times 10^{-27}}\right)^{1/2} = 1095 \text{ m/s}.$$

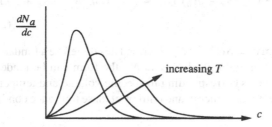

Figure 2.6

The same line of analysis applied to species b yields the relation

$$c_{\mathrm{mp},b} = \left(\frac{2k_B T}{m_b}\right)^{1/2}.$$

It is clear from the relations for $c_{\mathrm{mp},a}$ and $c_{\mathrm{mp},b}$ that the most probable speed will differ if the mass of the two particle types is different. Thus, in a mixture of neon ($m = 3.35 \times 10^{-26}$ kg) and helium ($m = 6.64 \times 10^{-27}$ kg), the most probable speeds for the two types of atoms will differ by more than a factor of 2.

Having evaluated the partition function, we are now in a position to evaluate other thermodynamic properties for the system we are considering. In Section 2.3 we obtained the following relation for the system internal energy:

$$U = \frac{N_a k_B T^2}{Z_a}\left(\frac{\partial Z_a}{\partial T}\right)_{N_a, N_b, V} + \frac{N_b k_B T^2}{Z_b}\left(\frac{\partial Z_b}{\partial T}\right)_{N_a, N_b, V}. \tag{2.52}$$

Substituting the partition function relations (2.109a,b) into the above equation, and rearranging somewhat, the relation for internal energy becomes

$$U = \tfrac{3}{2}(N_a + N_b)k_B T. \tag{2.120}$$

We also derived the following relation for the entropy of the system in Section 2.4:

$$\frac{S}{k_B} = \frac{U}{k_B T} + N_a \ln\left\{\frac{Z_a}{N_a}\right\} + N_b \ln\left\{\frac{Z_b}{N_b}\right\} + N_a + N_b. \tag{2.78}$$

Substituting Eq. (2.120) and the partition function relations (2.114a,b) into the above equation yields, after a little rearranging,

$$\frac{S}{k_B} = \frac{5}{2}(N_a + N_b) + \frac{3}{2}N_a \ln\left\{\frac{2\pi m_a V^{2/3} k_B T}{h^2 N_a^{2/3}}\right\} + \frac{3}{2}N_b \ln\left\{\frac{2\pi m_b V^{2/3} k_B T}{h^2 N_b^{2/3}}\right\}. \tag{2.121}$$

We now solve Eq. (2.120) for T and substitute into Eq. (2.121) to obtain

$$\frac{S}{k_B} = \frac{5}{2}(N_a + N_b) + \frac{3}{2}N_a \ln\left\{\frac{4\pi m_a V^{2/3} U}{3h^2 N_a^{2/3}(N_a + N_b)}\right\} + \frac{3}{2}N_b \ln\left\{\frac{4\pi m_b V^{2/3} U}{3h^2 N_b^{2/3}(N_a + N_b)}\right\}. \tag{2.122}$$

Note that this equation is of the form $S = S(U, V, N_a, N_b)$, and it is therefore a fundamental equation for the system. For a system with fixed N_a and N_b, the form can be made a bit cleaner if we define a reference state as corresponding to specific volume and temperature values V_0 and T_0. It follows that the internal energy and entropy at these reference conditions are given by

$$U_0 = \tfrac{3}{2}(N_a + N_b)k_B T_0, \tag{2.123}$$

$$\frac{S_0}{k_B} = \frac{5}{2}(N_a + N_b) + \frac{3}{2}N_a \ln\left\{\frac{4\pi m_a V_0^{2/3} U_0}{3h^2 N_a^{2/3}(N_a + N_b)}\right\} + \frac{3}{2}N_b \ln\left\{\frac{4\pi m_b V_0^{2/3} U_0}{3h^2 N_b^{2/3}(N_a + N_b)}\right\}.$$

(2.124)

The fundamental equation can then be cast in terms of the reference state values as

$$\frac{S - S_0}{k_B} = \frac{3}{2}(N_a + N_b) \ln\left\{\left(\frac{V}{V_0}\right)^{2/3}\left(\frac{U}{U_0}\right)\right\}.$$

(2.125)

This relation is plotted in Figure 2.7.

In engineering analysis, generally only the change of system entropy is of interest. For such cases, the reference values for S, V, and U cancel from the calculations when using the above equation. Hence the reference state is arbitrary and can be chosen for convenience. Regardless of the choice, the change in entropy between two states 1 and 2 is given by

$$\frac{S_2 - S_1}{k_B} = \frac{3}{2}(N_a + N_b) \ln\left\{\left(\frac{V_2}{V_1}\right)^{2/3}\left(\frac{U_2}{U_1}\right)\right\}$$

$$= \frac{3}{2}(N_a + N_b) \ln\left\{\left(\frac{V_2}{V_1}\right)^{2/3}\left(\frac{T_2}{T_1}\right)\right\}.$$

(2.126)

One additional interesting result can be obtained from this analysis. Solving the fundamental Eq. (2.125) for U yields the relation

$$U = U_0 \left(\frac{V}{V_0}\right)^{-2/3} \exp\left\{\frac{2(S - S_0)}{3(N_a + N_b)k_B}\right\}.$$

(2.127)

Also, by definition, we know that

$$P \equiv -\left(\frac{\partial U}{\partial V}\right)_{N_a, N_b, S}.$$

(2.101)

Evaluating the derivative above using Eq. (2.127) yields

$$P \equiv -\left(\frac{\partial U}{\partial V}\right)_{N_a, N_b, S} = \frac{2}{3}U_0\frac{V^{-5/3}}{V_0^{-2/3}} \exp\left\{\frac{2(S - S_0)}{3(N_a + N_b)k_B}\right\},$$

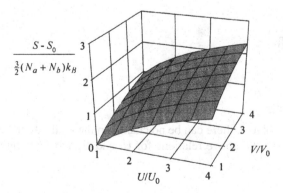

Figure 2.7

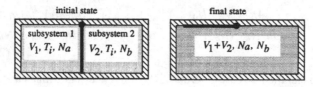

Figure 2.8

which, using Eq. (2.127), can be rewritten as

$$P = \frac{2}{3}\left(\frac{U}{V}\right). \tag{2.128}$$

Finally, combining Eqs. (2.120) and (2.128) to eliminate U, we obtain

$$P = (N_a + N_b)\frac{k_B T}{V}, \tag{2.129}$$

which the reader may recognize as the ideal gas equation of state. Thus, our model predicts that this system of boson particles that store energy only by translational motion will obey the ideal gas equation of state.

Example 2.6 Initially a composite system consists of two subsystems as shown in Figure 2.8. One has volume V_1 and contains N_a particles of species a and the other has volume V_2 and has N_b particle of species b. Initially the two subsystems are in equilibrium at the same temperature T_i. The partition separating the subsystems is removed, allowing the particles to mix. Find the change in temperature and entropy for the composite system if these particles behave as bosons that store energy by translation alone. You may assume that dilute occupancy applies.

Solution Using Eq. (2.120), we have initially

$$U_1 = \tfrac{3}{2}N_a k_B T_i, \qquad U_2 = \tfrac{3}{2}N_b k_B T_i$$

and therefore

$$U_i = \tfrac{3}{2}(N_a + N_b)\,k_B T_i.$$

In the final mixed state

$$U_f = \tfrac{3}{2}(N_a + N_b)\,k_B T_f.$$

Since the composite system is isolated, there can be no change in the total internal energy, implying that $U_f = U_i$. Equating the above relations for U_i and U_f and rearranging, we obtain

$$T_f - T_i = 0.$$

Thus, there is no change in temperature for the system. Using Eq. (2.121), we can obtain the following relations for the initial entropy of subsystems 1 and 2:

$$\frac{S_1}{k_B} = \frac{5}{2}N_a + \frac{3}{2}N_a \ln\left\{\frac{2\pi m_a V_1^{2/3} k_B T_i}{h^2 N_a^{2/3}}\right\},$$

$$\frac{S_2}{k_B} = \frac{5}{2}N_b + \frac{3}{2}N_b \ln\left\{\frac{2\pi m_b V_2^{2/3} k_B T_i}{h^2 N_b^{2/3}}\right\}.$$

Additivity requires that

$$S_i = S_1 + S_2$$

and for the final state Eq. (2.121) applies:

$$\frac{S_f}{k_B} = \frac{5}{2}(N_a + N_b) + \frac{3}{2}N_a \ln\left\{\frac{2\pi m_a (V_1 + V_2)^{2/3} k_B T_i}{h^2 N_a^{2/3}}\right\}$$

$$+ \frac{3}{2}N_b \ln\left\{\frac{2\pi m_b (V_1 + V_2)^{2/3} k_B T_i}{h^2 N_b^{2/3}}\right\}.$$

The change in entropy of the composite systems is therefore given by

$$S_f - S_i = S_f - S_1 - S_2.$$

Substituting the relations for S_1, S_2, and S_f into the above equation and simplifying yields the following relation for the change in entropy:

$$S_f - S_i = N_a k_B \ln\left\{\frac{V_1 + V_2}{V_1}\right\} + N_b k_B \ln\left\{\frac{V_1 + V_2}{V_2}\right\}.$$

Note that this change of entropy due to the mixing of two pure gases is always positive since $V_1 + V_2$ is always larger than either V_1 or V_2. This increase in entropy due to mixing alone is sometimes referred to as the *entropy of mixing*.

The statistical analysis of systems in a microcanonical ensemble presented in this chapter has led us to several important conclusions regarding macroscopic properties and equilibrium conditions in such systems. The interested reader can find further discussion of these topics in the references [6]–[9] listed at the end of this chapter. In the next chapter we will more fully develop a framework for macroscopic thermodynamic analysis of systems.

Exercises

2.1 Determine the possible distributions among energy states of three simple harmonic oscillators such that their total energy is $13h\nu/2$. All three harmonic oscillators have the same characteristics frequency ν. Is the set of occupancy numbers $\{1, 0, 1, 1, 0, 0, \ldots\}$ an acceptable distribution?

2.2 Determine the number of ways of distributing six distinguishable atoms into four energy levels such that one atom is in the first level, one is in the second, three atoms are in the third, and one is in the fourth level. (Note that the order within an energy level does not matter.) Repeat the calculation for indistinguishable atoms.

2.3 A system contains five indistinguishable bosons of a single species that can occupy one of five quantum energy levels that all have the same degeneracy. Determine and plot the total number of microstates for this type of system for energy level degeneracy values of 1, 10, and 100 and plot the number of microstates as a function of the degeneracy. (Hint: You may want to work in terms of the log of the number of microstates.)

2.4 A system contains five indistinguishable fermions of a single species that can occupy one of four quantum energy levels that all have the same degeneracy. Determine and plot the total number of microstates for this type of system for energy level degeneracy values of 2, 5, 10, and 50 and plot the number of microstates as a function of the degeneracy.

2.5 A system contains two indistinguishable bosons of species a and two indistinguishable bosons of species b. For both species, each particle can occupy one of three quantum energy levels. All the energy levels for both species have the same degeneracy. Determine and plot the total number of microstates for this type of system for energy level degeneracy values of 1, 10, and 50 and plot the number of microstates as a function of the degeneracy. (Hint: You may want to work in terms of the log of the number of microstates.)

2.6 A system of two energy levels with energies ε_0 and ε_1 is populated by N particles at temperature T. The degeneracy of both levels is one. The particles populate the microstates according to Boltzmann statistics.

(a) Derive an expression for the average energy per particle.
(b) Determine the limiting behavior of the average energy per particle as $T \rightarrow 0$ and $T \rightarrow \infty$.

2.7 A system contains boson particles with two internal energy levels: a ground state of degeneracy g_1 and a low-level excited state of degeneracy g_2 and energy $\Delta \varepsilon$ above the ground state. Dilute occupancy is a good approximation for this system. Derive a relation for the average energy per particle in the system in terms of g_1, g_2, $\Delta \varepsilon$, and temperature T.

2.8 Gibbs theorem states that the entropy for a binary mixture ideal gas is the same as the sum of the entropies each gas species would have if it occupied the total system volume V alone at temperature T. Use the results of Section 2.6 to show that this is true for a mixture of two ideal monatomic gases.

2.9 The standard deviation of an energy distribution

$$\frac{dN_a}{N_a} = f_a(\varepsilon) \, d\varepsilon$$

is given by

$$\sigma_\varepsilon = [\langle \varepsilon^2 \rangle - \langle \varepsilon \rangle^2]^{1/2},$$

where the angle brackets denote the weighted average of the enclosed quantity,

$$\langle \gamma(\varepsilon) \rangle = \int_0^\infty \gamma(\varepsilon) f_a(\varepsilon) \, d\varepsilon.$$

Derive a relation for σ_ε for the translational energy of an atom in a monatomic gas at temperature T. What happens to σ_ε at low temperatures ($T \rightarrow 0$)?

2.10 For a molecular speed distribution

$$\frac{dN_a}{N_a} = f_a(c) \, dc$$

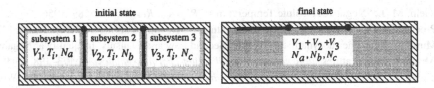

Figure 2.9

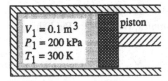

Figure 2.10

the root mean square (rms) speed $c_{\text{rms},a}$ is given by

$$c_{\text{rms},a} = \left[\int_0^\infty c^2 f_a(c)\, dc \right]^{1/2}.$$

Derive a relation for $c_{\text{rms},a}$ for a monatomic gas at temperature T. Use the result to evaluate $c_{\text{rms},a}$ for helium atoms at 290 K and compare $c_{\text{rms},a}$ to the most probable speed.

2.11 Initially a composite system consists of the three subsystems with different volumes as shown in Figure 2.9. Initially all three subsystems are in equilibrium with the same temperature T_i. Each subsystems contains a different monatomic gas species a, b, or c. The two partitions separating the systems are removed, allowing the different species to mix. Find the change in temperature and entropy of the composite system between the initial configuration and the final state. (Hint: Extend relations obtained in the two-subsystem problem discussed in Example 2.6 to a composite system having three subsystems.)

2.12 The cylinder shown in Figure 2.10 initially contains neon gas at 200 kPa and 300 K. The gas expands from its initial volume of 0.1 m³ in a constant-entropy process until its volume is 0.2 m³. Using results from Section 2.6, find the change in system internal energy for the process and the final system temperature.

2.13 For a monatomic gas that undergoes a constant entropy process from state 1 to state 2, use results of Section 2.6 to show that

$$P_1 V_1^{5/3} = P_2 V_2^{5/3}.$$

2.14 For a monatomic gas that undergoes a constant-entropy process from state 1 to state 2, use results of Section 2.6 to show that

$$\frac{T_1^{5/3}}{P_1} = \frac{T_2^{5/3}}{P_2}.$$

References

[1] Abragram, A. and Proctor, W. G., "Experiments on Spin Temperature," *Physical Review*, 106: 160, 1957.

[2] Ramsey, N. F., "Thermodynamics and Statistical Mechanics at Negative Absolute Temperature," *Physical Review*, 103: 20–8, 1956.

[3] Klein, M. J., "Negative Absolute Temperatures," *Physical Review*, 104: 589, 1956.

[4] Proctor, W. G., "Negative Absolute Temperatures," *Scientific American*, 239(2): 78–85, 1978.

[5] Muschik, W., "Thermodynamical Algebra, Second Law and Clausius' Inequality at Negative Absolute Temperatures," *Journal of Nonequilibrium Thermodynamics*, 14, 173–198, 1989.

[6] Callen, H. B., *Thermodynamics and an Introduction to Thermostatistics*, 2nd ed., John Wiley & Sons, New York, 1985.

[7] Kittel, C. and Kroemer, H., *Thermal Physics*, 2nd ed., W. H. Freeman and Company, New York, 1980.

[8] Robertson, H. S., *Statistical Thermophysics*, Prentice-Hall, Englewood Cliffs, NJ, 1993.

[9] Tien, C. L. and Lienhard, J. H., *Statistical Thermodynamics*, Hemisphere Publishing Corporation, New York, 1979.

CHAPTER 3

A Macroscopic Framework

The basic elements of statistical thermodynamics were developed in Chapter 2. In this chapter, we digress briefly from development of the statistical theory to expand the theoretical framework using mathematical tools and macroscopic analysis. By doing so we more strongly link the statistical theory to classical thermodynamics and set the stage for alternative statistical viewpoints considered in Chapter 4.

3.1   Necessary Conditions for Thermodynamic Equilibrium

In the previous chapter, we have derived several important pieces of information about thermodynamic systems. The goal of this chapter is to expand the framework of macroscopic thermodynamic theory so that it can be applied effectively to a variety of system types. We will begin by summarizing the important ideas developed in the last chapter.

So far, we have taken the volume V, internal energy U, and particle numbers N_a and N_b to be intrinsic properties for any system we may consider. We subsequently defined the properties entropy S, temperature T, pressure P, and chemical potentials μ_a and μ_b. Our analysis of the statistical characteristics of thermodynamic systems has led to the conclusion that for a system with fixed U, V, N_a, and N_b, equilibrium corresponds to a maximum value of the system entropy. This is referred to as the *entropy maximum principle*. The entropy of a composite system with an arbitrary number of subsystems is additive over the constituent subsystems. This is the additivity property of entropy. The value of entropy at equilibrium is a function of U, V, N_a, and N_b. This function is the fundamental relation for the system. We further concluded that entropy is a monotonically increasing function of energy. Finally, we considered two subsystems I and II separated by a boundary that is free to move and permits exchange of heat, species a, and species b. For such a system, we showed that the necessary conditions for equilibrium are:

$$T^{\mathrm{I}} = T^{\mathrm{II}}, \qquad P^{\mathrm{I}} = P^{\mathrm{II}}, \qquad \mu_a^{\mathrm{I}} = \mu_a^{\mathrm{II}}, \qquad \mu_b^{\mathrm{I}} = \mu_b^{\mathrm{II}}. \tag{3.1}$$

Observation of the physical environment around us reveals that matter exists in different forms that we call *phases*. A fluid phase is characterized by the free motion of molecules, atoms, or other particles within the system boundaries. In general, fluids are said to be gases or vapors when their density is low whereas at high density they are categorized as liquids. When the particles become bound together in such a way that their motion is limited to a very small region of space, the matter in the system is acknowledged to be in a solid phase. Because liquids and solids are characterized by high density, they are sometimes jointly referred to as *condensed phases*. The surface that defines the boundary between regions containing different phases is usually termed an *interface*.

We note that the boundary between the subsystems in Figure 2.4 has characteristics identical to those at the boundary between two different phases in a multiphase system. Specifically, this boundary is permeable to all species in the system, it permits heat interactions

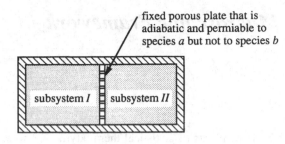

fixed porous plate that is
adiabatic and permiable to
species *a* but not to species *b*

subsystem *I*   subsystem *II*

Figure 3.1

between the phases, and the boundary can move in response to pressure changes on either side. The conditions specified in Eq. (3.1) are therefore the necessary conditions for equilibrium in a system containing two species and two phases with a flat interface between the phases. (The effects of interface curvature will be considered in a later chapter.) These equilibrium conditions are sometimes referred to as the *conditions for phase equilibrium*.

The necessary equilibrium conditions specified in Eq. (3.1) are specific to a system in which the boundary between the subsystems has the following characteristics:

(1) It is permeable to both species *a* and species *b* particles.
(2) It permits heat exchange between the subsystems through the boundary.
(3) The boundary can move in response to pressure changes within the subsystems.

It is important to note that the necessary equilibrium conditions will change if any of these characteristics are altered. As an example, we will consider the system shown in Figure 3.1 in which the boundary between the subsystems has the following characteristics:

(1) It is permeable to species *a* but not to species *b* particles.
(2) It permits no heat exchange between the subsystems through the boundary (the solid material in the porous wall is adiabatic).
(3) The boundary cannot move in response to pressure changes within the subsystems.

The first step of thermodynamic analysis for a system like that in Figure 3.1 must be to determine the necessary conditions for equilibrium. The analysis required to determine the necessary conditions for equilibrium begins with the fundamental relation for each subsystem:

$$S^{\mathrm{I}} = S^{\mathrm{I}}\left(U^{\mathrm{I}}, V^{\mathrm{I}}, N_a^{\mathrm{I}}, N_b^{\mathrm{I}}\right), \tag{3.2a}$$

$$S^{\mathrm{II}} = S^{\mathrm{II}}\left(U^{\mathrm{II}}, V^{\mathrm{II}}, N_a^{\mathrm{II}}, N_b^{\mathrm{II}}\right). \tag{3.2b}$$

Expanding each of these relations to first order in each variable and using the definitions of $T, P, \mu_a,$ and μ_b, we obtain

$$dS^{\mathrm{I}} = \frac{1}{T^{\mathrm{I}}}\, dU^{\mathrm{I}} + \frac{P^{\mathrm{I}}}{T^{\mathrm{I}}}\, dV^{\mathrm{I}} - \frac{\mu_a^{\mathrm{I}}}{T^{\mathrm{I}}}\, dN_a^{\mathrm{I}} - \frac{\mu_b^{\mathrm{I}}}{T^{\mathrm{I}}}\, dN_b^{\mathrm{I}}, \tag{3.3a}$$

$$dS^{\mathrm{II}} = \frac{1}{T^{\mathrm{II}}}\, dU^{\mathrm{II}} + \frac{P^{\mathrm{II}}}{T^{\mathrm{II}}}\, dV^{\mathrm{II}} - \frac{\mu_a^{\mathrm{II}}}{T^{\mathrm{II}}}\, dN_a^{\mathrm{II}} - \frac{\mu_b^{\mathrm{II}}}{T^{\mathrm{II}}}\, dN_b^{\mathrm{II}}. \tag{3.3b}$$

These relations dictate the small change in subsystem entropy that results from small changes in U, V, N_a, and N_b. Since at equilibrium

$$dS = dS^{\mathrm{I}} + dS^{\mathrm{II}} = 0 \tag{3.4}$$

it follows that, at equilibrium conditions, small changes in U, V, N_a, and N_b in the subsystems that result from microscopic fluctuations must satisfy

$$\frac{1}{T^{\mathrm{I}}}dU^{\mathrm{I}} + \frac{P^{\mathrm{I}}}{T^{\mathrm{I}}}dV^{\mathrm{I}} - \frac{\mu_a^{\mathrm{I}}}{T^{\mathrm{I}}}dN_a^{\mathrm{I}} - \frac{\mu_b^{\mathrm{I}}}{T^{\mathrm{I}}}dN_b^{\mathrm{I}} + \frac{1}{T^{\mathrm{II}}}dU^{\mathrm{II}} + \frac{P^{\mathrm{II}}}{T^{\mathrm{II}}}dV^{\mathrm{II}} - \frac{\mu_a^{\mathrm{II}}}{T^{\mathrm{II}}}dN_a^{\mathrm{II}} - \frac{\mu_b^{\mathrm{II}}}{T^{\mathrm{II}}}dN_b^{\mathrm{II}} = 0. \tag{3.5}$$

The analysis to this point is essentially identical to that in Section 2.5 for the specific systems shown in Figure 2.4. In fact, the above relation applies to any composite system with two subsystems and two species. The specific constraints associated with the boundary wall between the subsystems are invoked in the next step of the analysis. The nature of these constraints will differ for different boundary conditions.

For the system considered here, the location of the permeable wall separating the subsystems is fixed. It follows that the volumes of the subsystems are fixed and hence

$$dV^{\mathrm{I}} = dV^{\mathrm{II}} = 0. \tag{3.6}$$

Also, since the porous boundary between the subsystems is impermeable to species b, N_b^{II} is fixed and $N_b^{\mathrm{I}} = 0$. It follows that

$$dN_b^{\mathrm{I}} = dN_b^{\mathrm{II}} = 0. \tag{3.7}$$

Because the total number of species a particles in the system is fixed, if a fluctuation results in a transfer of such particles between the subsystems, it must be true that

$$dN_a^{\mathrm{I}} = -dN_a^{\mathrm{II}}. \tag{3.8}$$

The final constraint on the system is associated with energy transfer between the subsystems. The solid structure of the wall between the subsystems is specified as being adiabatic. It might appear at first glance that this implies that fluctuations of internal energy in the subsystems are not possible. This conclusion is not correct, however. Because species are exchanged between the subsystems, the internal energy of the subsystems can fluctuate. Nevertheless, conservation of energy does require that

$$dU^{\mathrm{I}} = -dU^{\mathrm{II}}. \tag{3.9}$$

Combining Eqs. (3.5)–(3.9), we obtain

$$\left(\frac{1}{T^{\mathrm{I}}} - \frac{1}{T^{\mathrm{II}}}\right)dU^{\mathrm{I}} - \left(\frac{\mu_a^{\mathrm{I}}}{T^{\mathrm{I}}} - \frac{\mu_a^{\mathrm{II}}}{T^{\mathrm{II}}}\right)dN_a^{\mathrm{I}} = 0. \tag{3.10}$$

Since this must be satisfied for arbitrary dU^{I} and dN_a^{I} at equilibrium, it follows that the necessary conditions for equilibrium are

$$T^{\mathrm{I}} = T^{\mathrm{II}}, \qquad \mu_a^{\mathrm{I}} = \mu_a^{\mathrm{II}}. \tag{3.11}$$

Note that, for this system, there are no requirements involving the pressure or the chemical potential for species b. These properties may take on whatever values are necessary to satisfy the specified values of N_b and V in the subsystems.

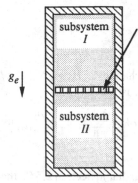

Figure 3.2

For an arbitrary system, determination of the necessary conditions for equilibrium requires simultaneous consideration of the $dS = 0$ relation and the constraints on the system. Equations (2.93) and (3.5) are examples of appropriate forms of the $dS = 0$ relation for this type of analysis. It should be noted, however, that sometimes a more complicated form is necessary to fully account for all the physics of the system. This circumstance is illustrated in Example 3.1. In the next section we examine equations of state that can be used together with the necessary conditions for equilibrium and other constraints to determine the equilibrium state of a composite system like that in Figure 3.1.

Example 3.1 Two parts of a completely isolated system are separated by a porous piston of mass m_p as in Figure 3.2. (Completely isolated implies no heat or work interactions with the surroundings are permitted.) The space on either side of the piston contains a binary gas mixture of species a and species b. The piston, which is free to move in the vertical direction, has a cross section A_p and is permeable to both species a and species b. Determine the necessary conditions for equilibrium.

Solution For a system with two subsystems and two species, the following $dS = 0$ relation must apply at equilibrium to satisfy the entropy maximum principle:

$$\left(\frac{\partial S}{\partial U}\right)^{\mathrm{I}}_{N_a,N_b,V} dU^{\mathrm{I}} + \left(\frac{\partial S}{\partial V}\right)^{\mathrm{I}}_{U,N_a,N_b} dV^{\mathrm{I}} + \left(\frac{\partial S}{\partial N_a}\right)^{\mathrm{I}}_{N_b,U,V} dN_a^{\mathrm{I}}$$

$$+ \left(\frac{\partial S}{\partial N_b}\right)^{\mathrm{I}}_{N_a,U,V} dN_b^{\mathrm{I}} + \left(\frac{\partial S}{\partial U}\right)^{\mathrm{II}}_{N_a,N_b,V} dU^{\mathrm{II}} + \left(\frac{\partial S}{\partial V}\right)^{\mathrm{II}}_{U,N_a,N_b} dV^{\mathrm{II}}$$

$$+ \left(\frac{\partial S}{\partial N_a}\right)^{\mathrm{II}}_{N_b,U,V} dN_a^{\mathrm{II}} + \left(\frac{\partial S}{\partial N_b}\right)^{\mathrm{II}}_{N_a,U,V} dN_b^{\mathrm{II}} = 0.$$

Because both the total volume and the total quantity of each species are fixed, any random perturbation must satisfy

$$dV^{\mathrm{I}} = -dV^{\mathrm{II}}, \qquad dN_a^{\mathrm{I}} = -dN_a^{\mathrm{II}}, \qquad dN_b^{\mathrm{I}} = -dN_b^{\mathrm{II}}.$$

Using these relations and the definitions of P, μ_a, and μ_b, we can manipulate the above

$dS = 0$ relation into the form

$$\left(\frac{\partial S}{\partial U}\right)^{\mathrm{I}}_{N_a, N_b, V} dU^{\mathrm{I}} + \left(\frac{\partial S}{\partial U}\right)^{\mathrm{II}}_{N_a, N_b, V} dU^{\mathrm{II}} + \left(\frac{P^{\mathrm{I}}}{T^{\mathrm{I}}} - \frac{P^{\mathrm{II}}}{T^{\mathrm{II}}}\right) dV^{\mathrm{I}}$$

$$- \left(\frac{\mu_a^{\mathrm{I}}}{T^{\mathrm{I}}} - \frac{\mu_a^{\mathrm{II}}}{T^{\mathrm{II}}}\right) dN_a^{\mathrm{I}} - \left(\frac{\mu_b^{\mathrm{I}}}{T^{\mathrm{I}}} - \frac{\mu_b^{\mathrm{II}}}{T^{\mathrm{II}}}\right) dN_b^{\mathrm{I}} = 0.$$

Energy in this system can be stored as internal energy in the gas in each subsystem and as potential energy associated with the vertical position of the piston. For any energy perturbation of the subsystems, conservation of energy requires

$$dU^{\mathrm{I}} + dU^{\mathrm{II}} = -m_{\mathrm{p}} g_{\mathrm{e}} \, dV^{\mathrm{I}} / A_{\mathrm{p}},$$

where g_{e} is the gravitational acceleration at the earth's surface. Combining this relation and the definition of T with the relation above yields

$$\left(\frac{1}{T^{\mathrm{I}}} - \frac{1}{T^{\mathrm{II}}}\right) dU^{\mathrm{I}} + \left(\frac{P^{\mathrm{I}}}{T^{\mathrm{I}}} - \frac{P^{\mathrm{II}}}{T^{\mathrm{II}}} + \frac{m_{\mathrm{p}} g_{\mathrm{e}}}{A_{\mathrm{p}} T^{\mathrm{II}}}\right) dV^{\mathrm{I}}$$

$$- \left(\frac{\mu_a^{\mathrm{I}}}{T^{\mathrm{I}}} - \frac{\mu_a^{\mathrm{II}}}{T^{\mathrm{II}}}\right) dN_a^{\mathrm{I}} - \left(\frac{\mu_b^{\mathrm{I}}}{T^{\mathrm{I}}} - \frac{\mu_b^{\mathrm{II}}}{T^{\mathrm{II}}}\right) dN_b^{\mathrm{I}} = 0.$$

Since this must hold for arbitrary perturbations, each of the prefactors of the differentials must be zero, which implies the following necessary conditions for equilibrium:

$$T^{\mathrm{I}} = T^{\mathrm{II}}, \qquad P^{\mathrm{II}} = P^{\mathrm{I}} + m_{\mathrm{p}} g_{\mathrm{e}} / A_{\mathrm{p}}, \qquad \mu_a^{\mathrm{I}} = \mu_a^{\mathrm{II}}, \qquad \mu_b^{\mathrm{I}} = \mu_b^{\mathrm{II}}.$$

3.2 The Fundamental Equation and Equations of State

The results obtained in the previous chapter implicitly embody the fact that macroscopically distinguishable states exist that are characterized completely by the internal energy U, volume V, and the number of particles N_a and N_b. In this chapter we will consider the more general case of a system containing r species of particles. The extension of results from the two-species case considered in the previous chapter to the case of an arbitrary number of components is generally quite transparent. In the general case, the fundamental equation has the generic form

$$S = S(U, V, N_1, N_2, \ldots, N_r). \tag{3.12}$$

Note that conservation of mass, energy, and physical space (for nonrelativistic systems) dictates that additivity applies to the particle numbers, internal energy, and volume, as well as entropy. In examining thermodynamic properties of a system at equilibrium, those whose values are proportional to the number of particles in the system are termed *extensive* properties, while those whose values are independent of the mass in the system are said to be *intensive* properties. For additivity to apply to entropy, S must be a *homogeneous first-order* function of the extensive variables $U, V, N_1, N_2, \ldots, N_r$. This implies that for an arbitrary constant ζ

$$S(\zeta U, \zeta V, \zeta N_1, \zeta N_2, \ldots, \zeta N_r) = \zeta S(U, V, N_1, N_2, \ldots, N_r). \tag{3.13}$$

The fact that S is a monotonically increasing function of U ensures that the absolute temperature T is positive. If we postulate that S is continuous and differentiable in addition to being a monotonic function of U, we can, in principle, invert the fundamental relation to get U as a function of $S, V, N_1, N_2, \ldots, N_r$. Thus

$$U = U(S, V, N_1, N_2, \ldots, N_r),$$

$$S = S(U, V, N_1, N_2, \ldots, N_r)$$

are equivalent forms of the fundamental relation. Because the fundamental relation is homogeneous and of first-order, it follows that for a system containing a single species of particle

$$S(U, V, N) = (N/N_A)S(U/(N/N_A), V/(N/N_A), N_A),$$

where N_A is Avogadro's number. For such a system, if we define molar specific properties as

$$\hat{u} = U/(N/N_A), \tag{3.14}$$

$$\hat{v} = V/(N/N_A), \tag{3.15}$$

$$\hat{s} = S/(N/N_A), \tag{3.16}$$

then the fundamental relation can be written in the form

$$\hat{s} = \hat{s}(\hat{u}, \hat{v}, N_A) = \hat{s}(\hat{u}, \hat{v}). \tag{3.17}$$

By convention, the basic principles of thermodynamics are organized in terms of three laws of thermodynamics. The first law of thermodynamics embodies basic conservation of energy. If we limit our considerations to simple systems that are macroscopically homogeneous, isotropic, and not subject to the effects of electric charge, chemical reactions, electromagnetic fields, or surface effects, then conservation of energy requires that changes in internal energy of a system must equal the net energy exchange across the boundary of the system. Generally we categorize system energy exchanges as either being work or heat. In differential form, the first law of thermodynamics for a simple closed system can therefore be stated as

$$dU = \delta Q - \delta W, \tag{3.18}$$

where δQ and δW represent differential quantities of energy crossing the system boundary as heat and work, respectively. Here, heat (δQ) is taken to be positive when it transfers energy into the system and work (δW) is taken to be positive when it transfers energy out of the system. Work and heat are not properties and hence are not exact differentials. From experience, we know that it is possible to make the system walls from materials that poorly transmit heat. In the limit of zero heat exchange with the system, the system walls are said to be *adiabatic*. If the walls of the system are virtually adiabatic, changes in U between states can be experimentally determined by measuring the work required to achieve the change of state.

The second law of thermodynamics embodies the extremum principle. There are several ways of stating the second law, but for our purposes we can state it as follows:

◆   *For a closed isolated system, the values of the extensive properties established at equilibrium are those that correspond to the macrostate having the maximum S of all macrostates accessible to the system.*

We will develop other equivalent ways of stating the second law in later sections of this text.

The third law of thermodynamics states that the entropy of a pure crystalline solid system vanishes as the absolute temperature approaches zero:

$$\left(\lim_{T \to 0} S = 0 \right)_{\text{pure solid crystal}}. \tag{3.19}$$

The full justification for the third law is best understood from the perspective of quantum statistical mechanics. We will return to discuss this point when we consider the thermodynamics of crystalline solids in Chapter 7. For the present we will simply adopt this as a plausible means of setting the zero point of entropy for pure crystalline solids.

We now will further explore the relationships among thermodynamic properties that evolve from the fundamental equation. Expanding the fundamental relation

$$U = U(S, V, N_1, N_2, \ldots, N_r) \tag{3.20}$$

in a Taylor series and retaining only first-order terms, we obtain

$$dU = \left(\frac{\partial U}{\partial S} \right)_{V, N_1, N_2, \ldots, N_r} dS + \left(\frac{\partial U}{\partial V} \right)_{S, N_1, N_2, \ldots, N_r} dV + \sum_{i=1}^{r} \left(\frac{\partial U}{\partial N_i} \right)_{S, V, N_{j \neq i}} dN_i. \tag{3.21}$$

Using generalized versions of the definitions of T, P, and the chemical potential in Eq. (2.47), (2.97), and (2.102) in the Chapter 2,

$$T = \left(\frac{\partial U}{\partial S} \right)_{V, N_1, N_2, \ldots, N_r}, \tag{3.22}$$

$$P = -\left(\frac{\partial U}{\partial V} \right)_{S, N_1, N_2, \ldots, N_r}, \tag{3.23}$$

$$\mu_i = \left(\frac{\partial U}{\partial N_i} \right)_{S, V, N_{j \neq i}}, \tag{3.24}$$

Eq. (3.21) can be written

$$dU = T\, dS - P\, dV + \sum_{i=1}^{r} \mu_i\, dN_i. \tag{3.25}$$

The above equation for dU applies to changes between two equilibrium states and hence we can integrate it only along a path composed of a sequence of equilibrium states. Such a path is termed a *quasiequilibrium* path. A process that follows a quasiequilibrium path is referred to as a *quasistatic process*. The properties T, P, and μ_i are all intensive properties. This is demonstrated as follows. Consider a system consisting of ζ identical subsystems. By definition, the temperature of the composite system is given by

$$T_{\text{comp}} = \frac{\partial}{\partial(\zeta S)} U(\zeta S, \zeta V, \zeta N_1, \ldots, \zeta N_r). \tag{3.26}$$

In the above relation $S, V, N_1, \ldots, N_r$ are the properties of the identical subsystems and, by additivity, ζS equals the entropy for the composite system. Since U is also additive over the subsystems,

$$U(\zeta S, \zeta V, \zeta N_1, \ldots, \zeta N_r) = \zeta U(S, V, N_1, \ldots, N_r) \tag{3.27}$$

and therefore

$$T_{\text{comp}} = \frac{\partial}{\partial(\zeta S)} \zeta U(S, V, N_1, \ldots, N_r). \tag{3.28}$$

The factors of ζ cancel to yield

$$T_{\text{comp}} = \frac{\partial}{\partial S} U(S, V, N_1, \ldots, N_r). \tag{3.29}$$

The right side is just the definition of the temperature for a subsystem. Thus the temperature in each subsystem is equal to that for the system as a whole. Temperature is therefore independent of the quantity of mass in the system, and it is therefore an intensive property. It is left as an exercise to show that the same line of analysis demonstrates that P and μ_i are also intensive. For a system containing a single species of particle, the definitions of the intensive properties P and T can be cast in the form

$$P = -\left(\frac{\partial \hat{u}}{\partial \hat{v}}\right)_{\hat{s}}, \tag{3.30}$$

$$T = \left(\frac{\partial \hat{u}}{\partial \hat{s}}\right)_{\hat{v}}. \tag{3.31}$$

For a closed system of fixed composition, no changes in particle numbers can occur, and for any changes in state $dN_i = 0$. Equation (3.25) then reduces to

$$dU = T\,dS - P\,dV. \tag{3.32}$$

Since by definition P, T, and μ_i are partial derivatives of a function of $S, V, N_1, \ldots, N_r$, they themselves are functions of these variables:

$$T = T(S, V, N_1, \ldots, N_r), \tag{3.33}$$

$$P = P(S, V, N_1, \ldots, N_r), \tag{3.34}$$

$$\mu_i = \mu_i(S, V, N_1, \ldots, N_r). \tag{3.35}$$

Relations of this type, expressing intensive properties as functions of independent extensive properties, are called *equations of state*. Knowledge of an equation of state does not imply knowledge of all properties of a system. Since a fundamental equation is homogeneous and of first-order, it follows that equations of state are *homogeneous and of zeroth order*. Consequently, multiplication of each independent extensive property by a scalar constant ζ leaves the function unchanged. Thus, for example

$$T(\zeta S, \zeta V, \zeta N_1, \ldots, \zeta N_r) = T(S, V, N_1, \ldots, N_r). \tag{3.36}$$

The equation of state for a pure ideal gas,

$$T = \frac{PV}{Nk_{\text{B}}}, \tag{3.37}$$

is homogeneous and of zeroth order since multiplication of both V and N by the same constant leaves the resulting value of T unchanged. It can also be written as

$$T = \frac{P\hat{v}}{R},\tag{3.38}$$

where $R = N_A k_B$ is the universal gas constant.

As an equation of state for real gases, the ideal gas equation of state is inadequate in two regards. First, the equation of state implies that $\hat{v} = RT/P$. But if $P \to \infty$ or $T \to 0$, $\hat{v}$ cannot go to zero as the ideal gas relation predicts because the molecules themselves occupy some space. We can patch up the formula by introducing a constant b_v representing this residual volume per molecule. It follows that

$$\hat{v} = RT/P + N_A b_v,\tag{3.39}$$

which can be rearranged to give

$$P = \frac{RT}{\hat{v} - N_A b_v}.\tag{3.40}$$

The second inadequacy is that attractive forces in a real gas reduce the pressure exerted by gas on walls of the containment below that which would exist in the absence of such forces. Attractive forces are often proportional to r^{-6}, where r is the separation distance between molecules. In a gas the mean separation distance is proportional to $\hat{v}^{-1/3}$ and we therefore expect that the reduction in pressure due to such forces will be proportional to $\hat{v}^{-2}$. This suggests that we account for this effect with a relation of the form

$$P = \frac{RT}{\hat{v} - N_A b_v} - \frac{N_A^2 a_v}{\hat{v}^2},\tag{3.41}$$

where a_v is a constant. Equation (3.41) is *van der Waal's equation*, a prototype equation of state for liquids as well as dense gases. Different constants must be used for different gases as the size and attractive forces for each molecular species will be different. Although it is not quantitatively exact over wide ranges for some gases, van der Waal's qualitatively represents the behavior of the liquid and gas phases of many pure substances.

Example 3.2 For the monatomic gas model (indistinguishable bosons with dilute occupancy) discussed in Section 2.6, derive the equations of state for the chemical potential.

Solution The entropy of the gas is given by Eq. (2.122) as

$$\frac{S}{k_B} = \frac{5}{2}(N_a + N_b) + \frac{3}{2}N_a \ln\left\{\frac{4\pi m_a V^{2/3} U}{3h^2 N_a^{2/3}(N_a + N_b)}\right\} + \frac{3}{2}N_b \ln\left\{\frac{4\pi m_b V^{2/3} U}{3h^2 N_b^{2/3}(N_a + N_b)}\right\}.$$

Since by definition $\mu_a = (\partial U/\partial N_a)_{V,S,N_b}$, we could invert this relation to get U as a function of the other variables and differentiate to determine μ_a. However, a somewhat easier alternative is to note that for a function y of x and z we have

$$\left(\frac{\partial x}{\partial z}\right)_y = -\left(\frac{\partial x}{\partial y}\right)_z \left(\frac{\partial y}{\partial z}\right)_x.$$

and therefore we can write that

$$\left(\frac{\partial S}{\partial N_a}\right)_{V,U,N_b} = -\left(\frac{\partial S}{\partial U}\right)_{V,N_a,N_b}\left(\frac{\partial U}{\partial N_a}\right)_{V,S,N_b}.$$

But by definition the first partial derivative on the right side above is just $1/T$ and the second is μ_a, and so

$$\left(\frac{\partial S}{\partial N_a}\right)_{V,U,N_b} = -\frac{\mu_a}{T}.$$

Differentiating the entropy relation above with respect to N_a and setting the resulting equation to $-\mu_a/T$ and rearranging, we obtain

$$\mu_a = -\left(\frac{3k_BT}{2}\right)\ln\left\{\frac{4\pi m_a V^{2/3} U}{3h^2 N_a^{2/3}(N_a+N_b)}\right\}.$$

Since the internal energy of the gas is given by

$$U = \frac{3}{2}(N_a+N_b)k_BT$$

we can solve this relation for T and substitute into the relation for μ_a. This yields the following equation of state:

$$\mu_a = -\left(\frac{U}{N_a+N_b}\right)\ln\left\{\frac{4\pi m_a V^{2/3} U}{3h^2 N_a^{2/3}(N_a+N_b)}\right\}.$$

The identical reasoning can be applied to species b to obtain

$$\mu_b = -\left(\frac{U}{N_a+N_b}\right)\ln\left\{\frac{4\pi m_b V^{2/3} U}{3h^2 N_b^{2/3}(N_a+N_b)}\right\}.$$

3.3   The Euler Equation and the Gibbs–Duhem Equation

The fundamental relation in terms of U is homogenous and of first order, implying that

$$U(\zeta S, \zeta V, \zeta N_1, \ldots, \zeta N_r) = \zeta U(S, V, N_1, \ldots, N_r). \tag{3.42}$$

Differentiating both sides of this relation with respect to ζ and applying the chain rule, we obtain

$$\frac{\partial U}{\partial(\zeta S)}\left(\frac{\partial(\zeta S)}{\partial \zeta}\right) + \frac{\partial U}{\partial(\zeta V)}\left(\frac{\partial(\zeta V)}{\partial \zeta}\right) + \sum_{i=1}^{r}\frac{\partial U}{\partial(\zeta N_i)}\left(\frac{\partial(\zeta N_i)}{\partial \zeta}\right) = U, \tag{3.43}$$

which reduces to

$$\frac{\partial U}{\partial(\zeta S)}S + \frac{\partial U}{\partial(\zeta V)}V + \sum_{i=1}^{r}\frac{\partial U}{\partial(\zeta N_i)}N_i = U. \tag{3.44}$$

Since Eq. (3.44) must be valid for any choice of ζ, we take $\zeta = 1$:

$$\frac{\partial U}{\partial S}S + \frac{\partial U}{\partial V}V + \sum_{i=1}^{r}\frac{\partial U}{\partial N_i}N_i = U. \tag{3.45}$$

Using relations (3.22)–(3.24), the partial derivatives can be replaced with the corresponding intensive properties, converting the Eq. (3.45) to

$$U = TS - PV + \sum_{i=1}^{r} \mu_i N_i. \tag{3.46}$$

Equation (3.46) is referred to as the *Euler equation*. Differentiating both sides of Eq. (3.46) yields

$$dU = T\,dS + S\,dT - P\,dV - V\,dP + \sum_{i=1}^{r} \mu_i\,dN_i + \sum_{i=1}^{r} N_i\,d\mu_i. \tag{3.47}$$

Replacing dU in the above equation with the right side of Eq. (3.25) and eliminating terms that cancel, we obtain the *Gibbs–Duhem equation*

$$S\,dT - V\,dP + \sum_{i=1}^{r} N_i\,d\mu_i = 0. \tag{3.48}$$

For a single-component system this reduces to

$$S\,dT - V\,dP + N\,d\mu = 0. \tag{3.49}$$

Equation (3.49) can also be written in terms of a molar chemical potential $\hat{\mu} = N_A \mu$:

$$d\hat{\mu} = -\hat{s}\,dT + \hat{v}\,dP, \tag{3.50}$$

which is the commonly cited form of the single-component Gibbs–Duhem equation. For a single-component system $\hat{\mu} = \hat{\mu}(\hat{s}, \hat{v})$ and it follows that $T = T(\hat{s}, \hat{v})$, $P = P(\hat{s}, \hat{v})$, $\hat{\mu} = \hat{\mu}(\hat{s}, \hat{v})$. We may use these three relations to eliminate $\hat{s}$ and $\hat{v}$, yielding one relation among three variables T, P, and $\hat{\mu}$. For a binary system $U/N_2 = U(S/N_2, V/N_2, N_1/N_2)$, from which we can derive four equations of state:

$$T = T(S/N_2, V/N_2, N_1/N_2),$$

$$P = P(S/N_2, V/N_2, N_1/N_2),$$

$$\mu_1 = \mu_1(S/N_2, V/N_2, N_1/N_2),$$

$$\mu_2 = \mu_2(S/N_2, V/N_2, N_1/N_2).$$

These four relations can be used to eliminate three variables S/N_2, V/N_2, and N_1/N_2, thereby obtaining a relation among T, P, μ_1, and μ_2. If we add one more species, we get one more variable to eliminate and one more equation of state to use to eliminate it. Hence, in principle, it is always possible to get a functional relation among the intensive properties T, P, and chemical potentials μ_i. The Gibbs–Duhem equation is the differential form of the relation among these properties.

Example 3.3 For a binary mixture of monatomic gases, derive the relation among T, P, and the chemical potentials μ_a and μ_b.

Solution The internal energy of the gas is given by Eq. (2.120) as

$$U = \tfrac{3}{2}(N_a + N_b)k_B T.$$

We can solve this relation for T and substitute into the relation for μ_a obtained in Example 3.2. This yields the following relation:

$$\mu_a = -\left(\frac{3}{2}\right)k_{\mathrm{B}}T \ln\left\{\frac{2\pi m_a k_{\mathrm{B}}T}{h^2}\left(\frac{V}{N_a}\right)^{2/3}\right\}.$$

The identical reasoning can be applied to species b to obtain

$$\mu_b = -\left(\frac{3}{2}\right)k_{\mathrm{B}}T \ln\left\{\frac{2\pi m_b k_{\mathrm{B}}T}{h^2}\left(\frac{V}{N_b}\right)^{2/3}\right\}.$$

This system also obeys the ideal gas law

$$PV = (N_a + N_b)k_{\mathrm{B}}T.$$

Rearranging, we can write this relation in the form

$$\frac{V}{N_a} = \left(\frac{k_{\mathrm{B}}T}{P}\right)\frac{(N_a + N_b)/N_{\mathrm{A}}}{N_a/N_{\mathrm{A}}} = \frac{k_{\mathrm{B}}T}{P X_a},$$

where X_a is the mole fraction of species a, defined as

$$X_a = \frac{N_a/N_{\mathrm{A}}}{(N_a + N_b)/N_{\mathrm{A}}}.$$

Substituting the right side of the above equation for V/N_a into the relation for μ_a, and rearranging, we obtain

$$\mu_a = -\left(\frac{3}{2}\right)k_{\mathrm{B}}T \ln\left\{\frac{2\pi m_a (k_{\mathrm{B}}T)^{5/3}}{h^2 P^{2/3}}\right\} + k_{\mathrm{B}}T \ln X_a.$$

Applying the same analysis to species b yields

$$\mu_b = -\left(\frac{3}{2}\right)k_{\mathrm{B}}T \ln\left\{\frac{2\pi m_b (k_{\mathrm{B}}T)^{5/3}}{h^2 P^{2/3}}\right\} + k_{\mathrm{B}}T \ln X_b.$$

Since $X_b = 1 - X_a$, the above relation for μ_b can be written

$$\mu_b = -\left(\frac{3}{2}\right)k_{\mathrm{B}}T \ln\left\{\frac{2\pi m_b (k_{\mathrm{B}}T)^{5/3}}{h^2 P^{2/3}}\right\} + k_{\mathrm{B}}T \ln\{1 - X_a\}.$$

This relation for μ_b can be combined with the above relation for μ_a to eliminate X_a. Doing so yields the following relation among T, P, μ_a, and μ_b:

$$\frac{\mu_a}{k_{\mathrm{B}}T} = -\left(\frac{3}{2}\right)\ln\left\{\frac{2\pi m_a (k_{\mathrm{B}}T)^{5/3}}{h^2 P^{2/3}}\right\}$$

$$+ \ln\left\{1 - \exp\left\{\frac{\mu_b}{k_{\mathrm{B}}T} - \left(\frac{3}{2}\right)\ln\left\{\frac{2\pi m_b (k_{\mathrm{B}}T)^{5/3}}{h^2 P^{2/3}}\right\}\right\}\right\}.$$

It is also noteworthy that the relation

$$\mu_a = -\left(\frac{3}{2}\right) k_B T \ln\left\{\frac{2\pi m_a (k_B T)^{5/3}}{h^2 P^{2/3}}\right\} + k_B T \ln X_a$$

can be rearranged to the form

$$\mu_a = -\left(\frac{5}{2}\right) k_B T \ln\left\{\frac{T}{\theta_{\mu,a}}\right\} + k_B T \ln\left\{\frac{P_a}{P_{a,\text{ref}}}\right\},$$

where P_a is the *partial pressure* of species a, defined as

$$P_a = X_a P,$$

$P_{a,\text{ref}}$ is an arbitrary reference value for P_a, and

$$\theta_{\mu,a} = -\left(\frac{P_{a,\text{ref}}^{3/2} h^2}{2\pi m_a k_B^{5/3}}\right)^{3/5}.$$

Corresponding relations (with the as changed to bs) apply to species b. This form of the relation for μ_a illustrates that, for a monatomic ideal gas at constant temperature, there is a direct functional relation between the partial pressure of the species and its chemical potential.

3.4   Legendre Transforms and Thermodynamic Functions

Legendre Transforms

The original extremum principle that has emerged from our statistical analysis indicates that the equilibrium state is that which maximizes S for a specified system total internal energy. We now want to develop other formulations of the extremum principle. To emphasize the mathematics, we generalize the fundamental relation

$$S = S(U, V, N_1, \ldots, N_r)$$

to the form

$$Y = Y(X_0, X_1, X_2, \ldots, X_t) \tag{3.51}$$

and we define

$$\xi_k = \frac{\partial Y}{\partial X_k}. \tag{3.52}$$

We want to express the information contained in the fundamental relation in terms of derivatives ξ_k, which in our thermodynamic analysis are intensive properties. We shall first consider the simplest case where Y is a function of one variable X:

$$Y = Y(X), \qquad \xi = \frac{\partial Y}{\partial X} = \frac{dY}{dX}.$$

The derivative ξ indicates the slope of the tangent lines to the curve $Y = Y(X)$ as shown in Figure 3.3.

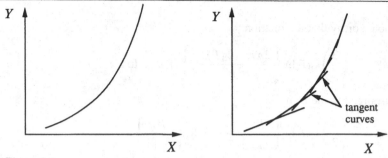

Figure 3.3

The tangent line at each point is uniquely defined by its slope, ξ, and its intercept of the Y axis, ψ. The family of tangent lines defined by $\psi(\xi)$ defines an envelope to the original curve $Y(X)$ and thus contains the same information. Both can be considered to be representations of the fundamental relation.

How do we determine $\psi(\xi)$ when given $Y(X)$? This type of manipulation is known as a *Legendre transformation*. For the simple one-variable case, since

$$\xi = \frac{Y - \psi}{X - 0} \tag{3.53}$$

it follows that

$$\psi = Y - \xi X. \tag{3.54}$$

To execute the transform, we therefore proceed as follows:

(1)   Differentiate $Y(X)$ to get $\xi(X)$.
(2)   Use $\psi = Y(X) - \xi(X)X$ to get $\psi(X)$.
(3)   Eliminate X to get $\psi(\xi)$.

The function $\psi(\xi)$ is termed the *Legendre transform* of $Y(X)$. The inverse process is also of interest. Since $\psi = Y - \xi X$, differentiating yields

$$d\psi = dY - \xi\, dX - X\, d\xi. \tag{3.55}$$

Using the fact that $\xi = dY/dX$ it follows that $d\psi = -X\, d\xi$, or equivalently

$$\frac{d\psi}{d\xi} = -X. \tag{3.56}$$

So if we start with $\psi(\xi)$, we can invert the transformation in the following steps:

(1)   Differentiate $\psi(\xi)$ to get $X(\xi)$.
(2)   Use $Y = \xi X(\xi) - \psi(\xi)$ to get $Y(\xi)$.
(3)   Eliminate ξ to get $Y(X)$.

The symmetry of the transform and its inverse can be seen in the comparison below:

Legendre Transform of $Y(X)$

$$\left.\begin{array}{l} Y = Y(X) \\ \xi = dY/dX \\ \psi(X) = -\xi X + Y \end{array}\right\} \text{ Eliminate } X \text{ and } Y \text{ to get } \psi(\xi)$$

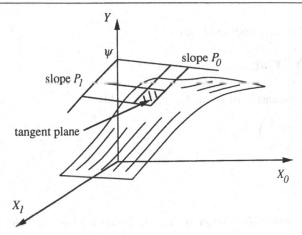

Figure 3.4

Inverse Transform

$$
\left.
\begin{aligned}
\psi &= \psi(\xi) \\
X(\xi) &= -d\psi/d\xi \\
Y &= X\xi + \psi
\end{aligned}
\right\} \text{ Eliminate } \xi \text{ and } \psi \text{ to get } Y(X)
$$

Generalization of the transform to three variables $Y = Y(X_0, X_1)$ requires that we consider a surface as being defined by an envelope of tangent planes. A plane is defined by the intercept of the plane on the Y axis, denoted as ψ, together with ξ_0 and ξ_1, which are the slopes of traces of the plane on the X_0–Y and X_1–Y planes (see Figure 3.4). The function $\psi = \psi(\xi_0, \xi_1)$ thus defines a family of planes that form an envelope about the surface.

We extrapolate the transform process to an arbitrary number of variables as follows. The relation

$$ Y = Y(X_0, X_1, X_2, \ldots, X_t) \tag{3.57} $$

represents a hypersurface in a $(t + 2)$-dimensional space and

$$ \xi_k(X) = \frac{\partial Y}{\partial X_k} \tag{3.58} $$

represents the partial slope of an associated hypersurface. The family of tangent hyperplanes that defines the hypersurface is mathematically specified as

$$ \psi = Y - \sum_{k=0}^{t} \xi_k X_k. \tag{3.59} $$

Taking the differential yields

$$ d\psi = dY - \sum_{k=0}^{t} (\xi_k \, dX_k + X_k \, d\xi_k). \tag{3.60} $$

Using the definition (3.58) of ξ_k and a Taylor expansion of Y,

$$ dY = \sum_{k=0}^{t} \left(\frac{\partial Y}{\partial X_k} \right) dX_k = \sum_{k=0}^{t} \xi_k \, dX_k, \tag{3.61} $$

we can replace dY in Eq. (3.60) to obtain

$$d\psi = -\sum_{k=0}^{t} X_k \, d\xi_k. \tag{3.62}$$

And since a Taylor expansion of $\psi(\xi_0, \xi_1, \ldots, \xi_t)$ yields

$$d\psi = \sum_{k=0}^{t} \left(\frac{\partial \psi}{\partial \xi_k} \right) d\xi_k, \tag{3.63}$$

it must be true that

$$-X_k = \frac{\partial \psi}{\partial \xi_k}. \tag{3.64}$$

To execute a Legendre transformation, we use the following $t + 3$ equations:

$$Y = Y(X_0, X_1, X_2, \ldots, X_t), \tag{3.57}$$

$$\xi_k(X) = \frac{\partial Y}{\partial X_k} \quad (k = 0, 1, 2, \ldots, t), \tag{3.58}$$

$$\psi = Y - \sum_{k=0}^{t} \xi_k X_k \tag{3.59}$$

to eliminate the variables Y and $X_0, X_1, X_2, \ldots, X_t$ ($t + 2$ unknowns) and obtain $\psi = \psi(\xi_0, \xi_1, \ldots, \xi_t)$. To execute the inverse transformation, we use the $t + 3$ equations

$$\psi = \psi(\xi_0, \xi_1, \ldots, \xi_t), \tag{3.65}$$

$$-X_k = \frac{\partial \psi}{\partial \xi_k} \quad (k = 0, 1, 2, \ldots, t), \tag{3.64}$$

$$Y = \psi + \sum_{k=0}^{t} \xi_k X_k \tag{3.66}$$

to eliminate the variables ψ and $\xi_0, \xi_1, \xi_2, \ldots, \xi_t$ ($t + 2$ unknowns) and obtain $Y = Y(X_0, X_1, X_2, \ldots, X_t)$.

We need not transform a function $Y = Y(X_0, X_1, X_2, \ldots, X_t)$ with respect to all variables $X_0, X_1, X_2, \ldots, X_t$. We can treat some as fixed constants. This is termed a *partial transform*. We denote the partial transform with respect to $X_0, X_1, X_2, \ldots, X_n$ on function $Y = Y(X_0, X_1, X_2, \ldots, X_t)$ as $Y = Y[\xi_0, \xi_1, \xi_2, \ldots, \xi_n, X_{n+1}, \ldots, X_t]$. To execute a partial transform we use the $n + 3$ relations

$$Y = Y(X_0, X_1, X_2, \ldots, X_t), \tag{3.67}$$

$$\xi_k(X) = \frac{\partial Y}{\partial X_k} \quad (k = 0, 1, 2, \ldots, n), \tag{3.68}$$

$$Y[\xi_0, \xi_1, \xi_2, \ldots, \xi_n, X_{n+1}, \ldots, X_t] = Y(X_0, \ldots, X_t) - \sum_{k=0}^{t} \xi_k X_k \tag{3.69}$$

to eliminate the variables Y and $X_0, X_1, X_2, \ldots, X_n$ ($n + 2$ unknowns) and obtain $Y = Y[\xi_0, \xi_1, \xi_2, \ldots, \xi_n, X_{n+1}, \ldots, X_t]$. To execute the inverse transformation, we use the $n + 3$

equations

$$Y = Y[\xi_0, \xi_1, \xi_2, \ldots, \xi_n, X_{n+1}, \ldots, X_t], \tag{3.70}$$

$$X_k = -\partial Y[\xi_0, \ldots, \xi_n, X_{n+1}, \ldots, X_t]/\partial \xi_k \quad (k = 0, 1, 2, \ldots, n), \tag{3.71}$$

$$Y(X_0, \ldots, X_t) = Y[\xi_0, \ldots, X_t] + \sum_{k=0}^{t} \xi_k X_k \tag{3.72}$$

to eliminate the variables $Y[\xi_0, \ldots, X_t]$ and $\xi_0, \xi_1, \xi_2, \ldots, \xi_n$ ($n + 2$ unknowns) and obtain $Y(X_0, X_1, \ldots, X_t)$.

The Legendre transformation is also useful in formulating the mechanics of multibody systems. As described in Chapter 1, the Lagrangian

$$L = L(v_1, v_2, \ldots, v_r, q_1, q_2, \ldots, q_r)$$

characterizes the dynamic state of a system in terms of generalized velocities v_i and generalized coordinates q_i. Another formulation of dynamics characterizes the state of the system in terms of the Hamiltonian

$$H = H(p_1, p_2, \ldots, p_r, q_1, q_2, \ldots, q_r),$$

which is a function of generalized momenta p_i and coordinates q_i. In Chapter 1, it was also noted that

$$p_i = \frac{\partial L}{\partial v_i}.$$

It follows directly that the Lagrangian for the system can be converted to the Hamiltonian by executing a partial Legendre transformation with respect to the generalized velocities.

Thermodynamic Functions

We will now explore the application of Legendre transforms to thermodynamic relations. For the fundamental relation $U = U(S, V, N_1, \ldots, N_r)$ the derivatives of U with respect to the independent variables correspond to the intensive parameters T, $-P$, and $\mu_1, \ldots, \mu_r$. Functions obtained by using a Legendre transform to replace one or more extensive properties with an intensive parameter are called *thermodynamic potentials*.

The *Helmholtz free energy* denoted as F (or A) is obtained by the partial Legendre transform replacing S by T:

$$Y = U = U(S, V, N_1, \ldots, N_r),$$

$$\xi_0 = T = \frac{\partial U}{\partial S},$$

$$F = Y - \sum_{k=0}^{t} \xi_k X_k.$$

As the only ξ_k is T and the only X_k is S, the above relation becomes

$$F = U - TS. \tag{3.73}$$

Differentiating gives

$$dF = dU - T\,dS - S\,dT. \tag{3.74}$$

Using the following relation obtained previously,

$$dU = T\,dS - P\,dV + \sum_{i=1}^{r} \mu_i\,dN_i, \tag{3.25}$$

and substituting to eliminate dU, yields the following relation for dF:

$$dF = -S\,dT - P\,dV + \sum_{i=1}^{r} \mu_i\,dN_i. \tag{3.75}$$

The *enthalpy*, denoted as H, is obtained by the transform replacing the volume with the pressure:

$$Y = U = U(S, V, N_1, \dots, N_r),$$

$$\xi_0 = -P = \frac{\partial U}{\partial V},$$

$$H = U - (-P)V,$$

which simplifies to

$$H = U + PV. \tag{3.76}$$

Differentiation yields

$$dH = dU + P\,dV + V\,dP. \tag{3.77}$$

Combining Eq. (3.77) with Eq. (3.25), we obtain

$$dH = T\,dS + V\,dP + \sum_{i=1}^{r} \mu_i\,dN_i. \tag{3.78}$$

The *Gibbs function*, denoted as G, is obtained by the transform replacing S by T and V by P:

$$U = U(S, V, N_1, \dots, N_r),$$

$$-P = \frac{\partial U}{\partial V}, \qquad T = \frac{\partial U}{\partial S},$$

$$G = U - TS - (-P)V.$$

The relation for G simplifies to

$$G = U - TS + PV. \tag{3.79}$$

Differentiating and using Eq. (3.25) to replace dU yields

$$dG = -S\,dT + V\,dP + \sum_{i=1}^{r} \mu_i\,dN_i. \tag{3.80}$$

For a system containing a single particle type, we can obtain a function referred to as the *grand canonical potential*, denoted as $U[T, V, \mu]$, by executing the transform that replaces S by T and N by μ:

$$U = U(S, V, N),$$
$$T = \frac{\partial U}{\partial S}, \qquad \mu = \frac{\partial U}{\partial N}, \tag{3.81}$$
$$U[T, V, \mu] = U - TS - \mu N.$$

If we replace S, V, and all the N_is with the corresponding intensive properties, we obtain the complete Legendre transform of U, given by

$$U[T, P, \mu_1, \ldots, \mu_r] = U - TS + PV - \sum_{i=1}^{r} \mu_i N_i.$$

If we use the Euler equation (3.46) to evaluate U, the above equation reduces to

$$U[T, P, \mu_1, \ldots, \mu_r] = 0. \tag{3.82}$$

Thus, the complete Legendre transform of U is identically zero.

Although we shall not do so here, it is possible to perform analogous transforms to the entropy form of the fundamental relation $S = S(U, V, N_1, \ldots, N_r)$. The resulting properties are referred to as *Massieu functions*. Massieu functions provide an alternate, and equivalent, means of formulating equilibrium thermodynamic theory.

Example 3.4 Prove that for a system containing a single species of particle, the molar specific Gibbs function is equal to the chemical potential per mole of particles.

Solution For a system with only one species of particle, the Euler equation (3.46) requires that

$$U = TS - PV + \mu_a N_a.$$

Solving this relation for μ_a, dividing by N_a, and multiplying by Avogadro's number, we obtain

$$\mu_a N_A = \frac{U + PV - TS}{N_a/N_A} = \hat{u} + P\hat{v} - T\hat{s}.$$

Since by definition $G = U + PV - TS$, the right side of the above equation is just the molar specific Gibbs function $\hat{g}$ and the left side of the above equation is the chemical potential per mole of particles, $\hat{\mu}_a$. Thus

$$\hat{\mu}_a = \hat{g}.$$

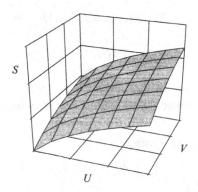

Figure 3.5 $S(U, V)$ for a single component system of fixed N.

3.5   Quasistatic and Reversible Processes

The fundamental equation of a simple system $S = S(U, V, N_1, \ldots, N_r)$ or $U = U(S, V, N_1, \ldots, N_r)$ defines a surface in a thermodynamic configuration space. For the simple case of a system containing N particles of a single species, N is fixed and $S = S(U, V)$ and the fundamental relation is a surface in three space similar to that shown in Figure 3.5.

For a composite system consisting of two or more subsystems, a similar interpretation applies, except that S is a function of more variables. Each point on the hypersurface represents an equilibrium state. An arbitrary curve drawn from state A to state B on the surface represents a sequence of equilibrium states. Such a process could occur only if, at any instant as it traversed from A to B, the system deviated from an equilibrium state by only an infinitesimal amount. Processes that conform to this idealization are termed *quasistatic processes* or *quasiequilibrium processes*. In most systems it is possible to identify a characteristic *relaxation time* required to reestablish equilibrium when the system is perturbed away from equilibrium by a small amount. In a quasistatic process, the relaxation time of a system is short compared to the time over which the process occurs. The differentials dS, dU, etc., considered previously represent changes along a quasistatic path.

Consider now a real system in equilibrium state A. Very often the release of an internal constraint allows a system to access a new state B of higher entropy than the state it was in at the instant the constraint was released. Since our extremum principle mandates that the system entropy must be the highest possible at equilibrium, we expect that release of the constraint will initiate a spontaneous process terminating in state B with higher entropy than state A. Perhaps the simplest example of a situation of this type is the free expansion of a gas into an evacuated subspace of a system when removal of a barrier allows the particles to enter the previously empty space (see Figure 3.6).

For a system containing N particles of a pure monatomic gas, it is a simple matter to show that the change in entropy from state A to state B is given by

$$S_B - S_A = N k_B \ln \left\{ \frac{V_B}{V_A} \right\}. \tag{3.83}$$

From the above relation, it is clear that if $V_B > V_A$ then $S_B > S_A$. Obviously it is also possible to similarly release an internal constraint on energy or number of particles.

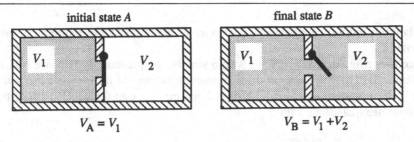

Figure 3.6 Free expansion of a gas.

To reverse the process resulting from the removal of the constraint, the system would have to spontaneously reduce its entropy, achieving an S value less than the original value and hence less than its maximum. This would violate the extremum principle. Hence, the transition from A to B resulting from the release of an internal constraint is an *irreversible process*. The idealized limiting case in which the increase in S is vanishingly small can be reversed without violating the extremum principle. It is therefore termed a *reversible process*.

It is possible to reverse state from B to A if the system is coupled appropriately to another system. However, to invert such a process and to decrease the entropy of a system, a corresponding or larger increase in S must occur in some coupled system. Why is this so? This must be true because if we consider the original system and the coupled system together to be one system, the interaction that reduces S in the original system is the spontaneous result of removing an impediment to such interactions. We have already argued that such spontaneous processes occur because the system is undergoing a transition to a state of higher entropy. Thus the entropy of the combined system must increase, or in the limit, remain the same. This directly implies that if the entropy of the original system decreases by an amount $S_B - S_A$, the entropy of the coupled system must increase by that amount or more.

An increase in S is an indication that the spontaneous process is irreversible, and an irreversible process always results in an increase in S. Hence, for any real process in a closed system

$$dS \geq 0. \tag{3.84}$$

For a closed simple compressible system with a fixed number of particles of each species, Eq. (3.32) indicates that

$$dU = T\,dS - P\,dV. \tag{3.32}$$

The first law requires that for differential changes of state

$$dU = \delta Q - \delta W. \tag{3.18}$$

Note that Eq. (3.32) is valid for changes along a quasiequilibrium path. It is therefore restricted to quasistatic processes. If we combine Eqs. (3.32) and (3.18) to eliminate dU, the following relation is obtained:

$$T\,dS = \delta Q - \delta W + P\,dV. \tag{3.85}$$

For systems of fixed volume and no work interaction, it follows that

$$\delta Q = T\,dS. \tag{3.86}$$

This relation is valid for quasisteady heat exchange at constant temperature for a system of fixed volume.

By considering energy exchanges with a simple compressible system, we can deduce several important consequences of the extremum principle. We first define a *thermal reservoir* as being a system enclosed by a rigid impermeable wall for which all heat transfer interactions are quasistatic with

$$dU = \delta Q = T_{\text{res}} \, dS_{\text{res}} \tag{3.87}$$

and transfer of heat to or from the reservoir has a negligible effect on its temperature. We now consider a quasistatic process in which heat is transferred from a thermal reservoir to a system held at constant volume so that there is no work interaction. For the system, Eq. (3.86) requires that

$$dS_{\text{sys}} = \delta Q / T_{\text{sys}}. \tag{3.88}$$

By additivity, the change of entropy of the composite system composed of the original system and the reservoir is given by

$$dS_{\text{comp sys}} = dS_{\text{sys}} + dS_{\text{res}} = \frac{\delta Q}{T_{\text{sys}}} - \frac{\delta Q}{T_{\text{res}}}. \tag{3.89}$$

Note that here we have adopted the convention that heat exchange is positive when it flows into the system. For this spontaneous process within the composite system, Eq. (3.84) implies that $dS_{\text{comp sys}} \geq 0$. Combining this requirement with Eq. (3.89) it is clear that for such a process to occur, we must have

$$T_{\text{res}} \geq T_{\text{sys}}. \tag{3.90}$$

Thus heat exchange spontaneously occurs between two systems only if their temperatures differ, and heat spontaneously flows from a high temperature system to a lower temperature one. It also follows that reversible transfer of heat can only be achieved if the temperatures of the two systems are the same.

Reversible heat transfer therefore requires that the transfer occur quasistatically and at constant temperature. These are exactly the requirements for the validity of Eq. (3.86). We can therefore state that

$$dS = \frac{\delta Q}{T} \tag{3.91}$$

for a reversible heat addition process when $\delta W = P \, dV$. Note that this requires perfect conversion of the energy associated with the $P \, dV$ effect to mechanical work, or vice versa, depending on the direction of the volume change. Because of friction and other effects, such conversion is never perfect. For work done on the system, δW and $P \, dV$ are both negative and $P \, dV$ is greater (less negative) than δW owing to imperfect conversion. For work done by the system, δW and $P \, dV$ are both positive and δW is less than $P \, dV$ owing to imperfect conversion. In either case, mathematically we can state that for any work interaction

$$P \, dV - \delta W \geq 0. \tag{3.92}$$

In the above relation the equal sign applies to the limiting case of a quasistatic work interaction with perfect conversion. Such a process would be reversible. The equal sign in Eq. (3.92) therefore applies to a reversible work process.

Solving Eq. (3.85) for $P\,dV - \delta W$ and substituting into the above inequality (3.92), we obtain

$$T\,dS - \delta Q \geq 0. \tag{3.93}$$

Solving this relation for $T\,dS$ yields the following relation, which applies to any real process:

$$dS \geq \frac{\delta Q}{T} \quad \text{(for any real process)}. \tag{3.94}$$

In Eq. (3.94), the equal sign applies in the limit of a reversible process. Note that if a process is reversible and adiabatic, Eq. (3.94) implies that $dS = 0$. Thus, a reversible and adiabatic process for a simple compressible system of fixed composition is a constant-entropy process. Such processes are also said to be *isentropic*.

This line of analysis naturally leads to a discussion of heat engines, heat pumps, and the Carnot cycle. The interested reader may wish to consult the references by Howell and Buckius [1] and Modell and Reid [2] for further discussion of these topics. At this point, however, we will explore other aspects of the theoretical framework of macroscopic thermodynamics.

3.6   Alternate Forms of the Extremum Principle

The extremum principle on which our development of macroscopic theory has been based has thus far been stated in terms of entropy. This statement of the principle arose from our consideration of a system held at fixed volume and energy. We can, however, obtain other, completely equivalent statements of the extremum principle by considering other system constraints. All these statements are equivalent forms of the second law of thermodynamics. The entropy maximum form of the second law can be succinctly stated as

◆   *For a system held at constant V, U, and N_i, at equilibrium, unconstrained parameters take on values that maximize the entropy of the system.*

A more mathematical statement is

$$dS = 0, \quad d^2S < 0 \quad \text{(at equilibrium for a system at fixed } V, U, \text{ and } N_i). \tag{3.95}$$

We now consider the more general case of an isolated system of fixed volume containing a pure substance. For any spontaneous process, the first and second laws in the entropy maximum formulation require that

$$dU = \delta Q - \delta W, \tag{3.96}$$

$$dS \geq \frac{\delta Q}{T}. \tag{3.97}$$

Combining these relations yields

$$T\,dS \geq dU + \delta W. \tag{3.98}$$

If the only possible work mode is pressure–volume work, then $\delta W = P\,dV$. Since V is fixed, $\delta W = 0$ and upon rearranging (3.98) becomes

$$dU \leq T\,dS. \tag{3.99}$$

Now rather than imposing a condition of fixed energy, we allow heat exchange so as to keep S fixed. It follows that $dS = 0$ and Eq. (3.99) reduces to

$$dU \leq 0. \tag{3.100}$$

Thus, for a system with fixed V and S, spontaneous processes always reduce the system energy. When the system reaches a condition of minimum U, spontaneous processes will produce zero change in U. The equilibrium condition thus corresponds to minimum U for a system at constant V and S:

$$dU = 0, \quad d^2U > 0 \quad \text{(at equilibrium for a system at fixed } V, S, \text{ and } N_i). \tag{3.101}$$

Instead of the entropy maximum principle, we have an *energy minimum principle:*

◆   *For a system held at constant V, S, and N_i, at equilibrium, unconstrained parameters take on values that minimize the internal energy of the system.*

We now consider an isolated system held at constant V, T, and N_i. For any spontaneous internal process the first and second laws take the form of Eqs. (3.96) and (3.97). Combining these relations to eliminate δQ gives

$$T\,dS \geq dU + \delta W.$$

Because V is fixed, $\delta W = P\,dV = 0$, which reduces the above equation to

$$T\,dS \geq dU. \tag{3.102}$$

Differentiating $F = U - TS$ and rearranging, we obtain

$$dU = dF + T\,dS + S\,dT. \tag{3.103}$$

Substituting this result into the $T\,dS$ inequality (3.102) yields, after a little rearranging,

$$dF \leq -S\,dT. \tag{3.104}$$

For fixed T, $dT = 0$, whereupon Eq. (3.104) reduces to

$$dF \leq 0. \tag{3.105}$$

Hence any spontaneous process must result in a decrease in F. It follows that at equilibrium, F must be a minimum. Mathematically we state this condition as

$$dF = 0, \qquad d^2F > 0. \tag{3.106}$$

We generalize this as the *Helmholtz potential minimum principle:*

◆   *For a system held at constant V, T, and N_i, at equilibrium, unconstrained parameters take on values that minimize the Helmholtz free energy of the system.*

We can similarly develop an extremum principle for a system held at constant P and S. With these constraints, the first and second laws require that for a spontaneous process

$$dH \leq 0. \tag{3.107}$$

It follows that any spontaneous process reduces H. At equilibrium, H must therefore be a minimum. We state this condition mathematically as

$$dH = 0, \qquad d^2H > 0. \tag{3.108}$$

This leads us to the *Enthalpy minimum principle:*

♦   *For a system held at constant P, S, and N_i, at equilibrium, unconstrained parameters take on values that minimize the enthalpy of the system.*

A final circumstance of interest is a system held at constant T and constant P. For these constraints, the first and second laws require that for any spontaneous process

$$dG \leq 0. \tag{3.109}$$

Any spontaneous process thus reduces G, implying that G must be a minimum at equilibrium with

$$dG = 0, \qquad d^2G > 0. \tag{3.110}$$

This can be stated as the *Gibbs function minimum principle:*

♦   *For a system held at constant P, T, and N_i, at equilibrium, unconstrained parameters take on values that minimize the Gibbs function of the system.*

It is important to note that all forms of the extremum principle are equivalent. Generally it is most advantageous to select the form that corresponds to the imposed boundary conditions on the system of interest.

3.7   Maxwell Relations

Additional relations among thermodynamics properties can be obtained purely from mathematical requirements. We consider here the fundamental equation in the energy form for a system containing a binary mixture:

$$U = U(S, V, N_a, N_b). \tag{3.111}$$

Expanding U in a Taylor series and truncating after the first-order terms, we obtain

$$dU = \left(\frac{\partial U}{\partial S}\right)_{N_a,N_b,V} dS + \left(\frac{\partial U}{\partial V}\right)_{S,N_a,N_b} dV + \left(\frac{\partial U}{\partial N_a}\right)_{S,V,N_b} dN_a + \left(\frac{\partial U}{\partial N_b}\right)_{S,V,N_a} dN_b. \tag{3.112}$$

We take the derivative in the first term of Eq. (3.112) and differentiate with respect to V and take the derivative in the second term and differentiate with respect to S, which yields

$$\left(\frac{\partial}{\partial V}\left(\frac{\partial U}{\partial S}\right)_{N_a,N_b,V}\right)_{S,N_a,N_b} = \frac{\partial^2 U}{\partial V \partial S}, \tag{3.113}$$

$$\left(\frac{\partial}{\partial S}\left(\frac{\partial U}{\partial V}\right)_{S,N_a,N_b}\right)_{N_a,N_b,V} = \frac{\partial^2 U}{\partial S \partial V}. \tag{3.114}$$

But since the order of differentiation does not matter, the cross derivatives in Eqs. (3.113) and (3.114) are equal. Consequently,

$$\left(\frac{\partial}{\partial V}\left(\frac{\partial U}{\partial S}\right)_{N_a,N_b,V}\right)_{S,N_a,N_b} = \left(\frac{\partial}{\partial S}\left(\frac{\partial U}{\partial V}\right)_{S,N_a,N_b}\right)_{N_a,N_b,V}. \tag{3.115}$$

However, by definition

$$\left(\frac{\partial U}{\partial S}\right)_{N_a,N_b,V} = T, \tag{3.116a}$$

$$\left(\frac{\partial U}{\partial V}\right)_{S,N_a,N_b} = -P, \tag{3.116b}$$

$$\left(\frac{\partial U}{\partial N_a}\right)_{N_b,S,V} = \mu_a, \tag{3.116c}$$

$$\left(\frac{\partial U}{\partial N_b}\right)_{N_a,S,V} = \mu_b, \tag{3.116d}$$

and Eq. (3.112) can be written as

$$dU = T\,dS - P\,dV + \mu_a\,dN_a + \mu_b\,dN_b. \tag{3.117}$$

Using Eqs. (3.116a) and (3.116b), we can write Eq. (3.115) as

$$\left(\frac{\partial T}{\partial V}\right)_{S,N_a,N_b} = -\left(\frac{\partial P}{\partial S}\right)_{N_a,N_b,V}. \tag{3.118}$$

Note that the above equation is a relation among thermodynamic properties known as a Maxwell relation. This line of analysis can be applied to any two of the partial derivatives that appear in Eq. (3.112). Doing so, the following additional relations among properties can be obtained:

$$\left(\frac{\partial T}{\partial N_a}\right)_{S,V,N_b} = \left(\frac{\partial \mu_a}{\partial S}\right)_{N_a,V,N_b}, \tag{3.119}$$

$$\left(\frac{\partial T}{\partial N_b}\right)_{S,V,N_a} = \left(\frac{\partial \mu_b}{\partial S}\right)_{N_b,V,N_b}, \tag{3.120}$$

$$-\left(\frac{\partial P}{\partial N_a}\right)_{S,V,N_b} = \left(\frac{\partial \mu_a}{\partial V}\right)_{N_a,S,N_b}, \tag{3.121}$$

$$-\left(\frac{\partial P}{\partial N_b}\right)_{S,V,N_a} = \left(\frac{\partial \mu_b}{\partial V}\right)_{N_b,S,N_a}, \tag{3.122}$$

$$\left(\frac{\partial \mu_a}{\partial N_b}\right)_{S,V,N_a} = \left(\frac{\partial \mu_b}{\partial N_a}\right)_{S,V,N_b}. \tag{3.123}$$

Other Maxwell relations can be derived by considering the Legendre transformed fundamental relations. In Section 3.3 we showed that the Helmholtz free energy F could be related to T, V, N_a, and N_b by executing a partial Legendre transform that replaces S by T in the fundamental relation $U = U(S, V, N_a, N_b)$. The resulting relation is of the form

$$F = F(T, V, N_a, N_b) \tag{3.124}$$

and the Legendre transform yielded the relation

$$dF = -S \, dT - P \, dV + \mu_a \, dN_a + \mu_b \, dN_b. \tag{3.125}$$

Expanding the relation (3.124) for the Helmholtz function in a Taylor series and retaining only first-order terms, we obtain

$$dF = \left(\frac{\partial F}{\partial T}\right)_{V,N_a,N_b} dT + \left(\frac{\partial F}{\partial V}\right)_{T,N_a,N_b} dV + \left(\frac{\partial F}{\partial N_a}\right)_{T,V,N_b} dN_a + \left(\frac{\partial F}{\partial N_b}\right)_{T,V,N_a} dN_b. \tag{3.126}$$

A comparison of Eqs. (3.125) and (3.126) clearly shows that

$$\left(\frac{\partial F}{\partial T}\right)_{V,N_a,N_b} = -S, \tag{3.127}$$

$$\left(\frac{\partial F}{\partial V}\right)_{T,N_a,N_b} = -P, \tag{3.128}$$

$$\left(\frac{\partial F}{\partial N_a}\right)_{T,V,N_b} = \mu_a, \tag{3.129}$$

$$\left(\frac{\partial F}{\partial N_b}\right)_{T,V,N_a} = \mu_b. \tag{3.130}$$

We now can proceed in the manner described above to obtain an additional set of Maxwell relations. For each relation, we select a pair of partial derivatives in Eq. (3.126), differentiate each with respect to the differentiation variable of the other, equate the cross derivatives, and substitute appropriate definitions from among Eqs. (3.127)–(3.130). This process yields the following six additional Maxwell relations:

$$\left(\frac{\partial S}{\partial V}\right)_{T,N_a,N_b} = \left(\frac{\partial P}{\partial T}\right)_{V,N_a,N_b}, \tag{3.131}$$

$$-\left(\frac{\partial S}{\partial N_a}\right)_{T,V,N_b} = \left(\frac{\partial \mu_a}{\partial T}\right)_{V,N_a,N_b}, \tag{3.132}$$

$$-\left(\frac{\partial S}{\partial N_b}\right)_{T,V,N_a} = \left(\frac{\partial \mu_b}{\partial T}\right)_{V,N_b,N_a}, \tag{3.133}$$

$$-\left(\frac{\partial P}{\partial N_a}\right)_{T,V,N_b} = \left(\frac{\partial \mu_a}{\partial V}\right)_{V,N_a,N_b}, \tag{3.134}$$

$$-\left(\frac{\partial P}{\partial N_b}\right)_{T,V,N_a} = \left(\frac{\partial \mu_b}{\partial V}\right)_{V,N_b,N_a}, \tag{3.135}$$

$$\left(\frac{\partial \mu_a}{\partial N_b}\right)_{T,V,N_a} = \left(\frac{\partial \mu_b}{\partial N_a}\right)_{T,V,N_b}. \tag{3.136}$$

The partial Legendre transform that replaces volume with pressure provides a relation for enthalpy,

$$H = H(U, P, N_a, N_b), \tag{3.137}$$

and subsequent differentiation yielded the relation

$$dH = T \, dS + V \, dP + \mu_a \, dN_a + \mu_b \, dN_b. \tag{3.138}$$

Using the same procedure as in the case of the Helmholtz function gives the following Maxwell relations:

$$\left(\frac{\partial T}{\partial P} \right)_{S,N_a,N_b} = \left(\frac{\partial V}{\partial S} \right)_{P,N_a,N_b}, \tag{3.139}$$

$$\left(\frac{\partial T}{\partial N_a} \right)_{S,P,N_b} = \left(\frac{\partial \mu_a}{\partial S} \right)_{P,N_a,N_b}, \tag{3.140}$$

$$\left(\frac{\partial T}{\partial N_b} \right)_{S,P,N_a} = \left(\frac{\partial \mu_b}{\partial S} \right)_{P,N_b,N_a}, \tag{3.141}$$

$$\left(\frac{\partial V}{\partial N_a} \right)_{S,P,N_b} = \left(\frac{\partial \mu_a}{\partial P} \right)_{S,N_a,N_b}, \tag{3.142}$$

$$\left(\frac{\partial V}{\partial N_b} \right)_{S,P,N_a} = \left(\frac{\partial \mu_b}{\partial P} \right)_{S,N_b,N_a}, \tag{3.143}$$

$$\left(\frac{\partial \mu_a}{\partial N_b} \right)_{S,P,N_a} = \left(\frac{\partial \mu_b}{\partial N_a} \right)_{S,P,N_b}. \tag{3.144}$$

The partial Legendre transform that replaces entropy by temperature and volume with pressure yields a relation for the Gibb's function

$$G = G(T, P, N_a, N_b), \tag{3.145}$$

for which

$$dG = -S \, dT + V \, dP + \mu_a \, dN_a + \mu_b \, dN_b. \tag{3.146}$$

Using the same procedure as in the case of the Helmholtz function yields the following Maxwell relations:

$$-\left(\frac{\partial S}{\partial P} \right)_{T,N_a,N_b} = \left(\frac{\partial V}{\partial T} \right)_{P,N_a,N_b}, \tag{3.147}$$

$$-\left(\frac{\partial S}{\partial N_a} \right)_{T,P,N_b} = \left(\frac{\partial \mu_a}{\partial T} \right)_{P,N_a,N_b}, \tag{3.148}$$

$$-\left(\frac{\partial S}{\partial N_b} \right)_{T,P,N_a} = \left(\frac{\partial \mu_b}{\partial T} \right)_{P,N_b,N_a}, \tag{3.149}$$

$$\left(\frac{\partial V}{\partial N_a} \right)_{T,P,N_b} = \left(\frac{\partial \mu_a}{\partial P} \right)_{T,N_a,N_b}, \tag{3.150}$$

$$\left(\frac{\partial V}{\partial N_b} \right)_{T,P,N_a} = \left(\frac{\partial \mu_b}{\partial P} \right)_{T,N_b,N_a}, \tag{3.151}$$

$$\left(\frac{\partial \mu_a}{\partial N_b} \right)_{T,P,N_a} = \left(\frac{\partial \mu_b}{\partial N_a} \right)_{T,P,N_b}. \tag{3.152}$$

Using the definition of the grand canonical potential

$$U[T, V, \mu_a, \mu_b] = U - TS - \mu_a N_a - \mu_b N_b, \tag{3.153}$$

for which it can be shown that

$$dU[T, V, \mu_a, \mu_b] = -S\,dT - P\,dV - N_a\,d\mu_a - N_b\,d\mu_b, \tag{3.154}$$

we can obtain the following Maxwell relations:

$$\left(\frac{\partial S}{\partial V}\right)_{T,\mu_a,\mu_b} = \left(\frac{\partial P}{\partial T}\right)_{V,\mu_a,\mu_b}, \tag{3.155}$$

$$\left(\frac{\partial S}{\partial \mu_a}\right)_{T,V,\mu_b} = \left(\frac{\partial N_a}{\partial T}\right)_{V,\mu_a,\mu_b}, \tag{3.156}$$

$$\left(\frac{\partial S}{\partial \mu_b}\right)_{T,V,\mu_a} = \left(\frac{\partial N_b}{\partial T}\right)_{V,\mu_b,\mu_a}, \tag{3.157}$$

$$\left(\frac{\partial P}{\partial \mu_a}\right)_{T,V,\mu_b} = \left(\frac{\partial N_a}{\partial V}\right)_{T,\mu_a,\mu_b}, \tag{3.158}$$

$$\left(\frac{\partial P}{\partial \mu_b}\right)_{T,V,\mu_a} = \left(\frac{\partial N_b}{\partial V}\right)_{T,\mu_b,\mu_a}, \tag{3.159}$$

$$\left(\frac{\partial N_a}{\partial \mu_b}\right)_{T,V,\mu_a} = \left(\frac{\partial N_b}{\partial \mu_a}\right)_{T,V,\mu_b}. \tag{3.160}$$

It should be clear at this point that this methodology can be applied to any of the Legendre transformed forms of the fundamental equation. Other Maxwell relations can be derived in this manner by considering the partial transform that replaces only N_a by μ_a ($U[S, V, \mu_a, N_b]$), the partial transform that replaces N_a by μ_a and N_b by μ_b ($U[S, V, \mu_a, \mu_b]$), and the partial transform that replaces V by P, N_a by μ_a, and N_b by μ_b ($U[S, P, \mu_a, \mu_b]$).

3.8 Other Properties

Several other properties are of interest in thermodynamic analysis of specific two-component systems. The *isothermal compressibility* κ_T of a two-component system is defined as

$$\kappa_T = -\left(\frac{1}{V}\right)\left(\frac{\partial V}{\partial P}\right)_{T,N_a,N_b}. \tag{3.161}$$

Another property of interest is the *coefficient of thermal expansion* β_T defined as

$$\beta_T = \left(\frac{1}{V}\right)\left(\frac{\partial V}{\partial T}\right)_{P,N_a,N_b}. \tag{3.162}$$

Two additional properties that are frequently of use in thermodynamic analysis are the molar specific heat at constant volume $\hat{c}_V$ and the molar specific heat at constant pressure $\hat{c}_P$. These are defined as

$$\hat{c}_V \equiv \left(\frac{N_A}{N_a + N_b}\right)\left(\frac{\partial U}{\partial T}\right)_{V,N_a,N_b}, \tag{3.163}$$

$$\hat{c}_P \equiv \left(\frac{N_A}{N_a + N_b}\right)\left(\frac{\partial H}{\partial T}\right)_{P,N_a,N_b}, \tag{3.164}$$

where N_A in the above relations is Avogadro's number. These defintions are equivalent to

$$\hat{c}_V = \left(\frac{N_A T}{N_a + N_b}\right)\left(\frac{\partial S}{\partial T}\right)_{V,N_a,N_b}, \tag{3.165}$$

$$\hat{c}_P = \left(\frac{N_A T}{N_a + N_b}\right)\left(\frac{\partial S}{\partial T}\right)_{P,N_a,N_b}. \tag{3.166}$$

The equivalence of (3.164) and (3.166) is demonstrated in Example 3.5. Note that setting either N_a or N_b to zero one recovers the definitions of the above parameters for a pure substance.

Example 3.5   Show that the definitions (3.164) and (3.166) for $\hat{c}_P$ are equivalent.

Solution   For a differential change in properties, Eq. (3.32) indicates that, for a system with fixed N_a and N_b,

$$dU = T\,dS - P\,dV.$$

By defintion, $H = U + PV$ and it follows that $dH = dU + P\,dV + V\,dP$. Substituting the above relation for dU and simplifying yields

$$dH = T\,dS + V\,dP.$$

Dividing both sides by dT, this relation becomes

$$\frac{dH}{dT} = T\left(\frac{dS}{dT}\right) + V\left(\frac{dP}{dT}\right).$$

At constant P, N_a, and N_b, the last term on the right vanishes and we can interpret dH/dT and dS/dT as partial derivatives. The above relation then becomes

$$\left(\frac{\partial H}{\partial T}\right)_{P,N_a,N_b} = T\left(\frac{\partial S}{\partial T}\right)_{P,N_a,N_b}.$$

The above result indicates that the enthalpy derivative in Eq. (3.164) can be replaced by the right side of the above relation. Doing so yields Eq. (3.166). This implies that the defintions (3.164) and (3.166) for $\hat{c}_P$ are equivalent. A similar line of reasoning can be used to demonstrate that the defintions (3.163) and (3.165) for $\hat{c}_V$ are equivalent.

Example 3.6   Derive a general relation for the difference $\hat{c}_P - \hat{c}_V$.

Solution Using a partial Legengdre transform that replaces U with T, a fundamental equation in the form $S = S(U, V, N_a, N_b)$ can be converted to $S = S(T, V, N_a, N_b)$. For a system with fixed N_a and N_b, it follows from basic calculus that

$$dS = \left(\frac{\partial S}{\partial T}\right)_{V,N_a,N_b} dT + \left(\frac{\partial S}{\partial V}\right)_{T,N_a,N_b} dV.$$

Using the Maxwell relation (3.131), this equation can be written in the form

$$dS = \left(\frac{\partial S}{\partial T}\right)_{V,N_a,N_b} dT + \left(\frac{\partial P}{\partial T}\right)_{V,N_a,N_b} dV.$$

We can similarly consider a partial Legendre transform that replaces U with T and V with P, thus converting the fundamental equation from the form $S = S(U, V, N_a, N_b)$ to $S = S(T, P, N_a, N_b)$. For fixed N_a and N_b, basic calculus dictates that the corresponding relation for dS is

$$dS = \left(\frac{\partial S}{\partial T}\right)_{P,N_a,N_b} dT + \left(\frac{\partial S}{\partial P}\right)_{T,N_a,N_b} dP.$$

Using the Maxwell relation (3.147), this equation becomes

$$dS = \left(\frac{\partial S}{\partial T}\right)_{P,N_a,N_b} dT - \left(\frac{\partial V}{\partial T}\right)_{P,N_a,N_b} dP.$$

Equating the right side of the above equation to the right side of the equation for dS obtained above, we obtain

$$\left(\frac{\partial S}{\partial T}\right)_{V,N_a,N_b} dT + \left(\frac{\partial P}{\partial T}\right)_{V,N_a,N_b} dV = \left(\frac{\partial S}{\partial T}\right)_{P,N_a,N_b} dT - \left(\frac{\partial V}{\partial T}\right)_{P,N_a,N_b} dP.$$

Using the defintions (3.165) for $\hat{c}_V$ and (3.166) for $\hat{c}_P$, this equation can be written in the form

$$\left[\frac{N_a + N_b}{N_A T}\right] \hat{c}_V \, dT + \left(\frac{\partial P}{\partial T}\right)_{V,N_a,N_b} dV = \left[\frac{N_a + N_b}{N_A T}\right] \hat{c}_P \, dT - \left(\frac{\partial V}{\partial T}\right)_{P,N_a,N_b} dP.$$

Solving for dT, we find

$$dT = \left[\frac{N_A T (\partial P/\partial T)_{V,N_a,N_b}}{(N_a + N_b)(\hat{c}_P - \hat{c}_V)}\right] dV + \left[\frac{N_A T (\partial V/\partial T)_{P,N_a,N_b}}{(N_a + N_b)(\hat{c}_P - \hat{c}_V)}\right] dP.$$

For a system with fixed N_a and N_b, there exists an equation of state of the form $P = P(T, V)$, or equivalently $T = T(P, V)$. Hence, it follows from basic calculus that

$$dT = \left(\frac{\partial T}{\partial V}\right)_{P,N_a,N_b} dV + \left(\frac{\partial T}{\partial P}\right)_{V,N_a,N_b} dP.$$

For this equation and the one above to both be valid for arbitrary differential changes, the respective coefficients of dV and dP must be equal. Equating the coefficients of either dV

or dP and solving for $\hat{c}_P - \hat{c}_V$ yields

$$\hat{c}_P - \hat{c}_V = \left(\frac{N_A T}{N_a + N_b}\right)\left(\frac{\partial P}{\partial T}\right)_{V,N_a,N_b}\left(\frac{\partial V}{\partial T}\right)_{P,N_a,N_b}.$$

As discussed in Appendix I, for a function $z = z(x, y)$

$$\left(\frac{\partial z}{\partial y}\right)_x = -\left(\frac{\partial z}{\partial x}\right)_y\left(\frac{\partial x}{\partial y}\right)_z.$$

Therefore

$$\left(\frac{\partial P}{\partial T}\right)_{V,N_a,N_b} = -\left(\frac{\partial V}{\partial T}\right)_{P,N_a,N_b}\left(\frac{\partial P}{\partial V}\right)_{T,N_a,N_b}.$$

Substituing the right side of the above equation for $(\partial P/\partial T)_{V,N_a,N_b}$ in the relation for $\hat{c}_P - \hat{c}_V$, we obtain

$$\hat{c}_P - \hat{c}_V = -\left(\frac{N_A T}{N_a + N_b}\right)\left(\frac{\partial V}{\partial T}\right)^2_{P,N_a,N_b}\left(\frac{\partial P}{\partial V}\right)_{T,N_a,N_b}.$$

Using the definitions (3.161) and (3.162), this relation can be written more compactly as

$$\hat{c}_P - \hat{c}_V = \frac{N_A T V \beta_T^2}{(N_a + N_b)\kappa_T}.$$

The above general equation relates $\hat{c}_P - \hat{c}_V$ to the isothermal compressibility and the coefficient of thermal expansion.

Exercises

3.1

$v = 1.67$		$v = 1.80$		$v = 1.92$	
$s = 7.35$	$u = 2506$	$s = 7.48$	$u = 2544$	$s = 7.61$	$u = 2582$
$s = 8.37$	$u = 3048$	$s = 8.52$	$u = 3131$	$s = 8.70$	$u = 3241$
$s = 9.06$	$u = 3665$	$s = 9.22$	$u = 3807$	$s = 9.45$	$u = 4005$

(Units: $v \sim$ m³/kg, $s \sim$ kJ/(kg K), $u \sim$ kJ/kg.)

Consider the tabulated properties above for steam. Note that u, v, and s are defined on a per unit mass basis and that $u = \hat{u}/M$, $v = \hat{v}/M$, and $s = \hat{s}/M$, where M is the molecular mass.

(a) For each of the three values of specific volume, curve fit the relation $u(s)$ with a function of the form $u = u_2 s^2 + u_1 s + u_0$, where u_0, u_1, and u_2 are constants. (Note that the constants will differ for each v value.)

(b) For the variations of u_0, u_1, and u_2 with v, develop curve fits of the form $u_2 = a_2 v^2 + a_1 v + a_3$, $u_1 = b_2 v^2 + b_1 v + b_3$, $u_0 = c_2 v^2 + c_1 v + c_3$.

(c) Assemble the results of parts (a) and (b) together to obtain a single function $u(s, v)$. Is this function a proper fundamental equation? Explain your answer.

(d) Use the relation obtained in part (c) to predict the temperature and pressure for $s = 8.52$ J/kg K and $v = 1.80$ m³/kg.

3.2 A system is divided into three subsystems by two rigid walls that are permeable to both species of molecules in the system. The walls are fixed so that they cannot move in response to pressure changes. The system is isolated so that no work or heat interactions with the surrounding are possible. The three subsystems have different volumes and initially contain different amounts of species a and species b molecules. After a long time, the overall system reaches equilibrium. Determine the necessary conditions for equilibrium in this composite system.

3.3 The system to be considered in this problem is identical to that in Figure 3.1 except that the porous plate separating the two subsystems is free to move horizontally in response to pressure changes in the subsystems. Determine the necessary conditions for equilibrium in this system.

3.4 The system to be considered in this problem is identical to that described in Exercise 3.1 except that the piston is permeable to species a and species b molecules. Initially there are no species b molecules in subsystem I and $N_{b,i}$ molecules of species b in subsystem II. Determine the necessary conditions for equilibrium in this system.

3.5 The system to be considered in this problem is identical to that in Figure 3.1 except that the plate separating the subsystems is adiabatic, is impermeable to both species, and is free to move horizontally in response to pressure changes in the subsystems. Find the one necessary condition for equilibrium in this system and explain why there is only one necessary condition.

3.6 Three tanks containing helium gas each have their own shut-off valve. Initially the valves are closed. Tank I has a volume of 0.010 m^3 and an initial pressure of 200 kPa. Tank II has a volume of 0.005 m^3 and an initial pressure of 500 kPa, and tank III has a volume of 0.015 m^3 and an initial pressure of 50 kPa. The tanks are connected to a common pipe and the valves are opened, allowing the pressure to equalize. Find the final pressure in the tanks. You may neglect the volume of the common pipe relative to the tank volumes.

3.7 For a single-component system, liquid and vapor may coexist under specific pressure and temperature conditions. Such circumstances are said to be *saturation conditions*. As the pressure is increased, the difference in density and other properties for the liquid and vapor phase decreases until a pressure is reached where the difference between the two phases vanishes. This is known as the *critical point* for the substance. The van der Waals equation exhibits this type of behavior and is therefore an appropriate prototype equation of state for a system that undergoes liquid–vapor phase change.

    (a) If P_c, T_c, and $\hat{v}_c$ are the pressure, temperature, and volume per kmol at the critical point, show that the van der Waals equation can be written in the form

$$P_r = \frac{8T_r}{3\hat{v}_r - 1} - \frac{3}{\hat{v}_r^2},$$

where $P_r = P/P_c$, $T_r = T/T_c$, and $\hat{v}_r = \hat{v}/\hat{v}_c$.

    (b) Determine the values of $\hat{v}_r$ that correspond to $P_r = 0.03$ and $T_r = 0.85$. What is the physical interpretation of each value?

3.8 In Example 3.2 it was shown that the relation for chemical potential in a binary mixture of monatomic gases can be written

$$\mu_a = -\left(\frac{5}{2}\right)k_BT \ln\left\{\frac{T}{\theta_{\mu,a}}\right\} + k_BT \ln\left\{\frac{P_a}{P_{a,\text{ref}}}\right\},$$

where P_a is the partial pressure of species a defined as

$$P_a = X_a P,$$

$P_{a,\text{ref}}$ is an arbitrary reference value for P_a, and

$$\theta_{\mu,a} = -\left(\frac{P_{a,\text{ref}}^{3/2}h^2}{2\pi m_a k_B^{5/3}}\right)^{3/5}.$$

(a) For $P_{a,\text{ref}} = 101$ kPa, determine the value of $\theta_{\mu,a}$ for helium, neon, and argon.

(b) For $P_a = 101$ kPa, determine and plot the variation of μ_a with T for helium, neon, and argon. What happens to μ_a as T approaches zero? (Note that we will show in Chapter 4 that the dilute occupancy assumption from which these results were derived is inaccurate at very low T, and so this behavior is not expected to be physically realistic.)

3.9 Show that the equations of state for a binary mixture of monatomic ideal gases satisfy the Euler equation (3.46).

3.10 We will demonstrate in Chapter 6 that the fundamental equation for a van der Waals fluid can be written in the form

$$\hat{s} = R\ln\left\{(\hat{v} - N_A b_v)\left(\hat{u} + N_A^2 a_v/\hat{v}\right)^{\xi_v}\right\} + \hat{s}_0,$$

where ξ_v and $\hat{s}_0$ are fixed constants for a given fluid.

(a) Show that the van der Waals equation of state (3.41) can be derived from this relation.

(b) Show that for a van der Waals fluid the enthalpy per kmol is given by

$$\hat{h} = RT\left(\xi_v + \frac{\hat{v}}{\hat{v} - N_A b_v}\right) - \frac{2N_A^2 a_v}{\hat{v}}.$$

3.11 A piston and cylinder device contains an ideal monatomic gas. The cross-sectional area of the piston is 3.14 cm^2. Initially the distance between the piston and the end of the cylinder is 2 cm, the temperature of the gas is 300 K, and the pressure is 100 kPa. The gas is compressed reversibly and adiabatically until the distance between the piston and the cylinder is 1 mm. Find the pressure and temperature of the gas at the end of the process and determine the work done on the gas.

3.12 The piston and cylinder system in Figure 3.7 contains a binary mixture of two monatomic ideal gases with $\hat{N}_a = 0.2$ kmol of species a and $\hat{N}_b = 0.5$ kmol of species b. Initially the system has volume $V_1 = 0.001$ m^3 and pressure $P_1 = 101$ kPa. Heat interaction between the gas and the piston transfers energy into the gas and the piston slowly moves, increasing the volume to $V_2 = 0.003$ m^3. The resulting constant-pressure (isobaric) expansion process in the gas is reversible. Determine the work done by the gas and the energy transferred as heat during the process.

3.13 An insulated rigid cylinder with a total volume of 0.015 m^3 is divided into two chambers (side 1 and side 2), each containing a binary mixture of monatomic gases. The volume of side 1 is twice that of side 2. The chambers are separated by a rigid membrane that is permeable to component a but impermeable to component b. Heat transfer may also occur across the membrane. Initially, on side 1: $N_a = 0.6$ moles, $N_b = 0.8$ moles, and $T = 400$ K,

Figure 3.7

and on side 2: $N_a = 1.2$ mole, $N_b = 0.6$ moles, and $T = 300$ K. After equilibrium is established, what are the values of T, N_a, and P on sides 1 and 2?

3.14 A simple compressible system undergoes a reversible constant-pressure (isobaric) process from state 1 to state 2. In general the process may involve simultaneous work and heat interactions with the system. Show that the heat exchanged during the process is equal to the change in enthalpy $H_2 - H_1$.

3.15 Determine $\kappa_T(P)$ and $\beta_T(T)$ for a mixture of three monatomic ideal gases.

3.16 Determine the value of $\hat{c}_P / \hat{c}_V$ for a binary mixture of two monatomic ideal gases.

3.17 Show that the definition (3.165) for $\hat{c}_V$ is equivalent to Eq. (3.163).

References

[1] Howell, J. R. and Buckius, R. O., *Fundamentals of Engineering Thermodynamics*, 2nd ed., Chapter 9, McGraw-Hill, Inc., New York, 1985.

[2] Modell, M. and Reid, R. C., *Thermodynamics and Its Applications*, 2nd ed., Chapters 5–7, Prentice-Hall, Inc., Englewood Cliffs, NJ, 1980.

CHAPTER 4

Other Ensemble Formulations

In this chapter, we explore alternate statistical viewpoints of equilibrium systems. We do so by defining alternative system ensembles with different combinations of system boundary constraints. We also examine information on system property fluctuations that can be extracted from the statistical analysis of the ensembles. This information on fluctuations provides important insight into phase transitions considered later in Chapter 8.

4.1   Microstates and Energy Levels

The purpose of this first section is to clarify the relationship between microstates and energy levels and to emphasize how they are represented in the statistical thermodynamics theory developed in this text. In the Boltzmann statistics analysis presented in Chapter 2, we considered the set of occupation numbers that indicated how the particles in the system were distributed over the available energy levels. This led to the following relations for the partition function:

$$Z_a = \sum_{i=0}^{\infty} g_i e^{-\varepsilon_i/k_B T}, \tag{2.50a}$$

$$Z_b = \sum_{j=0}^{\infty} g_j e^{-\varepsilon_j/k_B T}. \tag{2.50b}$$

The analysis can also be formulated in terms of occupation numbers that indicate how the particles are distributed over the accessible microstates. The summations in the partition functions would then be over all microstates rather than over the energy levels, and the summation would be written as

$$Z_a = \sum_{i'=0}^{\infty} e^{-\varepsilon_{i'}/k_B T}, \tag{4.1a}$$

$$Z_b = \sum_{j'=0}^{\infty} e^{-\varepsilon_{j'}/k_B T}, \tag{4.1b}$$

where each i' or j' designates a different microstate. If the degeneracy of energy level i is g_i, then there will be g_i identical terms of $e^{-\varepsilon_i/k_B T}$ in the summation in Eq. (4.1a). In the energy level formulation, these terms are represented by a single term $g_i e^{-\varepsilon_i/k_B T}$. The microstate summation indicated in Eqs. (4.1a,b) is therefore identical to the corresponding energy level summation in Eqs. (2.50a,b). The only difference is the method of bookkeeping.

For the most part, in the statistical models in this text we will consider distributions over energy levels. The reason for this choice is that distributions over energy levels can often be macroscopically detected, whereas it is impossible to detect how a system is distributed over its many microstates. In particular, spectroscopic measurements can reveal how molecules in a system are distributed over the accessible energy levels. Distributions over energy levels determined from statistical thermodynamics are valuable in interpreting the results of such measurements. Despite this preference for the energy level formulation, it will sometimes be convenient to consider the statistics of the system in terms of microstates. When we elect to do so, primed indices such as i' or j' will be used to emphasize that we are doing the bookkeeping in terms of microstates. Unprimed indices will indicate energy levels. The reader should keep in mind that the physics is the same regardless of the way we do the bookkeeping.

4.2   The Canonical Ensemble

In his pioneering work on statistical mechanics, J. Willard Gibbs [1] explored the statistical analysis of many-particle systems in terms of different ensemble types. One of these was the microcanonical ensemble, which we have examined in Chapter 2. Two other ensemble types that are particularly useful are the *canonical ensemble* and the *grand canonical ensemble*. The former we will examine in detail in this section and the latter we will consider in the next section.

As indicated schematically in Figure 4.1, a canonical ensemble is a collection of systems that all have the same volume, the same number of particles of species a, and the same number of particles of species b. (We are again considering systems that contain a binary mixture of particles to explicitly include mixture effects.) The systems are all assumed to be in perfect thermal contact with one another so that at all times they are at the same temperature. The ensemble is perfectly insulated from the rest of the universe so that the total internal energy of the ensemble is fixed at U_E. Each system may be in any system microstate consistent with the specified V, N_a, N_b, and T for each system in the ensemble. The total number of systems in the ensemble, n_E, is presumed to be so large that essentially every possible microstate is represented in the ensemble. The energy levels accessible to the systems in the ensemble, E_j, and the degeneracies, g_j, are assumed to be a function of N_a, N_b, and V:

$$E_j = E_j(N_a, N_b, V),$$

$$g_j = g_j(N_a, N_b, V).$$

Before proceeding to the canonical ensemble analysis, some comments regarding the advantages of the canonical ensemble are in order. First, note that for the canonical ensemble we fix the temperature of the systems whereas for the microcanonical ensemble we specified

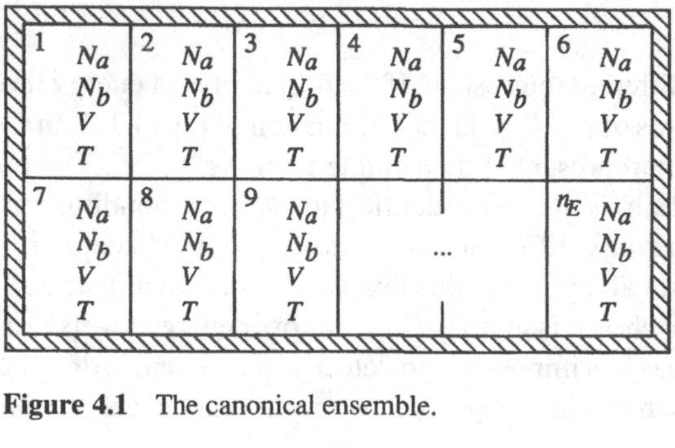

Figure 4.1   The canonical ensemble.

the energy of the systems. At a macroscopic level we can closely approximate isolating a system so that its energy is fixed. However, even the best insulation can allow microscale energy exchanges with the system. In that sense, we can never exactly construct a completely isolated system with energy fixed at an exact value. On the other hand, it is often easy to envelop a system in a larger body at a fixed temperature. One example would be to immerse the system in a large well-stirred pool of water. The system and the pool will reach an equilibrium at a temperature essentially equal to that of the pool. Microscale energy fluctuations may momentarily transfer energy to or from the pool and lower or raise the energy of the system, but over time, microscale fluctuations will keep the temperature of the system fixed to a high level of precision. In addition, it is often easier to quantify the temperature and volume for a system of interest than its energy and volume. As a result it is sometimes more useful to have a statistical thermodynamics model for a system held at fixed volume and temperature than one held at fixed volume and energy. These arguments provide some hints that the canonical ensemble formulation of statistical thermodynamics theory may be particularly useful in some applications. We will see later that this is indeed the case.

We begin by considering the set of occupation numbers $\{n_j\}$ that indicates how the member systems in the ensemble are distributed over the system energy levels. Each n_j in the set indicates the number of systems in energy level ε_j. It follows directly that the distribution must satisfy

$$\sum_{j=0}^{\infty} n_j = n_{\mathrm{E}}, \tag{4.2}$$

$$\sum_{j=0}^{\infty} n_j E_j = U_{\mathrm{E}}, \tag{4.3}$$

where n_{E} is the total number of systems in the ensemble and U_{E} is the total energy in the ensemble. Based on the ergodic hypothesis (see Section 2.2), we assume that the average properties of the ensemble are equivalent to the time average properties of the system being considered. We therefore seek to determine the set of occupation numbers for the ensemble, which we can then use to evaluate the statistical properties of the ensemble. For fixed N_a, N_b, T, and V, the energy of a real system at equilibrium is observed to be virtually constant with time. Since particle collisions are continuously changing the microstate of the system, the virtually constant system energy implies that, in the equilibrium distribution, the majority of the ensemble members are in microstates with energies very close to a single value. In terms of our ensemble, this implies that equilibrium corresponds to an occupation number set that produces the greatest number of ensemble microstates. We are therefore seeking the occupation number distribution that maximizes the number of ensemble microstates W subject to the ensemble constraints indicated in Eqs. (4.2) and (4.3).

The systems in the ensemble are distinguishable. The number of ways of dividing the n_{E} distinguishable systems in the ensemble among the system energy levels as dictated by the occupation number set $\{n_j\}$ is equal to the number of way of dividing n_{E} distinguishable objects into groups of $n_0, n_1, \ldots, n_j, \ldots$. From basic combinatorial analysis (see Appendix I), this is given by

$$\left\{ \begin{array}{l} \textit{number of ways of dividing systems} \\ \textit{among energy levels as dictated by } \{n_j\} \end{array} \right\} = n_{\mathrm{E}}! \prod_{j=0}^{\infty} \frac{1}{n_j!}. \tag{4.4}$$

However, each system energy level E_j has degeneracy g_j, which means that the n_j systems in that energy level may be in any of g_j microstates. To get a relation for the number of microstates corresponding to an occupancy number set, we must multiply the contribution of each energy level in the above equation by the number of ways that the n_j systems can be distributed among the g_j microstates for that energy level. As described in Appendix I, the number of ways of distributing n_j distinguishable systems over g_j microstates is equal to $g_j^{n_j}$. Multiplying this term for each energy level by the corresponding factor on the right side of Eq. (4.4), we obtain the following relation for the total number of ensemble microstates for a specified occupancy number set:

$$W(\{n_j\}) = n_E! \prod_{j=0}^{\infty} \frac{g_j^{n_j}}{n_j!}. \tag{4.5}$$

We will denote the probability of a system being in the jth energy level as $\tilde{P}_j$. From the above observations it follows that

$$\tilde{P}_j = \frac{n_j}{n_E} = \frac{\sum_{\{n_j\}} n_j W(\{n_j\})}{n_E \sum_{\{n_j\}} W(\{n_j\})}. \tag{4.6}$$

For the ensemble considered here, we therefore seek to find the occupation number set that maximizes W subject to the constraints (4.2) and (4.3). We expect that this set of occupation numbers are those that characterize the statistics of the real system at equilibrium. Again, for convenience, we elect to maximize $\ln W$ subject to the constraints. Taking the natural log of both sides of Eq. (4.5) yields

$$\ln W = \ln n_E! + \sum_{j=0}^{\infty} n_j \ln g_j - \sum_{j=0}^{\infty} \ln n_j!. \tag{4.7}$$

Since n_E and n_j are typically very large, we make use of Stirling's approximation (see Appendix I) to evaluate the $\ln n_E!$ and $\ln n_j!$ terms. The relation (4.7) then becomes

$$\ln W = n_E \ln n_E - n_E + \sum_{j=0}^{\infty} n_j \ln g_j - \sum_{j=0}^{\infty} (n_j \ln n_j - n_j). \tag{4.8}$$

Using constraint (4.2), this reduces to

$$\ln W = n_E \ln n_E + \sum_{j=0}^{\infty} n_j \ln g_j - \sum_{j=0}^{\infty} n_j \ln n_j. \tag{4.9}$$

Differentiating both sides of the above equation gives

$$d(\ln W) = dn_E \ln n_E + dn_E + \sum_{j=0}^{\infty} dn_j \ln g_j - \sum_{j=0}^{\infty} (dn_j \ln n_j + dn_j). \tag{4.10}$$

Since constraint (4.2) implies that

$$\sum_{j=0}^{\infty} dn_j = dn_E = 0,$$

Eq. (4.10) reduces to

$$d(\ln W) = \sum_{j=0}^{\infty} dn_j \ln g_j - \sum_{j=0}^{\infty} dn_j \ln n_j. \tag{4.11}$$

The maximum in $\ln W$ implies that $d(\ln W) = 0$. Imposing this requirement on Eq. (4.11) yields

$$\sum_{j=0}^{\infty} dn_j \ln g_j - \sum_{j=0}^{\infty} dn_j \ln n_j = 0. \tag{4.12}$$

Differentiating the constraint relations then gives

$$\sum_{j=0}^{\infty} dn_j = dn_E = 0 \tag{4.13a}$$

and

$$\sum_{j=0}^{\infty} E_j dn_j = dU_E = 0. \tag{4.13b}$$

We now make use of the method of Lagrange multipliers. (See Appendix 1 for a description of this technique.) We multiply the differentiated constraint equations by Lagrange multipliers α and β respectively:

$$\alpha \sum_{j=0}^{\infty} dn_j = 0, \tag{4.14a}$$

$$\beta \sum_{j=0}^{\infty} E_j dn_j = 0. \tag{4.14b}$$

Adding these equations, subtracting Eq. (4.12), and collecting terms within a single summation yields

$$\sum_{j=0}^{\infty} (\alpha + \beta E_j - \ln g_j + \ln n_j) \, dn_j = 0. \tag{4.15}$$

At the maximum, this relation must hold for arbitrary choices of dn_j, which implies that the coefficients of the differential terms must all be zero:

$$\alpha + \beta E_j - \ln g_j + \ln n_j = 0. \tag{4.16}$$

We can now solve this relation for n_j. In doing so, we will designate the resulting n_j distribution as n_j^* to serve as a reminder that this distribution applies to a system at equilibrium. The resulting relation is

$$n_j^* = g_j e^{-\alpha} e^{-\beta E_j}. \tag{4.17}$$

Using the constraint relation (4.2) together with Eq. (4.17) yields

$$n_E = \sum_{j=0}^{\infty} n_j^* = \sum_{j=0}^{\infty} g_j e^{-\alpha} e^{-\beta E_j} = e^{-\alpha} \sum_{j=0}^{\infty} g_j e^{-\beta E_j}. \tag{4.18}$$

Combining Eqs. (4.17) and (4.18) leads to

$$\frac{n_j^*}{n_{\mathrm{E}}} = \frac{g_j e^{-\alpha} e^{-\beta E_j}}{e^{-\alpha} \sum_{j=0}^{\infty} g_j e^{-\beta E_j}} = \frac{g_j e^{-\beta E_j}}{\sum_{j=0}^{\infty} g_j e^{-\beta E_j}}. \tag{4.19}$$

The distribution above is generally written in the form

$$\frac{n_j^*}{n_{\mathrm{E}}} = \frac{g_j e^{-\beta E_j}}{Q}, \tag{4.20}$$

where

$$Q = \sum_{j=0}^{\infty} g_j e^{-\beta E_j}. \tag{4.21}$$

In the above relations, Q is termed the *canonical partition function*. Note that the summation in Eq. (4.21) is over all system energy levels.

Two tasks remain to complete the canonical ensemble formulation: We must evaluate the multiplier β and we must relate the partition function to other thermodynamic properties. To accomplish these tasks, we will take advantage of information already obtained for systems in a microcanonical ensemble. We first note that for the canonical ensemble as a whole, the total energy and the total number of systems are fixed.

Furthermore, since the volume of each system is fixed and the number of systems is fixed, the volume of the ensemble is fixed. These fixed conditions are the same as those imposed on systems in a microcanonical ensemble. Thus, if we consider the systems in the canonical ensembles to be just a special kind of "particle," with energy levels and degeneracies equal to those for the canonical ensemble system, the canonical ensemble as a whole can be treated as if it is one system in a microcanonical ensemble (see Figure 4.2). Note that the systems in the canonical ensemble are free to exchange energy through random fluctuations just as the particles could in the microcanonical ensemble. The results of microcanonical

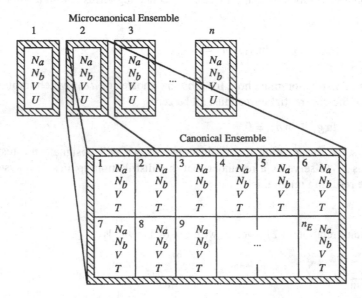

Figure 4.2   A canonical ensemble as one system in a microcanonical ensemble.

analysis in Sections 2.2 and 2.3 apply to the canonical ensemble as a whole if we assign variables as follows:

> N_a is set to zero since we consider only one type of system.
>
> N_b is set equal to n_E.
>
> Particle energy level ε_j with degeneracy g_j corresponds to system energy level E_j with degeneracy g_j.
>
> The particle occupation numbers $N_{b,j}$ correspond to the system occupation numbers n_j.

Invoking these correspondence arguments, we can modify Eq. (2.31) accordingly to obtain the following relation for S_E, the total entropy of the canonical ensemble:

$$\frac{S_E}{k_B} = n_E \ln n_E + \sum_{j=0}^{\infty} n_j \ln g_j - \sum_{j=0}^{\infty} n_j \ln n_j. \tag{4.22}$$

Consistent with the usual assumptions for the microcanonical ensemble, this relation should apply to an ensemble of distinguishable systems that have reached equilibrium by achieving a system energy distribution that maximizes the number of ensemble microstates. Into Eq. (4.22) we substitute the equilibrium distribution for the ensemble given by Eq. (4.20) to obtain

$$\frac{S_E}{k_B} = n_E \ln n_E + \sum_{j=0}^{\infty} \frac{n_E g_j e^{-\beta E_j}}{Q} \ln g_j - \sum_{j=0}^{\infty} \frac{n_E g_j e^{-\beta E_j}}{Q} \ln \left\{ \frac{n_E g_j e^{-\beta E_j}}{Q} \right\}. \tag{4.23}$$

Rearranging, this becomes

$$\frac{S_E}{k_B} = n_E \ln n_E - \left(\frac{n_E}{Q}\right) \sum_{j=0}^{\infty} g_j e^{-\beta E_j} \ln g_j + \left(\frac{n_E}{Q}\right) \sum_{j=0}^{\infty} g_j e^{-\beta E_j} \ln n_E$$

$$+ \left(\frac{n_E}{Q}\right) \sum_{j=0}^{\infty} g_j e^{-\beta E_j} \ln g_j - \beta \left(\frac{n_E}{Q}\right) \sum_{j=0}^{\infty} E_j g_j e^{-\beta E_j}$$

$$+ \left(\frac{n_E}{Q}\right) \sum_{j=0}^{\infty} g_j e^{-\beta E_j} \ln Q. \tag{4.24}$$

Canceling terms and using Eq. (4.21) to evaluate some of the summation terms, we can simplify this relation to

$$\frac{S_E}{k_B} = -\beta \left(\frac{n_E}{Q}\right) \sum_{j=0}^{\infty} E_j g_j e^{-\beta E_j} + n_E \ln Q. \tag{4.25}$$

We define the ensemble average of any property as being the summation over all energy levels of the value of that property, for the system energy level, weighted by the probability that the system exists in that energy level. Designating an arbitrary property as Y, the definition of its ensemble average of $\langle Y \rangle$ is

$$\langle Y \rangle = \sum_{j=0}^{\infty} Y(E_j, V, N_a, N_b) \tilde{P}(E_j) = \sum_{j=0}^{\infty} Y(E_j, V, N_a, N_b) \frac{g_j e^{-\beta E_j}}{Q}. \tag{4.26}$$

The canonical ensemble analysis presumes that the average property value computed using definition (4.26) is the value of that property that would be observed macroscopically for a system at equilibrium at the N_a, N_b, V, and T values specified for the ensemble. The internal energy and entropy values observed at equilibrium would therefore be given by

$$U = \langle U \rangle = \sum_{j=0}^{\infty} E_j \frac{g_j e^{-\beta E_j}}{Q}, \tag{4.27}$$

$$S = \langle S \rangle = \sum_{j=0}^{\infty} S(E_j, V, N_a, N_b) \frac{g_j e^{-\beta E_j}}{Q}. \tag{4.28}$$

Substituting Eq. (4.27) into Eq. (4.25) we can write

$$\frac{S_E}{k_B} = -\beta n_E \langle U \rangle + n_E \ln Q. \tag{4.29}$$

The total entropy for the canonical ensemble, by additivity, is just the sum of the entropies of the member systems. We write this sum in the form

$$S_E = \sum_{j=0}^{\infty} S(E_j, V, N_a, N_b) \, n_j^*. \tag{4.30}$$

In the above relation, the entropy for each level is multiplied by the number of systems in each level, n_j^*, and the terms for all levels are added, thus accounting for the contribution of all systems in the ensemble. Using the occupation number distribution (4.20) to evaluate n_j^*, Eq. (4.30) becomes

$$S_E = \sum_{j=0}^{\infty} S(E_j, V, N_a, N_b) \frac{n_E g_j e^{-\beta E_j}}{Q}. \tag{4.31}$$

Combining Eqs. (4.28) and (4.31) we obtain

$$S_E = n_E \langle S \rangle. \tag{4.32}$$

Substituting the right side of Eq. (4.31) for S_E converts Eq. (4.28) into

$$\frac{\langle S \rangle}{k_B} = -\beta \langle U \rangle + \ln Q. \tag{4.33}$$

Since the ensemble average values of U and S are taken to be equal to the equilibrium properties for the system considered, Eq. (4.33) can be written

$$\frac{S}{k_B} = -\beta U + \ln Q. \tag{4.34}$$

We next differentiate (4.34) with respect to β at constant N_a, N_b, and V:

$$\frac{1}{k_B} \left(\frac{\partial S}{\partial \beta} \right)_{N_a, N_b, V} = U + \beta k_B \left(\frac{\partial U}{\partial \beta} \right)_{N_a, N_b, V} + \frac{1}{Q} \left(\frac{\partial Q}{\partial \beta} \right)_{N_a, N_b, V}. \tag{4.35}$$

The last partial derivative on the right side can be evaluated using the definition of Q, Eq. (4.21):

$$\frac{1}{Q} \left(\frac{\partial Q}{\partial \beta} \right)_{N_a, N_b, V} = \frac{1}{Q} \left(\frac{\partial}{\partial \beta} \sum_{j=0}^{\infty} g_j e^{-\beta E_j} \right)_{N_a, N_b, V}. \tag{4.36}$$

As discussed in Section 2.3, the energy levels and the degeneracies for the particles in the systems in the ensemble are expected to be functions of the volume of the system and the number of particles of each species. It follows directly that the energy levels and respective degeneracies for the systems are also functions of those variables. Since N_a, N_b, and V are fixed for these systems, the E_j and g_j are fixed and can be considered as constants in taking the partial derivative on the right side of (4.36). Evaluation of the derivative in Eq. (4.36) then yields

$$\frac{1}{Q}\left(\frac{\partial Q}{\partial \beta}\right)_{N_a,N_b,V} = \frac{1}{Q}\sum_{j=0}^{\infty} -E_j g_j e^{-\beta E_j} = -U. \tag{4.37}$$

Substituting Eq. (4.37) into Eq. (4.35) and solving for β gives

$$\beta = \frac{(\partial S/\partial \beta)_{N_a,N_b,V}}{k_B(\partial U/\partial \beta)_{N_a,N_b,V}} = \frac{1}{k_B(\partial U/\partial S)_{N_a,N_b,V}}. \tag{4.38}$$

Since, by definition, $T = (\partial U/\partial S)_{N_a,N_b,V}$, the relation for β becomes

$$\beta = \frac{1}{k_B T}. \tag{4.39}$$

Having evaluated β, we can substitute this result into Eqs. (4.21) and (4.34) to obtain the following relations for the partition function and entropy:

$$Q = \sum_{j=0}^{\infty} g_j e^{-E_j/k_B T}, \tag{4.40}$$

$$\frac{S}{k_B} = \frac{U}{k_B T} + \ln Q. \tag{4.41}$$

Upon examining Eq. (4.27), it can easily be verified that

$$U = -\frac{1}{Q}\left(\frac{\partial Q}{\partial \beta}\right)_{V,N_a,N_b}. \tag{4.42}$$

Upon using Eq. (4.39) and the chain rule to convert the derivative, we obtain the following relation for U:

$$U = \frac{k_B T^2}{Q}\left(\frac{\partial Q}{\partial T}\right)_{V,N_a,N_b}. \tag{4.43}$$

Relations for other properties are derived as follows. Solving Eq. (4.41), for $\ln Q$, we obtain

$$\ln Q = \frac{S}{k_B} - \frac{U}{k_B T}. \tag{4.44}$$

Differentiating this relation yields

$$d(\ln Q) = \frac{dS}{k_B} - \frac{dU}{k_B T} + \frac{U\,dT}{k_B T^2}. \tag{4.45}$$

But in Chapter 3 we showed that for any system containing a binary mixture of two species of particles

$$dU = T\,dS - P\,dV + \mu_a dN_a + \mu_b dN_b. \tag{4.46}$$

Substituting the right side of Eq. (4.46) for dU in Eq. (4.45), we obtain

$$d(\ln Q) = \frac{P\,dV}{k_B T} - \frac{\mu_a dN_a}{k_B T} - \frac{\mu_b dN_b}{k_B T} + \frac{U\,dT}{k_B T^2}. \tag{4.47}$$

It is clear, however, from Eq. (4.40) that since g_j and E_j are functions of V, N_A, and N_b, $\ln Q$ is a function of T, V, N_a, and N_b. Mathematically expanding $\ln Q$ to first order in these variables yields

$$d(\ln Q) = \left(\frac{\partial(\ln Q)}{\partial T}\right)_{V,N_a,N_b} dT + \left(\frac{\partial(\ln Q)}{\partial V}\right)_{T,N_a,N_b} dV$$

$$+ \left(\frac{\partial(\ln Q)}{\partial N_a}\right)_{T,V,N_b} dN_a + \left(\frac{\partial(\ln Q)}{\partial N_b}\right)_{T,V,N_a} dN_b. \tag{4.48}$$

For both (4.47) and (4.48) to be valid for all differential changes, the corresponding coefficients of the differential terms must be equal. Equating the coefficients of the dT terms leads to Eq. (4.43). Equating the coefficients for the other differential terms gives the following relations for the pressure and the chemical potentials:

$$P = k_B T \left(\frac{\partial(\ln Q)}{\partial V}\right)_{T,N_a,N_b}, \tag{4.49}$$

$$\mu_a = -k_B T \left(\frac{\partial(\ln Q)}{\partial N_a}\right)_{T,V,N_b}, \tag{4.50}$$

$$\mu_b = -k_B T \cdot \left(\frac{\partial(\ln Q)}{\partial N_b}\right)_{T,V,N_a}. \tag{4.51}$$

The relations (4.40), (4.41), (4.43), and (4.49)–(4.51) provide the link between microscale energy storage in the system and macroscopic thermodynamic properties. In addition, by substituting Eq. (4.39) into Eq. (4.20), we also obtain the following relation for the probability of finding a system in system energy level j:

$$\tilde{P}_j = \frac{n_j^*}{n_E} = \frac{g_j e^{-E_j/k_B T}}{Q}. \tag{4.52}$$

Example 4.1   Evaluate the canonical partition function Q for a monatomic ideal gas and show that Eq. (4.49) is consistent with the ideal gas equation of state.

Solution   Equation (4.44) links the partition function to entropy, temperature, and internal energy:

$$\ln Q = \frac{S}{k_B} - \frac{U}{k_B T}.$$

For a monatomic gas, the internal energy and entropy are given by Eqs. (2.119) and (2.120). Substituting these relations for S and U into the above relation yields

$$\ln Q = \frac{3}{2} N_a \ln\left\{\frac{2\pi m_a V^{2/3} k_B T}{h^2 N_a^{2/3}}\right\} + \frac{3}{2} N_b \ln\left\{\frac{2\pi m_b V^{2/3} k_B T}{h^2 N_b^{2/3}}\right\} + N_a + N_b.$$

Differentiating with respect to V, we obtain

$$\left(\frac{\partial(\ln Q)}{\partial V}\right)_{T,N_a,N_b} = \frac{N_a}{V} + \frac{N_b}{V}.$$

Finally, we substitute the right side of the above relation into Eq. (4.49) to get

$$P = k_\mathrm{B}T\left(\frac{\partial(\ln Q)}{\partial V}\right)_{T,N_a,N_b} = \frac{(N_a + N_b)}{V}k_\mathrm{B}T.$$

The final result above demonstrates that Eq. (4.49) is consistent with the ideal gas equation of state.

4.3   The Grand Canonical Ensemble

We will now consider the statistical behavior of systems in a grand canonical ensemble.

As indicated schematically in Figure 4.3, a grand canonical ensemble is a collection of systems that all have the same volume. We are again considering systems that contain a binary mixture of species a and species b particles to explicitly include mixture effects. The systems are all assumed to be in perfect thermal contact with one another so that they are all at the same temperature T. The ensemble is perfectly insulated from the rest of the universe so that the total internal energy of the ensemble is fixed at U_E. The walls of each system are permeable so that both species of particles may freely move among the systems. The systems will thus reach mass transfer equilibrium, with the result that the chemical potential for species a and species b will be the same in all systems. The boundary of the ensemble is impermeable to both species and the total number of species a and species b particles in the ensemble are fixed at $N_{a,\mathrm{E}}$ and $N_{b,\mathrm{E}}$, respectively.

Each system may be in any system microstate consistent with the specified V and the values of N_a and N_b for that specific system. The total number of systems in the ensemble, n_E, is presumed to be so large that microstates spanning all possible values of N_a and N_b are represented in the ensemble. The energy levels accessible to the systems in the ensemble and the associated degeneracies are assumed to be a function of N_a, N_b, and V. For systems with a specific pair of N_a and N_b values, $E_{N_a,N_b,j}$ and $g_{N_a,N_b,j}$ are a function of V:

$$E_{N_a,N_b,j} = E_{N_a,N_b,j}(V),$$

$$g_{N_a,N_b,j} = g_{N_a,N_b,j}(V).$$

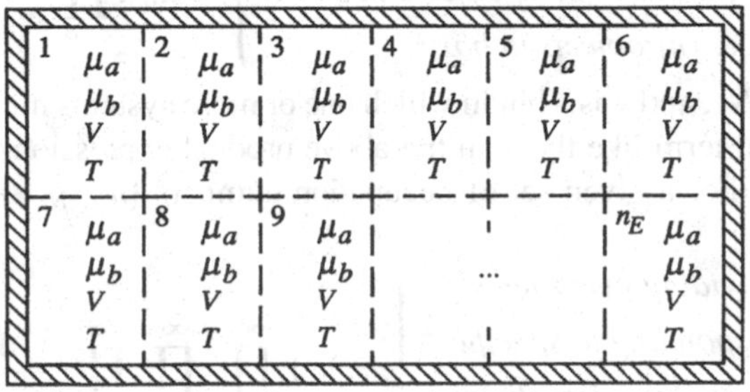

Figure 4.3   The grand canonical ensemble.

We begin by considering the set of occupation numbers $\{n_{N_a,N_b,j}\}$ that indicates how the member systems in the ensemble are distributed over the system energy levels and the ranges of particle numbers. Each $n_{N_a,N_b,j}$ in the set indicates the number of systems in energy level $E_{N_a,N_b,j}$ with particle number values of N_a and N_b. It follows directly that the distribution must satisfy

$$\sum_{N_a=0}^{\infty} \sum_{N_b=0}^{\infty} \sum_{j=0}^{\infty} n_{N_a,N_b,j} = n_E, \qquad (4.53)$$

$$\sum_{N_a=0}^{\infty} \sum_{N_b=0}^{\infty} \sum_{j=0}^{\infty} n_{N_a,N_b,j} E_{N_a,N_b,j} = U_E, \qquad (4.54)$$

$$\sum_{N_a=0}^{\infty} \sum_{N_b=0}^{\infty} \sum_{j=0}^{\infty} n_{N_a,N_b,j} N_a = N_{a,E}, \qquad (4.55)$$

$$\sum_{N_a=0}^{\infty} \sum_{N_b=0}^{\infty} \sum_{j=0}^{\infty} n_{N_a,N_b,j} N_b = N_{b,E}, \qquad (4.56)$$

where n_E is the total number of systems in the ensemble and U_E is the total energy in the ensemble.

Based on the ergodic hypothesis (see Section 2.2), we assume that the average properties of the ensemble are equivalent to the time average properties of the system being considered. We therefore seek to determine the set of occupation numbers for the ensemble so that we evaluate the statistical properties of the ensemble. For fixed μ_a, μ_b, T, and V, the N_a, N_b, and U values for a real system at equilibrium are observed to be virtually constant with time. Since the microstate of the system is continuously changing, the virtually constant values of N_a, N_b, and U imply that, in the equilibrium distribution, the vast majority of the ensemble members are in microstates corresponding to one value of energy and one combination of N_a and N_b values. In terms of our ensemble, this implies that equilibrium corresponds to an occupation number distribution that produces the greatest number of ensemble microstates. We are therefore seeking the occupation number distribution that maximizes the number of ensemble microstates W subject to the ensemble constraints indicated in Eqs. (4.53)–(4.56).

The systems in the ensemble are distinguishable. As discussed in Appendix I, from basic combinatorial analysis, we know that

$$\left\{ \begin{array}{l} \textit{number of ways of putting } n_{do} \textit{ distinguishable objects} \\ \textit{in boxes with } n_0 \textit{ objects in box 0, } n_1 \textit{ objects in box 1,} \\ n_2 \textit{ objects in box 2, } \dots, n_i \textit{ objects in box } i, \dots \end{array} \right\} = n_{do}! \prod_{i=0}^{\infty} \frac{1}{n_i!}.$$

Here, each combination of N_a, N_b, and j is a bin in which one or more systems may reside. Hence, there must be a factorial term like those in the above product expression for each combination of N_a, N_b, and j. For a given set of occupation numbers $\{n_{N_a,N_b,j}\}$ we can therefore write

$$\left\{ \begin{array}{l} \textit{number of ways of dividing } n_E \textit{ systems} \\ \textit{among groups corresponding to specific} \\ \textit{combinations of } N_a, N_b, \textit{ and } j \textit{ as dictated} \\ \textit{by } \{n_{N_a,N_b,j}\} \end{array} \right\} = n_E! \prod_{N_a=0}^{\infty} \prod_{N_b=0}^{\infty} \prod_{j=0}^{\infty} \frac{1}{n_{N_a,N_b,j}!}.$$

$$(4.57)$$

The triple product above indicates that we multiply the factorial terms for all combinations of N_a, N_b, and j. Each system energy level $E_{N_a,N_b,j}$ has degeneracy $g_{N_a,N_b,j}$, which means that the $n_{N_a,N_b,j}$ systems in that energy level may be in any of $g_{N_a,N_b,j}$ microstates. Note that, in general, the degeneracy may depend on N_a, N_b, and j. For simplicity in notation, we will denote the degeneracy as $g_{N_a,N_b,j}$, keeping in mind that it implicitly depends on N_a, N_b, as well as j. To get a relation for the number of microstates corresponding to an occupancy number set, we must multiply the contribution of each energy level in the above equation by the number of ways that the $n_{N_a,N_b,j}$ systems can be distributed among the $g_{N_a,N_b,j}$ microstates for that energy level. As described in Appendix I, the number of ways of distributing $n_{N_a,N_b,j}$ distinguishable systems over $g_{N_a,N_b,j}$ microstates is equal to $g_{N_a,N_b,j}^{n_{N_a,N_b,j}}$. Multiplying this term for each energy level by the corresponding factor on the right side of Eq. (4.57), we obtain the following relation for the total number of ensemble microstates for a specified occupancy number set $\{n_{N_a,N_b,j}\}$:

$$W(\{n_{N_a,N_b,j}\}) = n_E! \prod_{N_a=0}^{\infty} \prod_{N_b=0}^{\infty} \prod_{j=0}^{\infty} \frac{g_{N_a,N_b,j}^{n_{N_a,N_b,j}}}{n_{N_a,N_b,j}!}. \tag{4.58}$$

We will denote the probability of a system being in the jth energy level with particle numbers N_a and N_b as $\tilde{P}_{N_a,N_b,j}$. The above observations imply that

$$\tilde{P}_{N_a,N_b,j} = \frac{n_{N_a,N_b,j}}{n_E} = \frac{\sum_{\{n_{N_a,N_b,j}\}} n_{N_a,N_b,j} W(\{n_{N_a,N_b,j}\})}{n_E \sum_{\{n_{N_a,N_b,j}\}} W(\{n_{N_a,N_b,j}\})}. \tag{4.59}$$

For the ensemble considered here, we therefore seek to find the occupation number set that maximizes W subject to the constraints (4.53)–(4.56). We expect that this set of occupation numbers are those that characterize the real system at equilibrium. Again, for convenience, we elect to maximize $\ln W$ subject to the constraints. Taking the natural log of both sides of Eq. (4.58) yields

$$\ln W = \sum_{N_a=0}^{\infty} \sum_{N_b=0}^{\infty} \sum_{j=0}^{\infty} \left(n_{N_a,N_b,j} \ln g_{N_a,N_b,j} - \ln n_{N_a,N_b,j}! \right) + \ln n_E!. \tag{4.60}$$

Since n_E and $n_{N_a,N_b,j}$ are typically very large, we make use of Stirling's approximation (see Appendix I) to evaluate the $\ln n_E!$ and $\ln n_{N_a,N_b,j}!$ terms. The relation (4.60) for $\ln W$ then becomes

$$\ln W = \sum_{N_a=0}^{\infty} \sum_{N_b=0}^{\infty} \sum_{j=0}^{\infty} \left(n_{N_a,N_b,j} \ln g_{N_a,N_b,j} - n_{N_a,N_b,j} \ln n_{N_a,N_b,j} + n_{N_a,N_b,j} \right) + n_E \ln n_E - n_E. \tag{4.61}$$

Using constraint (4.53), this reduces to

$$\ln W = \sum_{N_a=0}^{\infty} \sum_{N_b=0}^{\infty} \sum_{j=0}^{\infty} \left(n_{N_a,N_b,j} \ln g_{N_a,N_b,j} - n_{N_a,N_b,j} \ln n_{N_a,N_b,j} \right) + n_E \ln n_E. \tag{4.62}$$

Differentiating Eq. (4.53) yields

$$\sum_{N_a=0}^{\infty} \sum_{N_b=0}^{\infty} \sum_{j=0}^{\infty} dn_{N_a,N_b,j} = dn_E = 0. \tag{4.63}$$

Differentiating both sides of Eq. (4.62) and using Eq. (4.63), we obtain

$$d(\ln W) = \sum_{N_a=0}^{\infty} \sum_{N_b=0}^{\infty} \sum_{j=0}^{\infty} \left(dn_{N_a,N_b,j} \ln g_{N_a,N_b,j} - dn_{N_a,N_b,j} \ln n_{N_a,N_b,j} \right). \qquad (4.64)$$

Since $d(\ln W) = 0$ at the maximum we seek, Eq. (4.64) further reduces to

$$\sum_{N_a=0}^{\infty} \sum_{N_b=0}^{\infty} \sum_{j=0}^{\infty} \left(dn_{N_a,N_b,j} \ln g_{N_a,N_b,j} - dn_{N_a,N_b,j} \ln n_{N_a,N_b,j} \right) = 0. \qquad (4.65)$$

We once again employ the method of Lagrange multipliers. To do so, we first differentiate the constraint relations and multiply both sides of each by a Lagrange multiplier to obtain

$$\sum_{N_a=0}^{\infty} \sum_{N_b=0}^{\infty} \sum_{j=0}^{\infty} \alpha \, dn_{N_a,N_b,j} = 0, \qquad (4.66)$$

$$\sum_{N_a=0}^{\infty} \sum_{N_b=0}^{\infty} \sum_{j=0}^{\infty} \beta E_{N_a,N_b,j} dn_{N_a,N_b,j} = 0, \qquad (4.67)$$

$$\sum_{N_a=0}^{\infty} \sum_{N_b=0}^{\infty} \sum_{j=0}^{\infty} \gamma_a N_a dn_{N_a,N_b,j} = 0, \qquad (4.68)$$

$$\sum_{N_a=0}^{\infty} \sum_{N_b=0}^{\infty} \sum_{j=0}^{\infty} \gamma_b N_b dn_{N_a,N_b,j} = 0. \qquad (4.69)$$

We then add Eqs. (4.66)–(4.69) and subtract Eq. (4.65). After a little rearranging, the resulting relation takes the form

$$\sum_{N_a=0}^{\infty} \sum_{N_b=0}^{\infty} \sum_{j=0}^{\infty} \left(\alpha + \beta E_{N_a,N_b,j} + \gamma_a N_a + \gamma_b N_b - \ln g_{N_a,N_b,j} + \ln n_{N_a,N_b,j} \right) dn_{N_a,N_b,j} = 0.$$
$$(4.70)$$

At the maximum, this relation must hold for arbitrary $dn_{N_a,N_b,j}$, which can only be true if the coefficients of the differential terms are all zero:

$$\alpha + \beta E_{N_a,N_b,j} + \gamma_a N_a + \gamma_b N_b - \ln g_{N_a,N_b,j} + \ln n_{N_a,N_b,j} = 0. \qquad (4.71)$$

The distribution that we obtain by solving Eq. (4.71) will be designated as $n^*_{N_a,N_b,j}$ to denote that it is expected to apply to a system at equilibrium. The resulting relation for $n^*_{N_a,N_b,j}$ is

$$n^*_{N_a,N_b,j} = g_{N_a,N_b,j} \exp\left(-\alpha - \gamma_a N_a - \gamma_b N_b - \beta E_{N_a,N_b,j} \right). \qquad (4.72)$$

Applying the constraint relation (4.53) we find

$$n_{\mathrm{E}} = \sum_{N_a=0}^{\infty} \sum_{N_b=0}^{\infty} \sum_{j=0}^{\infty} n^*_{N_a,N_b,j}$$

$$= e^{-\alpha} \sum_{N_a=0}^{\infty} \sum_{N_b=0}^{\infty} \sum_{j=0}^{\infty} g_{N_a,N_b,j} \exp\left(-\gamma_a N_a - \gamma_b N_b - \beta E_{N_a,N_b,j} \right). \qquad (4.73)$$

Combining Eqs. (4.72) and (4.73) yields the following relation for the distribution:

$$\frac{n^*_{N_a,N_b,j}}{n_{\mathrm{E}}} = \frac{g_{N_a,N_b,j} \exp\left(-\gamma_a N_a - \gamma_b N_b - \beta E_{N_a,N_b,j} \right)}{\Xi}, \qquad (4.74)$$

where Ξ is the grand canonical partition function defined as

$$\Xi = \sum_{N_a=0}^{\infty} \sum_{N_b=0}^{\infty} \sum_{j=0}^{\infty} g_{N_a,N_b,j} \exp\left(-\gamma_a N_a - \gamma_b N_b - \beta E_{N_a,N_b,j}\right). \tag{4.75}$$

Our next task is to determine the unknown multipliers β, γ_a, and γ_b. We begin by noting that each subset of systems within the grand canonical ensemble that have the same N_a and N_b values are, taken together, a canonical ensemble. In addition to having the same N_a and N_b values, they all also have the same volume and are in thermal equilibrium with each other at the same temperature. Within this subset, all energy microstates should be represented, and thus this subset fulfills all the requirements to be considered as a canonical ensemble. The probability that a system within this subset is in system energy level j is given by

$$\tilde{P}_{N_a,N_b,j} = \frac{n^*_{N_a,N_b,j}}{\sum_{j=0}^{\infty} n^*_{N_a,N_b,j}}. \tag{4.76}$$

Using Eq. (4.74) to evaluate $n^*_{N_a,N_b,j}$ in the above relation, we obtain

$$\tilde{P}_{N_a,N_b,j} = \frac{\left(\frac{1}{\Xi}\right) n_E g_{N_a,N_b,j} \exp\left(-\gamma_a N_a - \gamma_b N_b - \beta E_{N_a,N_b,j}\right)}{\sum_{j=0}^{\infty} \left(\frac{1}{\Xi}\right) n_E g_{N_a,N_b,j} \exp\left(-\gamma_a N_a - \gamma_b N_b - \beta E_{N_a,N_b,j}\right)}, \tag{4.77}$$

which reduces to

$$\tilde{P}_{N_a,N_b,j} = \frac{g_{N_a,N_b,j} e^{-\beta E_{N_a,N_b,j}}}{\sum_{j=0}^{\infty} g_{N_a,N_b,j} e^{-\beta E_{N_a,N_b,j}}}. \tag{4.78}$$

But we have shown in the previous section that for a canonical ensemble with specified N_a and N_b, the probability that a system is in energy level j is given by

$$\tilde{P}_j = \frac{g_j e^{-E_j/k_B T}}{Q}, \tag{4.79}$$

where

$$Q = \sum_{j=0}^{\infty} g_j e^{-E_j/k_B T}. \tag{4.40}$$

Clearly, Eqs. (4.78) and (4.79) are equivalent only if $\beta = 1/k_B T$. Since this must be true for all the canonical ensemble subsets within the grand canonical ensemble, we therefore conclude that for the grand canonical ensemble as a whole

$$\beta = \frac{1}{k_B T}. \tag{4.80}$$

To evaluate γ_a and γ_b, we return to the definition of the grand canonical partition function, Eq. (4.75). For convenience, we define

$$\hat{\gamma}_a = \gamma_a k_B T, \tag{4.81}$$

$$\hat{\gamma}_b = \gamma_b k_B T. \tag{4.82}$$

Substituting (4.80)–(4.82) into Eq. (4.75), the definition of the grand canonical ensemble becomes

$$\Xi = \sum_{N_a=0}^{\infty} \sum_{N_b=0}^{\infty} \sum_{j=0}^{\infty} g_{N_a,N_b,j} \exp\left[-\left(\hat{\gamma}_a N_a + \hat{\gamma}_b N_b + E_{N_a,N_b,j}\right)/k_B T\right]. \tag{4.83}$$

We note that for specified N_a, N_b, and j, the energy levels and degeneracies are expected to be at most functions of the volume V. The other parameters that appear in the definition of Ξ are T, $\hat{\gamma}_a$, and $\hat{\gamma}_b$. If we consider the natural log of Ξ, it too would be a function of V, T, $\hat{\gamma}_a$, and $\hat{\gamma}_b$. Expanding $\ln \Xi$ to first order in these variables, we obtain

$$
d(\ln \Xi) = \left(\frac{\partial(\ln \Xi)}{\partial T}\right)_{V,\hat{\gamma}_a,\hat{\gamma}_b} dT + \left(\frac{\partial(\ln \Xi)}{\partial V}\right)_{T,\hat{\gamma}_a,\hat{\gamma}_b} dV
$$

$$
+ \left(\frac{\partial(\ln \Xi)}{\partial \hat{\gamma}_a}\right)_{T,V,\hat{\gamma}_b} d\hat{\gamma}_a + \left(\frac{\partial(\ln \Xi)}{\partial \hat{\gamma}_b}\right)_{T,V,\hat{\gamma}_a} d\hat{\gamma}_b. \tag{4.84}
$$

We must bring to this analysis some additional information on properties. The ensemble average of any property is computed by multiplying the value of the property at each combination of N_a, N_b, and energy level j by the probability that a system has that combination and summing such terms for all possible combinations of N_a, N_b, and j. Designating an arbitrary property as Y, this definition of its ensemble average $\langle Y \rangle$ is written as

$$
\langle Y \rangle = \sum_{N_a=0}^{\infty} \sum_{N_b=0}^{\infty} \sum_{j=0}^{\infty} Y\left(E_{N_a,N_b,j}, V, N_a, N_b\right) \tilde{P}\left(E_{N_a,N_b,j}, N_a, N_b\right)
$$

$$
= \sum_{N_a=0}^{\infty} \sum_{N_b=0}^{\infty} \sum_{j=0}^{\infty} Y\left(E_{N_a,N_b,j}, V, N_a, N_b\right)
$$

$$
\times \frac{g_{N_a,N_b,j} \exp\left[-\left(\hat{\gamma}_a N_a + \hat{\gamma}_b N_b + E_{N_a,N_b,j}\right)/k_B T\right]}{\Xi}. \tag{4.85}
$$

The grand canonical ensemble analysis is based on the premise that the average property value computed using definition (4.85) is the value of that property that would be observed macroscopically for a system at equilibrium at the μ_a, μ_b, V, and T values specified for the ensemble. The ensemble average values of N_a, N_b, internal energy U, and pressure P are therefore given by

$$
\langle N_a \rangle = \sum_{N_a=0}^{\infty} \sum_{N_b=0}^{\infty} \sum_{j=0}^{\infty} N_a \frac{g_{N_a,N_b,j} \exp\left[-\left(\hat{\gamma}_a N_a + \hat{\gamma}_b N_b + E_{N_a,N_b,j}\right)/k_B T\right]}{\Xi},
$$
$$
\tag{4.86}
$$

$$
\langle N_b \rangle = \sum_{N_a=0}^{\infty} \sum_{N_b=0}^{\infty} \sum_{j=0}^{\infty} N_b \frac{g_{N_a,N_b,j} \exp\left[-\left(\hat{\gamma}_a N_a + \hat{\gamma}_b N_b + E_{N_a,N_b,j}\right)/k_B T\right]}{\Xi},
$$
$$
\tag{4.87}
$$

$$
\langle U \rangle = \sum_{N_a=0}^{\infty} \sum_{N_b=0}^{\infty} \sum_{j=0}^{\infty} E_{N_a,N_b,j} \frac{g_{N_a,N_b,j} \exp\left[-\left(\hat{\gamma}_a N_a + \hat{\gamma}_b N_b + E_{N_a,N_b,j}\right)/k_B T\right]}{\Xi},
$$
$$
\tag{4.88}
$$

$$
\langle P \rangle = \sum_{N_a=0}^{\infty} \sum_{N_b=0}^{\infty} \sum_{j=0}^{\infty} P\left(E_{N_a,N_b,j}, V, N_a, N_b\right)
$$

$$
\times \frac{g_{N_a,N_b,j} \exp\left[-\left(\hat{\gamma}_a N_a + \hat{\gamma}_b N_b + E_{N_a,N_b,j}\right)/k_B T\right]}{\Xi}. \tag{4.89}
$$

We will now use Eq. (4.83) together with Eqs. (4.86)–(4.89) to evaluate the partial derivatives in Eq. (4.84). The derivatives with respect to T, $\hat{\gamma}_a$, and $\hat{\gamma}_b$ are fairly straightforward to evaluate.

$$\left(\frac{\partial(\ln\Xi)}{\partial T}\right)_{V,\hat{\gamma}_a,\hat{\gamma}_b} = \frac{1}{\Xi}\left(\frac{\partial\Xi}{\partial T}\right)_{V,\hat{\gamma}_a,\hat{\gamma}_b}$$

$$= \frac{1}{k_B T^2}\sum_{N_a=0}^{\infty}\sum_{N_b=0}^{\infty}\sum_{j=0}^{\infty}\left(\hat{\gamma}_a N_a + \hat{\gamma}_b N_b + E_{N_a,N_b,j}\right)$$

$$\times \frac{g_{N_a,N_b,j}\exp\left[-\left(\hat{\gamma}_a N_a + \hat{\gamma}_b N_b + E_{N_a,N_b,j}\right)/k_B T\right]}{\Xi}$$

$$= \frac{\hat{\gamma}_a\langle N_a\rangle}{k_B T^2} + \frac{\hat{\gamma}_b\langle N_b\rangle}{k_B T^2} + \frac{\langle U\rangle}{k_B T^2}, \tag{4.90}$$

$$\left(\frac{\partial(\ln\Xi)}{\partial\hat{\gamma}_a}\right)_{V,T,\hat{\gamma}_b} = \frac{1}{\Xi}\left(\frac{\partial\Xi}{\partial\hat{\gamma}_a}\right)_{V,T,\hat{\gamma}_b}$$

$$= \sum_{N_a=0}^{\infty}\sum_{N_b=0}^{\infty}\sum_{j=0}^{\infty} -\left(\frac{N_a}{k_B T}\right)$$

$$\times \frac{g_{N_a,N_b,j}\exp\left[-\left(\hat{\gamma}_a N_a + \hat{\gamma}_b N_b + E_{N_a,N_b,j}\right)/k_B T\right]}{\Xi}$$

$$= -\frac{\langle N_a\rangle}{k_B T}, \tag{4.91}$$

$$\left(\frac{\partial(\ln\Xi)}{\partial\hat{\gamma}_b}\right)_{V,T,\hat{\gamma}_a} = \frac{1}{\Xi}\left(\frac{\partial\Xi}{\partial\hat{\gamma}_b}\right)_{V,T,\hat{\gamma}_a}$$

$$= \sum_{N_a=0}^{\infty}\sum_{N_b=0}^{\infty}\sum_{j=0}^{\infty} -\left(\frac{N_b}{k_B T}\right)$$

$$\times \frac{g_{N_a,N_b,j}\exp\left[-\left(\hat{\gamma}_a N_a + \hat{\gamma}_b N_b + E_{N_a,N_b,j}\right)/k_B T\right]}{\Xi}$$

$$= -\frac{\langle N_b\rangle}{k_B T}. \tag{4.92}$$

The derivative with respect to V is a bit trickier to evaluate:

$$\left(\frac{\partial(\ln\Xi)}{\partial V}\right)_{T,\hat{\gamma}_a,\hat{\gamma}_b} = \frac{1}{\Xi}\left(\frac{\partial\Xi}{\partial V}\right)_{T,\hat{\gamma}_a,\hat{\gamma}_b}$$

$$= \frac{1}{\Xi}\frac{\partial}{\partial V}\sum_{N_a=0}^{\infty}\sum_{N_b=0}^{\infty}\sum_{j=0}^{\infty} g_{N_a,N_b,j}$$

$$\times \exp\left[-\left(\hat{\gamma}_a N_a + \hat{\gamma}_b N_b + E_{N_a,N_b,j}\right)/k_B T\right]$$

$$= \frac{1}{\Xi}\sum_{N_a=0}^{\infty}\sum_{N_b=0}^{\infty} e^{-(\hat{\gamma}_a N_a + \hat{\gamma}_b N_b)/k_B T}\frac{\partial}{\partial V}\sum_{j=0}^{\infty} g_{N_a,N_b,j}e^{-E_{N_a,N_b,j}/k_B T}$$

$$= \frac{1}{\Xi}\sum_{N_a=0}^{\infty}\sum_{N_b=0}^{\infty} e^{-(\hat{\gamma}_a N_a + \hat{\gamma}_b N_b)/k_B T}\frac{\partial Q(N_a, N_b, V, T)}{\partial V}. \tag{4.93}$$

At this point we note that Eq. (4.49) from our analysis of the canonical ensemble implies that the V derivative of Q in the above equation is just Q/k_BT times the pressure for a system at specified N_a, N_b, V, and T. We can therefore write the relation for the V derivative of $\ln \Xi$ as

$$\left(\frac{\partial(\ln \Xi)}{\partial V}\right)_{T,\hat{\gamma}_a,\hat{\gamma}_b} = \frac{1}{\Xi}\sum_{N_a=0}^{\infty}\sum_{N_b=0}^{\infty}\frac{P(N_a,N_b,V,T)Q(N_a,N_b,V,T)}{k_BT}e^{-(\hat{\gamma}_aN_a+\hat{\gamma}_bN_b)/k_BT}.$$

(4.94)

The factor $P(N_a, N_b, V, T)$ in the above relation is equal to the ensemble average pressure for a canonical ensemble at the specified N_a, N_b, V, and T. It follows from the definition of the canonical ensemble average properties (4.26) that

$$P(N_a,N_b,V,T) = \sum_{j=0}^{\infty} P\left(E_{N_a,N_b,j},V,N_a,N_b\right)\frac{g_je^{-E_{N_a,N_b,j}/k_BT}}{Q}.$$

(4.95)

Substituting this result into Eq. (4.94) yields

$$\left(\frac{\partial(\ln \Xi)}{\partial V}\right)_{T,\hat{\gamma}_a,\hat{\gamma}_b} = \sum_{N_a=0}^{\infty}\sum_{N_b=0}^{\infty}\sum_{j=0}^{\infty}\left[P(E_{N_a,N_b,j},V,N_a,N_b)\right.$$
$$\left.\times\frac{g_{N_a,N_b,j}\exp\left[-\left(\hat{\gamma}_aN_a+\hat{\gamma}_bN_b+E_{N_a,N_b,j}\right)/k_BT\right]}{\Xi k_BT}\right].$$

(4.96)

Comparing the right side of the above equation with Eq. (4.89), we see that the triple summation in Eq. (4.96) is just the ensemble average pressure $\langle P \rangle$ for the grand canonical ensemble. Replacing the triple summation with $\langle P \rangle$ simplifies Eq. (4.96) to

$$\left(\frac{\partial(\ln \Xi)}{\partial V}\right)_{T,\hat{\gamma}_a,\hat{\gamma}_b} = \frac{\langle P \rangle}{k_BT}.$$

(4.97)

Substituting Eqs. (4.90)–(4.92) and (4.97) for the partial derivatives in Eq. (4.84), we obtain the following relation:

$$d(\ln \Xi) = \left(\frac{\hat{\gamma}_a\langle N_a\rangle}{k_BT^2}+\frac{\hat{\gamma}_b\langle N_b\rangle}{k_BT^2}+\frac{\langle U\rangle}{k_BT^2}\right)dT + \frac{\langle P\rangle}{k_BT}dV - \frac{\langle N_a\rangle}{k_BT}d\hat{\gamma}_a - \frac{\langle N_b\rangle}{k_BT}d\hat{\gamma}_b.$$

(4.98)

We add $d(\langle U\rangle/k_BT)$, $d(\hat{\gamma}_a\langle N_a\rangle/k_BT)$, and $d(\hat{\gamma}_b\langle N_b\rangle/k_BT)$ to both sides and rearrange to get

$$d\left(\ln \Xi + \frac{\langle U\rangle}{k_BT}+\frac{\hat{\gamma}_a\langle N_a\rangle}{k_BT}+\frac{\hat{\gamma}_b\langle N_b\rangle}{k_BT}\right) = \frac{d\langle U\rangle}{k_BT}+\frac{\langle P\rangle}{k_BT}dV+\frac{\hat{\gamma}_a}{k_BT}d\langle N_a\rangle+\frac{\hat{\gamma}_b}{k_BT}d\langle N_b\rangle.$$

(4.99)

As a final step, we drop the ensemble average brackets, acknowledging that the ensemble average properties are expected to equal the real physical properties for the system of interest:

$$d\left(\ln \Xi + \frac{U}{k_BT}+\frac{\hat{\gamma}_aN_a}{k_BT}+\frac{\hat{\gamma}_bN_b}{k_BT}\right) = \frac{dU}{k_BT}+\frac{P}{k_BT}dV+\frac{\hat{\gamma}_a}{k_BT}dN_a+\frac{\hat{\gamma}_b}{k_BT}dN_b.$$

(4.100)

We now compare Eq. (4.100) with the following equation, developed in Chapter 3, which applies to a binary mixture of two species:

$$T dS = dU + P dV - \mu_a dN_a - \mu_b dN_b. \tag{4.101}$$

Equation (4.101) can be written in the form

$$d\left(\frac{S}{k_B}\right) = \frac{dU}{k_B T} + \frac{P dV}{k_B T} - \frac{\mu_a}{k_B T} dN_a - \frac{\mu_b}{k_B T} dN_b. \tag{4.102}$$

Equations (4.100) and (4.102) both must be valid for any choices of the differential terms. If we set dN_a and dN_b both to zero, the right sides of both equations are equal and it follows directly that

$$d\left(\frac{S}{k_B}\right) = d\left(\ln \Xi + \frac{U}{k_B T} + \frac{\hat{\gamma}_a N_a}{k_B T} + \frac{\hat{\gamma}_b N_b}{k_B T}\right). \tag{4.103}$$

Since the differentials on the left side of Eqs. (4.100) and (4.102) must be equal, the right sides must also be equal for any choices of the differentials. For this to be true, the coefficients of corresponding differentials on the right sides must be equal. Equality of the coefficients for the dN_a and dN_b terms requires that

$$\hat{\gamma}_a = -\mu_a, \tag{4.104}$$

$$\hat{\gamma}_b = -\mu_b. \tag{4.105}$$

Equations (4.104) and (4.105) can then be used with Eqs. (4.81) and (4.82) to evaluate the Lagrange multipliers γ_a and γ_b:

$$\gamma_a = -\mu_a/k_B T, \tag{4.106}$$

$$\gamma_b = -\mu_b/k_B T. \tag{4.107}$$

Substituting Eqs. (4.80), (4.106), and (4.107) into Eqs. (4.74) and (4.75), we obtain the following relations for the occupancy distribution and the grand canonical partition function:

$$\frac{n^*_{N_a,N_b,j}}{n_E} = \frac{g_{N_a,N_b,j} \exp\left[\left(\mu_a N_a + \mu_b N_b - E_{N_a,N_b,j}\right)/k_B T\right]}{\Xi}, \tag{4.108}$$

where Ξ is the grand canonical partition function defined as

$$\Xi = \sum_{N_a=0}^{\infty} \sum_{N_b=0}^{\infty} \sum_{j=0}^{\infty} g_{N_a,N_b,j} \exp\left[\left(\mu_a N_a + \mu_b N_b - E_{N_a,N_b,j}\right)/k_B T\right]. \tag{4.109}$$

The distribution (4.108) can also be interpreted as the probability $\tilde{P}_{N_a,N_b,j}$ that a system at equilibrium with specified T, V, μ_a, and μ_b will, at an arbitrarily chosen time, contain N_a species a particles, N_b species b particles and be in energy level j:

$$\tilde{P}_{N_a,N_b,j} = \frac{g_{N_a,N_b,j} \exp\left[\left(\mu_a N_a + \mu_b N_b - E_{N_a,N_b,j}\right)/k_B T\right]}{\Xi}. \tag{4.110}$$

Having evaluated all of the Lagrange multipliers, we now will turn to the problem of obtaining relations for thermodynamic properties. Using Eqs. (4.104) and (4.105) to evaluate $\hat{\gamma}_a$ and $\hat{\gamma}_b$, the differential relation (4.103) can be integrated to obtain

$$S = k_B \ln \Xi + \frac{U}{T} - \frac{\mu_a N_a}{T} - \frac{\mu_b N_b}{T} + S_0.$$

In the above relation S_0 is a constant of integration. We note, however, that in the limit of N_a and N_b both going to zero, the system has only one accessible microstate: that with zero internal energy and zero particles. In that limit, the partition function defined by Eq. (4.109) goes to one and $\ln \Xi$ is zero. With the exception of the S_0 term, the other terms on the right side of the above equation also vanish as N_a and N_b approach zero, implying that S_0 is the limiting value of entropy as the total number of particles approaches zero. In general our definition of entropy has been based on the notion that it is an indicator of the number of accessible microstates for the system. This notion suggests that a perfectly ordered system with only one microstate should have zero entropy. We therefore set the integration constant S_0 equal to zero. Our relation for entropy then becomes

$$S = k_B \ln \Xi + \frac{U}{T} - \frac{\mu_a N_a}{T} - \frac{\mu_b N_b}{T}. \tag{4.111}$$

Replacing the ensemble average properties with the actual physical properties and using Eqs. (4.104) and (4.105) to evaluate $\hat{\gamma}_a$ and $\hat{\gamma}_b$, Eq. (4.90) becomes

$$k_B T \left(\frac{\partial (\ln \Xi)}{\partial T} \right)_{V, \mu_a, \mu_b} = \frac{U}{T} - \frac{\mu_a N_a}{T} - \frac{\mu_b N_b}{T}. \tag{4.112}$$

The right side of Eq. (4.112) is identical to the last three terms on the right of Eq. (4.111). We replace those three terms by the left side of Eq. (4.112) to obtain

$$S = k_B \ln \Xi + k_B T \left(\frac{\partial (\ln \Xi)}{\partial T} \right)_{V, \mu_a, \mu_b}. \tag{4.113}$$

Equation (4.113) explicitly relates the system entropy to the grand canonical partition function.

In Eqs. (4.91), (4.92), and (4.97), we similarly replace $\hat{\gamma}_a$ with $-\mu_a$, replace $\hat{\gamma}_b$ with $-\mu_b$, and replace each ensemble average property with the corresponding system physical property. Rearranging the resulting equations a bit yields the following additional property relations:

$$N_a = k_B T \left(\frac{\partial (\ln \Xi)}{\partial \mu_a} \right)_{V, T, \mu_b}, \tag{4.114}$$

$$N_b = k_B T \left(\frac{\partial (\ln \Xi)}{\partial \mu_b} \right)_{V, T, \mu_a}, \tag{4.115}$$

$$P = k_B T \left(\frac{\partial (\ln \Xi)}{\partial V} \right)_{T, \mu_a, \mu_b}. \tag{4.116}$$

Solving for U, Eq. (4.111) becomes

$$U = TS - k_B T \ln \Xi + \mu_a N_a + \mu_b N_b. \tag{4.117}$$

In Chapter 3, we derived the Euler equation, which takes the following form for a system containing a binary mixture of components a and b:

$$U = TS - PV + \mu_a N_a + \mu_b N_b. \tag{4.118}$$

Comparing Eqs. (4.117) and (4.118), we see that both relations can be valid only if

$$PV = k_B T \ln \Xi. \tag{4.119}$$

Solving Eq. (4.112) for U and using (4.114) and (4.115) to evaluate N_a and N_b, we also obtain

$$U = k_B T \left[\mu_a \left(\frac{\partial (\ln \Xi)}{\partial \mu_a} \right)_{V,T,\mu_b} + \mu_b \left(\frac{\partial (\ln \Xi)}{\partial \mu_b} \right)_{V,T,\mu_a} + T \left(\frac{\partial (\ln \Xi)}{\partial T} \right)_{V,\mu_a,\mu_b} \right].$$

(4.120)

Equations (4.113)–(4.116), (4.119), and (4.120) provide the link between the grand canonical ensemble statistics and macroscopic thermodynamic properties.

Example 4.2   Determine the value of the grand canonical partition function Ξ for a binary mixture of monatomic ideal gases containing one kmol of each species.

Solution   Since the mixture must obey Eq. (4.119),

$$PV = k_B T \ln \Xi,$$

and the ideal gas equation of state,

$$PV = (N_a + N_b) k_B T,$$

we can combine these relations to obtain

$$\ln \Xi = N_a + N_b.$$

This relation can be solved for Ξ, yielding

$$\Xi = e^{N_a + N_b}.$$

Substituting for N_a and N_b, we obtain

$$\Xi = e^{(1+1)6.02 \times 10^{26}} = 10^{5.2 \times 10^{26}}.$$

Clearly the numerical value of Ξ is extraordinarily large, reflecting the fact that the number of accessible system energy microstates with energies comparable to $k_B T$ is enormous.

4.4   Fluctuations

In this section we are going to examine microscale fluctuations in a system at macroscopic equilibrium. We will consider fluctuations in a system that contains a binary mixture of species a and species b particles. The degree to which a system fluctuates about the mean value of a thermodynamic property Y can be quantified in terms of its *variance* σ_Y^2 (where σ_Y is the *standard deviation*):

$$\sigma_Y^2 = \langle Y - \langle Y \rangle \rangle^2.$$

(4.121)

Here we will specifically consider fluctuations of N_a in a grand canonical ensemble (for which V, T, μ_a, and μ_b are held fixed). Applying Eq. (4.121) to fluctuations in N_a yields

$$\sigma_{N_a}^2 = \langle N_a - \langle N_a \rangle \rangle^2 = \langle N_a^2 - 2N_a \langle N_a \rangle + \langle N_a \rangle^2 \rangle = \langle N_a^2 \rangle - 2\langle N_a \rangle \langle N_a \rangle + \langle N_a \rangle^2$$

$$= \langle N_a^2 \rangle - \langle N_a \rangle^2.$$

(4.122)

The ensemble average of a property Y for the grand canonical ensemble for a two-species system is

$$\langle Y \rangle = \sum_{N_a=0}^{\infty} \sum_{N_b=0}^{\infty} \sum_{j=0}^{\infty} Y\left(N_a, N_b, E_{N_a,N_b,j}\right) \tilde{P}_{N_a,N_b,j}$$

$$= \sum_{N_a=0}^{\infty} \sum_{N_b=0}^{\infty} \sum_{j=0}^{\infty} Y\left(N_a, N_b, E_{N_a,N_b,j}\right)$$

$$\times \frac{g_{N_a,N_b,j} \exp\left[\left(\mu_a N_a + \mu_b N_b - E_{N_a,N_b,j}\right)/k_{\mathrm{B}}T\right]}{\Xi},$$

where the partition function for the two-species system is

$$\Xi = \sum_{N_a=0}^{\infty} \sum_{N_b=0}^{\infty} \sum_{j=0}^{\infty} g_{N_a,N_b,j} \exp\left[\left(\mu_a N_a + \mu_b N_b - E_{N_a,N_b,j}\right)/k_{\mathrm{B}}T\right]. \qquad (4.123)$$

Treating N_a^2 as the property and applying this relation, we have

$$\langle N_a^2 \rangle = \sum_{N_a=0}^{\infty} \sum_{N_b=0}^{\infty} \sum_{j=0}^{\infty} N_a^2 \frac{g_{N_a,N_b,j} \exp\left[\left(\mu_a N_a + \mu_b N_b - E_{N_a,N_b,j}\right)/k_{\mathrm{B}}T\right]}{\Xi}$$

$$= \sum_{N_a=0}^{\infty} \sum_{N_b=0}^{\infty} \sum_{j=0}^{\infty} \frac{k_{\mathrm{B}}T}{\Xi}$$

$$\times \frac{\partial}{\partial \mu_a} \left(N_a g_{N_a,N_b,j} \exp\left[\left(\mu_a N_a + \mu_b N_b - E_{N_a,N_b,j}\right)/k_{\mathrm{B}}T\right]\right).$$

We simplify this relation by manipulating the right side as follows:

$$\langle N_a^2 \rangle = \sum_{N_a=0}^{\infty} \sum_{N_b=0}^{\infty} \sum_{j=0}^{\infty} k_{\mathrm{B}}T$$

$$\times \left[\frac{\partial}{\partial \mu_a} \left(\frac{N_a g_{N_a,N_b,j} \exp\left[\left(\mu_a N_a + \mu_b N_b - E_{N_a,N_b,j}\right)/k_{\mathrm{B}}T\right]}{\Xi} \right) \right.$$

$$\left. + \frac{\partial \Xi}{\partial \mu_a} \left(\frac{N_a g_{N_a,N_b,j} \exp\left[\left(\mu_a N_a + \mu_b N_b - E_{N_a,N_b,j}\right)/k_{\mathrm{B}}T\right]}{\Xi^2} \right) \right]$$

$$= k_{\mathrm{B}}T \frac{\partial}{\partial \mu_a} \sum_{N_a=0}^{\infty} \sum_{N_b=0}^{\infty} \sum_{j=0}^{\infty} N_a$$

$$\times \frac{g_{N_a,N_b,j} \exp\left[\left(\mu_a N_a + \mu_b N_b - E_{N_a,N_b,j}\right)/k_{\mathrm{B}}T\right]}{\Xi} + \frac{k_{\mathrm{B}}T}{\Xi} \frac{\partial \Xi}{\partial \mu_a}$$

$$\times \sum_{N_a=0}^{\infty} \sum_{N_b=0}^{\infty} \sum_{j=0}^{\infty} \frac{N_a g_{N_a,N_b,j} \exp\left[\left(\mu_a N_a + \mu_b N_b - E_{N_a,N_b,j}\right)/k_{\mathrm{B}}T\right]}{\Xi}$$

$$= k_{\mathrm{B}}T \frac{\partial \langle N_a \rangle}{\partial \mu_a} + \frac{k_{\mathrm{B}}T}{\Xi} \frac{\partial \Xi}{\partial \mu_a} \langle N_a \rangle$$

$$= k_{\mathrm{B}}T \frac{\partial \langle N_a \rangle}{\partial \mu_a} + k_{\mathrm{B}}T \frac{\partial (\ln \Xi)}{\partial \mu_a} \langle N_a \rangle. \qquad (4.124)$$

Using Eq. (4.114) from the previous section to simplify the second term on the right of Eq. (4.124), we obtain

$$\langle N_a^2 \rangle = k_B T \left(\frac{\partial \langle N_a \rangle}{\partial \mu_a} \right)_{V,T,\mu_b} + \langle N_a \rangle^2.$$

Substituting the right side of the above equation for $\langle N_a^2 \rangle$ in Eq. (4.122) yields

$$\sigma_{N_a}^2 = k_B T \left(\frac{\partial \langle N_a \rangle}{\partial \mu_a} \right)_{V,T,\mu_b} + \langle N_a \rangle^2 - \langle N_a \rangle^2,$$

which reduces to

$$\sigma_{N_a}^2 = k_B T \left(\frac{\partial \langle N_a \rangle}{\partial \mu_a} \right)_{V,T,\mu_b}. \tag{4.125}$$

It should be obvious that a similar relation applies for the variance for species b, which can be obtained by switching the roles of a and b in Eq. (4.125). Equation (4.125) links the mean magnitude of the fluctuations in N_a to the derivative $(\partial N_a / \partial \mu_a)_{V,T,\mu_b}$. This provides information about the fluctuations in the number of particles for one species in the system. We may also be interested in fluctuations in the total number of particles in the system $N = N_a + N_b$. With some additional effort, the line of analysis above can be extended to show that the variance in the total number of particles in the binary system is given by

$$\sigma_N^2 = k_B T \left[\left(\frac{\partial \langle N_a \rangle}{\partial \mu_a} \right)_{V,T,\mu_b} + \left(\frac{\partial \langle N_b \rangle}{\partial \mu_b} \right)_{V,T,\mu_a} + \left(\frac{\partial \langle N_a \rangle}{\partial \mu_b} \right)_{V,T,\mu_a} + \left(\frac{\partial \langle N_b \rangle}{\partial \mu_a} \right)_{V,T,\mu_b} \right]. \tag{4.126}$$

With a similar (but longer) line of analysis, it can be shown that the variance in the system internal energy for the binary system is

$$\sigma_U^2 = k_B T^2 \left(\frac{\partial U}{\partial T} \right)_{V,N_a,N_b} + \Phi \sigma_{N_a}^2 \left(\frac{\partial U}{\partial N_a} \right)_{V,T,\mu_b}^2 + \Phi \sigma_{N_b}^2 \left(\frac{\partial U}{\partial N_b} \right)_{V,T,\mu_a}^2, \tag{4.127a}$$

where

$$\Phi = \frac{\left(\frac{\partial \langle N_a \rangle}{\partial \mu_a} \right)_{V,T,\mu_b} \left(\frac{\partial \langle N_b \rangle}{\partial \mu_b} \right)_{V,T,\mu_a}}{\left(\frac{\partial \langle N_a \rangle}{\partial \mu_a} \right)_{V,T,\mu_b} \left(\frac{\partial \langle N_b \rangle}{\partial \mu_b} \right)_{V,T,\mu_a} - \left(\frac{\partial \langle N_a \rangle}{\partial \mu_b} \right)_{V,T,\mu_a} \left(\frac{\partial \langle N_b \rangle}{\partial \mu_a} \right)_{V,T,\mu_b}}. \tag{4.127b}$$

We will now explore the interpretation of Eq. (4.125) in terms of other thermodynamic properties. From basic calculus, for a function $z = z(x, y)$ we know that (see Appendix I)

$$\left(\frac{\partial z}{\partial y} \right)_x = - \left(\frac{\partial z}{\partial x} \right)_y \left(\frac{\partial x}{\partial y} \right)_z.$$

Holding T and μ_b fixed, we can therefore write

$$\left(\frac{\partial \langle N_a \rangle}{\partial \mu_a} \right)_{V,T,\mu_b} = - \left(\frac{\partial \langle N_a \rangle}{\partial V} \right)_{\mu_a,T,\mu_b} \left(\frac{\partial V}{\partial \mu_a} \right)_{\langle N_a \rangle,T,\mu_b}. \tag{4.128}$$

At constant $\langle N_a \rangle$ and T, we expand $(\partial V/\partial \mu_a)_{\langle N_a \rangle, T, \mu_b}$ using the chain rule

$$\left(\frac{\partial V}{\partial \mu_a}\right)_{\langle N_a \rangle, T, \mu_b} = \left(\frac{\partial V}{\partial P}\right)_{\langle N_a \rangle, T, \mu_b} \left(\frac{\partial P}{\partial \mu_a}\right)_{\langle N_a \rangle, T, \mu_b}. \tag{4.129}$$

Substituting the right side of Eq. (4.129) for $(\partial V/\partial \mu_a)_{\langle N_a \rangle, T, \mu_b}$ in Eq. (4.128) gives

$$\left(\frac{\partial \langle N_a \rangle}{\partial \mu_a}\right)_{V, T, \mu_b} = -\left(\frac{\partial \langle N_a \rangle}{\partial V}\right)_{\mu_a, T, \mu_b} \left(\frac{\partial V}{\partial P}\right)_{\langle N_a \rangle, T, \mu_b} \left(\frac{\partial P}{\partial \mu_a}\right)_{\langle N_a \rangle, T, \mu_b}. \tag{4.130}$$

The first and third partial derivatives on the right side of the above relation can be evaluated using the following two Maxwell relations:

$$\left(\frac{\partial P}{\partial \mu_a}\right)_{\langle N_a \rangle, T, \mu_b} = \left(\frac{\partial \langle N_a \rangle}{\partial V}\right)_{P, T, \mu_b}, \tag{4.131}$$

$$\left(\frac{\partial \langle N_a \rangle}{\partial V}\right)_{\mu_a, T, \mu_b} = \left(\frac{\partial P}{\partial \mu_a}\right)_{V, T, \mu_b}. \tag{4.132}$$

Replacing the first and third derivatives in Eq. (4.130) with the derivatives indicated in Eqs. (4.131) and (4.132), we obtain

$$\left(\frac{\partial \langle N_a \rangle}{\partial \mu_a}\right)_{V, T, \mu_b} = -\left(\frac{\partial P}{\partial \mu_a}\right)_{V, T, \mu_b} \left(\frac{\partial V}{\partial P}\right)_{\langle N_a \rangle, T, \mu_b} \left(\frac{\partial \langle N_a \rangle}{\partial V}\right)_{P, T, \mu_b}. \tag{4.133}$$

Solving Eq. (4.119) for $\ln \Xi$ and substituting the resulting expression for $\ln \Xi$ in Eq. (4.114) yields

$$N_a = k_B T \left(\frac{\partial (PV/k_B T)}{\partial \mu_a}\right)_{V, T, \mu_b}$$

$$= V \left(\frac{\partial P}{\partial \mu_a}\right)_{V, T, \mu_b}. \tag{4.134}$$

It is clear from this result that the first partial derivative in Eq. (4.133) is just N_a/V. Using this result to replace this derivative in Eq. (4.133) leads to

$$\left(\frac{\partial \langle N_a \rangle}{\partial \mu_a}\right)_{V, T, \mu_b} = -\frac{N_a}{V} \left(\frac{\partial V}{\partial P}\right)_{\langle N_a \rangle, T, \mu_b} \left(\frac{\partial \langle N_a \rangle}{\partial V}\right)_{P, T, \mu_b}. \tag{4.135}$$

To evaluate the rightmost derivative in the above equation, we proceed as follows. First, we rewrite Eq. (4.134) in the form

$$N_a = \left(\frac{\partial (PV)}{\partial \mu_a}\right)_{V, T, \mu_b}. \tag{4.136}$$

Differentiating both sides of Eq. (4.136) with respect to V, we obtain

$$\left(\frac{\partial N_a}{\partial V}\right)_{P, T, \mu_b} = \left(\frac{\partial}{\partial V}\left[\left(\frac{\partial (PV)}{\partial \mu_a}\right)_{V, T, \mu_b}\right]\right)_{P, T, \mu_b}. \tag{4.137}$$

Switching the order of differentiation yields

$$\left(\frac{\partial N_a}{\partial V}\right)_{P,T,\mu_b} = \left(\frac{\partial}{\partial \mu_a}\left[\left(\frac{\partial(PV)}{\partial V}\right)_{P,T,\mu_b}\right]\right)_{V,T,\mu_b}$$

$$= \left(\frac{\partial}{\partial \mu_a}[P]\right)_{V,T,\mu_b} = \left(\frac{\partial P}{\partial \mu_a}\right)_{V,T,\mu_b}. \tag{4.138}$$

Since Eq. (4.134) implies that $(\partial P/\partial \mu_a)_{V,T,\mu_b} = N_a/V$, the above equation can be written

$$\left(\frac{\partial N_a}{\partial V}\right)_{P,T,\mu_b} = \frac{N_a}{V}. \tag{4.139}$$

Substituting the right side of Eq. (4.139) for the rightmost derivative on the right side of (4.135), we obtain

$$\left(\frac{\partial \langle N_a \rangle}{\partial \mu_a}\right)_{V,T,\mu_b} = -\frac{N_a^2}{V^2}\left(\frac{\partial V}{\partial P}\right)_{\langle N_a \rangle, T, \mu_b}. \tag{4.140}$$

Replacing the derivative in Eq. (4.125) with the right side of Eq. (4.140) and replacing the ensemble average properties with system physical properties, the relation for the variance of N_a becomes

$$\sigma_{N_a}^2 = -\frac{k_BTN_a^2}{V^2}\left(\frac{\partial V}{\partial P}\right)_{N_a, T, \mu_b}. \tag{4.141}$$

Dividing both side by N_a^2 and taking the square root, we obtain the following relation for the fractional standard deviation of N_a from its mean value:

$$\frac{\sigma_{N_a}}{N_a} = \left[\frac{k_BT}{-(\partial P/\partial V)_{N_a, T, \mu_b}V^2}\right]^{1/2}. \tag{4.142}$$

The relations obtained for a binary system apply to single-species system if we set N_b to zero and drop μ_b as a parameter. For a single-species system, here we will drop the a subscript and designate the number of particles and chemical potential simply as N and μ, respectively. Conversion of relations derived earlier thus requires that we set N_b to zero and replace N_a and μ_a with N and μ respectively. It follows that for a single-species system, the relation for the fractional standard deviation in the number of particles N is given by

$$\frac{\sigma_N}{N} = \left[\frac{k_BT}{-(\partial P/\partial V)_{N,T}V^2}\right]^{1/2}. \tag{4.143}$$

Using a similar strategy to simplify Eq. (4.127), it follows that for a single-species system the variance of the internal energy about its mean value is given by

$$\sigma_U^2 = k_BT^2\left(\frac{\partial U}{\partial T}\right)_{V,N} + \sigma_N^2\left(\frac{\partial U}{\partial N}\right)_{T,V}^2, \tag{4.144}$$

where σ_N^2 is the variance of N, which can be computed using Eq. (4.143). With some minor manipulation of Eq. (4.144), the following relation for the fractional standard deviation of

U can be obtained:

$$\frac{\sigma_U}{U} = \left[\frac{k_B T^2}{U^2} \left(\frac{\partial U}{\partial T} \right)_{V,N} + \frac{\sigma_N^2}{U^2} \left(\frac{\partial U}{\partial N} \right)_{T,V}^2 \right]^{1/2}. \tag{4.145}$$

At this point, a comment on the significance of these results is warranted. Note that the analysis of fluctuations presented in this section was based on results for the grand canonical ensemble formulation. The grand canonical ensemble formulation applies to an open system in thermal and mass exchange communication with a host of other systems subject to the same constraints. Within in a large body of fluid, such as the air in a typical room, if we define a subsystem with a specified volume that is small compared to the overall size of the body of fluid, that subsystem is subject to the constraints imposed on systems in a grand canonical ensemble. For that reason, the results of the statistical analysis of the grand canonical ensemble provide useful tools for thermodynamic analysis of fluid elements within a larger body of fluid.

Our previous analysis of the grand canonical ensemble led to relations for the number of ensemble members for each particle–number–energy level combination. The distribution we found for a single-component system has a sharp maximum at a particular combination of N and U, which we interpret as being the equilibrium values for a system at the volume, temperature, and chemical potential values for the ensemble. The shape of the distribution is shown schematically in Figure 4.4. The relations for σ_N / N and σ_U / U derived in this section directly indicate the sharpness of the peak in the distribution. If the standard deviation from the mean property value is very small, the peak is very sharp, and the probability that the system will be observed with a value of U or N other than the mean value is very remote.

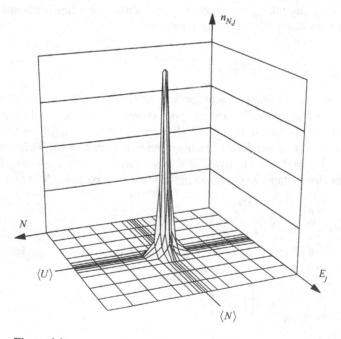

Figure 4.4

However, if the fractional standard deviation is significantly greater than zero, there is a nonnegligible chance that the system may be observed to wander away from the mean U or N value.

The sharpness of the distribution for N can be easily assessed for an ideal gas. Since the equation of state for an ideal gas is $PV = Nk_BT$, it follows that

$$-(\partial P / \partial V)_{N,T} = \frac{Nk_BT}{V^2}. \tag{4.146}$$

Substituting the right side of Eq. (4.146) for $-(\partial P / \partial V)_{N,T}$ in Eq. (4.142), we obtain

$$\frac{\sigma_N}{N} = \frac{1}{N^{1/2}}. \tag{4.147}$$

Thus for an ideal gas, as the number of particles increases, the fractional level of fluctuations decreases. For a system containing Avogadro's number of particles, the fractional deviation from the mean number of particles is of the order of 10^{-13}. It is therefore not surprising that any attempts to measure the number of particles in a system of macroscopic size will always give a result in close agreement with the mean value of N, and the measured value for an equilibrium system will not measurably change with time. Furthermore, to detect deviations from the mean in macroscopic systems, we must have instrumentation that can resolve such measurements to better than one part in 10^{13}. This level of precision is generally far beyond common instrumentation. These results imply, however, that if the system is small so that N is not large, the fluctuations may be a significant fraction of the mean value.

Example 4.3   Estimate the level of density fluctuations in a cubic centimeter of air at normal atmospheric pressure and room temperature.

Solution   For normal atmospheric pressure and room temperature we take $P = 101$ kPa and $T = 293$ K. Using the ideal gas equation of state we find

$$N = \frac{PV}{k_BT} = \frac{(101{,}000)(10^{-2})^3}{1.38 \times 10^{-23}\,(293)} = 2.50 \times 10^{19} \text{ molecules},$$

$$\frac{\sigma_N}{N} = \frac{1}{\sqrt{N}} = \frac{1}{\sqrt{2.50 \times 10^{19}}} = 2.00 \times 10^{-10}.$$

This implies that the standard deviation in the density in the 1 cm^3 volume is only two parts in 10^{10}, an amount so small it is virtually impossible to measure.

Example 4.4   Estimate the level of density fluctuations in a spherical control volume with a diameter of 200 Å at conditions typical of the earth's atmosphere at 3,000 meters altitude.

Solution   For the specified altitude, typical temperature and pressure conditions are $P = 70$ kPa and $T = 268$ K. The volume of the control volume is given by

$$V = \frac{4\pi r^3}{3} = \frac{4\pi(100 \times 10^{-10})^3}{3} = 4.2 \times 10^{-24} \text{ m}^3.$$

Using the ideal gas equation of state we find

$$N = \frac{PV}{k_BT} = \frac{(70,000)(4.2 \times 10^{-24})}{1.38 \times 10^{-23}\,(268)} = 79 \text{ molecules},$$

$$\frac{\sigma_N}{N} = \frac{1}{\sqrt{N}} = \frac{1}{\sqrt{79}} = 0.11.$$

Thus, density shifts of 11% over regions of this size are common in the atmosphere at this altitude. These fluctuations are significant because the variation in density results in a variation in the dielectric constant, and the variation of the dielectric constant causes the region to scatter light passing through the atmosphere. Visible light corresponds to values of wavelength λ between 4,000 and 7,000 Å. For scattering objects that are small compared to the wavelength of the radiation, the fraction of the incident radiation that is scattered is strongly wavelength dependent, being proportional to λ^{-4}. This regime is known as *Rayleigh scattering*. Rayleigh scattering does occur from regions of anomalous density in the atmosphere that result from density fluctuations. Because of its strong wavelength dependence, shorter wavelength blue light is scattered to a greater degree than the other colors in the visible range, making the sky appear blue. The blue sky is a result of the combined effects of density fluctuations and Rayleigh scattering.

Equation (4.142) also implies that fluctuations may be significant if $-(\partial P/\partial V)_{N,T}$ is small. For pure fluids, $-(\partial P/\partial V)_{N,T}$ becomes small in the vicinity of the critical point and for metastable superheated liquid or supersaturated vapor. We will discuss metastable fluids in more detail in a Chapter 8. However, it is worth noting here that the large density fluctuations that occur when $-(\partial P/\partial V)_{N,T}$ becomes small may raise the density of a vapor nearly to that of a liquid phase or may lower the density of a liquid near that for a vapor phase. Large density fluctuations that may initiate a change of phase are sometimes referred to as *heterophase fluctuations*.

Example 4.5   A system contains saturated argon vapor at atmospheric pressure and 87.3 K. Treating the vapor as an ideal gas, estimate the probability that a heterophase fluctuation may occur within a spherical region of the system having a diameter of 5 nanometers.

Solution   We consider subsystems of the overall system that permit mass and energy to cross their boundaries with each subsystem held at the temperature and chemical potential of the overall system. The subsystems are modeled as members of a grand canonical ensemble. The probability that a subsystem has N_a argon atoms is obtained by setting $N_b = 0$ in Eq. (4.110) and summing over all energy levels j. Doing so yields

$$\tilde{P}_{N_a} = \sum_{j=0}^{\infty} \frac{g_{N_a,j} e^{(\mu_a N_a - E_{N_a,j})/k_BT}}{\Xi},$$

where Ξ, the grand canonical partition function defined by Eq. (4.109) with N_b set to zero, is

$$\Xi = \sum_{N_a=0}^{\infty} \sum_{j=0}^{\infty} g_{N_a,j} e^{(\mu_a N_a - E_{N_a,N_b,j})/k_BT}.$$

Since the summation in the relation for $\tilde{P}_{N_a}$ is over j, it can be rewritten as

$$\tilde{P}_{N_a} = \frac{e^{\mu_a N_a / k_B T}}{\Xi} \sum_{j=0}^{\infty} g_{N_a,j} e^{-E_{N_a,j}/k_B T}.$$

The summation in the above relation is equal to the canonical partition function for an ensemble of systems having volume V, N_a particles, and temperature T. The above relation is therefore equivalent to

$$\tilde{P}_{N_a} = e^{\mu_a N_a / k_B T} \frac{Q(T, V, N_a)}{\Xi(T, V, \mu_a)}.$$

In this equation, the dependencies of the partition functions have been indicated explicitly. Taking the natural log of both sides converts the relation to

$$\ln \tilde{P}_{N_a} = \frac{\mu_a N_a}{k_B T} + \ln Q - \ln \Xi.$$

In this model analysis, we will treat the vapor as if it is an ideal gas. We will therefore use relations derived previously for a monatomic ideal gas to evaluate μ_a, $\ln Q$, and $\ln \Xi$ in the above relations. Equation (4.119) is used to evaluate $\ln \Xi$:

$$\ln \Xi = PV / k_B T.$$

For $\ln Q$, we use the relation derived in Example 4.1 with N_b set to zero,

$$\ln Q = \left(\frac{3}{2}\right) N_a \ln \left\{ \frac{2\pi m_a k_B T}{h^2} \left(\frac{V}{N_a}\right)^{2/3} \right\} + N_a,$$

and the relation obtained in Example 3.3 for a monatomic gas is used to evaluate the chemical potential,

$$\mu_a = -\left(\frac{3}{2}\right) k_B T \ln \left\{ \frac{2\pi m_a k_B T}{h^2} \left(\frac{V}{N_a}\right)^{2/3} \right\}.$$

In this analysis, the subsystem is a control volume with volume V. The values of μ_a and T for the subsystem are taken to be the same as those in the overall bulk system. In calculating μ_a we therefore use T and V/N_a values associated with the bulk system. The overall system is assumed to obey the ideal gas equation of state $PV = N_a k_B T$, and to get the μ_a value for the bulk, we can replace V/N_a in the above relation with $k_B T / P$ since T and P are specified for the bulk system:

$$\mu_a = -\left(\frac{3}{2}\right) k_B T \ln \left\{ \frac{2\pi m_a k_B T}{h^2} \left(\frac{k_B T}{P}\right)^{2/3} \right\}.$$

Substituting this relation for the chemical potential and the equations above for $\ln \Xi$ and $\ln Q$ into the equation for $\tilde{P}_{N_a}$, we get, after some simplification,

$$\ln \tilde{P}_{N_a} = N_a \ln \left(\frac{PV}{N_a k_B T}\right) + N_a - \left(\frac{PV}{k_B T}\right).$$

In the subsystem, the chemical potential and the temperature are taken to be fixed at the bulk system values, but we expect that the number of atoms in the volume may fluctuate with time about a mean, which is the equilibrium value of N_a. Since the gas in the subsystem volume obeys the ideal gas law, and the mean pressure in the volume is expected to equal the bulk system pressure P, we interpret PV/k_BT as being equal to the mean number of atoms in the subsystem $\langle N_a \rangle$. We can therefore rewrite the above relation as

$$\ln \tilde{P}_{N_a} = N_a \ln \left(\frac{\langle N_a \rangle}{N_a} \right) + N_a - \langle N_a \rangle.$$

Taking the exponential function (inverse natural log) of each side of the above equation, we get the following relation for the probability that the subsystem volume will contain N_a atoms:

$$\tilde{P}_{N_a} = e^{N_a - \langle N_a \rangle} \left(\frac{\langle N_a \rangle}{N_a} \right)^{N_a}.$$

In a spherical volume with a diameter of 5 nm, the mean number of argon atoms in saturated vapor at atmospheric pressure and 87.3 K is computed as

$$\langle N_a \rangle = PV/k_BT = (101,000)(\pi/6)(5 \times 10^{-9})^3/(1.38 \times 10^{-23} \times 87.3)$$

$$= 5.50 \text{ molecules.}$$

The density of saturated argon liquid at these conditions is 1,394 kg/m^3 (see Appendix III). Dividing this mass density by the molecular mass of 39.9 kg/kmol for argon and multiplying by Avogadro's number yields a density of 2.10×10^{28} molecules/m^3. To fill the subsystem with molecules at the liquid density would require a fluctuation such that

$$N_a = (2.10 \times 10^{28})(\pi/6)(5 \times 10^{-9})^3 = 1,380 \text{ molecules.}$$

Putting the calculated values of $\langle N_a \rangle$ and N_a into the relation for $\tilde{P}_{N_a}$, we find that the probability that such a fluctuation would occur is

$$\tilde{P}_{N_a} = e^{1,380 - 5.50} \left(\frac{5.50}{1,380} \right)^{1,380} \approx 10^{-2,714}.$$

Thus, the probability that random fluctuations in the vapor will produce near liquid density in a region nanometers in dimension is virtually zero. The accuracy of this prediction is limited because we treat the vapor as an ideal gas, and its behavior will deviate from that for an ideal gas as it approaches saturation. The result nevertheless suggests that formation of a liquid phase in a tiny subregion of the system by random fluctuations in the vapor is very unlikely, even at saturation conditions. This observation is qualitatively correct for most fluids. We will return to consider this issue in more detail in Chapter 8. In that chapter, we will show that if the system is pushed into a supersaturated state, heterophase fluctuations may have high enough probability to be likely initiators of a phase transition.

The ensemble formulations discussed so far in this chapter have considered system energy levels. To apply the results of these formulations to real systems of atoms, molecules, and other particles, we need to establish the link between molecular or particle energy levels and system energy levels. We will examine this issue in detail in the next section.

4.5   Distinguishability and Evaluation of the Partition Function

We now wish to turn our attention to the problem of evaluating the canonical and grand canonical partition function for systems of atoms, molecules, or other particles. As discussed in Chapter 2, particles commonly encountered in real systems fall into one of two categories: fermions and bosons. For bosons, there are no restrictions on the manner in which the system of particles distributes itself over available energy microstates. The behavior of fermions is limited in that no two identical fermions can occupy the same particle energy microstate. Analysis of fermions gives rise to *Fermi–Dirac statistics*, whereas consideration of bosons leads to *Bose–Einstein statistics*.

To apply the ensemble statistical framework to a system of N particles, we must first determine the complete set of energy microstates accessible to the system. If the system is governed by quantum theory, this would generally require the solution of the N-body Schrödinger equation, which is a virtually impossible task. Fortunately, there are many systems for which the N-body Hamiltonian operator can be written as a sum of the individual Hamiltonians. The total energy can then be written as a sum of individual energies for all N particles:

$$\hat{H} = \sum_{i=1}^{N} \hat{H}_i. \qquad (4.148)$$

The most common example of a system that conforms to this behavior is an ideal gas. Another example is the decomposition of the Hamiltonian of a polyatomic molecule into the contributions for its various degrees of freedom:

$$\hat{H} = \hat{H}_{\text{translation}} + \hat{H}_{\text{rotation}} + \hat{H}_{\text{vibration}} + \hat{H}_{\text{electronic}}. \qquad (4.149)$$

For many systems amenable to analysis by classical mechanics and quantum mechanics treatments, $\hat{H}$, by a proper selection of variables, can be written as a sum of individual terms. Pseudoparticles are sometimes associated with terms in this sum even if there are no real particles. These are sometimes also called *quasiparticles*. These include *photons, phonons, plasmons, magnons*, and *rotons*. For a systems of many distinguishable particles that behave this way, the total system energy can be computed as

$$\hat{H} = \sum_{i=1}^{N} \hat{H}_{\text{particle } i}. \qquad (4.150)$$

If the system is held at fixed V and T, the canonical partition function is given by

$$Q(N, V, T) = \sum_{j'=0}^{\infty} e^{-E_{j'}/k_{\text{B}}T} = \sum_{j'_a, j'_b, j'_c, \dots} e^{-(\varepsilon_{j'_a} + \varepsilon_{j'_b} + \varepsilon_{j'_c} + \cdots)/k_{\text{B}}T}, \qquad (4.151)$$

where $a, b, c, \dots$ designate the different molecules and $\varepsilon_{j'_a}$ indicates energy microstate j'_a for particle a, $\varepsilon_{j'_b}$ indicates energy microstate j'_b for particle b, etc. Note that we have elected, at least initially, to write the relation for the partition function as a sum over all microstates rather than over energy levels. The notation $j'_a, j'_b, j'_c, \dots$ in the summation denotes a summation over all possible combinations of these indices. Equation (4.151) can be rewritten as

$$Q = \sum_{j_a'=0}^{\infty} e^{-\varepsilon_{j_a'}/k_B T} \sum_{j_b'=0}^{\infty} e^{-\varepsilon_{j_b'}/k_B T} \sum_{j_c'=0}^{\infty} e^{-\varepsilon_{j_c'}/k_B T} \cdots . \tag{4.152}$$

Note that we have cast this relation as a product of summations over all the energy microstates for each particle in the system. If the degeneracy of each energy level for particle $a, b, c, \ldots$ is $g_{j_a}, g_{j_b}, g_{j_c}, \ldots$ respectively, we can write Eq. (4.151) as the product of summations over all energy levels for the particles:

$$Q = \sum_{j_a=0}^{\infty} g_{j_a} e^{-\varepsilon_{j_a}/k_B T} \sum_{j_b=0}^{\infty} g_{j_b} e^{-\varepsilon_{j_a}/k_B T} \sum_{j_c=0}^{\infty} g_{j_c} e^{-\varepsilon_{j_c}/k_B T} \cdots . \tag{4.153}$$

Each summation in the above relation is, in fact, a partition function for a specific particle. Since we will most often apply this concept to molecules, we will define a *molecular partition function q* as

$$q = \sum_{j=0}^{\infty} g_j e^{-\varepsilon_j/k_B T} . \tag{4.154}$$

The above definition is written as a sum over the energy levels of the particle or molecule. When it is useful, we can also write it in terms of a summation over energy microstates:

$$q = \sum_{j'=0}^{\infty} e^{-\varepsilon_{j'}/k_B T} . \tag{4.155}$$

With this definition we can write the relation (4.153) for Q as

$$Q = q_a q_b q_c \cdots q_N \quad \text{(for N particles)}, \tag{4.156}$$

where the subscripts denote the particles for which the molecular partition function is defined. Note that this implicitly assumes that the molecules are distinguishable. If the N molecules or other particles are all of same type, this relation reduces to

$$Q(N, V, T) = [q(V, T)]^N . \tag{4.157}$$

If the particles are of two types, a and b, and the system contains N_a particles of type a and N_b of type b, the relation for Q becomes

$$Q(N_a, N_b, V, T) = [q_a(V, T)]^{N_a} [q_b(V, T)]^{N_b} . \tag{4.158}$$

If we now specifically consider molecules for which the molecular Hamiltonian can be approximated by a sum of Hamiltonians for the various modes of energy storage in the molecule, by the same line of reasoning outlined above, we can break the molecular partition function into factors associated with each mode of energy storage,

$$q_{\text{molecular}} = q_{\text{tr}} q_{\text{rot}} q_{\text{vib}} q_e , \tag{4.159}$$

where $q_{\text{tr}}, q_{\text{rot}}, q_{\text{vib}},$ and q_e are the partition functions for translational, rotational, vibrational, and electronic energy storage, respectively. These are defined as

$$q_{\text{tr}} = \sum_{j=0}^{\infty} g_{\text{tr}, j} e^{-\varepsilon_{\text{tr}, j}/k_B T} , \tag{4.160}$$

$$q_{rot} = \sum_{j=0}^{\infty} g_{rot,j} e^{-\varepsilon_{rot,j}/k_B T}, \tag{4.161}$$

$$q_{vib} = \sum_{j=0}^{\infty} g_{vib,j} e^{-\varepsilon_{vib,j}/k_B T}, \tag{4.162}$$

$$q_e = \sum_{j=0}^{\infty} g_{e,j} e^{-\varepsilon_{e,j}/k_B T}. \tag{4.163}$$

Note that the above definitions are written as summations over energy levels. Equivalent definitions in terms of a summation over energy microstates also apply.

In this manner, not only can we reduce an N-particle problem to a one-particle problem, but we can reduce it further to the individual energy storage modes of a single particle. While this is an attractive result, there is one major problem with it: Atoms and molecules, in general, are not distinguishable. One exception is a system of atoms in a crystal lattice. Because the mean position of each atom is fixed in space, each particle is distinguishable. The results obtained above for distinguishable particles are therefore directly applicable to solid crystals, and we will explore how we can apply them to solid crystals in a later section.

In a liquid or gas, the particles are free to move about within the system and any of the particles may occupy a given location. For a given microstate of a system containing a fluid, the particles have specific positions, velocities, and internal energy microstates. If we exchanged two particles of the same species, the microstates would be different if the particles were distinguishable, but we would not be able to tell them apart if the particles are indistinguishable. If we treated the particles as distinguishable when in fact they are not, we would overcount the number of microstates. Thus, the relation between Q and the molecular partition function obtained for distinguishable particles is not correct for systems containing a gas or liquid.

Without a means to link the canonical partition function to energy storage in indistinguishable particles or molecules, the canonical ensemble formulation of statistical thermodynamics would be of little value as an analysis tool. Fortunately, there is a fairly straightforward way to establish such a link. In our analysis of the canonical ensemble, we derived the following relation, which links the partition function Q to the internal energy, entropy, and temperature.

$$\frac{S}{k_B} = \frac{U}{k_B T} + \ln Q. \tag{4.41}$$

In Chapter 2 we also derived the following equation for a system in a microcanonical ensemble that contains a binary mixture of distinguishable particles:

$$\frac{S}{k_B} = \frac{U}{k_B T} + N_a \ln Z_a + N_b \ln Z_b. \tag{2.51}$$

Comparing Eqs. (4.41) and (2.51), it is clear that both can be valid for a system containing a binary mixture of distinguishable particles only if

$$\ln Q = N_a \ln Z_a + N_b \ln Z_b. \tag{4.164}$$

The definitions of the partition functions

$$Z_a = \sum_{i=0}^{\infty} g_i e^{-\varepsilon_i / k_B T},$$ (2.50a)

$$Z_b = \sum_{j=0}^{\infty} g_j e^{-\varepsilon_j / k_B T}$$ (2.50b)

are identical to that for the molecular partition functions defined by Eq. (4.154). Note that both are summations over all accessible particle or molecular energy levels. It follows directly that

$$Z_a = q_a,$$ (4.165)

$$Z_b = q_b.$$ (4.166)

Substituting Eqs. (4.165) and (4.166) into Eq. (4.164) and solving for Q yields

$$Q = q_a^{N_a} q_b^{N_b} \quad \text{(for distinguishable particles)}.$$ (4.167)

This result is identical to Eq. (4.158) obtained earlier in this section by writing the total system energy in the canonical partition function as the sum of individual energies for all particles in the system.

In Chapter 2 we also derived the following relation, which is valid for indistinguishable bosons or fermions in the limit of dilute occupancy:

$$\frac{S}{k_B} = \frac{U}{k_B T} + N_a \ln \left\{ \frac{Z_a}{N_a} \right\} + N_b \ln \left\{ \frac{Z_b}{N_b} \right\} + N_a + N_b.$$ (2.78)

Comparing Eqs. (4.41) and (2.78), it is clear that both can be valid for a system containing a binary mixture of indistinguishable particles only if

$$\ln Q = N_a \ln \left\{ \frac{Z_a}{N_a} \right\} + N_b \ln \left\{ \frac{Z_b}{N_b} \right\} + N_a + N_b.$$ (4.168)

Rearranging the right side, Eq. (4.168) can be written in the form

$$\ln Q = N_a \ln Z_a + N_b \ln Z_b - (N_a \ln N_a - N_a) - (N_b \ln N_b - N_b).$$ (4.169)

Using Stirling's approximation in reverse, the terms in parentheses on the right side of Eq. (4.169) are given by

$$N_a \ln N_a - N_a = \ln N_a!,$$ (4.170)

$$N_b \ln N_b - N_b = \ln N_b!.$$ (4.171)

Substituting the right sides of Eqs. (4.170) and (4.171) into (4.169) yields

$$\ln Q = N_a \ln Z_a + N_b \ln Z_b - \ln N_a! - \ln N_b!.$$ (4.172)

Since $Z_a = q_a$ and $Z_b = q_b$, we can substitute and solve Eq. (4.172) to obtain the following relation for Q:

$$Q = \left(\frac{q_a^{N_a}}{N_a!} \right) \left(\frac{q_b^{N_b}}{N_b!} \right) \quad \left(\begin{array}{l} \text{for indistinguishable particles} \\ \text{in the limit of dilute occupancy} \end{array} \right).$$ (4.173)

Thus, by requiring consistency between the results for the microcanonical ensemble and the canonical ensemble we have derived a relation between the molecular partition function for distinguishable bosons and for indistinguishable bosons or fermions in the limit of dilute occupancy. Furthermore, if we compare Eq. (4.167) for distinguishable particles with Eq. (4.173), we note that Q for the binary mixture of indistinguishable particles in the high temperature limit is lower by a factor of $1/(N_a!N_b!)$. This reflects the reduced number of distinguishable system microstates when the particles are indistinguishable.

It should be transparent from the form of Eq. (4.173) that if the system contains r species, the relation for Q becomes

$$Q = \prod_{i=1}^{r} \frac{q_i^{N_i}}{N_i!} \quad \begin{pmatrix} \text{for indistinguishable particles} \\ \text{in the limit of dilute occupancy} \end{pmatrix}, \qquad (4.174)$$

where q_i and N_i are the molecular partition function and number of particles for the ith species.

The importance of Eq. (4.173) cannot be overstated. It provides the link between molecular energy storage characteristics and the system partition function. We now have a clear sequence of steps to determine system thermodynamic properties from particle energy storage characteristics in the high temperature limit:

(1) Determine the partition functions for the individual modes of energy storage for each species of molecule or other particle.
(2) Construct the molecular partition function for each species as $q_{\text{molecular}} = q_{\text{tr}}q_{\text{rot}}q_{\text{vib}}q_e$.
(3) Construct the system canonical partition function using Eq. (4.173) (or (4.174)).
(4) Use the relations derived in Section 4.2 to determine thermodynamic properties from the canonical partition function Q.

It should be noted that Eq. (4.173) and the methodology defined in the steps above are applicable if the system meets the following criteria:

(1) The particles are indistinguishable.
(2) The total system energy can be written as the sum of the energies associated with each particle.
(3) The system exhibits dilute occupancy.

Chapter 5 of this text will be devoted to applying the methodology defined in the steps above to systems containing different ideal gases.

In closing this section, we take on the task of defining more precisely the high temperature limit in which dilute occupancy is attained. Based on the quantum analysis of a single particle in a box, in Chapter 1 we found that at moderate to high particle energies, the number of translational states with energy less than a specified value ε was given by

$$n_\varepsilon = \frac{\pi}{6}\left(\frac{8mV^{2/3}\varepsilon}{h^2}\right)^{3/2}. \qquad (4.175)$$

If the mean translational energy of a particle in a system is $\langle \varepsilon \rangle$, then $n_{\langle \varepsilon \rangle}$ is a good estimate of the number of microstates accessible to the particle. Furthermore, in Section 2.6 we showed that in a system of particles that store energy only by translation, the mean energy

Table 4.1 *Assessment of the Dilute Occupancy Criterion for Several System Types*

Fluid	T (K)	$\dfrac{6N}{\pi V}\left(\dfrac{h^2}{12mk_BT}\right)^{3/2}$
Liquid helium	4	1.6
Gaseous helium	4	0.10
Gaseous helium	20	0.0020
Gaseous helium	100	4.0×10^{-5}
Liquid neon	27	0.011
Gaseous neon	100	3.0×10^{-6}
Liquid argon	86	5.1×10^{-4}
Gaseous argon	86	2.0×10^{-6}
Electrons in sodium	300	1,500

per particle is related to the temperature as

$$\langle \varepsilon \rangle = \tfrac{3}{2}k_BT.$$

Dilute occupancy implies that

$$n_\varepsilon \gg N. \tag{4.176}$$

Taking $\varepsilon = (3/2)k_BT$ in Eq. (4.175) and combining Eqs. (4.175) and (4.176) yields the following criterion for dilute occupancy:

$$\frac{6N}{\pi V}\left(\frac{h^2}{12mk_BT}\right)^{3/2} \ll 1. \tag{4.177}$$

This condition is favored by high molecular mass, high temperature, and low particle density (N/V). It turns out that this condition is satisfied for all but the lightest molecules at low temperatures (see the typical values in Table 4.1).

Strictly speaking, the criterion described above was derived for systems containing particles that store energy by translation only. However, the results are also reasonably accurate for most polyatomic molecules, since translational states usually account for much of their energy. In any case, additional internal storage modes only increase the number of accessible microstates and so if a system satisfies the inequality condition (4.177), additional internal modes will only further dilute the occupancy. Thus (4.177) is a sufficient condition for dilute occupancy as long as the particles are free to translate within the system.

At high system temperatures, the mean energy per particle will be high and the separation of the quantum energy states will be small compared to the energy of the particle. As a result, the energy can be considered to be a continuous variable, and the behavior of the system is accurately predicted by classical mechanics models. At high temperatures, the statistical behavior of such systems can also be treated by the statistical mechanics model developed by Boltzmann, which is based on a classical mechanics view of energy storage in particles. For that reason, in the limit of dilute occupancy, the particles are said to obey *Boltzmann statistics* (for indistinguishable particles). Because their original formulation was based on a classical mechanics view of energy storage, the Boltzmann statistics behavior at high temperature is also sometimes called the *classical limit*. Further discussion of ensemble theory and the dilute occupancy limit can be found in References [2]–[5].

Exercises

4.1 The following relation defines an approximate partition function for a pure dense gas:

$$Q(N, V, T) = \frac{1}{N!} \left(\frac{2\pi m k_B T}{h^2} \right)^{3N/2} (V - Nb)^N e^{aN^2/V k_B T}.$$

In this relation, a and b are different constants for each molecular species. From this partition function, derive an equation of state of the form $P = P(V, N, T)$.

4.2 For a single-species system we can take $N_b = 0$, drop μ_b as a parameter, and drop the a subscripts on N_a and μ_a. Equation (4.125) then becomes

$$\sigma_N^2 = k_B T \left(\frac{\partial \langle N \rangle}{\partial \mu} \right)_{V,T}.$$

Use this result to derive Eq. (4.145) for a single-species system held at constant V, μ, and T.

4.3 Show that Eq. (4.143) can be written in the form

$$\frac{\sigma_N}{N} = \left[\frac{k_B T}{-(\partial P / \partial \hat{v})_T (N/N_A) \hat{v}^2} \right]^{1/2},$$

where $\hat{v}$ is the molar specific volume in m^3/kmol.

4.4 Show that for a single-species system held at constant V, N, and T, the variance of the energy is given by

$$\sigma_U^2 = k_B T^2 \left(\frac{\partial U}{\partial T} \right)_{V,N}.$$

4.5 As noted in Chapter 3, for a single-component system, the van der Waals equation of state can be written in the form

$$P_r = \frac{8 T_r}{3 \hat{v}_r - 1} - \frac{3}{\hat{v}_r^2},$$

where $P_r = P/P_c$, $T_r = T/T_c$, and $\hat{v}_r = \hat{v}/\hat{v}_c$, with P_c, T_c, and $\hat{v}_c$ being the pressure, temperature, and volume per kmol at the critical point. For $\hat{v}_r = 1$, plot the variation of $(\sigma_N/N)(P_c \hat{v}_c N/N_A k_B T_c)^{1/2}$ with T_r predicted by the van der Waals equation of state for $1 \leq T_r \leq 2$. What is the physical interpretation for the behavior of the fluid as $T_r \to 1$?

4.6

(a) At constant T and V, the chemical potentials for a binary mixture are functions of the number of particles of each species:

$$\mu_a = \mu_a(N_a, N_b), \qquad \mu_b = \mu_b(N_a, N_b) \quad \text{(at constant } T \text{ and } V\text{)}.$$

Show that this implies that

$$\left(\frac{\partial N_a}{\partial \mu_a} \right)_{V,T,\mu_b} = \frac{\left(\frac{\partial \mu_b}{\partial N_b} \right)_{V,T,N_a}}{\left(\frac{\partial \mu_a}{\partial N_a} \right)_{V,T,N_b} \left(\frac{\partial \mu_b}{\partial N_b} \right)_{V,T,N_a} + \left(\frac{\partial \mu_a}{\partial N_b} \right)_{V,T,N_a} \left(\frac{\partial \mu_b}{\partial N_a} \right)_{V,T,N_b}}.$$

(b)   A binary mixture of two monatomic ideal gases a and b has temperature of 300 K, a pressure of 50 kPa, and a concentration of species a of one part per trillion (i.e., a mole fraction of 10^{-12}). You take a small sample of the binary mixture at one instant of time that has a volume of one cubic millimeter. Assuming that the fractional uncertainty in the measured concentration from this sample is equal to σ_{N_a}/N_a, use the results of part (a) to assess the accuracy of bulk concentration measurements from a sample this small.

4.7   Derive Eq. (4.126) for the variance of $N = N_a + N_b$ in a binary mixture of two particle species a and b.

4.8   For 0.018 kg of steam at 100°C, use the van der Waals equation to estimate how far the pressure must rise above the equilibrium saturation pressure before σ_N/N is of order one.

4.9   For 0.001 kmol of liquid nitrogen at 120 K, use the van der Waals equation to estimate how far the pressure must fall below the equilibrium saturation pressure before σ_N/N is of order one.

4.10   For saturated neon vapor at 27.1 K and atmospheric pressure, estimate the probability that a heterophase fluctuation will raise the density to that of saturated liquid in a spherical region with a diameter of 5 nm. Note that neon is a monatomic substance with a molecular mass of 20.18 kg/kmol and a saturated liquid density of 1,205 kg/m^3 at 27.1 K.

4.11   Evaluate the parameter

$$\frac{6N}{\pi V} \left(\frac{h^2}{12 m k_\mathrm{B} T} \right)^{3/2}$$

for steam at 100°C and assess the validity of the dilute occupancy assumption for this system.

4.12   In a gas of protons the mean kinetic energy of each proton is 1.04×10^{-19} J. The mass of a proton is 1.66×10^{-27} kg. Evaluate the dilute occupancy assumption for this system.

4.13   Evaluate the parameter

$$\frac{6N}{\pi V} \left(\frac{h^2}{12 m k_\mathrm{B} T} \right)^{3/2}$$

for electrons in a copper wire at room temperature and assess the validity of the dilute occupancy assumption for this system.

References

[1]   Gibbs, J. W., *Elementary Principles in Statistical Mechanics*, reprinted by Ox Bow Press, Woodbridge, CT, 1981.
[2]   Hill, T. L., *An Introduction to Statistical Thermodynamics*, Dover Publications, New York, 1986.
[3]   Kittel, C. and Kroemer, H., *Thermal Physics*, 2nd ed., W. H. Freeman and Company, New York, 1980.
[4]   McQuarrie, D. A., *Statistical Mechanics*, Harper and Row, New York, 1976.
[5]   Robertson, H. S., *Statistical Thermophysics*, Prentice-Hall, Englewood Cliffs, NJ, 1993.

Ideal Gases

In Chapter 5 we apply the statistical theory developed in earlier chapters to ideal gases. We consider a binary mixture of two gas species and note that pure component relations can be recovered by setting the number of particles of one species to zero. The necessary conditions for chemical equilibrium in reacting gas mixtures are also examined.

5.1 Energy Storage and the Molecular Partition Function

We now wish to turn our attention to the problem of evaluating the partition function for a system containing a pure ideal gas. In doing so we will make use of the methodology developed in Section 4.5 that relates the canonical partition function to the molecular partition function for indistinguishable particles under conditions of dilute occupancy. The major task in applying this analysis to systems of ideal gas molecules is to derive relations for the molecular partition function from available information about energy storage in the molecules.

In general, energy can be stored in a molecule as translational, rotational, or vibrational motion, and as a result of electronic or nuclear transitions. For the gases considered in this chapter, these energy storage mechanisms will be assumed to be independent. We therefore expect that we can decompose the Hamiltonian for a polyatomic molecule into its various degrees of freedom

$$\hat{H} \cong \hat{H}_{\text{translational}} + \hat{H}_{\text{rotational}} + \hat{H}_{\text{vibrational}} + \hat{H}_{\text{electronic}} + \hat{H}_{\text{nuclear}} \tag{5.1}$$

and, as a result, the molecular partition function can be written as the product of factors associated with each of the energy storage modes,

$$q(V, T) = q_{\text{tr}} q_{\text{rot}} q_{\text{vib}} q_{\text{e}} q_{\text{nucl}}. \tag{5.2}$$

To evaluate each of the factors on the right side of Eq. (5.2), we will analyze each of the energy storage modes that are active for a given type of molecule. We will begin in the next section by considering monatomic gases. We then consider more complicated diatomic and polyatomic molecules in subsequent sections.

5.2 Ideal Monatomic Gases

We now will apply the analysis tools developed in the previous chapter to an ideal monatomic gas. "Ideal" implies that the gas is dilute enough that intermolecular interactions can be neglected except for their effect during very brief collision processes between molecules. This is generally true for monatomic gases at pressures below a few atmospheres and temperatures greater than 20°C.

The number of independent energy storage modes is sometimes referred to as the number of *degrees of freedom* for the molecule. A monatomic gas atom has translational, electronic, and nuclear energy storage modes. To a very good approximation, the translational, nuclear,

145

and electronic Hamiltonians are separable. Based on the results obtained in Section 4.5, this implies that

$$q(V, T) = q_{tr}q_e q_{nucl}. \tag{5.3}$$

The Translational Partition Function

In Section 2.6, we determined the partition function for a particle that stores energy by translation alone. The definition of the partition function Z_a in that section is identical to the definition of q_{tr}:

$$q_{tr} = \sum_{i=0}^{\infty} g_{tr,i} e^{-\varepsilon_{tr,i}/k_B T}. \tag{5.4}$$

It follows directly from the analysis presented in Section 2.6 that the translational partition function for a monatomic gas is given by

$$q_{tr} = \left(\frac{2\pi m k_B T}{h^2}\right)^{3/2} V. \tag{5.5}$$

The above relation can be written in the form

$$q_{tr} = \frac{V}{\Lambda^3}, \tag{5.6}$$

where

$$\Lambda = \left(\frac{h^2}{2\pi m k_B T}\right)^{1/2}. \tag{5.7}$$

The factor Λ that occurs in the translational partition function has units of length. The usual interpretation of Λ is based on the following. The mean translational energy of a particle is computed as

$$\langle \varepsilon_{tr} \rangle = \sum_{i=0}^{\infty} \varepsilon_{tr,i} \frac{g_{tr,i} e^{-\varepsilon_{tr,i}/k_B T}}{q_{tr}}. \tag{5.8}$$

With a little rearranging, it is easy to show that

$$\langle \varepsilon_{tr} \rangle = k_B T^2 \left(\frac{\partial (\ln q_{tr})}{\partial T}\right). \tag{5.9}$$

Substituting the above relation (5.5) for q_{tr} yields

$$\langle \varepsilon_{tr} \rangle = \left(\frac{3}{2}\right) k_B T. \tag{5.10}$$

Since $\varepsilon_{tr} = p^2/2m$, where p is the particle momentum, the average momentum is proportional to $(m k_B T)^{1/2}$. Thus Λ is essentially h/p, which is equal to the de Broglie wavelength associated with the thermal motion of the particle. As a result, Λ is termed the *thermal de Broglie wavelength*. Note also that the condition for the validity of dilute occupancy, Eq. (4.177), can be stated in terms of Λ as

$$\frac{\Lambda^3}{V/N} \ll \sqrt{\frac{\pi}{6}}. \tag{5.11}$$

Since the fraction on the right side is close to one (0.72), the criterion can be more simply stated as

$$\frac{\Lambda^3}{V/N} \ll 1. \tag{5.12}$$

This implies that the thermal de Broglie wavelength must be small compared to the mean volume per molecule in the system containment. Note that this is similar to the condition that quantum effects decrease as the de Broglie wavelength becomes small compared to the physical system dimensions.

Example 5.1   Calculate the thermal de Broglie wavelength and assess the assumption of dilute occupancy for nitrogen gas at 300 K and 101 kPa.

Solution   The mass of a nitrogen molecule (N_2) is 4.65×10^{-26} kg. Substituting into Eq. (5.7), we obtain

$$\Lambda = \left(\frac{h^2}{2\pi m k_B T}\right)^{1/2} = \left(\frac{(6.63 \times 10^{-34})^2}{2\pi (4.65 \times 10^{-26})(1.38 \times 10^{-23})300}\right)^{1/2}$$
$$= 1.91 \times 10^{-11} \text{ m}.$$

Treating the gas as ideal, we can evaluate V/N using the ideal gas law

$$\frac{V}{N} = \frac{k_B T}{P}.$$

Substituting the system temperature and pressure yields

$$\frac{V}{N} = \frac{(1.38 \times 10^{-23})300}{101,000} = 4.10 \times 10^{-26} \text{ m}^3 \text{ (per molecule)}.$$

It follows from these results that

$$\frac{\Lambda^3}{V/N} = \frac{(1.91 \times 10^{-11})^3}{4.10 \times 10^{-26}} = 1.70 \times 10^{-7}.$$

This ratio being much less than one implies that the dilute occupancy idealization is appropriate for this system.

The Electronic Partition Function

For our analysis here, we write the electronic partition function as a sum over energy levels:

$$q_e = \sum_{i=0}^{\infty} g_{e,i} e^{-\varepsilon_{e,i}/k_B T}, \tag{5.13}$$

where $g_{e,i}$ is the degeneracy and $\varepsilon_{e,i}$ is the energy of the ith electronic level. We fix the arbitrary zero of the energy such that $\varepsilon_{e,0} = 0$. In doing so we effectively measure all electronic energy levels relative to the ground state:

$$q_e = g_{e,0} + g_{e,1} e^{-\Delta\varepsilon_{e,10}/k_B T} + \cdots, \tag{5.14}$$

where $\Delta\varepsilon_{e,i0}$ is the energy of electronic level i relative to the lowest energy level (ground state). These $\Delta\varepsilon_{e,i0}$ terms are typically of the order of electron volts. In general, $\Delta\varepsilon_{e,i0}/k_B T$ is quite large at ordinary temperatures and only the first term in the summation is significantly different from zero. Consequently, we need retain only the first term in most cases. In a few cases, such as for halogen gases, the second and third levels are close to the ground state. For such cases, the second- and third-level terms should also be retained.

For all monatomic gases, the electronic energy levels and degeneracies are known from measurements and/or quantum theory models. Hence, we write the electronic partition in the form

$$q_e \cong g_{e,0} + g_{e,1}e^{-\Delta\varepsilon_{e,10}/k_B T}. \tag{5.15}$$

Possible inclusion of additional higher energy level terms needs to be considered only at very small changes $\Delta\varepsilon_{e,i0}$ or at extremely high temperatures.

The Nuclear Partition Function

Because nuclear energy levels are separated by millions of electron volts, at room temperature only the first term in a summation for q_{nucl} need be considered:

$$q_{nucl} \cong g_{nucl,0}, \tag{5.16}$$

where $g_{nucl,0}$ is the degeneracy of the lowest energy level.

Properties

Assembling the parts, we have the following relation for the molecular partition function for a monatomic ideal gas:

$$q = q_{tr}q_e q_{nucl}$$

$$= \left(\frac{2\pi m k_B T}{h^2}\right)^{3/2} V\left(g_{e,0} + g_{e,1}e^{-\Delta\varepsilon_{e,10}/k_B T}\right)g_{nucl,0}. \tag{5.17}$$

Applying Eq. (4.174) we obtain the following relation for the canonical partition function:

$$Q = \frac{1}{N!}\left[\left(\frac{2\pi m k_B T}{h^2}\right)^{3/2} V\left(g_{e,0} + g_{e,1}e^{-\Delta\varepsilon_{e,10}/k_B T}\right)g_{nucl,0}\right]^N. \tag{5.18}$$

We will usually find that it is appropriate to take $g_{nucl,0} = 1$. We will therefore do so in evaluating the properties of the gas. It follows directly from our analysis of the canonical ensemble that

$$U = k_B T^2\left(\frac{\partial(\ln Q)}{\partial T}\right)_{N,V} = \frac{3}{2}N k_B T + \frac{N g_{e,1}\Delta\varepsilon_{e,10}e^{-\Delta\varepsilon_{e,10}/k_B T}}{g_{e,0} + g_{e,1}e^{-\Delta\varepsilon_{e,10}/k_B T}}, \tag{5.19}$$

$$S = k_B \ln Q + k_B T\left(\frac{\partial(\ln Q)}{\partial T}\right)_{N,V}$$

$$= N k_B \ln\left\{\left(\frac{2\pi m k_B T}{h^2}\right)^{3/2}\frac{V e^{5/2}}{N}\right\} + S_e, \tag{5.20}$$

where

$$S_e = Nk_B \ln\{g_{e,0} + g_{e,1}e^{-\Delta\varepsilon_{e,10}/k_BT}\} + \frac{(N/T)g_{e,1}\Delta\varepsilon_{e,10}e^{-\Delta\varepsilon_{e,10}/k_BT}}{g_{e,0} + g_{e,1}e^{-\Delta\varepsilon_{e,10}/k_BT}}. \tag{5.21}$$

Equation (5.20) is referred to as the *Sackur–Tetrode equation*. We also can obtain a relation for the pressure

$$P = k_B T \left(\frac{\partial(\ln Q)}{\partial T}\right)_{N,T} = \frac{Nk_B T}{V}, \tag{5.22}$$

which, not surprisingly, is the ideal gas equation of state. Note that inclusion of the electronic storage terms does not affect this equation of state.

5.3   Ideal Diatomic Gases

As a next step, we will consider an ideal gas composed of diatomic molecules. In addition to translational and electronic degrees of freedom, diatomic molecules possess vibrational and rotational degrees of freedom as well. Diatomic molecules represent the next step upward in complexity above a monatomic gas. Many technologically and environmentally important gases are diatomic. Examples include H_2, N_2, O_2, CO, NO, Cl_2, and HCl.

A general procedure to determine the energy levels of a diatomic molecule would be to set up the Schrödinger equation for the two nuclei and n electrons and solve it for the set of eigenvalues of the diatomic molecule. Exact application of this method has been done for H_2, but it becomes increasingly difficult as the number of electrons in the molecule increases. Approximate methodologies have been developed to circumvent difficulties associated with full solution of the Schrödinger equation. One way of determining electron orbitals in the molecule is the so-called LCAO–MO method. The acronym "LCAO–MO" stands for linear combination of atomic orbitals–molecular orbital. This method is based on the notion that orbitals for the molecule can be predicted as linear combinations of orbitals in the individual atoms. Using this method (or variations of it), total electron density maps can be generated for virtually all diatomic molecules.

The simpler model that we will adopt here makes use of the Born–Oppenheimer approximation. This approximation is based on the fact that the nuclei are much more massive than the electrons. The electrons are therefore considered to move in a field produced by the nuclei fixed at some internuclear separation. The motion of the nuclei in the electronic potential field established by the surrounding electron cloud varies with the electronic state of the molecule. Calculation of the nuclei interaction force potential $\phi_j(r)$ for even the ground state is a difficult n-electron calculation. As a result, semiempirical approximate relations for $\phi_j(r)$ are often used (where r is the internuclear separation distance).

It is a simple matter to show that the motion of two masses in a spherically symmetric potential can be rigorously separated into two separate problems by the introduction of center-of-mass and relative coordinates. For example, for the analogous 2-D case of two masses m_1 and m_2, moving in the x–y plane, the total energy is given by

$$E = \frac{m_1}{2}\left(\dot{x}_1^2 + \dot{y}_1^2\right) + \frac{m_2}{2}\left(\dot{x}_2^2 + \dot{y}_2^2\right) + \phi(x_1 - x_2, y_1 - y_2). \tag{5.23}$$

Introducing the new variables

$$X = \frac{m_1 x_1 + m_2 x_2}{m_1 + m_2}, \qquad Y = \frac{m_1 y_1 + m_2 y_2}{m_1 + m_2}, \tag{5.24}$$

$$x_{12} = x_1 - x_2, \qquad y_{12} = y_1 - y_2, \tag{5.25}$$

we can reduce this two-body problem to two one-body problems. Substituting to execute the above transformation of variables, we can show that the Hamiltonian for this system is given by

$$\hat{H} = E = \frac{m_1 + m_2}{2}(\dot{X}^2 + \dot{Y}^2) + \frac{m_r}{2}\left(\dot{x}_{12}^2 + \dot{y}_{12}^2\right) + \phi(x_{12}, y_{12}), \tag{5.26}$$

where m_r, the *reduced mass* of the two particles, is

$$m_r = \frac{m_1 m_2}{m_1 + m_2}. \tag{5.27}$$

Using

$$\frac{\partial \hat{H}}{\partial p_j} = \dot{q}_j, \qquad \frac{\partial \hat{H}}{\partial q_j} = -\dot{p}_j \tag{5.28}$$

equations of motion for this system can be derived that can be divided into two groups: (1) those that govern the translation of the center of mass having effective mass $m_1 + m_2$ and (2) equations that govern kinetic and potential energy interactions resulting from the relative motion of the two particles. Note that this transformation of the problem can also easily be extended to the 3-D case. If we lump the relative motion modes of storage together as internal storage, the Hamiltonian can be written as

$$\hat{H} = \hat{H}_{tr} + \hat{H}_{int}, \tag{5.29}$$

where $\hat{H}_{tr}$ is the contribution associated with translation of the center of mass and $\hat{H}_{int}$ is the contribution associated with relative motion of the atomic nuclei in the molecule. The latter contribution includes vibrational and rotational energy storage. From quantum analysis, this decomposition of the Hamiltonian will result in energy eigenvalues that also separate into two independent contributions,

$$\varepsilon = \varepsilon_{tr} + \varepsilon_{int}. \tag{5.30}$$

The partition function of a diatomic molecule is therefore expected to be given by

$$q = q_{tr} q_{int}. \tag{5.31}$$

Since the center-of-mass motion is analytically the same as that of a monatomic particle, we also expect that

$$q_{tr} = \left(\frac{2\pi(m_1 + m_2)k_B T}{h^2}\right)^{3/2} V. \tag{5.32}$$

The density of the translational states alone is so high that we expect dilute occupancy. It follows that

$$Q(N, V, T) = \frac{q_{tr}^N q_{int}^N}{N!}. \tag{5.33}$$

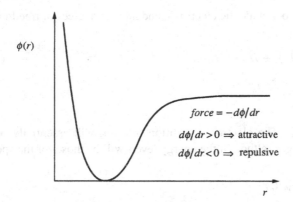

Figure 5.1 Potential function for model system.

When decoupled from translation, as described above, the relative motion of the two bodies can be interpreted as one body of reduced mass $m_r = m_1 m_2/(m_1 + m_2)$ moving about the origin. The potential function representing the force acting on the particle of reduced mass $\phi(r)$ is of the form shown in Figure 5.1. The relative motion of the nuclei in the molecule consists of rotary motion about the center of mass and relative vibrational motion of the two nuclei. This is mathematically represented by the rotary motion and vibrational motion of the body of reduced mass in the potential $\phi(r)$. Because the amplitude of the vibrational motion is very small it is a good approximation to consider it as representing the angular motion of the effective mass at a fixed internuclear distance r_e. The equilibrium internuclear distance is presumed to correspond to the zero force ($d\phi/dr = 0$) condition at the bottom of the potential well. Expanding $\phi(r)$ about r_e we get

$$\phi(r) = \phi(r_e) + \left(\frac{d\phi}{dr}\right)_{r=r_e} (r - r_e) + \frac{1}{2}\left(\frac{d^2\phi}{dr^2}\right)_{r=r_e} (r - r_e)^2 + \cdots. \tag{5.34}$$

The linear term in the $\phi(r)$ expansion vanishes because $d\phi/dr = 0$ at the bottom of the well. The expansion can then be written as

$$\phi(r) = \phi(r_e) + \tfrac{1}{2}k(r - r_e)^2 + \cdots, \tag{5.35}$$

with k being the effective force constant defined as

$$k = \left(\frac{d^2\phi}{dr^2}\right)_{r=r_e}. \tag{5.36}$$

The approximation embodied in this treatment of the relative motion storage modes is called the *rigid-rotor–harmonic-oscillator approximation*. It follows directly from these arguments that we can write

$$\hat{H}_{\mathrm{int}} = \hat{H}_{\mathrm{rot}} + \hat{H}_{\mathrm{vib}} \tag{5.37}$$

and

$$\varepsilon_{\mathrm{int}} = \varepsilon_{\mathrm{rot}} + \varepsilon_{\mathrm{vib}}, \tag{5.38}$$

$$q_{\mathrm{int}} = q_{\mathrm{rot}}q_{\mathrm{vib}}. \tag{5.39}$$

If we extend this treatment to include the electronic and nuclear degrees of freedom, for the complete molecule we have

$$\hat{H} = \hat{H}_{tr} + \hat{H}_{rot} + \hat{H}_{vib} + \hat{H}_e + \hat{H}_{nucl}, \tag{5.40}$$

$$\varepsilon = \varepsilon_{tr} + \varepsilon_{rot} + \varepsilon_{vib} + \varepsilon_e + \varepsilon_{nucl}, \tag{5.41}$$

$$q = q_{tr}q_{rot}q_{vib}q_eq_{nucl}. \tag{5.42}$$

We have already evaluated q_{tr}. The electronic partition function will essentially be the same as that for a monatomic gas (although the energy levels will be those for the specific molecule):

$$q_e \cong g_{e,0} + g_{e,1}e^{-\Delta\varepsilon_{e,10}/k_BT}. \tag{5.43}$$

As for the monatomic gas, we will adopt the approximation that $q_{nucl} = 1$ (i.e., the molecule remains in the nuclear ground state, which presumably has a degeneracy of one). Note that the implication in the above relations that these modes are independent is an idealization that is not exactly correct. It does, however, serve as a useful approximate model. Within this approximation, the partition function is given by

$$Q(N, V, T) = \frac{(q_{tr}q_{rot}q_{vib}q_eq_{nucl})^N}{N!}. \tag{5.44}$$

To fully evaluate Q, we therefore must develop a means of evaluating q_{rot} and q_{vib}.

The Vibrational Partition Function

Modeling the vibrational mode as a harmonic oscillator, our previous quantum analysis indicated that the energy levels and degeneracies are given by

$$\left.\begin{array}{r} \varepsilon_{vib,n} = h\nu\left(n + \tfrac{1}{2}\right) \\ g_{vib,n} = 1 \end{array}\right\} \quad \text{for } n = 0, 1, 2, \ldots, \tag{5.45}$$

where the characteristic frequency ν is related to the reduced mass and force constant as

$$\nu = \frac{1}{2\pi}\left(\frac{k}{m_r}\right)^{1/2} \tag{5.46}$$

Since the above relation predicts that the lowest (ground) state has finite energy, assigning zero energy to that state would be inconsistent. Instead, we therefore assign zero energy to the bottom of the potential well for the lowest-energy (ground) electronic state. The oscillator will then not actually be able to reach the bottom of the well but will have a lowest energy of $(1/2)h\nu$ as a ground state (see Figure 5.2).

As discussed in Chapter 1, transitions from one vibrational level to another can be induced by electromagnetic radiation. For a molecule to change its vibrational state by absorbing radiation it must (1) change its dipole moment when vibrating and (2) obey the selection rule: $\Delta n = \pm 1$. Note that the frequency of the absorbed or emitted radiation for a transition between levels is

$$\nu_e = \frac{\varepsilon_{n+1} - \varepsilon_n}{h} = \frac{1}{2\pi}\left(\frac{k}{m_r}\right)^{1/2}. \tag{5.47}$$

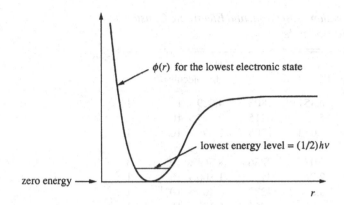

Figure 5.2 Lowest energy electronic ground state.

Consequently, the vibration spectrum for a diatomic molecule will consist of one line. Vibration lines typically occur in the infrared at wave numbers $(1/\lambda = \nu_e/c_l)$ near 1,000 cm^{-1}. Based on the above arguments, the vibrational partition function becomes

$$q_{vib} = \sum_{n=0}^{\infty} e^{-\varepsilon_{vib,n}/k_B T} = e^{-h\nu/2k_B T} \sum_{n=0}^{\infty} e^{-nh\nu/k_B T}. \tag{5.48}$$

In this case, the summation converges to a result that can be expressed in closed form:

$$\sum_{n=0}^{\infty} z^{-n} = \frac{1}{1 - z^{-1}} \quad \text{for } z^{-1} < 1, \tag{5.49}$$

and hence

$$q_{vib} = \frac{e^{-h\nu/2k_B T}}{1 - e^{-h\nu/k_B T}}. \tag{5.50}$$

The constraint on convergence of the sum is satisfied for any value of absolute temperature. The quantity $h\nu/k_B T$ is larger than 1 at low to moderate temperatures. However, if the temperature is high enough, $h\nu/k_B T$ may be much less than 1, and because the energy levels are close together, the summation for q_{vib} can be replaced with an integral:

$$q_{vib} = e^{-h\nu/2k_B T} \int_0^{\infty} e^{-nh\nu/k_B T} \, dn = \frac{k_B T}{h\nu}. \tag{5.51}$$

This result is identical to the limit of the closed-form relation for q_{vib} for $h\nu/k_B T \rightarrow 0$. Using $q_{vib}(T)$ we can evaluate the vibrational contribution to the gas internal energy,

$$U_{vib} = Nk_B T^2 \frac{\partial \ln\{q_{vib}\}}{\partial T} = Nk_B \left(\frac{\theta_{vib}}{2} + \frac{\theta_{vib}}{e^{\theta_{vib}/T} - 1} \right), \tag{5.52}$$

where the *vibrational temperature* θ_{vib} is defined as

$$\theta_{vib} = \frac{h\nu}{k_B}. \tag{5.53}$$

The vibrational contribution to the molar heat capacity is

$$\hat{c}_{V,vib} = \left(\frac{N_A}{N} \right) \left(\frac{\partial U_{vib}}{\partial T} \right)_V = N_A k_B \left(\frac{\theta_{vib}}{T} \right)^2 \frac{e^{\theta_{vib}/T}}{(e^{\theta_{vib}/T} - 1)^2}. \tag{5.54}$$

Table 5.1 *Rotation, Vibration, and Electronic Constants for Diatomic Molecules*

	σ_s	θ_{rot} (K)	θ_{vib} (K)	D_e (J/molecule)	$g_{e,0}$
Cl_2	2	0.351	808	4.029×10^{-19}	1
H_2	2	85.3	6215	7.607×10^{-19}	1
I_2	2	0.054	308	2.497×10^{-19}	1
N_2	2	2.88	3374	1.589×10^{-18}	1
O_2	2	2.07	2256	8.363×10^{-19}	3
CO	1	2.77	3103	1.800×10^{-18}	1
HCl	1	15.2	4227	7.400×10^{-19}	1
HI	1	9.06	3266	5.128×10^{-19}	1

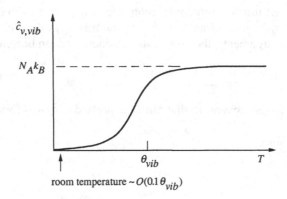

Figure 5.3   Vibrational component of the specific heat for a diatomic gas.

Note that as $T \to \infty$, $U_{vib} \to N k_B T$ and $\hat{c}_{V,vib} \to N_A k_B$. The variation of $\hat{c}_{V,vib}$ with temperature has the form shown in Figure 5.3. Vibration-related constants for some diatomic molecules are listed in Table 5.1.

The Rotational Partition Function for a Heteronuclear Diatomic Molecule

Modeling the rotational mode as a rigid rotor, our previous quantum analysis indicated that the energy levels and degeneracy are

$$\left.\begin{array}{l} \varepsilon_j = \frac{\hbar^2 j(j+1)}{2I} \\[2mm] g_j = 2j+1 \end{array}\right\} \quad \text{for } j = 0, 1, 2, \ldots . \tag{5.55}$$

Consistent with the above result, the zero of the rotational energy is taken to be the $j = 0$ state. As discussed in Chapter 1, transitions from one rotational level to another can be induced by electromagnetic radiation if (1) the molecule has a permanent dipole moment and (2) $\Delta j = \pm 1$. The frequency of radiation emitted in the process of changing from level $j + 1$ to j is given by

$$\nu_e = \frac{\varepsilon_{j+1} - \varepsilon_j}{h} = \frac{h}{4\pi^2 I}(j+1) \quad \text{for } j = 0, 1, 2, \ldots . \tag{5.56}$$

This implies that radiation is absorbed at frequencies given by multiples of $h/4\pi^2 I$, resulting in a set of equally spaced spectral lines, often found in the microwave region.

Based on the above results, the rotational partition function is given by

$$q_{rot} = \sum_{j=0}^{\infty} g_j e^{-\varepsilon_{rot,n}/k_B T} = \sum_{n=0}^{\infty} (2j+1) e^{-\bar{B} j(j+1) h c_1/k_B T}, \tag{5.57}$$

where

$$\bar{B} = \frac{h}{8\pi^2 I c_1} \tag{5.58}$$

is the *rotational constant* for the molecule. We further define

$$\theta_{rot} = \frac{h\bar{B}c_1}{k_B} = \frac{h^2}{8\pi^2 I k_B} \tag{5.59}$$

as the characteristic temperature for rotation. Unlike the vibrational case, the summation for q_{rot} cannot be written in closed form. We can replace the summation with an integral if the energy levels are close together ($(\varepsilon_{j+1} - \varepsilon_j)/k_B T \ll 1$). This condition can be equivalently stated as

$$2(j+1)\theta_{rot}/k_B T \ll 1. \tag{5.60}$$

This condition is satisfied at moderate temperatures except for large j values. However, the terms with large j are so small compared to those for low j values that slight inaccuracy in evaluating them has a negligible effect on the overall result. For high temperatures, we therefore write

$$q_{rot} = \int_0^{\infty} (2j+1) e^{-j(j+1)\theta_{rot}/T} \, dj = \frac{T}{\theta_{rot}}. \tag{5.61}$$

Thus for $\theta_{rot} \ll T$

$$q_{rot} = \frac{8\pi^2 I k_B T}{h^2}. \tag{5.62}$$

The accuracy of this result increases as the temperature increases. Noting that the summation for q_{rot} can be written as

$$q_{rot} = \sum_{n=0}^{\infty} (2j+1) e^{-j(j+1)\theta_{rot}/T}, \tag{5.63}$$

it is clear that if T is small compared to θ_{rot} the first few terms should suffice to evaluate q_{rot}:

$$q_{rot} = 1 + 3e^{-2\theta_{rot}/T} + 5e^{-6\theta_{rot}/T} + 7e^{-12\theta_{rot}/T} + \cdots. \tag{5.64}$$

The number of terms shown above is sufficient to determine the sum to within 0.1% for $\theta_{rot} > 0.7T$. For $\theta_{rot} < 0.7T$, but not small enough for the integral to give a good approximation, an intermediate approximation is needed. Using the integral result as the lowest-order approximation, higher-order correction terms can be generated using a series expansion technique. The following expansion can be obtained with this method:

$$q_{rot} = \frac{T}{\theta_{rot}} \left[1 + \frac{1}{3}\left(\frac{\theta_{rot}}{T}\right) + \frac{1}{15}\left(\frac{\theta_{rot}}{T}\right)^2 + \frac{4}{315}\left(\frac{\theta_{rot}}{T}\right)^3 + \cdots \right]. \tag{5.65}$$

For most substances, however, θ_{rot} is very low and the high-temperature approximation will be accurate. For such gases the rotational contribution to the internal energy is

$$U_{rot} = Nk_BT^2\frac{\partial \ln\{q_{rot}\}}{\partial T} = Nk_BT + \cdots. \qquad (5.66)$$

It follows that the contribution to the specific heat is

$$\hat{c}_{V,rot} = \left(\frac{N_A}{N}\right)\left(\frac{\partial U_{rot}}{\partial T}\right)_V = N_Ak_B + \cdots. \qquad (5.67)$$

Another feature of interest is the fraction of molecules in the jth rotational energy level

$$\frac{N_j}{N} = \frac{(2j+1)e^{-j(j+1)\theta_{rot}/T}}{q_{rot}}. \qquad (5.68)$$

This variation exhibits a maximum that can be determined by differentiating with respect to j and setting the result equal to zero. This yields

$$j_{max} = \left(\frac{T}{2\theta_{rot}}\right)^{1/2}. \qquad (5.69)$$

The variation of N_j/N with j for CO at 300 K is shown in Figure 5.4. The plot clearly indicates that most molecules are in excited rotational levels at ordinary temperatures. Note that all of the above analysis of the rotational storage mode presumes that we can treat the atomic nuclei as distinguishable particles. In fact, this is not appropriate for homonuclear molecules.

The Rotational Partition Function for a Homonuclear Diatomic Molecule

The wave function of a homonuclear diatomic molecule must have a certain symmetry with respect to the interchange of the two identical nuclei in the molecule. The nature of the required symmetry is dictated by the spins of the nuclei. If the two nuclei have integral spins, the wave function must be symmetric under interchange of the nuclei (they are bosons). If the two nuclei have half-integral spins, the wave function must be antisymmetric (they are fermions). Quantum analysis of the effect of these restrictions is somewhat complicated. However, in the high temperature limit, the net effects are easily quantified. Of particular note is that the restrictions on the overall wave function link the nuclear and

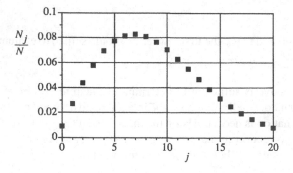

Figure 5.4   Variation of N_j/N with j for CO at 300 K.

rotational effects in such a way that the combined nuclear and rotational partition function does not factor into $q_{\text{nucl}}q_{\text{rot}}$. (This is true, in general, for either the boson or fermion case.) For either the half-integral or integral spin case, it can be shown that for the high temperature limit the rotational–nuclear partition function reduces to

$$q_{\text{rot,nucl}} = (2i + 1)^2 \left(\frac{T}{2\theta_{\text{rot}}} \right), \tag{5.70}$$

where i is the nuclear spin quantum number. In the high temperature limit, we can therefore consider

$$q_{\text{rot,nucl}} = q_{\text{nucl}}q_{\text{rot}}, \tag{5.71}$$

where

$$q_{\text{nucl}} = (2i + 1)^2, \tag{5.72}$$

$$q_{\text{rot}} = \frac{T}{2\theta_{\text{rot}}}. \tag{5.73}$$

For these relations to be valid, T must be greater than about $5\theta_{\text{rot}}$. The high temperature result for the heteronuclear and homonuclear cases are often represented together as

$$q_{\text{rot}} = \frac{T}{\sigma_{\text{s}}\theta_{\text{rot}}}, \tag{5.74}$$

where $\sigma_{\text{s}} = 1$ for the heteronuclear case and $\sigma_{\text{s}} = 2$ for the homonuclear case. Because its deviation from 1 is linked to symmetry considerations, σ_{s} is termed the *symmetry number*. It is frequently assumed that the same symmetry number correction applies to the expansion relation that can be developed for small θ_{rot}/T:

$$q_{\text{rot}} = \frac{T}{\sigma_{\text{s}}\theta_{\text{rot}}} \left[1 + \frac{1}{3} \left(\frac{\theta_{\text{rot}}}{T} \right) + \frac{1}{15} \left(\frac{\theta_{\text{rot}}}{T} \right)^2 + \frac{4}{315} \left(\frac{\theta_{\text{rot}}}{T} \right)^3 + \cdots \right]. \tag{5.75}$$

Note from Table 5.1 that θ_{rot} for virtually all molecules except H_2 is so low that almost all circumstances of practical interest correspond to the high temperature limit. For H_2, two nuclear spin configurations are possible: (1) parallel nuclear spins, which is termed ortho-hydrogen, and (2) opposed nuclear spins, which is para-hydrogen. The strong nuclear spin effect for hydrogen at low temperatures causes the thermodynamic properties of ortho-hydrogen to be distinctly different from those of para-hydrogen.

Using the general relation above for q_{rot}, the rotational contributions to U, $\hat{c}_V$, and S are determined to be

$$U_{\text{rot}} = Nk_{\text{B}}T \left[1 - \frac{1}{3} \left(\frac{\theta_{\text{rot}}}{T} \right) - \frac{1}{45} \left(\frac{\theta_{\text{rot}}}{T} \right)^2 + \cdots \right], \tag{5.76}$$

$$\hat{c}_{V,\text{rot}} = N_{\text{A}}k_{\text{B}} \left[1 + \frac{1}{45} \left(\frac{\theta_{\text{rot}}}{T} \right)^2 + \cdots \right], \tag{5.77}$$

$$S_{\text{rot}} = Nk_{\text{B}} \left[1 - \ln \left(\frac{\sigma_{\text{s}}\theta_{\text{rot}}}{T} \right) + \cdots \right]. \tag{5.78}$$

Before putting together the pieces of the overall partition function, it is useful to reexamine the electronic contribution. We argued previously that the electronic partition function for a diatomic molecule should essentially be of the same form as a monatomic one,

$$q_e = g_{e,0} + g_{e,1}e^{-\Delta\varepsilon_{e,10}/k_BT} + \cdots. \tag{5.79}$$

In treating diatomic molecules it is common, however, to take the zero of the electronic energy to be the separated, electronically unexcited atoms at rest. When the atoms are brought together to the bonded configuration, the electronic energy decreases (following the same potential energy curve that governs the vibrational mode). If the depth of the electronic ground state potential well is D_e then the energy of the electronic ground state for the molecule is $-D_e$.

Note that D_e is the energy required to break the molecule into two atoms (i.e., the energy of dissociation). We also define the quantity D_0 as the energy difference between the lowest vibrational state and the dissociated molecule:

$$D_0 = D_e - \frac{1}{2}h\nu. \tag{5.80}$$

As indicated in Figure 5.5, if the electronic state is altered to the second level, the potential well in which the nuclei vibrate may, in general, also change. Based on the above-defined zero point for the electronic state, the electronic partition function is given by

$$q_e = g_{e,0}e^{D_e/k_BT} + g_{e,1}e^{-\varepsilon_{e,1}/k_BT} + \cdots. \tag{5.81}$$

Having considered each contribution to the total partition function q, we can now assemble the parts to obtain the harmonic-oscillator–rigid-rotor approximation for q:

$$q = \left(\frac{2\pi(m_1 + m_2)k_BT}{h^2}\right)^{3/2} V \left(\frac{8\pi^2 I k_BT}{\sigma_s h^2}\right) e^{-h\nu/2k_BT} \frac{g_{e,0}e^{D_e/k_BT}}{1 - e^{-h\nu/k_BT}}. \tag{5.82}$$

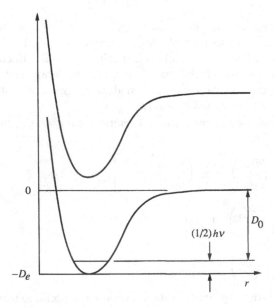

Figure 5.5   Definition of the electronic ground state.

This relation applies provided that $\theta_{\text{rot}} \ll T$, only the ground electronic state is important, and the energy zero points are defined as described above. Using this relation for q, we can obtain thermodynamic properties for the ideal diatomic gas:

$$U = Nk_BT\left[\frac{5}{2} + \frac{h\nu}{2k_BT} + \frac{h\nu/k_BT}{e^{h\nu/k_BT} - 1} - \frac{D_e}{k_BT}\right], \tag{5.83}$$

$$\hat{c}_V = N_Ak_B\left[\frac{5}{2} + \left(\frac{h\nu}{k_BT}\right)^2 \frac{e^{h\nu/k_BT}}{\left(e^{h\nu/k_BT} - 1\right)^2}\right], \tag{5.84}$$

$$S = Nk_B\left[\ln\left\{\left(\frac{2\pi(m_1 + m_2)k_BT}{h^2}\right)^{3/2} \frac{V e^{5/2}}{N}\right\} + \ln\left\{\left(\frac{8\pi^2 I k_BT}{\sigma_s h^2}\right)\right\}\right.$$
$$\left. + \frac{h\nu/k_BT}{(e^{h\nu/k_BT} - 1)} - \ln\{1 - e^{-h\nu/k_BT}\} + \ln g_{e0}\right], \tag{5.85}$$

$$PV = Nk_BT. \tag{5.86}$$

Although beyond the scope of the present discussions, this treatment can be extended to account for mode coupling, centrifugal distortion, and anharmonic effects. In spite of its simplicity, the relations obtained from the harmonic-oscillator–rigid-rotor model generally agree well with measured properties for most diatomic gases. The additional effects mentioned above are generally important only at very high temperatures.

Example 5.2 A piston and cylinder device contains one kmol of N_2 gas initially at $T_1 = 300\,K$ with initial volume $V_1 = 0.002\,m^3$. The gas is compressed reversibly and adiabatically to a volume of $1.0 \times 10^{-5}\,m^3$. Determine the final temperature T_2 of the gas.

Solution The reversible and adiabatic compression is a constant entropy process. Setting $S_2 = S_1$ and using Eq. (5.85) to evaluate the entropy values, we obtain

$$Nk_B\left[\ln\left\{\left(\frac{2\pi(m_1 + m_2)k_BT_1}{h^2}\right)^{3/2} \frac{V_1 e^{5/2}}{N}\right\} + \ln\left\{\left(\frac{8\pi^2 I k_BT_1}{\sigma_s h^2}\right)\right\}\right.$$
$$\left. + \frac{h\nu/k_BT_1}{(e^{h\nu/k_BT_1} - 1)} - \ln\{1 - e^{-h\nu/k_BT_1}\} + \ln g_{e0}\right]$$
$$= Nk_B\left[\ln\left\{\left(\frac{2\pi(m_1 + m_2)k_BT_2}{h^2}\right)^{3/2} \frac{V_2 e^{5/2}}{N}\right\} + \ln\left\{\left(\frac{8\pi^2 I k_BT_2}{\sigma_s h^2}\right)\right\}\right.$$
$$\left. + \frac{h\nu/k_BT_2}{(e^{h\nu/k_BT_2} - 1)} - \ln\{1 - e^{-h\nu/k_BT_2}\} + \ln g_{e0}\right].$$

Substituting for T_1, V_1, and V_2 and simplifying, we can reduce this relation to

$$\frac{5}{2}\ln T_2 + \frac{\theta_{\text{vib}}/T_2}{e^{\theta_{\text{vib}}/T_2} - 1} - \ln\{1 - e^{-\theta_{\text{vib}}/T_2}\} = 20.66,$$

where θ_{vib} is the vibrational temperature. $\theta_{vib} = h\nu/k_B = 3{,}392$ K for nitrogen. Iteratively solving this equation yields

$$T_2 = 2750 \text{ K}.$$

5.4 Polyatomic Gases

Treatment of polyatomic molecules makes use of the same analytical tools used to analyze diatomic molecules. Where advantageous and appropriate, we may use the classical limit results to evaluate contributions to the partition function. As in the case of diatomic molecules, the Schrödinger equation that describes the motion of the nuclei in the potential field of the molecule can be transformed into center-of-mass and relative coordinates. This allows us to break down the Hamiltonian into a component associated with translation of the center of mass and a component associated with other internal modes,

$$\hat{H} = \hat{H}_{tr} + \hat{H}_{int}, \tag{5.87}$$

from which it follows that

$$\varepsilon = \varepsilon_{tr} + \varepsilon_{int}, \tag{5.88}$$

$$q = q_{tr}q_{int}, \tag{5.89}$$

where

$$q_{tr} = \left(\frac{2\pi M k_B T}{h^2}\right)^{3/2} V \tag{5.90}$$

and M is the total mass of the molecule. We again expect that the density of translational energy states alone is sufficiently large that the dilute occupancy approximation is valid, and hence

$$Q = \frac{q_{tr}^N q_{int}^N}{N!}. \tag{5.91}$$

We further expect that

$$q_{int} = q_{rot}q_{vib}q_e q_{nucl}. \tag{5.92}$$

Each of the n atoms in the molecule requires three coordinates to locate. The total required for the molecule is therefore $3n$ total coordinates, which, in the modified treatment, fall into three groups:

Number of Coordinates	To Specify
3	Center-of-mass location
2 or 3 (2–linear, 3–nonlinear)	Molecule orientation
$3n - 5$ or $3n - 6$	Relative locations of n nuclei

We will again make use of the rigid-rotor–harmonic-oscillator approximation to allow us to separately evaluate q_{rot} and q_{vib}. We choose, as the zero of energy, all n atoms completely separated in their ground electronic states. It follows that the energy of the ground electronic state for the molecule is $-D_e$ and the electronic partition function is

$$q_e = g_{e0}e^{D_e/k_BT} + \cdots. \tag{5.93}$$

The two or three coordinates specifying orientation of the molecular represent rotational degrees of freedom. The $3n - 5$ or $3n - 6$ internal coordinates represent vibrational degrees of freedom. By selectively defining the internal coordinates, the oscillatory behavior can be represented as a combination of independent harmonic oscillators. Internal coordinates so defined are termed *normal coordinates*. The Hamiltonian can then be written as the sum of contributions for each oscillator. The total energy is then given by

$$\varepsilon_{vib} = \sum_{j=1}^{\alpha} \left(n_j + \frac{1}{2} \right) h\nu_j, \tag{5.94}$$

where

$$\nu_j = \frac{1}{2\pi} \left(\frac{k_j}{m_{r,j}} \right)^{1/2} \tag{5.95}$$

and $\alpha = 3n - 5$ for a linear molecules and $\alpha = 3n - 6$ for a nonlinear molecule. The α fundamental frequencies ν_j can be obtained from a normal coordinates analysis of the molecule. In practice they are often determined spectroscopically. The oscillatory modes for a linear triatomic molecule such as CO_2 are shown in Figure 5.6.

For a nonlinear triatomic molecule such as water, there are three modes. The modes for the water molecule are shown in Figure 5.7. Each normal mode of vibration makes an independent contribution to q_{vib}. Treating each as a harmonic oscillator yields

$$q_{vib} = \prod_{j=1}^{\alpha} \frac{e^{-\theta_{vib,j}/2T}}{1 - e^{-\theta_{vib,j}/T}}, \tag{5.96}$$

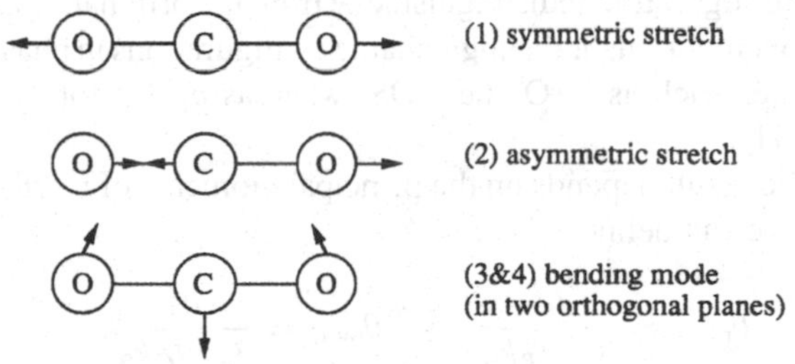

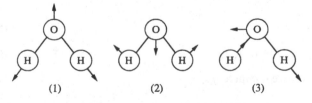

(1) symmetric stretch

(2) asymmetric stretch

(3&4) bending mode
(in two orthogonal planes)

Figure 5.6 Vibrational modes for CO_2.

(1) (2) (3)

Figure 5.7 Vibrational modes for water molecules.

where

$$\theta_{\text{vib},j} = h\nu_j / k_B. \tag{5.97}$$

The rotational properties of a polyatomic molecule depend on the general shape of the molecule. If the molecule is linear (e.g., CO_2, C_2H_2) the problem is the same as for a diatomic molecule. The energy levels are given by

$$\left.\begin{array}{l} \varepsilon_j = \frac{\hbar^2 j(j+1)}{2I} \\[2mm] g_j = 2j + 1 \end{array}\right\} \quad \text{for } j = 0, 1, 2, \ldots, \tag{5.55}$$

where for a polyatomic molecule

$$I = \sum_{j=1}^{n} m_j s_j^2, \tag{5.98}$$

with

$s_j = $ the distance of the jth nucleus from the center of mass of the molecule,

$m_j = $ the mass of the jth atom in the molecule.

In the high temperature limit, the rotational partition function for a linear polyatomic molecule becomes

$$q_{\text{rot}} = \frac{8\pi^2 I k_B T}{\sigma_s h^2} = \frac{T}{\sigma_s \theta_{\text{rot}}}, \tag{5.99}$$

where

$$\theta_{\text{rot}} = \frac{h^2}{8\pi^2 I k_B}. \tag{5.100}$$

In Eq. (5.99), σ_s is a symmetry number, which is equal to the number of ways that the molecule can be rotated into a configuration indistinguishable from the original. Classically, σ_s is a factor included to avoid overcounting indistinguishable configurations in phase space; $\sigma_s = 1$ for asymmetric molecules such as N_2O and COS, whereas $\sigma_s = 2$ for symmetric molecules such as CO_2 and C_2H_2.

For a nonlinear molecule, the result depends on the principal moments of inertia I_A, I_B, and I_C. With respect to these, we can define

$$\theta_{\text{rot},A} = \frac{h^2}{8\pi^2 I_A k_B}, \qquad \theta_{\text{rot},B} = \frac{h^2}{8\pi^2 I_B k_B}, \qquad \theta_{\text{rot},C} = \frac{h^2}{8\pi^2 I_C k_B}. \tag{5.101}$$

The simplest case is the spherical top, corresponding to $I_A = I_B = I_C$. For this case, the energy levels and degeneracy are

$$\varepsilon_j = \frac{\hbar^2 j(j+1)}{2I}, \tag{5.102}$$

$$g_j = (2j + 1)^2, \qquad j = 0, 1, 2, \ldots. \tag{5.103}$$

In the high temperature limit, we compute q_{rot} as

$$q_{\text{rot}} = \frac{1}{\sigma_s} \int_0^\infty (2j + 1)^2 e^{-j(j+1)\hbar^2/2I k_B T} \, dj, \tag{5.104}$$

where σ_s is the symmetry number for a polyatomic molecule (the number of ways the molecule can be rotated "into itself"). $\sigma_s = 2$ for H_2O, $\sigma_s = 3$ for NH_3, etc. In the high temperature limit, the integrand in Eq. (5.104) is small except at j values that are large compared to 1. The relation for q_{rot} then becomes

$$q_{rot} = \frac{1}{\sigma_s} \int_0^\infty 4j^2 e^{-j^2 \hbar^2 / 2I k_B T} \, dj = \frac{\pi^{1/2}}{\sigma_s} \left(\frac{8\pi^2 I k_B T}{h^2} \right)^{3/2}. \tag{5.105}$$

With a bit more analytical effort, it can be shown that in the high temperature limit for the general case of $I_A \neq I_B \neq I_C$

$$q_{rot} = \frac{\pi^{1/2}}{\sigma_s} \left(\frac{T^3}{\theta_{rot,A} \theta_{rot,B} \theta_{rot,C}} \right)^{1/2}. \tag{5.106}$$

Assuming that the nuclear storage mode is always in the nondegenerate ground state, $q_{nucl} = 1$ and complete relations for q can be assembled as follows:

for a linear polyatomic molecule:

$$q = \left(\frac{2\pi M k_B T}{h^2} \right)^{3/2} V \left(\frac{T}{\sigma_s \theta_{rot}} \right) \left(\prod_{j-1}^{3n-5} \frac{e^{-\theta_{vib,j}/2T}}{1 - e^{-\theta_{vib,j}/T}} \right) g_{e,0} e^{D_e/k_B T}, \tag{5.107}$$

for a nonlinear polyatomic molecule:

$$q = \left(\frac{2\pi M k_B T}{h^2} \right)^{3/2} V \frac{\pi^{1/2}}{\sigma_s} \left(\frac{T^3}{\theta_{rot,A} \theta_{rot,B} \theta_{rot,C}} \right)^{1/2} \left(\prod_{j=1}^{3n-6} \frac{e^{-\theta_{vib,j}/2T}}{1 - e^{-\theta_{vib,j}/T}} \right) g_{e,0} e^{D_e/k_B T}, \tag{5.108}$$

where the vibrational and rotational temperatures are given by Eqs. (5.97) and (5.100) or (5.101).

Using the property relations for the canonical ensemble, the following relations can be derived from the above relations for q:

for linear polyatomic molecules:

$$\frac{U}{Nk_B T} = \frac{5}{2} + \sum_{j=1}^{3n-5} \left[\frac{\theta_{vib,j}}{2T} + \frac{\theta_{vib,j}/T}{e^{\theta_{vib,j}/T} - 1} \right] - \frac{D_e}{k_B T}, \tag{5.109}$$

$$\frac{\hat{c}_V}{N_A k_B} = \frac{5}{2} + \sum_{j=1}^{3n-5} \left[\left(\frac{\theta_{vib,j}}{T} \right)^2 \frac{e^{\theta_{vib,j}/T}}{(e^{\theta_{vib,j}/T} - 1)^2} \right], \tag{5.110}$$

$$\frac{S}{Nk_B} = \ln \left\{ \left(\frac{2\pi M k_B T}{h^2} \right)^{3/2} \frac{V e^{5/2}}{N} \right\} + \ln \left\{ \frac{Te}{\sigma_s \theta_{rot}} \right\}$$

$$+ \sum_{j=1}^{3n-5} \left[\frac{\theta_{vib,j}/T}{e^{\theta_{vib,j}/T} - 1} - \ln\{1 - e^{-\theta_{vib,j}/T}\} \right] + \ln g_{e,0}, \tag{5.111}$$

$$PV = Nk_B T; \tag{5.112}$$

Table 5.2 *Constants for Polyatomic Molecules*

Molecule	σ_s	θ_{rot} (K)	θ_{vib} (K)	D_e (J/molecule)	$g_{e,0}$
CCl_4	12	0.0823 (3)	310 (2), 450 (3), 660, 1120 (3)	2.189×10^{-18}	1
CH_4	12	7.54 (3)	1870 (3), 2180 (2), 4170, 4320 (3)	2.914×10^{-18}	1
CO_2	2	0.561	954 (2), 1890, 3360	2.702×10^{-18}	1
H_2O	2	13.4, 20.9, 40.1	2290, 5160, 5360	1.613×10^{-18}	1
NH_3	3	8.92, 13.6, 13.6	1360, 4800, 4880 (2), 2330 (2)	2.067×10^{-18}	1
NO_2	2	0.603	850 (2), 1840, 3200	1.586×10^{-18}	2
SO_2	2	0.422, 0.495, 2.92	750, 1660, 1960	1.797×10^{-18}	1

for nonlinear polyatomic molecules:

$$\frac{U}{N k_B T} = 3 + \sum_{j=1}^{3n-6} \left[\frac{\theta_{vib,j}}{2T} + \frac{\theta_{vib,j}/T}{e^{\theta_{vib,j}/T} - 1} \right] - \frac{D_e}{k_B T}, \tag{5.113}$$

$$\frac{\hat{c}_V}{N_A k_B} = 3 + \sum_{j=1}^{3n-6} \left[\left(\frac{\theta_{vib,j}}{T} \right)^2 \frac{e^{\theta_{vib,j}/T}}{(e^{\theta_{vib,j}/T} - 1)^2} \right], \tag{5.114}$$

$$\frac{S}{N k_B} = \ln \left\{ \left(\frac{2\pi M k_B T}{h^2} \right)^{3/2} \frac{V e^{5/2}}{N} \right\} + \ln \left\{ \frac{\pi^{1/2} e^{3/2}}{\sigma_s} \left(\frac{T^3}{\theta_{rot,A} \theta_{rot,B} \theta_{rot,C}} \right)^{1/2} \right\}$$

$$+ \sum_{j=1}^{3n-6} \left[\frac{\theta_{vib,j}/T}{e^{\theta_{vib,j}/T} - 1} - \ln\{1 - e^{-\theta_{vib,j}/T}\} \right] + \ln g_{e,0}, \tag{5.115}$$

$$PV = N k_B T. \tag{5.116}$$

Values of rotational and vibrational temperatures for various molecules are listed in Table 5.2. Note that when two or more modes have the same θ_{rot} or θ_{vib} value, the multiplicity is indicated in parentheses.

5.5   Equipartition of Energy

We now will examine how energy is distributed among modes of molecular energy storage for a particular common system type. To do so, we will begin by considering each molecule as a system in a canonical ensemble. This is consistent with the fact that in a real system each molecule is confined to a specified volume and collisions with other molecules essentially keep it in equilibrium with a system of fixed temperature. Suppose the molecular energy microstate is specified by a relation of the form

$$\varepsilon_{i',j'} = \gamma \xi_{j'}^2 + \sum_{l=0}^{\alpha} \varepsilon_{i',l}. \tag{5.117}$$

The first term on the right side of Eq. (5.117) is the contribution associated with energy storage that depends quadratically on a spatial coordinate or generalized velocity ξ. The second term on the right side of Eq. (5.117) is the sum total effect of all other modes of energy storage in the molecule. The quadratic ξ storage mode is assumed to be independent of all other modes, from which it follows that

$$q = q_\xi q_\alpha, \tag{5.118}$$

where

$$q_\xi = \sum_{j'=0}^{\infty} e^{-\gamma \xi_{j'}^2 / k_B T}, \tag{5.119}$$

$$q_\alpha = \sum_{i'=0}^{\infty} \exp\left(-\sum_{l=0}^{\alpha} \varepsilon_{i',l} / k_B T\right). \tag{5.120}$$

From the definition of the partition function, it follows that the probability of a particular microstate i', j' is given by

$$\tilde{P}_{i',j'} = \frac{N_{i',j'}}{N} = \frac{\exp\left(-\gamma \xi_{j'}^2 / k_B T - \sum_{l=0}^{\alpha} \varepsilon_{i',l} / k_B T\right)}{q}. \tag{5.121}$$

To get the probability of a specific $\xi_{j'}$ value, irrespective of i', we sum $\tilde{P}_{i',j'}$ over all i':

$$\tilde{P}_{j'} = \sum_{i'=0}^{\infty} \tilde{P}_{i',j'} = \sum_{i'=0}^{\infty} \frac{\exp\left(-\gamma \xi_{j'}^2 / k_B T - \sum_{l=0}^{\alpha} \varepsilon_{i',l} / k_B T\right)}{q}. \tag{5.122}$$

Using Eqs. (5.118) and (5.120) and interpreting $\tilde{P}_{j'}$ as $N_{j'}/N$, we can rewrite Eq. (5.122) as

$$\frac{N_{j'}}{N} = \frac{e^{-\gamma \xi_{j'}^2 / k_B T}}{q_\xi}. \tag{5.123}$$

Passing to the continuum limit for energy levels close together, $\xi_{j'}$ passes to ξ and $N_{j'}/N$ becomes $(1/N)dN/d\xi$. In this limit, the equivalent form of Eq. (5.123) becomes

$$\frac{dN}{d\xi} = \frac{N e^{-\gamma \xi^2 / k_B T}}{q_\xi}, \tag{5.124}$$

where the relation for q_ξ can be written in integral form,

$$q_\xi = \int_0^\infty e^{-\gamma \xi^2 / k_B T} \, d\xi. \tag{5.125}$$

In terms of the continuum formulation, the mean energy of a molecule associated with the generalized coordinate or velocity ξ is given by

$$\langle \varepsilon_\xi \rangle = \gamma \langle \xi^2 \rangle = \frac{\int_0^\infty \gamma \xi^2 \, dN}{\int_0^\infty dN}. \tag{5.126}$$

Substituting Eq. (5.124) yields

$$\langle \varepsilon_\xi \rangle = \frac{\int_0^\infty \frac{N \gamma \xi^2}{q_\xi} e^{-\gamma \xi^2 / k_B T} \, d\xi}{\int_0^\infty \frac{N}{q_\xi} e^{-\gamma \xi^2 / k_B T} \, d\xi}, \tag{5.127}$$

which, upon evaluating the integrals, becomes

$$\langle \varepsilon_\xi \rangle = \frac{k_B T}{2}. \tag{5.128}$$

We have thus demonstrated that each energy mode of a particle that is an independent quadratic function of one of the coordinates or velocity components of the particle contributes an average energy of $k_B T/2$ per particle to the system. This result is referred to as the *theorem of equipartition of energy* because it implies that system energy distributes itself equally, on the average, over quadratic energy storage modes.

Note that validity of the equipartition theorem requires that (1) the Boltzmann distribution is valid (the system is in equilibrium), (2) energy storage for each particle is independent of that for other particles, (3) the spacing between energy levels must be close enough that the continuum model is a reasonable representation, (4) the total energy of each particle can be split into energies for storage modes that are independent, and (5) the energy storage modes considered must be quadratic functions of position or velocity. If some particle energy storage modes are independent and quadratic while others are not, the equipartition theorem applies to the quadratic modes, if the other requirements listed above are met.

It follows from the above discussion that if particles in a system have B independent quadratic energy storage modes, the molar internal energy $\hat{u}$ is given by

$$\hat{u} = N_A B k_B T/2 \tag{5.129}$$

and the molar specific heat $\hat{c}_V$ is

$$\hat{c}_V = \left(\frac{\partial \hat{u}}{\partial T} \right)_{V,N} = N_A B k_B/2 = BR/2, \tag{5.130}$$

where $R = N_A k_B$ is the universal gas constant.

Example 5.3   Assuming that the proper requirements are met, use the equipartition theorem to determine $\hat{c}_V$ for a monatomic ideal gas and a diatomic ideal gas at moderate temperatures.

Solution   At moderate temperatures, molecules of a monatomic ideal gas store energy by translation only. The energy is therefore given by

$$\varepsilon = \frac{1}{2}mu^2 + \frac{1}{2}mv^2 + \frac{1}{2}mw^2.$$

The energy storage thus consists of three independent modes, each of which is a quadratic function of the velocity component u, v, or w. Thus $B = 3$ in Eq. (5.130) and $\hat{c}_V = 3R/2$. Since $\hat{c}_P = \hat{c}_V + R$ for an ideal gas, it follows that $\hat{c}_P = 5R/2$ and $\hat{c}_P/\hat{c}_V = 5/3$. These results agree well with measured values for monatomic gases such as argon.

Diatomic molecules have the "dumbbell" configuration shown schematically in Figure 5.8. Each molecule has three independent quadratic translational modes, just as

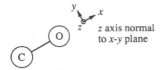

Figure 5.8

single atom species do. In addition, at moderate temperature, each molecule can store significant energy in rotation about the y and z axes. None is stored in rotation about the x axis because the atomic masses are located right on the axis. The rotational energies associated with rotation about the two orthogonal axes are quadratic functions of angular velocity and are independent. Hence there are a total of five independent quadratic energy storage modes, and the specific heat at moderate temperatures is $\hat{c}_V = 5R/2$. At high temperatures, diatomic molecules also store energy in vibrational motion of the bond between the atoms. As a first approximation, if the bond is considered elastic, the structure stores energy like a one-dimensional harmonic oscillator. There are actually two independent quadratic storage modes associated with this motion. One is the potential energy stored by compressing or stretching the bond. This mode is quadratic in the deviation of the interatomic spacing from its equilibrium value. The other mode is the kinetic energy associated with the relative motion of the atoms. This mode is quadratic in the relative velocity of the atoms with respect to each other. When energy storage in vibration is fully active, the molecule has seven independent quadratic energy storage modes: three in translation, two in rotation, and two in vibration. Equipartition then dictates that $\hat{c}_V = 7R/2$.

Example 5.4 Determine the final temperature of the gas for the compression process described in Example 5.2 assuming that the modes of energy storage obey the equipartition theorem.

Solution Classical theory for an ideal gas with constant specific heat predicts that for a constant entropy process

$$T_2 = T_1 \left(\frac{V_1}{V_2}\right)^{\gamma-1}, \qquad \gamma = \frac{\hat{c}_P}{\hat{c}_V} = \frac{\hat{c}_V + R}{\hat{c}_V}.$$

If we assume that three translational modes, two rotational modes, and two vibrational modes of energy storage are all fully quadratic and conform to the requirements of equipartition, $\hat{c}_V = 7R/2$, $\gamma = 1.40$, and the above relation predicts $T_2 = 2{,}498$ K for the values of T_1, V_1, and V_2 specified in Example 5.2. This is about 250 K lower than the prediction of fully statistical theory. Alternatively we could assume that the vibrational modes do not participate to a significant degree with the result that there are only five quadratic modes: three translational modes and two rotational modes. Then, $\hat{c}_V = 5R/2$, $\gamma = 1.29$, and the above relation predicts $T_2 = 1{,}395$. This prediction is obviously much lower than the prediction of the full statistical theory. This is clearly a case where the full statistical theory provides a better prediction than that provided by a constant specific heat model.

5.6 Ideal Gas Mixtures

To this point in this chapter we have focused on the properties of a system containing a single molecular species. However, many systems of scientific and engineering interest contain mixtures of ideal gases. Fortunately it is a relatively simple matter to extend the analysis framework developed thus far to systems containing mixtures. In Chapter 4 we showed that for a binary mixture the internal energy and entropy are related to the canonical partition function as

$$U = \frac{k_B T^2}{Q} \left(\frac{\partial Q}{\partial T} \right)_{V,N_a,N_b}, \tag{4.43}$$

$$\frac{S}{k_B} = \frac{U}{k_B T} + \ln Q. \tag{4.41}$$

These relations can be rearranged to obtain the forms

$$U = k_B T^2 \left(\frac{\partial (\ln Q)}{\partial T} \right)_{V,N_a,N_b}, \tag{5.131}$$

$$S = k_B T \left(\frac{\partial (\ln Q)}{\partial T} \right)_{V,N_a,N_b} + k_B \ln Q. \tag{5.132}$$

We also showed in Chapter 4 that for a binary mixture

$$Q = \left(\frac{q_a^{N_a}}{N_a!} \right) \left(\frac{q_b^{N_b}}{N_b!} \right). \tag{4.173}$$

Substituting the right side of Eq. (4.173) for Q in Eqs. (5.131) and (5.132) and simplifying by using Stirling's approximation ($\ln N! \cong N \ln N + N$) yields the following relations for U and S:

$$U = k_B T^2 N_a \left(\frac{\partial (\ln q_a)}{\partial T} \right)_{V,N_a,N_b} + k_B T^2 N_b \left(\frac{\partial (\ln q_b)}{\partial T} \right)_{V,N_a,N_b}, \tag{5.133}$$

$$S = k_B T N_a \left(\frac{\partial (\ln q_a)}{\partial T} \right)_{V,N_a,N_b} + k_B N_a \ln q_a - k_B N_a \ln N_a - k_B N_a$$

$$+ k_B T N_b \left(\frac{\partial (\ln q_b)}{\partial T} \right)_{V,N_a,N_b} + k_B N_b \ln q_b - k_B N_b \ln N_b - k_B N_b. \tag{5.134}$$

We define $\hat{u}_a$ and $\hat{u}_b$ as the internal energy per mole of species a and b, respectively, if they occupied the full system volume V at temperature T:

$$\hat{u}_a = k_B T^2 N_A \left(\frac{\partial [\ln q_a(V,T)]}{\partial T} \right)_V, \qquad \hat{u}_b = k_B T^2 N_A \left(\frac{\partial [\ln q_b(V,T)]}{\partial T} \right)_V. \tag{5.135}$$

With these definitions, the relation for the mixture internal energy becomes

$$U = (N_a/N_A)\hat{u}_a + (N_b/N_A)\hat{u}_b. \tag{5.136}$$

Note that this implies that the total internal energy is equal to the sum of the molar internal energies for each species, with each being weighted by the number of moles of that species in the mixture. Using the definition developed in Chapter 3, we can also derive a relation for the specific heat of the mixture:

$$\hat{c}_V = \frac{N_A}{(N_a + N_b)} \left(\frac{\partial U}{\partial T} \right)_{V,N_a,N_b}$$

$$= \frac{N_a}{(N_a + N_b)} \left(\frac{\partial \hat{u}_a}{\partial T} \right)_{V,N_a} + \frac{N_b}{(N_a + N_b)} \left(\frac{\partial \hat{u}_b}{\partial T} \right)_{V,N_b}. \tag{5.137}$$

Denoting the pure component molar specific heats as

$$\hat{c}_{V,a} = \left(\frac{\partial \hat{u}_a}{\partial T}\right)_{V,N_a}, \qquad \hat{c}_{V,b} = \left(\frac{\partial \hat{u}_b}{\partial T}\right)_{V,N_b} \tag{5.138}$$

we can write the relation for the mixture specific heat in the form

$$\hat{c}_V = \frac{N_a}{(N_a + N_b)}\hat{c}_{V,a} + \frac{N_b}{(N_a + N_b)}\hat{c}_{V,b}. \tag{5.139}$$

Thus, the molar specific heat of the mixture is sum of the molar specific heats for the constituents, with each being weighted by the mole fraction of that species in the mixture.

We also define an entropy per mole for each of the two species at the system volume and temperature:

$$\hat{s}_a = k_B T N_A \left(\frac{\partial [\ln q_a(V,T)]}{\partial T}\right)_V + k_B N_A \ln q_a(V,T) - k_B N_A \ln N_A - k_B N_A, \tag{5.140}$$

$$\hat{s}_b = k_B T N_A \left(\frac{\partial [\ln q_b(V,T)]}{\partial T}\right)_V + k_B N_A \ln q_b(V,T) - k_B N_A \ln N_A - k_B N_A. \tag{5.141}$$

In terms of these molar entropies, the entropy of the binary mixture can be written as

$$S = (N_a/N_A)[\hat{s}_a - R\ln(N_a/N_A)] + (N_b/N_A)[\hat{s}_b - R\ln(N_b/N_A)]. \tag{5.142}$$

The equivalence of Eqs. (5.134) and (5.142) can be verified by substitution of Eqs. (5.140) and (5.141) into Eq. (5.142). The usefulness of Eqs. (5.136), (5.139), and (5.142) is that they make it possible to use pure gas relations to evaluate the properties of mixtures. Their forms also clearly indicate how the relations must be extended for mixtures of more than two ideal gases.

Example 5.5 In a prototype combustion engine, at the end of the combustion process, the working fluid is a mixture of H_2O, CO_2, and N_2 gas at a pressure of 600 kPa and a temperature of 1,800 K. The mole fractions of H_2O, CO_2, and N_2 in the mixture are 0.190, 0.095, and 0.715, respectively. In the engine, this mixture expands until its pressure is 101 kPa. Modeling the expansion as reversible and adiabatic, determine the temperature of the gas at the end of the expansion process.

Solution The entropy of the mixture can be represented as

$$S = S_{H_2O}(N_{H_2O}, T, V) + S_{CO_2}(N_{CO_2}, T, V) + S_{N_2}(N_{N_2}, T, V),$$

where S_{H_2O} is evaluated using Eq. (5.115), S_{CO_2} is evaluated using (5.111), and S_{N_2} is evaluated using Eq. (5.85). The reversible and adiabatic expansion is a constant entropy process. Hence, for the mixture $S_2 - S_1 = 0$, which can be expressed as

$$\Delta S_{H_2O} + \Delta S_{CO_2} + \Delta S_{N_2} = 0,$$

where

$$\Delta S_{H_2O} = S_{H_2O}(N_{H_2O}, T_2, V_2) - S_{H_2O}(N_{H_2O}, T_1, V_1),$$

$$\Delta S_{CO_2} = S_{CO_2}(N_{CO_2}, T_2, V_2) - S_{CO_2}(N_{CO_2}, T_1, V_1),$$

$$\Delta S_{N_2} = S_{N_2}(N_{N_2}, T_2, V_2) - S_{N_2}(N_{N_2}, T_1, V_1).$$

Using Eqs. (5.85), (5.111), and (5.115) and the ideal gas law for the mixture ($PV = Nk_BT$) together with appropriate values of M, θ_{rot}, θ_{vib}, and σ_s for the three gases, we obtain the following relations:

$$
\Delta S_{H_2O} = N_{H_2O}k_B \left[\frac{3}{2}\ln(T_2/T_1) + \ln(T_2P_1/T_1P_2) + \frac{3}{2}\ln(T_2/T_1) + \frac{2290/T_2}{e^{2290/T_2} - 1} \right.
$$

$$
- \frac{2290/T_1}{e^{2290/T_1} - 1} + \frac{5160/T_2}{e^{5160/T_2} - 1} - \frac{5160/T_1}{e^{5160/T_1} - 1} + \frac{5360/T_2}{e^{5360/T_2} - 1}
$$

$$
- \frac{5360/T_1}{e^{5360/T_1} - 1} - \ln\left\{ \left(\frac{1 - e^{-2290/T_2}}{1 - e^{-2290/T_1}} \right)\left(\frac{1 - e^{-5160/T_2}}{1 - e^{-5160/T_1}} \right) \right.
$$

$$
\left. \left. \times \left(\frac{1 - e^{-5360/T_2}}{1 - e^{-5360/T_1}} \right) \right\} \right],
$$

$$
\Delta S_{CO_2} = N_{CO_2}k_B \left[\frac{3}{2}\ln(T_2/T_1) + \ln(T_2P_1/T_1P_2) + \ln(T_2/T_1) \right.
$$

$$
+ 2\left(\frac{954/T_2}{e^{954/T_2} - 1} - \frac{954/T_1}{e^{954/T_1} - 1} \right) + \frac{1890/T_2}{e^{1890/T_2} - 1} - \frac{1890/T_1}{e^{1890/T_1} - 1}
$$

$$
+ \frac{3360/T_2}{e^{3360/T_2} - 1} - \frac{3360/T_1}{e^{3360/T_1} - 1} - \ln\left\{ \left(\frac{1 - e^{-954/T_2}}{1 - e^{-954/T_1}} \right)^2 \left(\frac{1 - e^{-1890/T_2}}{1 - e^{-1890/T_1}} \right) \right.
$$

$$
\left. \left. \times \left(\frac{1 - e^{-3360/T_2}}{1 - e^{-3360/T_1}} \right) \right\} \right],
$$

$$
\Delta S_{N_2} = N_{N_2}k_B \left[\frac{3}{2}\ln(T_2/T_1) + \ln(T_2P_1/T_1P_2) + \ln(T_2/T_1) \right.
$$

$$
\left. + \frac{3374/T_2}{e^{3374/T_2} - 1} - \frac{3374/T_1}{e^{3374/T_1} - 1} - \ln\left\{ \frac{1 - e^{-3374/T_2}}{1 - e^{-3374/T_1}} \right\} \right].
$$

Per mole of the gas mixture, the number of molecules of each species can be calculated from the specified mole fractions as $N_{H_2O} = 0.190N_A$, $N_{CO_2} = 0.095N_A$, and $N_{N_2} = 0.715N_A$. Using these relations and the specified P_1, T_1, and P_2 values, we sum the right-hand sides of the three species ΔS relations and set the result equal to zero. The resulting equation can be iteratively solved for T_2. Doing so yields

$$
T_2 = 1230 \text{ K}.
$$

5.7   Chemical Equilibrium in Gas Mixtures

We showed in Chapter 4 that for a binary mixture

$$
Q = \left(\frac{q_a^{N_a}}{N_a!} \right)\left(\frac{q_b^{N_b}}{N_b!} \right). \tag{4.173}
$$

For a mixture of four gas species a, b, c, d, extension of the analysis in Chapter 4 yields

$$Q = \left(\frac{q_a^{N_a}}{N_a!}\right)\left(\frac{q_b^{N_b}}{N_b!}\right)\left(\frac{q_c^{N_c}}{N_c!}\right)\left(\frac{q_d^{N_d}}{N_d!}\right). \tag{5.143}$$

We will consider the following general gas phase reaction among the four species in the mixture:

$$v_a A + v_b B \Leftrightarrow v_c C + v_d D. \tag{5.144}$$

For a mixture in a vessel of fixed volume held at a specified temperature T, this reaction governs conversion of species until an equilibrium concentration of reactants and products is established. Suppose we somehow force the reaction in the forward direction by a small amount resulting in a change dN_a in species a. The reaction (5.144) requires that

$$dN_b = \frac{v_b}{v_a}dN_a, \qquad dN_c = -\frac{v_c}{v_a}dN_a, \qquad dN_d = -\frac{v_d}{v_a}dN_a. \tag{5.145}$$

We designate the ratio of the change in the amount of species a to its stoichiometric coefficient as

$$d\gamma = \frac{dN_a}{v_a}. \tag{5.146}$$

Equations (5.145) and (5.146) together require that

$$d\gamma = \frac{dN_a}{v_a} = \frac{dN_b}{v_b} = -\frac{dN_c}{v_c} = -\frac{dN_d}{v_d}. \tag{5.147}$$

Thus the parameter γ $(= N_a/v_a)$ can be interpreted as a reaction coordinate, changes in which indicate the extent to which reactants A and B have been converted to products C and D.

We showed in Chapter 3 that the Helmholtz free energy change for a multicomponent system can be written as

$$dF = -S\,dT - P\,dV + \sum_{\eta} \mu_\eta dN_\eta, \tag{5.148}$$

where the summation is taken over all species η in the system. In the four-species system of interest here, η can be any member of the set $\{a, b, c, d\}$. Because the system is held at fixed V and T, Eq. (5.148) simplifies to

$$dF = \mu_a dN_a + \mu_b dN_b + \mu_c dN_c + \mu_d dN_d. \tag{5.149}$$

Using Eq. (5.145), we can reorganize the right side of Eq. (5.149) to obtain

$$dF = (v_a \mu_a + v_b \mu_b - v_c \mu_c - v_d \mu_d)\,d\gamma. \tag{5.150}$$

For a system held at constant T and V, equilibrium corresponds to a minimum in the Helmholtz free energy with respect to all possible changes in the reaction coordinate $d\gamma$. At the minimum in F, $dF = 0$. This is true for arbitrary $d\gamma$ if and only if the collection of terms multiplying $d\gamma$ in Eq. (5.150) is identically zero:

$$v_a \mu_a + v_b \mu_b - v_c \mu_c - v_d \mu_d = 0. \tag{5.151}$$

Equation (5.151) is a necessary condition for chemical equilibrium in a system held at constant T and V. The chemical potential for each species in the gas is given by

$$\mu_\eta = -k_B T \left(\frac{\partial (\ln Q)}{\partial N_\eta} \right)_{N_{\gamma \neq \eta}, V, T}, \tag{5.152}$$

where η can be any member of the set $\{a, b, c, d\}$. Treating the system as a mixture of ideal gases, it follows from the extension of our binary mixture results that

$$Q = \left(\frac{q_a^{N_a}}{N_a!} \right) \left(\frac{q_b^{N_b}}{N_b!} \right) \left(\frac{q_c^{N_c}}{N_c!} \right) \left(\frac{q_d^{N_d}}{N_d!} \right).$$

Taking the natural log of both sides of this relation yields

$$\ln Q = N_a \ln q_a - \ln N_a! + N_b \ln q_b - \ln N_b! + N_c \ln q_c - \ln N_c! + N_d \ln q_d - \ln N_d!. \tag{5.153}$$

Using Stirling's approximation: $\ln N! \cong N \ln N + N$, this relation can be reorganized to

$$\ln Q = N_a \ln(q_a/N_a) + N_b \ln(q_b/N_b) + N_c \ln(q_c/N_c)$$
$$+ N_d \ln(q_d/N_d) + N_a + N_b + N_c + N_d. \tag{5.154}$$

Noting that $q_a = q_a(V, T)$, $q_b = q_b(V, T)$, $q_c = q_c(V, T)$, and $q_d = q_d(V, T)$, we can differentiate this relation for $\ln Q$ to obtain

$$\mu_a = -k_B T \left(\frac{\partial (\ln Q)}{\partial N_a} \right) = -k_B T \ln(q_a/N_a). \tag{5.155}$$

It similarly follows that

$$\mu_b = -k_B T \ln(q_b/N_b), \qquad \mu_c = -k_B T \ln(q_c/N_c), \qquad \mu_d = -k_B T \ln(q_d/N_d). \tag{5.156}$$

After substituting and some rearranging, we can write the requirement for chemical equilibrium (5.151) as

$$\frac{N_c^{\nu_c} N_d^{\nu_d}}{N_a^{\nu_a} N_b^{\nu_b}} = \frac{q_c^{\nu_c} q_d^{\nu_d}}{q_a^{\nu_a} q_b^{\nu_b}}. \tag{5.157}$$

Noting that the number concentration of species η is $\rho_{N,\eta} = N_\eta / V$ (in number of species η molecules per m³), we can write the above relation in the form

$$\frac{\rho_{N,c}^{\nu_c} \rho_{N,d}^{\nu_d}}{\rho_{N,a}^{\nu_a} \rho_{N,b}^{\nu_b}} = \frac{(q_c/V)^{\nu_c} (q_d/V)^{\nu_d}}{(q_a/V)^{\nu_a} (q_b/V)^{\nu_b}}. \tag{5.158}$$

The left side of Eq. (5.158) can be taken as the definition of the equilibrium constant $K_c(T)$ of the reaction

$$K_c(T) = \frac{\rho_{N,c}^{\nu_c} \rho_{N,d}^{\nu_d}}{\rho_{N,a}^{\nu_a} \rho_{N,b}^{\nu_b}} = \frac{(q_c/V)^{\nu_c} (q_d/V)^{\nu_d}}{(q_a/V)^{\nu_a} (q_b/V)^{\nu_b}}. \tag{5.159}$$

For ideal gas mixtures, it is sometimes convenient to define an equilibrium constant in terms of the partial pressures of the reactants and products at equilibrium. For the reaction considered here, this equilibrium constant is defined as

$$K_p(T) = \frac{(P_c/P_0)^{\nu_c}(P_d/P_0)^{\nu_d}}{(P_a/P_0)^{\nu_a}(P_b/P_0)^{\nu_b}},$$ (5.160)

where P_η is the equilibrium partial pressure of species η and P_0 is the standard reference pressure, which is usually taken to be 1 atmosphere or 1 bar. For each ideal gas species, the partial pressure is related to the number density of that species as

$$P_\eta = \rho_{N,\eta} k_B T.$$ (5.161)

Using Eqs. (5.161) and (5.159), it is easily shown that Eq. (5.160) implies that

$$K_p(T) = \frac{\rho_{N,c}^{\nu_c}\rho_{N,d}^{\nu_d}}{\rho_{N,a}^{\nu_a}\rho_{N,b}^{\nu_b}} \left(\frac{k_B T}{P_0}\right)^{\nu_c+\nu_d-\nu_a-\nu_b} = K_c(T)\left(\frac{k_B T}{P_0}\right)^{\nu_c+\nu_d-\nu_a-\nu_b}$$ (5.162)

and hence $K_p(T)$ can be evaluated as

$$K_p(T) = \frac{(P_c/P_0)^{\nu_c}(P_d/P_0)^{\nu_d}}{(P_a/P_0)^{\nu_a}(P_b/P_0)^{\nu_b}} = \frac{(q_c/V)^{\nu_c}(q_d/V)^{\nu_d}}{(q_a/V)^{\nu_a}(q_b/V)^{\nu_b}}\left(\frac{k_B T}{P_0}\right)^{\nu_c+\nu_d-\nu_a-\nu_b}.$$ (5.163)

For a reaction involving an arbitrary number of reactant (n_r) and product (n_p) species,

$$\nu_1 A_1 + \nu_2 A_2 + \cdots + \nu_{n_r} A_{n_r} \Leftrightarrow \nu_1' B_1 + \nu_2' B_2 + \cdots + \nu_{n_p}' A_{n_p},$$ (5.164)

the above analysis can be generalized to obtain

$$K_c(T) = \frac{\prod_{j=1}^{n_p} \rho_j^{\nu_j'}}{\prod_{i=1}^{n_r} \rho_i^{\nu_i}} = \frac{\prod_{j=1}^{n_p} (q_j/V)^{\nu_j'}}{\prod_{i=1}^{n_r} (q_i/V)^{\nu_i}},$$ (5.165)

$$K_p(T) = \frac{\prod_{j=1}^{n_p} (P_j/P_0)^{\nu_j'}}{\prod_{i=1}^{n_r} (P_i/P_0)^{\nu_i}} = \frac{\prod_{j=1}^{n_p} (q_j/V)^{\nu_j'}}{\prod_{i=1}^{n_r} (q_i/V)^{\nu_i}}\left(\frac{k_B T}{P_0}\right)^{\sum_{j=1}^{n_p} \nu_j' - \sum_{i=1}^{n_r} \nu_i}.$$ (5.166)

Equation (5.165) or (5.166) provides a theoretical link between the reaction equilibrium constant and the molecular partition functions for the species in the ideal gas mixture.

Example 5.6 Consider the reaction below involving diatomic gases in the temperature range 550 K to 1,000 K:

$$H_2 + I_2 \Leftrightarrow 2HI.$$

Derive a relation for the equilibrium constant for this reaction.

Solution The definition of the equilibrium constant (5.165) dictates that for this reaction

$$K_c(T) = \frac{(q_{HI}/V)^2}{(q_{H_2}/V)(q_{I_2}/V)} = \frac{q_{HI}^2}{q_{H_2}q_{I_2}}.$$

Substituting the molecular partition function relation for diatomic gases (5.82) with appropriate constants for the three gases involved, we obtain the following relation:

$$K_c(T) = \left(\frac{m_{HI}^2}{m_{H_2}m_{I_2}}\right)^{3/2}\left(\frac{4\theta_{rot,H_2}\theta_{rot,I_2}}{\theta_{rot,HI}^2}\right)\frac{(1-e^{-\theta_{vib,H_2}/T})(1-e^{-\theta_{vib,I_2}/T})}{1-e^{-\theta_{vib,HI}^2/T}}$$
$$\times e^{\{(2D_{0,HI}-D_{0,H_2}-D_{0,I_2})/k_B T\}}.$$

Note that this relation specifies the temperature dependence of the equilibrium constant for this reaction. We have implicitly assumed that translational, rotational, vibrational, and electronic storage modes are all important.

Additional insight into the application of statistical thermodynamics theory to ideal gases is provided by the treatment in References [1] and [2].

Exercises

5.1 The O—H bond length in water is 0.96 Å and the H—O—H bond angle is 104 degrees. Lay out a drawing of the water molecule graphically and determine the location of its center of mass and its three moments of inertia. From these results, verify the values of θ_{rot} given for water in Table 5.2.

5.2 What is the most probable value of the rotational quantum number for N_2 gas molecules at 300 K? What is the most probable vibration quantum number for this same situation?

5.3 Determine the value of $\hat{c}_V$ for NH_3 at 300 K. You may neglect electronic and nuclear storage effects. Does the value agree with that predicted by the equipartition theorem?

5.4 Use the equipartition theorem to determine the value of $\hat{c}_V$ for a mixture containing half water and half ammonia vapor by mole at 300 K. You may neglect electronic and nuclear storage effects.

5.5 Determine the value of $\hat{c}_V$ for a mixture of water vapor and nitrogen gas at 2,200 K. In the mixture the mole fraction of water vapor is 0.3. Electronic and nuclear storage effects may be neglected.

5.6 A somewhat more accurate expression for the vibrational energy of a diatomic molecule is

$$\varepsilon_n = \left(n + \frac{1}{2}\right)h\nu - x_e\left(n + \frac{1}{2}\right)^2 h\nu,$$

where x_e is called the *anharmonicity constant*. The additional term here represents the first correction for deviations from strictly harmonic behavior. Treating x_e as a small parameter, derive relations for P, U, and S that include the anharmonic effect at least to first order in x_e. Specifically explore the nature of such relations in the high temperature limit $(T/\theta_{vib} = O(1))$.

5.7 A cylinder in a compressor contains one kmol of O_2 gas initially at $T_1 = 300$ K with initial volume $V_1 = 0.002$ m^3. The gas is compressed reversibly and adiabatically to a volume of 1.0×10^{-5} m^3. Determine the final temperature T_2 of the gas assuming that the modes of energy storage obey the equipartition theorem.

5.8 In a prototype combustion engine, at the end of the combustion process, the working fluid is a mixture of H_2O, CO_2, and N_2 gas at a pressure of 800 kPa and a temperature of 2,000 K. The mole fractions of H_2O, CO_2, and N_2 in the mixture are 0.190, 0.095, and 0.715, respectively. In the engine, this mixture expands until its pressure is 120 kPa. Modeling the expansion as reversible and adiabatic, determine the temperature of the gas at the end of the expansion process.

5.9 Using the results of Example 5.6, plot the variation of the equilibrium constant with temperature for the reaction $H_2 + I_2 \Leftrightarrow 2HI$ over the temperature range from 500 to 1,000 K. Show that the variation is nearly a straight line when plotted as K_c versus $1/T$.

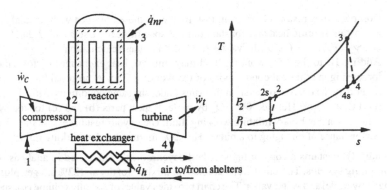

Figure 5.9

5.10 For a manned expedition to Mars, the Brayton cycle system shown schematically in Figure 5.9 is to be used to provide electrical power and heat. The atmosphere of Mars is essentially pure CO_2 gas at about 1 kPa. In its simplest embodiment, the system would use CO_2 gas from the atmosphere as a working fluid for the system. The gas would be compressed to fill the system. In its design operating mode, carbon dioxide gas at state 1 would be compressed to high pressure (state 2) and the gas would be heated using a small nuclear reactor to temperature T_3. The high temperature, high pressure gas is then expanded through a gas turbine to pressure P_4. The high temperature, low pressure gas is then sent through a heat exchanger to recover energy to heat structures in which the crew will live. The working fluid is then recycled to the inlet of the compressor. The system design operating conditions are summarized below:

$T_1 = 303$ K, $P_1 = 80$ kPa, $T_3 = 1{,}500$ K, $P_3 = 400$ kPa, $P_4 = 80$ kPa,

compressor efficiency $= (h_{2s} - h_1)/(h_2 - h_1) = 0.87$,

turbine efficiency $= (h_4 - h_3)/(h_{4s} - h_3) = 0.91$.

For this system design, analyze the following performance issues:

(a) One concern about this system is that by heating the CO_2 up to 1,500 K, the gas may partially decompose into CO and O_2 according to the equation

$$2CO_2 \Leftrightarrow 2CO + O_2.$$

Consider one cubic meter of CO_2 gas at equilibrium and assess whether significant decomposition is likely to occur at 1,500 K. Refer to the D_e values in Tables 5.1 and 5.2 in formulating your analysis. How hot would the CO_2 have to get before 1% of it has decomposed?

(b) Derive relations for the enthalpy and entropy of carbon dioxide that are valid over the range of temperatures encountered in this system.

(c) Write one or more computer programs to accomplish the following:

   (i) Use the property relations to determine the mass flow rate of CO_2 required to generate 100 kW of electrical power, assuming that the electrical generators are 85% efficient.

   (ii) For the resulting flow rate determined in (i) determine the rate of heat delivered to waste heat recovery unit $\dot{q}_h$.

   (iii) Determine how the efficiency of the system $(\dot{w}_t - \dot{w}_c)/\dot{q}_{nr}$ varies with T_3 for T_3 values between 1,300 and 1,700 K. Plot the resulting variation and assess the effect of reducing or increasing T_3 on the system performance.

(d) Compare your results obtained in task (c.i) with those obtained with an analysis for a constant specific heat and constant ratio of specific heats. Use $\hat{c}_P = 37.0$ kJ/kmol K and $\gamma = \hat{c}_P/\hat{c}_V = 1.289$ (the values for 300 K) in these calculations.

(e) Adding He to the CO_2 working fluid may improve heat exchanger performance by increasing the thermal conductivity of the working fluid. A proposal has been made to modify the design to a system that runs under the same conditions using a working fluid that is 50% He and 50% CO_2 by mole. Repeat parts (b) and (c) for the system running on this binary mixture working fluid. Using your results, assess the effect on performance of changing to a binary He–CO_2 mixture for the working fluid.

5.11 Appendix III contains a copy of the standard psychrometric chart used in analysis of air-conditioning systems. This chart is constructed for mixtures of dry air (22% oxygen plus 78% nitrogen by mole) and water vapor. The chart provides values of specific volume (v), specific enthalpy (h), and relative humidity (ϕ) as functions of humidity ratio (W) and temperature (T). Units to be used in this problem for these properties are specified in the table below:

Property	Units
v	m^3/(kg of dry air)
h	kJ/(kg of dry air)
W	(kg of water vapor)/(kg of dry air)
ϕ	%
T	°C

(a) Using results from Chapter 5, show that Dalton's law applies to mixtures of O_2, N_2, and H_2O if the mixture behaves as an ideal gas. Dalton's law states that the mole fraction of any species is equal to the ratio of its partial pressure to the total pressure. The partial pressure is the pressure the species would have if it occupied the system volume alone at the same temperature.

(b) Using results from Chapters 3–5, derive relations for specific enthalpy, relative humidity, and specific volume as functions of humidity ratio and temperature. Correct the enthalpy equation with appropriate constants so that in the limit of zero water vapor content, the enthalpy goes to zero at a temperature of 0°C and in the limit of no dry air content, the ratio h/W goes to 2,501 kJ/kg.

(c) Write a computer program to predict h, v, and ϕ for specified values of system pressure, temperature, and humidity ratio.

(d) Use the program developed in part (6) to construct a psychrometric chart for 1 atmosphere pressure. Your chart should show lines corresponding to

$h = 0, 20, 40, 60,$ and 80 kJ/kg dry air,

$v = 0.8, 0.85, 0.9,$ and 0.95 m^3/kg dry air,

$\phi = 20, 40, 60, 80,$ and 100%.

The chart should span temperatures from 0°C to 50°C and W values from 0 to 0.03.

(e) Repeat (d) at a pressure of 0.687 atmospheres (the pressure at 10,000 feet). Give an assessment of the differences between the charts. Are they large enough to be important?

5.12 A composite system consists of two chambers connected by a short pipe with an insulated valve in it. Each chamber has a volume of 0.5 m^3 and the composite system is insulated so it is thermally isolated from the surroundings. Initially the valve is closed and one chamber contains 0.5 moles of N_2 gas at 400 K while the other contains 0.5 moles of Br_2 gas at 800 K. The valve is opened and the gases are allowed to mix within the composite system. Determine the final temperature of the gas in the composite system after it reaches equilibrium. In these circumstances, you may neglect electronic and nuclear storage.

5.13 Four kmol of hydrogen (H_2) gas and 0.1 kmol of nitrogen (N_2) gas are fed into an initially evacuated chamber having a volume of 0.9 m^3. The gas in the chamber is held at 400°C by heating or cooling the chamber walls as required. The gases react according to

$$N_2 + 3H_2 \Leftrightarrow 2NH_3.$$

(a) Assuming the system reaches thermodynamic equilibrium, estimate the concentration of ammonia (NH_3) in the mixture.

(b) If we want to increase the concentration of ammonia in the mixture, should we increase or decrease the temperature? Justify your answer.

In analyzing this system, take the degeneracy of the electronic ground state to be one.

References

[1] McQuarrie, D. A., *Statistical Mechanics*, Chapters 5, 6, and 8, Harper and Row, New York, 1976.

[2] Tien, C. L. and Lienhard, J. H., *Statistical Thermodynamics*, Chapter 7, Hemisphere Publishing Corporation, New York, 1979.

CHAPTER 6

Dense Gases, Liquids, and Quantum Fluids

The ideal gases considered in Chapter 5 are arguably the simplest fluids encountered in real systems. The behavior of dense gases, liquids, and quantum fluids deviates strongly from that of an ideal gas. In this chapter, we examine how the thermodynamic framework developed in earlier chapters can be applied to such fluids. The van der Waals model for dense gases and liquids is explored in detail for pure and binary mixture systems. In doing so, we demonstrate that the statistical thermodynamic framework provides a link between microscopic characteristics of the molecules or particles and the macroscopic behavior of these fluids.

6.1 Behavior of Gases in the Classical Limit

In Chapter 5, we observed that if the temperature is high enough, we can replace the summation in the definition of the partition function with an integral to obtain the limiting form the partition function at high temperature. This reflects one of the fundamental characteristics of quantum theory, which is that classical behavior is attained in the limit of large quantum number. At high temperature, the average energy per molecule increases and this implies that the average quantum number for each energy storage mode is higher. Thus at higher temperatures, the mean behavior of the system is classical in nature.

In the previous sections of this text we have attacked the problem of determining the partition function by considering the problem from a quantum perspective. We now will explore how we can determine the partition function in the high temperature limit by considering the energy storage from a classical viewpoint.

The basic definition of the molecular partition function that we developed earlier can be written in terms of a summation over all quantum states j':

$$q = \sum_{j'=0}^{\infty} e^{-\beta \varepsilon_{j'}}, \tag{6.1}$$

where $\beta = 1/k_B T$. In the classical limit we expect that the summation can be replaced by an integral. Here we first consider independent particles for which the energy microstates are not affected by the presence of other particles in the system. For a single molecule the energy in a particular microstate is equal to the Hamiltonian evaluated at particular values of generalized momenta $\hat{p}_{j'}$ and position coordinates $\hat{q}_{j'}$ that specify the microstate. Thus

$$e^{-\beta \varepsilon_{j'}} = e^{-\beta \hat{H}(\hat{p}_1, \hat{p}_2, \dots, \hat{p}_\xi, \hat{q}_1, \hat{q}_2, \dots, \hat{q}_\xi)}, \tag{6.2}$$

where ξ is the number of momenta and coordinates necessary to completely specify the motion and position of the molecule in a particular microstate. Summing over all possible

179

microstates, when the microstates are very close together, must, in some sense, be equivalent
to integrating over all possible momenta and coordinate values:

$$q \sim \int\int\int \cdots \int e^{-\beta \hat{H}(\hat{p}_1, \hat{p}_2, \ldots, \hat{p}_\xi, \hat{q}_1, \hat{q}_2, \ldots, \hat{q}_\xi)} \, d\hat{p}_1 \, d\hat{p}_2 \, \ldots d\hat{p}_\xi \, d\hat{q}_1 \, d\hat{q}_2 \ldots d\hat{q}_\xi.$$

However, the integration above over the coordinates and momenta results in a dimensional
quantity and q is, by definition, dimensionless. We expect, therefore, that a proportionality
constant is required that renders the expression for q dimensionless. This suggests the
following definition for the molecular partition function in the classical limit:

$$q = C_q \int\int\int \cdots \int e^{-\beta \hat{H}(\hat{p}_1, \hat{p}_2, \ldots, \hat{p}_\xi, \hat{q}_1, \hat{q}_2, \ldots, \hat{q}_\xi)} \, d\hat{p}_1 \, d\hat{p}_2 \, \ldots d\hat{p}_\xi \, d\hat{q}_1 \, d\hat{q}_2 \ldots d\hat{q}_\xi. \tag{6.3}$$

Note that the integrals are taken over all accessible values of the parameters. The momenta
range from $-\infty$ to ∞ and the spatial coordinates span the volume of space accessible to
the particle.

To determine the proportionality constant C_q, we consider the specific case of the classical
limit for a monatomic ideal gas. In the absence of electronic and nuclear effects, molecules
of a monatomic ideal gas store energy by translation only. The Hamiltonian is therefore
given by

$$\hat{H} = \frac{1}{2m}\left(\hat{p}_x^2 + \hat{p}_y^2 + \hat{p}_z^2\right).$$

Substituting this relation into Eq. (6.3) yields

$$q_{\text{tr}} = C_q \int\int\int \ \int\int\int e^{-(\hat{p}_x^2 + \hat{p}_y^2 + \hat{p}_z^2)/2mk_B T} \, d\hat{p}_x \, d\hat{p}_y \, d\hat{p}_z \, dx \, dy \, dz.$$

In the above equation the spatial integrals are taken over all accessible values of x, y, and
z. The integration over these variables yields the volume of the container,

$$q_{\text{tr}} = C_q V \int\int\int e^{-(\hat{p}_x^2 + \hat{p}_y^2 + \hat{p}_z^2)/2mk_B T} \, d\hat{p}_x \, d\hat{p}_y \, d\hat{p}_z.$$

The momentum integrals are evaluated from $-\infty$ to ∞, which results in the following
relation:

$$q_{\text{tr}} = C_q V (2\pi m k_B T)^{3/2}. \tag{6.4}$$

In Chapters 2 and 5, we derived the following relation for the translational component of
the molecular partition function:

$$q_{\text{tr}} = \frac{V (2\pi m k_B T)^{3/2}}{h^3}. \tag{5.5}$$

Equivalence of Eq. (5.5) with the relation (6.4) obtained from the integral definition of q
requires that

$$C_q = \frac{1}{h^3}.$$

It can be easily shown that if the particle is restricted to translational energy storage in only two dimensions, consistency between the integral and summation definitions for q requires that $C_q = 1/h^2$. This implies that, in general,

$$C_q = \frac{1}{h^\xi}.$$

Thus, for the summation and integral definitions to be consistent,

$$q = \frac{1}{h^\xi} \int \int \int \cdots \int e^{-\beta \hat{H}(\hat{p}_1, \hat{p}_2, \ldots, \hat{p}_\xi, \hat{q}_1, \hat{q}_2, \ldots, \hat{q}_\xi)} d\hat{p}_1 \, d\hat{p}_2 \ldots d\hat{p}_\xi \, d\hat{q}_1 \, d\hat{q}_2 \ldots d\hat{q}_\xi.$$

$$(6.5)$$

Note that since Planck's constant h has units of momentum (kg m/s) multiplied by length (m), setting $C_q = 1/h^\xi$ makes the integral expression for q dimensionless.

We can extend the results above to a system of N indistinguishable particles using the Eq. (4.173) obtained in Chapter 4 for the partition function for the canonical ensemble. For a single-component system this relation reduces to

$$Q = \frac{q^N}{N!}.$$

$$(6.6)$$

Substituting the right side of Eq. (6.5) for the molecular partition function yields

$$Q = \frac{1}{N!} \prod_{j=1}^{N} \frac{1}{h^\xi} \int \int \int \cdots \int e^{-\beta \hat{H}_j(\hat{p}_{j,1}, \hat{p}_{j,2}, \ldots, \hat{p}_{j,\xi}, \hat{q}_{j,1}, \hat{q}_{j,2}, \ldots, \hat{q}_{j,\xi})} d\hat{p}_{j,1} \ldots d\hat{p}_{j,\xi} \, d\hat{q}_{j,1} \ldots d\hat{q}_{j,\xi}.$$

For independent particles, the coordinates and momenta in the integrals for each molecule are independent of one another. The N-fold product of 2ξ-dimensional integrals is therefore equivalent to a single integral over all $2\xi N$ momenta and coordinates:

$$Q = \frac{1}{N! h^{N\xi}} \int \int \int \cdots \int e^{-\beta \sum_{j=1}^{N} \hat{H}_j(\hat{p}_{j,1}, \hat{p}_{j,2}, \ldots, \hat{p}_{j,\xi}, \hat{q}_{j,1}, \hat{q}_{j,2}, \ldots, \hat{q}_{j,\xi})}$$

$$\times \prod_{j=1}^{N} d\hat{p}_{j,1} \ldots d\hat{p}_{j,\xi} d\hat{q}_{j,1} \ldots d\hat{q}_{j,\xi}.$$

Since the momenta and coordinates are independent, we replace the N sets of ξ momenta variables with a single set of $N\xi$ momenta: $\hat{p}_{1,1}$ to $\hat{p}_{1,\xi}$ becomes $\hat{p}_1$ to $\hat{p}_\xi$, $\hat{p}_{2,1}$ to $\hat{p}_{2,\xi}$ becomes $\hat{p}_{\xi+1}$ to $\hat{p}_{2\xi}$, etc. We similarly replace the N sets of ξ coordinates with a single set of coordinates ranging from 1 to $N\xi$. The above relation for Q then becomes

$$Q = \frac{1}{N! h^{N\xi}} \int \int \int \cdots \int e^{-\beta \sum_{j=1}^{N} \hat{H}_j(\hat{p}_{(j-1)\xi+1}, \ldots, \hat{p}_{j\xi}, \hat{q}_{(j-1)\xi+1}, \ldots, \hat{q}_{j\xi})} \prod_{j=1}^{N\xi} d\hat{p}_j \, d\hat{q}_j. \quad (6.7)$$

In the exponential term in the above equation, the sum of the $\hat{H}_j$ for all N particles is equal to the total system energy $\hat{H}(\hat{p}_1, \ldots, \hat{p}_{N\xi}, \hat{q}_1, \ldots, \hat{q}_{N\xi})$ for the system microstate corresponding to the specified set of $\hat{p}_j$ and $\hat{q}_j$ values:

$$Q = \frac{1}{N!h^{N\xi}} \int\int \cdots \int e^{-\beta\hat{H}(\hat{p}_1,\ldots,\hat{p}_{N\xi},\hat{q}_1,\ldots,\hat{q}_{N\xi})} \prod_{j=1}^{N\xi} d\hat{p}_j \, d\hat{q}_j. \tag{6.8}$$

This relation links the Hamiltonian $\hat{H}$ of the N molecule system to the canonical partition function for independent particles in the classical limit. As suggested by the above relation the system Hamiltonian will, in general, be a function of the position and momenta of all the particles in the system.

Ideal gases are an important category of substances that can be accurately modeled by assuming that the particles are independent. Dense gases and liquids do not conform to this model. The molecules in such systems interact with each other through long-range attractive forces and short-range repulsive forces. As a result, the Hamiltonian for the system is a much more complicated function of the positions and momenta of the particles. Equation (6.8) relates the total system Hamiltonian to the system partition function. Thus although this equation was derived for independent particles, it can also be applied to a system of interacting particles if the system Hamiltonian can be computed as a function of the coordinates and momenta of the particles.

Although we have not proven that Eq. (6.8) is a correct definition for the partition function in the classical limit for a system of interacting particles, this equation is, in fact, correct for a system of this type. Although we will not do so here, it is possible to start with the quantum sum definition for Q and derive Eq. (6.8) as the classical limiting behavior for $h \to 0$ (see the discussion in the reference by McQuarrie [1] cited at the end of this chapter).

6.2   Van der Waals Models of Dense Gases and Liquids

Single-Species Systems

In dense gases and liquids, the molecules are close enough that they are constantly interacting with neighboring molecules. This is fundamentally different from ideal gases in which molecules travel through space without interacting with other molecules until a brief (and infrequent) collision occurs. In this section we will first consider a system containing one species of interacting molecules. To evaluate the partition function for such a system using Eq. (6.8),

$$Q = \frac{1}{N!h^{N\xi}} \int\int \cdots \int e^{-\beta\hat{H}(\hat{p}_1,\ldots,\hat{p}_{N\xi},\hat{q}_1,\ldots,\hat{q}_{N\xi})} \prod_{j=1}^{N\xi} d\hat{p}_j \, d\hat{q}_j, \tag{6.8}$$

we must develop a relation for the system Hamiltonian, $\hat{H}$. For a system containing interacting particles, we write the system Hamiltonian as the sum of three components, as indicated below:

$$\hat{H} = \sum_{j=1}^{N} \hat{H}_{\text{trans},j}(\hat{p}_{x,j}, \hat{p}_{y,j}, \hat{p}_{z,j}) + \sum_{j=1}^{N} \hat{H}_{\text{rot},j}(\hat{p}_{\theta,j}, \ldots)$$
$$+ \Phi_N(\hat{q}_{1,1}, \ldots, \hat{q}_{1\xi}, \ldots, \hat{q}_{N,1}, \ldots, \hat{q}_{N,\xi}). \tag{6.9}$$

The first term on the right side of Eq. (6.9) is the sum of the translational energies of the molecules, which depends only on the momenta of their centers of mass. The second is the sum of the rotational energies of the molecules, which depends on rotational momenta of the molecules. The third term, Φ_N, is the potential energy due to intermolecular force interactions, which is a function only of the position coordinates of the molecules. Here we specifically will model the molecules as diatomic, linear polyatomic, or nonlinear poly-atomic with three different moments of inertia. The translational energy is a function of three Cartesian momenta and the rotational energy depends on two or three momenta. Hence, the total number of momenta in the Hamiltonian per molecule, ξ, is either five or six. The integral for Q in Eq. (6.8) is over all molecular momenta and coordinates, which are taken to be independent variables. We can therefore group the integrations into three factors, the first being integrals over the translational momenta, the second being integrals over the rotational coordinates and momenta, and the third being integrals over the translational coordinates:

$$Q = \frac{1}{N!} \left[\frac{1}{h^{3N}} \int\int\int \cdots \int e^{-\beta \Sigma_{j=1}^{N} \hat{H}_{\text{trans},j}(\hat{p}_{x,j}, \hat{p}_{y,j}, \hat{p}_{z,j})} \prod_{j=1}^{N} d\hat{p}_{x,j}\, d\hat{p}_{y,j}\, d\hat{p}_{z,j} \right]$$

$$\times \left[\frac{1}{h^{(\xi-3)N}} \int\int\int \cdots \int e^{-\beta \Sigma_{j=1}^{N} \hat{H}_{\text{rot},j}(\hat{p}_{\theta_1,j}, \ldots)} \prod_{j=1}^{N} \prod_{i=1}^{\xi-3} d\hat{p}_{\theta_i,j}\, d\hat{q}_{\theta_i,j} \right]$$

$$\times \left[\int\int\int \cdots \int e^{-\beta \Phi_N(x_1, y_1, z_1, \ldots, x_N, y_N, z_N)} \prod_{j=1}^{N} dx_j\, dy_j\, dz_j \right]. \tag{6.10}$$

Since the translational Hamiltonian for each molecule is the same and has the usual quadratic form, the factor inside the first set of square brackets in Eq. (6.10) is easily evaluated. Doing so, one finds that it equals the translational component of the molecular partition function divided by V to the Nth power, $(q_{\text{trans}}/V)^N = (2\pi M k_B T / h^2)^{3N/2}$. The rotational Hamiltonian is also assumed to be the same for each molecule. It follows that the factor in the second set of square brackets in Eq. (6.10) can be decomposed into the product of N sets of integrals, each of which is equal to the rotational component of the molecular partition function q_{rot}. The factor inside these square brackets therefore equals q_{rot}^N.

The factor inside the third set of square brackets in Eq. (6.10) is the classical *configuration integral* Z_N:

$$Z_N = \int\int\int \cdots \int e^{-\Phi(x_1, y_1, z_1, \ldots, x_N, y_N, z_N)/k_B T} \prod_{j=1}^{N} dx_j\, dy_j\, dz_j. \tag{6.11}$$

With this definition, the above relation for Q can be written as

$$Q = \frac{(2\pi M k_B T)^{3N/2}}{N! h^{3N}} q_{\text{rot}}^N Z_N. \tag{6.12}$$

As discussed in Chapter 5, q_{rot} varies with the molecular structure. For a linear diatomic or polyatomic molecule

$$q_{\text{rot}} = \frac{T}{\sigma_s \theta_{\text{rot}}}, \tag{6.13}$$

whereas for a nonlinear polyatomic molecule with three different moments of inertia

$$q_{\text{rot}} = \frac{\pi^{1/2}}{\sigma_{\text{s}}} \left(\frac{T}{\theta_{\text{rot},m}} \right)^{3/2},$$ (6.14)

where

$$\theta_{\text{rot},m} = (\theta_{\text{rot},A}\theta_{\text{rot},B}\theta_{\text{rot},C})^{1/3}.$$ (6.15)

The relation (6.12) for Q for either case can be written as

$$Q = \frac{(2\pi M k_{\text{B}} T)^{3N/2}}{N! h^{3N}} \frac{\pi^{(\xi-5)N/2}}{\sigma_{\text{s}}^{N}} \left(\frac{T}{\theta_{\text{rot},m}} \right)^{(\xi-3)N/2} Z_N,$$ (6.16)

where ξ is the total number of translational and rotational energy storage modes and it is understood that $\theta_{\text{rot},m} = \theta_{\text{rot}}$ for linear molecules.

To complete our model analysis we must evaluate Z_N. To do so using Eq. (6.11) we must develop a relation for the potential energy associated with particle force interactions, Φ_N. In dense fluids like liquids, one approach to dealing with the system potential energy associated with force interactions among neighboring molecules is to make use of a radially symmetric distribution function $g(r)$ defined such that

$$4\pi r^2 (N/V) g(r)\, dr = \left(\begin{array}{l} \text{the number density of molecules with centers between} \\ r \text{ and } r + dr \text{ measured relative to a specific molecule} \end{array} \right).$$

Qualitatively, the distribution function looks like the curve shown in Figure 6.1. The distribution function can be determined using Monte Carlo simulation methods or experimentally using x-ray scattering. The product $g(r)dr$ can be thought of as a correction factor that when multiplied by the system mean density gives the local (time-averaged) density.

We choose any molecule in the bulk fluid as the central molecule shown in Figure 6.2. In the type of model considered here, the intermolecular potential function $\phi(r)$ is idealized as being spherically symmetric. A spherically symmetric potential is likely to be a good

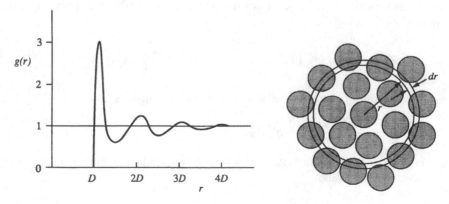

Figure 6.1   Figure 6.2

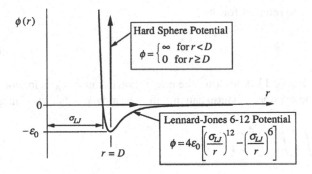

Figure 6.3

approximation for nonpolar uncharged molecules with a reasonably spherically symmetric structure. A more general formulation is required for molecules having a permanent dipole moment or grossly unsymmetric structure or electron charge. For a spherically symmetric potential function it follows that

$$4\pi r^2\phi(r)(N/V)g(r)dr = \left(\begin{array}{l}\text{the intermolecular potential energy between the}\\ \text{central molecule and other molecules with}\\ \text{centers at distances between } r \text{ and } r+dr\end{array}\right).$$

The intermolecular potential can be determined from quantum mechanical analysis or modeled empirically. Two common models, the hard sphere and Lennard–Jones 6-12 models, are shown in Figure 6.3.

The total potential energy of the fluid is obtained by integrating over all values of r, multiplying by N, since any of N molecules may be central, and multiplying by $1/2$ so that each interacting pair is counted only once. The total potential energy is therefore given by

$$\Phi_N = \frac{N}{2}\int_0^\infty 4\pi r^2\phi(r)(N/V)g(r)\,dr. \tag{6.17}$$

Considering Figure 6.1, we can make the following crude, but reasonable approximation for $g(r)$:

$$g(r) = \begin{cases} 0 & \text{for } r < D, \\ 1 & \text{for } r \geq D, \end{cases} \tag{6.18}$$

where D is the molecular diameter. With this idealization, the relation (6.17) for Φ_N can be rearranged to obtain

$$\Phi_N = \frac{2\pi N^2}{V}\int_D^\infty r^2\phi(r)\,dr. \tag{6.19}$$

Note that in adopting Eq. (6.17) we are implicitly assuming that the intermolecular potential is radially symmetric. It follows that if we define

$$a_v = -2\pi\int_D^\infty r^2\phi(r)\,dr, \tag{6.20}$$

we obtain the following simple relation for Φ_N:

$$\Phi_N = -\frac{a_v N^2}{V}.$$ (6.21)

Substituting Eq. (6.21) into Eq. (6.11) simplifies the integration because Φ_N is independent of any of the coordinates. The exponential term can therefore be moved outside the integrals:

$$Z_N = e^{a_v N^2 / V k_B T} \int\int\int \cdots \int \prod_{j=1}^{N} dx_j \, dy_j \, dz_j$$ (6.22)

The multiple integral is equal to the integral of unity over all accessible values of x, y, and z raised to the power N. At first glance it might appear that this multiple integral should equal V^N. However, in dense gases and liquids, the molecules may occupy a significant fraction of the physical system space. This occupied space is inaccessible to any given molecule and should not be included in the multiple integration. Consequently the quantity $(V - N b_v)^N$ is a better approximation for the multiple integral, where b_v is a constant representing the mean volume occupied by a molecule in the system. The resulting relation for Z_N is

$$Z_N = e^{a_v N^2 / v k_B T} (V - N b_v)^N.$$ (6.23)

Substituting the above relation for Z_N into Eq. (6.16), we obtain the following relation for the partition function:

$$Q = \frac{\pi^{(\xi-5)N/2}}{\sigma_s^N N!} \left(\frac{2\pi M k_B T (V - N b_v)^{2/3}}{h^2} \right)^{3N/2} \left(\frac{T}{\theta_{rot,m}} \right)^{(\xi-3)N/2} e^{a_v N^2 / V k_B T}.$$ (6.24)

Since most thermodynamic properties are related to the natural log of Q (see Chapter 4) it is useful to take the log of both sides of Eq. (6.24) and simplify using Stirling's approximation to obtain

$$\ln Q = N + \left(\frac{3N}{2} \right) \ln \left[\frac{2\pi M k_B T (V - N b_v)^{2/3}}{N^{2/3} h^2} \right] + N \left[\frac{(\xi - 5)}{2} \ln \pi - \ln \sigma_s \right]$$

$$+ \frac{(\xi - 3)N}{2} \ln \left(\frac{T}{\theta_{rot,m}} \right) + \frac{a_v N^2}{V k_B T}.$$ (6.25)

Substituting this result into Eq. (4.49), it is easily shown that for the model fluid in our analysis

$$P = \frac{N k_B T}{V - N b_v} - \frac{a_v N^2}{V^2}.$$ (6.26)

The above relation is the van der Waals equation of state discussed previously. Thus, our model analysis has led us to the partition function for a van der Waals fluid. Now that we have the partition function, we can generate relations for other properties as well. From

Eqs. (4.50) and (4.43) and the definition of enthalpy H, one can also obtain the following relations for our model fluid:

$$\frac{\mu}{k_{\mathrm{B}}T} = \frac{Nb_v}{V - Nb_v} - \left(\frac{3}{2}\right)\ln\left[\frac{2\pi Mk_{\mathrm{B}}T(V - Nb_v)^{2/3}}{N^{2/3}h^2}\right]$$

$$- \left[\frac{(\xi - 5)}{2}\ln\pi + \ln\sigma_s\right] - \frac{(\xi - 3)}{2}\ln\left(\frac{T}{\theta_{\mathrm{rot},m}}\right) - \frac{2a_v N}{Vk_{\mathrm{B}}T}, \tag{6.27}$$

$$U = \frac{\xi N}{2}k_{\mathrm{B}}T - \frac{a_v N^2}{V}, \tag{6.28}$$

$$H = \left(\frac{\xi}{2} + \frac{V}{V - Nb_v}\right)Nk_{\mathrm{B}}T - \frac{2a_v N^2}{V}. \tag{6.29}$$

Using Eq. (4.41) we can also obtain the following relation for the entropy:

$$S = Nk_{\mathrm{B}}\left(\frac{\xi}{2} + 1\right) + \frac{3Nk_{\mathrm{B}}}{2}\ln\left\{\frac{2\pi Mk_{\mathrm{B}}T(V - Nb_v)^{2/3}}{N^{2/3}h^2}\right\}$$

$$+ \frac{(\xi - 3)Nk_{\mathrm{B}}}{2}\ln\left\{\frac{T}{\theta_{\mathrm{rot},m}}\right\} + Nk_{\mathrm{B}}\ln\left\{\frac{\pi^{(\xi-5)/2}}{\sigma_s}\right\}. \tag{6.30}$$

Using Eq. (6.28) to eliminate T from the above equation, we obtain the fundamental equation for a van der Waals fluid:

$$S = Nk_{\mathrm{B}}\left(\frac{\xi}{2} + 1\right) + \frac{3Nk_{\mathrm{B}}}{2}\ln\left\{\frac{4\pi M(V - Nb_v)^{2/3}}{\xi N^{5/3}h^2}\left(U + \frac{a_v N^2}{V}\right)\right\}$$

$$+ \frac{(\xi - 3)Nk_{\mathrm{B}}}{2}\ln\left\{\frac{2}{\xi Nk_{\mathrm{B}}\theta_{\mathrm{rot},m}}\left(U + \frac{a_v N^2}{V}\right)\right\} + Nk_{\mathrm{B}}\ln\left\{\frac{\pi^{(\xi-5)/2}}{\sigma_s}\right\}. \tag{6.31}$$

Although it incorporates many idealizations, this model analysis provides a workable framework for predicting properties of dense gases and liquids that relates macroscopic thermodynamic properties to molecular characteristics. Note in particular that it relates the van der Waals constants a_v and b_v to the physical size of the molecules and the intermolecular potential function, which characterizes force interactions among molecules. Values of these constants for selected molecules are listed in Table 6.1. A plot of the fundamental equation for nitrogen obtained using Eq. (6.31) with constants from Table 6.1 is shown in Figure 6.4.

The model analysis for a fluid of interacting molecules presented in this section obviously incorporates several idealizations. It is based upon the central idealization in van der Waals' doctoral thesis in 1873 [2] that a vapor and a liquid are really just extremes of a continuous fluid phase. The van der Waals model described here can be improved by increasing the accuracy of the idealizations in the model. Specifically, more accurate approximations for the intermolecular potential function and the distribution function $g(r)$ could be incorporated into the model. For some molecules, an accurate equation of state requires that the

Table 6.1   *Van der Waals Constants Determined from Critical Data for selected Fluids*[†]

Molecule	$10^{49} \times a_v$ (Pa m^6/molecule^2)	$10^{29} \times b_v$ (m^3/molecule)	ξ
C_2H_2(acetylene)	12.2	8.47	5
NH_3	11.7	6.20	6
CO_2	10.1	7.09	5
CO	4.04	6.54	5
CF_4H_2(R134a)	27.7	15.9	6
H_2	0.682	4.40	5
CH_4	6.31	7.09	6
N_2	3.76	6.40	5
O_2	3.78	5.23	5
H_2O	15.2	5.05	6

[†] Adapted from Wark [5].

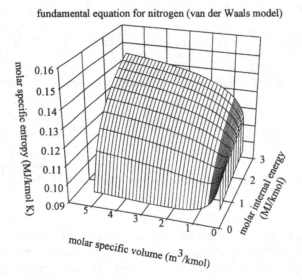

Figure 6.4

model include a nonsymmetric intermolecular potential function. Discussions of alternate intermolecular potential function models can be found in Hirschfelder, Curtis, and Bird [3], and Tien and Lienhard [4].

The property variations predicted by the van der Waals model exhibit all the features of those for real liquids and gases. As a predictive tool it is fairly good for low and moderate gas densities, but it generally becomes increasingly inaccurate as the density approaches that of the liquid phase. At liquid densities, it is not uncommon for the property predictions to be off by as much as 100%. The accuracy of the van der Waals property relations can be improved over a limited range of conditions if the constants are chosen to provide an optimal fit to data in that range.

The van der Waals model clearly has limited usefulness as a predictive tool. Yet it remains valuable because it illuminates the linkage between the molecular characteristics of the substance and its macroscopic thermodynamic properties, and it does so at high densities characteristic of liquids and low densities characteristic of vapors. Because it is applicable to both liquid and vapor phases, in Chapter 8 we will use the van der Waals model to explore thermodynamic features of phase transitions from liquid to vapor for a pure substance. We will see that the transformation from a low density to a high density fluid (or vice versa) is often not a simple continuous transition for real fluids and that the van der Waals model predicts such behavior.

Example 6.1 For nitrogen (N_2) $\sigma_{LJ} = 3.75 \times 10^{-10}$ m and $\varepsilon_0 = 1.31 \times 10^{-21}$ J are recommended values for the constants in the Lennard–Jones 6-12 intermolecular potential. Use these values to predict the van der Waals constants a_v and b_v for nitrogen.

Solution Substituting the Lennard–Jones 6-12 potential into Eq. (6.20) we obtain the following relation for a_v:

$$a_v = -2\pi \int_D^\infty r^2 4\varepsilon_0 \left[\left(\frac{\sigma_{LJ}}{r} \right)^{12} - \left(\frac{\sigma_{LJ}}{r} \right)^6 \right] dr.$$

Execution of the integral yields

$$a_v = \frac{8\pi \varepsilon_0 \sigma_{LJ}^6}{3D^3} \left[1 - \frac{\sigma_{LJ}^6}{3D^6} \right].$$

Here we take $D = \sigma_{LJ}$ where the potential becomes strongly repulsive. Substituting $\sigma_{LJ} = 3.75 \times 10^{-10}$ m and $\varepsilon_0 = 1.31 \times 10^{-21}$ J for nitrogen, we obtain

$$a_v = 3.86 \times 10^{-49} \text{ Pa m}^6/\text{molecule}^2.$$

The parameter b_v is interpreted as the volume of space made inaccessible by the presence of each of the other molecules. We therefore compute this as

$$b_v \cong 2\pi D^3/3.$$

Again we take $D = \sigma_{LJ}$ and substitute to obtain

$$b_v = 11.0 \times 10^{-29} \text{ m}^3/\text{molecule}.$$

The values of a_v and b_v obtained in this way do not exactly match those in Table 6.1 for nitrogen. Although the value for a_v is close, the b_v value in the table is only about 60% of that computed here.

Example 6.2 Using the relations for the van der Waals model, determine the value of the specific heat at constant pressure for saturated liquid nitrogen and saturated nitrogen vapor at 100 K and 778 kPa. Compare the results to the recommended values of 64.8 kJ/(kmol K) for liquid and 41.3 kJ/(kmol K) for vapor.

Solution By definition $\hat{c}_P = [(\partial H/\partial T)_{P,N}]_{N=N_A}$. For $H = H(V, N, T)$ and $P = P(V, N, T)$, it can be shown by considering changes in H along a line of constant P that

$$
\hat{c}_P = \left[\left(\frac{\partial H}{\partial T} \right)_{P,N} \right]_{N=N_A} = \left[\left(\frac{\partial H}{\partial T} \right)_{V,N} - \left(\frac{\partial H}{\partial V} \right)_{T,N} \frac{(\partial P/\partial T)_{V,N}}{(\partial P/\partial V)_{T,N}} \right]_{N=N_A}.
$$

Differentiating Eqs. (6.26) and (6.29) as required and substituting yields

$$
\hat{c}_P = \left(\frac{\xi}{2} + \frac{V/N_A}{V/N_A - b_v} \right) N_A k_B - \frac{\left[\frac{2a_v}{(V/N_A)^2} - \frac{b_v k_B T}{(V/N_A - b_v)^2} \right] \left(\frac{N_A k_B}{V/N_A - b_v} \right)}{\frac{2a_v}{(V/N_A)^3} - \frac{k_B T}{(V/N_A - b_v)^2}}.
$$

To compute the value of $\hat{c}_P$, the a_v, b_v, and ξ values for nitrogen from Table 6.1 are used. We first compute V/N_A for the liquid and vapor phases at the specified temperature and pressure conditions using Eq. (6.26). (These are the lowest and highest of the three solution values for V/N_A.) Doing so, we find that $V/N_A = 9.99 \times 10^{-29}$ m^3/molecule for the liquid phase and $V/N_A = 1.536 \times 10^{-27}$ m^3/molecule for the vapor phase. Substituting into the above equation then yields

$\hat{c}_P = 48.9$ kJ/(kmol K) for the liquid

and

$\hat{c}_P = 29.7$ kJ/(kmol K) for the vapor.

These values are within 30% of the corresponding recommended values cited above.

An Extended van der Waals Model for Binary Mixtures

We will now explore how the type of model discussed above for pure fluids can be extended to a binary mixture. For a binary mixture containing N_a molecules of species a and N_b molecules of species b, we showed in Chapter 4 that the canonical partition function is given by

$$
Q = \left(\frac{q_a^{N_a}}{N_a!} \right) \left(\frac{q_b^{N_b}}{N_b!} \right). \tag{6.32}
$$

If we substitute the right side of Eq. (6.3) for the molecular partition function for each species, the above relation becomes

$$
Q = \frac{1}{N_a!} \prod_{i=1}^{N_a} \frac{1}{h^{\xi_a}} \int \int \int \cdots \int e^{-\beta \hat{H}_i} d\hat{p}_{i,1} \ldots d\hat{p}_{i,\xi_a} d\hat{q}_{i,1} \ldots d\hat{q}_{i,\xi_a}
$$
$$
\times \frac{1}{N_b!} \prod_{j=1}^{N_b} \frac{1}{h^{\xi_b}} \int \int \int \cdots \int e^{-\beta \hat{H}_j} d\hat{p}_{j,1} \ldots d\hat{p}_{j,\xi_b} d\hat{q}_{j,1} \ldots d\hat{q}_{j,\xi_b}. \tag{6.33}
$$

In the above relation ξ_a and ξ_b are the number of momenta and coordinates required to specify the Hamiltonian of species a and species b molecules, respectively. Since the momenta and coordinates for each molecule (regardless of species) are independent and we are integrating over all possible values, we can combine the two lists into one and reorder them to run from 1 to $N_a\xi_a + N_b\xi_b$. Doing so yields

$$Q = \frac{1}{N_a!N_b!h^{N_a\xi_a+N_b\xi_b}} \int\int\int \cdots \int \exp\left\{-\beta\left(\sum_{i=1}^{N_a}\hat{H}_i + \sum_{j=1}^{N_b}\hat{H}_j\right)\right\} \prod_{l=1}^{N_a\xi_a+N_b\xi_b} d\hat{p}_l\, d\hat{q}_l. \tag{6.34}$$

But when combined, the two summations in the exponential term are just the total Hamiltonian for the system

$$\hat{H}_{\text{sys}} = \sum_{i=1}^{N_a}\hat{H}_i + \sum_{j=1}^{N_b}\hat{H}_j. \tag{6.35}$$

Substituting yields

$$Q = \frac{1}{N_a!N_b!h^{N_a\xi_a+N_b\xi_b}} \int\int\int \cdots \int e^{-\beta\hat{H}_{\text{sys}}} \prod_{l=1}^{N_a\xi_a+N_b\xi_b} d\hat{p}_l\, d\hat{q}_l. \tag{6.36}$$

As was the case in our initial treatment of a single-species system above, this relation has been derived using relations that were derived for particles for which energy storage is independent of the presence of the other particles. We have arrived at a result that depends only on the overall system energy storage. To explore the behavior of this mixture, we will assume here that this relation for Q also is valid for a binary system of interacting particles.

Continuing the analysis, we write the system Hamiltonian as the sum of five components. The first two are the sums of the translational energies of the molecules of each species, which depend only on the momenta of their centers of mass. The second two are the sums of the rotational energies of the two species, which depend on rotational momenta and coordinates of the molecules. The fifth component is the potential energy due to intermolecular force interactions, which is a function only of the position coordinates of the molecules. Here we specifically will model the molecules as diatomic, linear polyatomic, or nonlinear polyatomic with three different moments of inertia. The translational energy is a function of three Cartesian momenta and the rotational energy depends on two or three momenta. Hence, the total number of momenta in the Hamiltonian per molecule, ξ_a or ξ_b, is either five or six. The resulting relation for the Hamiltonian is

$$\hat{H}_{\text{sys}} = \sum_{i=1}^{N_a}\hat{H}_{\text{trans},i}(\hat{p}_{x,i}, \hat{p}_{y,i}, \hat{p}_{z,i}) + \sum_{j=1}^{N_b}\hat{H}_{\text{trans},j}(\hat{p}_{x,j}, \hat{p}_{y,j}, \hat{p}_{z,j})$$

$$+ \sum_{i=1}^{N_a}\hat{H}_{\text{rot},i}(\hat{p}_{\theta,i}, \ldots) + \sum_{j=1}^{N_b}\hat{H}_{\text{rot},j}(\hat{p}_{\theta,j}, \ldots)$$

$$+ \Phi(x_1, y_1, z_1, \ldots, x_{N_a+N_b}, y_{N_a+N_b}, z_{N_a+N_b}). \tag{6.37}$$

The integral for Q in Eq. (6.36) is over all molecular momenta and coordinates, which are taken to be independent variables. We can therefore group the integrations into five factors, the first two being integrals over the translational momenta, the third and fourth being integrals over the rotational coordinates and momenta, and the fifth being integrals over the translational coordinates. Thus we have

$$
Q = \frac{1}{N_a!} \left[\frac{1}{h^{3N_a}} \int\int\int \cdots \int e^{-\beta \Sigma_{i=1}^{N_a} \hat{H}_{\text{trans},i}(\hat{p}_{x,i}, \hat{p}_{y,i}, \hat{p}_{z,i})} \prod_{i=1}^{N_a} d\hat{p}_{x,i}\, d\hat{p}_{y,i}\, d\hat{p}_{z,i} \right]
$$

$$
\times \frac{1}{N_b!} \left[\frac{1}{h^{3N_b}} \int\int\int \cdots \int e^{-\beta \Sigma_{j=1}^{N_b} \hat{H}_{\text{trans},j}(\hat{p}_{x,j}, \hat{p}_{y,j}, \hat{p}_{z,j})} \prod_{j=1}^{N_b} d\hat{p}_{x,j}\, d\hat{p}_{y,j}\, d\hat{p}_{z,j} \right]
$$

$$
\times \left[\frac{1}{h^{(\xi_a-3)N_a}} \int\int\int \cdots \int e^{-\beta \Sigma_{j=1}^{N_a} \hat{H}_{\text{rot},j}(\hat{p}_{\theta_1,j},\ldots)} \prod_{j=1}^{N_a} \prod_{i=1}^{\xi_a-3} d\hat{p}_{\theta_i,j}\, d\hat{q}_{\theta_i,j} \right]
$$

$$
\times \left[\frac{1}{h^{(\xi_b-3)N_b}} \int\int\int \cdots \int e^{-\beta \Sigma_{j=1}^{N_b} \hat{H}_{\text{rot},j}(\hat{p}_{\theta_1,j},\ldots)} \prod_{j=1}^{N_b} \prod_{i=1}^{\xi_b-3} d\hat{p}_{\theta_i,j}\, d\hat{q}_{\theta_i,j} \right]
$$

$$
\times \left[\int\int\int \cdots \int \exp[-\beta \Phi(x_1, y_1, z_1, \ldots, x_{N_a+N_b}, y_{N_a+N_b}, z_{N_a+N_b})] \right.
$$

$$
\left. \times \prod_{j=1}^{N_a+N_b} dx_j, dy_j\, dz_j \right]. \tag{6.38}
$$

Since the translational Hamiltonian for the molecules of each species is the same and has the usual quadratic form, the factor inside the first square brackets in Eq. (6.38) is easily evaluated. Doing so one finds that it equals the translational component of the molecular partition function divided by V to the N_a power: $(q_{\text{trans}}/V)^{N_a} = (2\pi M_a k_B T / h^2)^{3N_a/2}$. Similar reasoning applied to the second integral indicates that it equals $(2\pi M_b k_B T / h^2)^{3N_b/2}$. Since the rotational Hamiltonian for the molecules of each type is also assumed to be the same, the factor in the third set of square brackets can be decomposed into the product of N_a sets of integrals for each molecule, each of which is equal to the rotational component of the molecular partition function for species a. The factor inside these square brackets therefore equals $q_{\text{rot},a}^{N_a}$. The same reasoning implies that the fourth group of integrals is $q_{\text{rot},b}^{N_b}$. The factor inside the fifth set of square brackets is the configuration integral for the binary mixture:

$$
Z_{N_a,N_b} = \int\int\int \cdots \int \exp[-\Phi(x_1, y_1, z_1, \ldots, x_{N_a+N_b}, y_{N_a+N_b}, z_{N_a+N_b})/k_B T]
$$

$$
\times \prod_{j=1}^{N_a+N_b} dx_j, dy_j\, dz_j. \tag{6.39}
$$

With this definition, the above relation for the partition function can be written as

$$
Q = \frac{(2\pi M_a k_B T)^{3N_a/2}(2\pi M_b k_B T)^{3N_b/2}}{N_a! N_b! h^{3N_a+3N_b}}\, q_{\text{rot},a}^{N_a} q_{\text{rot},b}^{N_b} Z_{N_a,N_b}. \tag{6.40}
$$

To complete our determination of the partition function, we need to develop a relation for the configuration integral. To do so, we first define the distribution functions $g_{aa}(r)$, $g_{ab}(r)$, $g_{ba}(r)$, and $g_{bb}(r)$ such that

$$4\pi r^2 (N_a/V) g_{aa}(r)\, dr = \begin{pmatrix} \text{the number density of type } a \text{ molecules with} \\ \text{centers between } r \text{ and } r + dr \text{ measured relative} \\ \text{to a specific } a \text{ molecule} \end{pmatrix},$$

$$4\pi r^2 (N_a/V) g_{ab}(r)\, dr = \begin{pmatrix} \text{the number density of type } a \text{ molecules with} \\ \text{centers between } r \text{ and } r + dr \text{ measured relative} \\ \text{to a specific } b \text{ molecule} \end{pmatrix},$$

$$4\pi r^2 (N_b/V) g_{bb}(r)\, dr = \begin{pmatrix} \text{the number density of type } b \text{ molecules with} \\ \text{centers between } r \text{ and } r + dr \text{ measured relative} \\ \text{to a specific } b \text{ molecule} \end{pmatrix},$$

$$4\pi r^2 (N_b/V) g_{ba}(r)\, dr = \begin{pmatrix} \text{the number density of type } b \text{ molecules with} \\ \text{centers between } r \text{ and } r + dr \text{ measured relative} \\ \text{to a specific } a \text{ molecule} \end{pmatrix}.$$

We also define $\phi_{aa}(r)$, $\phi_{bb}(r)$, and $\phi_{ab}(r)$ as the potential functions representing the force interaction between two molecules of type a, two molecules of type b, and one type a and one type b molecule, respectively. As in the pure fluid system considered above, we will idealize these potential functions as being radially symmetric. It follows that

$$4\pi r^2 \left[\phi_{aa}(r) g_{aa}(r) \left(\frac{N_a}{V} \right) + \phi_{ab}(r) g_{ba}(r) \left(\frac{N_b}{V} \right) \right] dr$$

$$= \begin{pmatrix} \text{the intermolecular potential energy} \\ \text{between a central type } a \text{ molecule} \\ \text{and other molecules with centers at} \\ \text{distances between } r \text{ and } r + dr \end{pmatrix},$$

$$4\pi r^2 \left[\phi_{bb}(r) g_{bb}(r) \left(\frac{N_b}{V} \right) + \phi_{ab}(r) g_{ab}(r) \left(\frac{N_a}{V} \right) \right] dr$$

$$= \begin{pmatrix} \text{the intermolecular potential energy} \\ \text{between a central type } b \text{ molecule} \\ \text{and other molecules with centers at} \\ \text{distances between } r \text{ and } r + dr \end{pmatrix}.$$

To get the total intermolecular potential energy for the system we integrate the above relations from $r = 0$ to infinity and multiply each integral by the number of corresponding molecules divided by 2 to properly account for the total effect of the molecules of that type.

Adding the results for the two species types together yields the relation

$$\Phi = \frac{N_a}{2} \int_0^\infty 4\pi r^2 \left[\phi_{aa}(r) g_{aa}(r) \left(\frac{N_a}{V} \right) + \phi_{ab}(r) g_{ba}(r) \left(\frac{N_b}{V} \right) \right] dr$$

$$+ \frac{N_b}{2} \int_0^\infty 4\pi r^2 \left[\phi_{bb}(r) g_{bb}(r) \left(\frac{N_b}{V} \right) + \phi_{ab}(r) g_{ab}(r) \left(\frac{N_a}{V} \right) \right] dr. \qquad (6.41)$$

Defining

$$a_{v,aa} = -2\pi \int_0^\infty r^2 \phi_{aa}(r) g_{aa}(r)\, dr, \qquad\qquad\qquad\qquad (6.42a)$$

$$a_{v,ab} = -\pi \int_0^\infty r^2 \phi_{ab}(r) [g_{ba}(r) + g_{ba}(r)]\, dr, \qquad\qquad\quad (6.42b)$$

$$a_{v,bb} = -2\pi \int_0^\infty r^2 \phi_{bb}(r) g_{bb}(r)\, dr, \qquad\qquad\qquad\qquad (6.42c)$$

we can rearrange this relation to obtain

$$\Phi = -\frac{a_{v,aa} N_a^2}{V} - \frac{2a_{v,ab} N_a N_b}{V} - \frac{a_{v,bb} N_b^2}{V}. \qquad\qquad (6.43)$$

Because Φ is independent of the integration variables, the relation (6.39) for the configuration integral can be written

$$Z_{N_a,N_b} = \exp\left[(a_{v,aa} N_a^2 + 2a_{v,ab} N_a N_b + a_{v,bb} N_b^2) / V k_B T \right]$$

$$\times \int\int\int \cdots \int \prod_{j=1}^{N_a+N_b} dx_j, dy_j\, dz_j. \qquad\qquad (6.44)$$

The multiple integral is equal to the integral of unity over all accessible values of x, y, and z raised to the power $N_a + N_b$. In dense gases and liquids, the molecules may occupy a significant fraction of the physical system space. If we ignored this effect, the multiple integral would equal $V^{N_a+N_b}$. Since this occupied space is inaccessible to any given molecule and should not be included in the multiple integration, a more accurate expression for this multiple integral would be $(V - N_a b_{v,a} - N_b b_{v,b})^{N_a+N_b}$, where $b_{v,a}$ and $b_{v,b}$ are constants representing the mean volume occupied by a species a and a species b molecule, respectively. The resulting relation for Z_{N_a,N_b} is

$$Z_{N_a,N_b} = \exp\left[(a_{v,aa} N_a^2 + 2a_{v,ab} N_a N_b + a_{v,bb} N_b^2) / V k_B T \right] (V - N_a b_{v,a} - N_b b_{v,b})^{N_a+N_b}. \qquad (6.45)$$

Consistent with our treatment of linear or nonlinear molecules described earlier in this section, we take

$$q_{\text{rot},a} = \frac{\pi^{(\xi_a-5)/2}}{\sigma_{s,a}} \left(\frac{T}{(\theta_{\text{rot},m})_a} \right)^{(\xi_a-3)/2}, \qquad\qquad (6.46)$$

$$q_{\text{rot},b} = \frac{\pi^{(\xi_b-5)/2}}{\sigma_{s,b}} \left(\frac{T}{(\theta_{\text{rot},m})_b} \right)^{(\xi_b-3)/2}. \qquad\qquad (6.47)$$

Substituting the above relations for $q_{\text{rot},a}$, $q_{\text{rot},b}$, and Z_{N_a,N_b} into Eq. (6.40), we obtain the following relation for the partition function:

$$Q = \left(\frac{1}{N_a!}\right)\left[\frac{2\pi M_a k_B T (V - N_a b_{v,a} - N_b b_{v,b})^{2/3}}{h^2}\right]^{3N_a/2}$$

$$\times \left(\frac{1}{N_b!}\right)\left[\frac{2\pi M_b k_B T (V - N_a b_{v,a} - N_b b_{v,b})^{2/3}}{h^2}\right]^{3N_b/2}$$

$$\times \frac{\pi^{(\xi_a-5)N_a/2}}{\sigma_{s,a}^{N_a}}\left(\frac{T}{(\theta_{\text{rot},m})_a}\right)^{(\xi_a-3)N_a/2} \frac{\pi^{(\xi_b-5)N_b/2}}{\sigma_{s,b}^{N_b}}\left(\frac{T}{(\theta_{\text{rot},m})_b}\right)^{(\xi_b-3)N_b/2}$$

$$\times \exp\left[\left(a_{v,aa}N_a^2 + 2a_{v,ab}N_aN_b + a_{v,bb}N_b^2\right)/Vk_BT\right]. \tag{6.48}$$

Taking the natural log of both sides and using Stirling's approximation yields

$$\ln Q = N_a + N_b + \left(\frac{3N_a}{2}\right)\ln\left[\frac{2\pi M_a k_B T (V - N_a b_{v,a} - N_b b_{v,b})^{2/3}}{N_a^{2/3}h^2}\right]$$

$$+ \left(\frac{3N_b}{2}\right)\ln\left[\frac{2\pi M_b k_B T (V - N_a b_{v,a} - N_b b_{v,b})^{2/3}}{N_b^{2/3}h^2}\right]$$

$$+ N_a\left[\frac{(\xi_a-5)}{2}\ln\pi - \ln\sigma_{s,a}\right] + N_b\left[\frac{(\xi_b-5)}{2}\ln\pi - \ln\sigma_{s,b}\right]$$

$$+ \frac{(\xi_a-3)N_a}{2}\ln\left(\frac{T}{(\theta_{\text{rot},m})_a}\right) + \frac{(\xi_b-3)N_b}{2}\ln\left(\frac{T}{(\theta_{\text{rot},m})_b}\right)$$

$$+ \frac{a_{v,aa}N_a^2 + 2a_{v,ab}N_aN_b + a_{v,bb}N_b^2}{Vk_BT}. \tag{6.49}$$

Using Eqs. (4.41), (4.43), and (4.49) with the above relation for $\ln Q$, we obtain the following property relations:

$$\frac{S}{k_B} = \left(\frac{\xi_a}{2}+1\right)N_a + \left(\frac{\xi_b}{2}+1\right)N_b$$

$$+ \left(\frac{3N_a}{2}\right)\ln\left[\frac{2\pi M_a k_B T (V - N_a b_{v,a} - N_b b_{v,b})^{2/3}}{N_a^{2/3}h^2}\right]$$

$$+ \left(\frac{3N_b}{2}\right)\ln\left[\frac{2\pi M_b k_B T (V - N_a b_{v,a} - N_b b_{v,b})^{2/3}}{N_b^{2/3}h^2}\right]$$

$$+ N_a\left[\frac{(\xi_a-5)}{2}\ln\pi - \ln\sigma_{s,a}\right] + N_b\left[\frac{(\xi_b-5)}{2}\ln\pi - \ln\sigma_{s,b}\right]$$

$$+ \frac{(\xi_a-3)N_a}{2}\ln\left(\frac{T}{(\theta_{\text{rot},m})_a}\right) + \frac{(\xi_b-3)N_b}{2}\ln\left(\frac{T}{(\theta_{\text{rot},m})_b}\right), \tag{6.50}$$

$$U = \frac{\xi_a}{2}N_a k_B T + \frac{\xi_b}{2}N_b k_B T - \left(\frac{a_{v,aa}N_a^2 + 2a_{v,ab}N_aN_b + a_{v,bb}N_b^2}{V}\right), \tag{6.51}$$

$$P = \frac{(N_a+N_b)k_B T}{V - N_a b_{v,a} - N_b b_{v,b}} - \frac{a_{v,aa}N_a^2 + 2a_{v,ab}N_aN_b + a_{v,bb}N_b^2}{V^2}, \tag{6.52}$$

$$H = \left[\frac{\xi_a}{2}N_a + \frac{\xi_b}{2}N_b + \frac{(N_a+N_b)V}{V - N_a b_{v,a} - N_b b_{v,b}}\right]k_B T$$

$$- 2\left(\frac{a_{v,aa}N_a^2 + 2a_{v,ab}N_aN_b + a_{v,bb}N_b^2}{V}\right). \tag{6.53}$$

Relations for the chemical potential for each species can also be obtained using Eqs. (4.50) and (4.51).

The van der Waals type model described here is obviously highly idealized. It grossly simplifies the interaction between the two species of molecules in the binary mixture. For some systems, this model is too idealized to be of much use as a general predictive tool. It is presented here primarily because it illustrates how a microscale model of the molecular interactions can be used to generate property relations for a binary mixture in either a liquid or vapor phase. Its main value is that it illuminates how the interdependencies among thermodynamic properties that occur in real binary fluid mixtures are related to microscopic features of the system. In Chapter 8 we will use this model to explore the thermophysics of liquid–vapor phase equilibrium conditions in a system containing a binary mixture. This type of binary mixture model may also be useful as a starting point for development of empirical tools for predicting thermodynamic properties.

Although it embodies idealized treatments of excluded volume effects and particle force interactions, the van der Waals model remains attractive because of its simplicity. Over the past thirty years, there have been numerous efforts to develop augmented or generalized van der Waals equations of state. These efforts have typically sought to develop improved treatments of excluded volume effects and molecular force interactions within the van der Waals framework. This has led to a series of generalized van der Waals models that are simple and accurate enough to be useful for engineering analysis. Further information on efforts to develop generalized van der Waals models can be found in references by Kac et al. [6], Alder and Hecht [7], Rigby [8], Hoover et al. [9], Lebowitz and Waisman [10], Song and Mason [11], Nordholm et al. [12–14], Hooper and Nordholm [15–17] and Penfold et al. [18]. The earliest of these efforts focused on equations of state for pure systems, whereas more recent efforts have included development of a van der Waals model for vapor–liquid equilibrium in simple binary mixtures. Other approaches to modeling dense gases and liquids are discussed in the next section.

6.3   Other Models of Dense Gases and Liquids

The Redlich–Kwong Equation of State

Since the late 1940s numerous efforts have been made to develop an equation of state for pressure having better accuracy than the van der Waals model. In 1949 Redlich and Kwong [19] proposed an improved equation of state that is substantially more accurate than the van der Waals equation. This was originally introduced as an empirical refinement of the van der Waals equation of state relating P, V, N, and T. To gain a broader perspective on this model we will reconsider the model analysis presented in the previous section.

The analytical development of the van der Waals model for a pure fluid in the previous section lead us to Eq. (6.17), which can be written as

$$\Phi_N = -\frac{N^2}{V}\left[-2\pi \int_0^\infty r^2 \phi(r) g(r)\, dr\right]. \tag{6.54}$$

We subsequently introduced idealizations regarding $\phi(r)$ and $g(r)$ that led to the conclusion that the factor in the square brackets was a constant for a given fluid. To improve the model, we now relax this idealization by admitting the possibility that the term in square brackets is a function of T and density. To do so, we postulate that

$$-2\pi \int_0^\infty r^2 \phi(r) g(r) \, dr = a_R \gamma(T) \eta(V/N),$$ (6.55)

where a_R is a constant for a given fluid and $\gamma(T)$ and $\eta(V/N)$ are unknown functions of temperature and inverse density, respectively. Note that for $\gamma = \eta = 1$ this treatment reverts to the van der Waals model. Replacing the term in square brackets, we can write the relation (6.54) for Φ_N in the form

$$\Phi_N = -\frac{a_R \gamma \eta N^2}{V}.$$ (6.56)

Following the same line of analysis as for the van der Waals model in the previous section, the relation for the configuration integral becomes

$$Z_N = e^{a_R \gamma \eta N^2/V k_B T}(V - N b_R)^N,$$ (6.57)

where now the mean effective volume occupied by a molecule is denoted as b_R. Substituting into Eq. (6.16) and taking the log of both sides yields

$$\ln Q = N + \left(\frac{3N}{2}\right) \ln \left[\frac{2\pi M k_B T (V - N b_R)^{2/3}}{N^{2/3} h^2}\right] + N \left[\frac{(\xi - 5)}{2} \ln \pi - \ln \sigma_s\right]$$
$$+ \frac{(\xi - 3)N}{2} \ln \left(\frac{T}{\theta_{rot,m}}\right) + \frac{a_R \gamma \eta N^2}{V k_B T}.$$ (6.58)

Since $P = k_B T (\partial \ln Q/\partial V)_{T,N}$, differentiating the above relation yields the following equation of state:

$$P = \frac{N k_B T}{V - N b_R} - \frac{a_R \gamma N^2}{V^2} \left[\frac{\eta}{V} - \frac{1}{N}\left(\frac{d\eta}{dV}\right)\right].$$ (6.59)

The following relation was proposed by Redlich and Kwong [19] as an improved equation of state:

$$P = \frac{N k_B T}{V - N b_R} - \frac{a_R N^2}{T^{1/2} V (V + N b_R)}.$$ (6.60)

Comparing Eqs. (6.59) and (6.60), it is clear that the fluid modeled in our analysis will obey the Redlich–Kwong equation of state if we require that

$$\gamma = T^{-1/2},$$ (6.61)

$$\frac{\eta}{V} - \frac{1}{N}\left(\frac{d\eta}{dV}\right) = \frac{1}{V + N b_R}.$$ (6.62)

The differential Eq. (6.62) is easily solved in closed form. In doing so, we set the integration constant to zero to ensure that as the mean separation distance between molecules becomes large $(V/(N b_R) \to \infty)$, the effect of Φ on the partition function vanishes so that the resulting asymptotic behavior of Q is consistent with ideal gas behavior. The resulting relation for η is

$$\eta = \left(\frac{V}{N b_R}\right) \ln \left(\frac{V + N b_R}{V}\right).$$ (6.63)

Substituting (6.61) and (6.63) into Eq. (6.58) yields a relation for the log of the partition function for a fluid that obeys the Redlich–Kwong equation of state:

$$\ln Q = N + \left(\frac{3N}{2}\right)\ln\left[\frac{2\pi M k_B T (V - N b_R)^{2/3}}{N^{2/3}h^2}\right] + N\left[\frac{(\xi - 5)}{2}\ln\pi - \ln\sigma_s\right]$$

$$+ \frac{(\xi - 3)N}{2}\ln\left(\frac{T}{\theta_{rot,m}}\right) + \frac{a_R N}{b_R k_B T^{3/2}}\ln\left(\frac{V + N b_R}{V}\right). \tag{6.64}$$

Using Eqs. (4.50) and (4.43) and the definition of enthalpy, we obtain the following additional property relations for a Redlich–Kwong fluid:

$$\frac{\mu}{k_B T} = \frac{N b_R}{V - N b_R} - \left(\frac{3}{2}\right)\ln\left[\frac{2\pi M k_B T (V - N b_R)^{2/3}}{N^{2/3}h^2}\right]$$

$$- \left[\frac{(\xi - 5)}{2}\ln\pi - \ln\sigma_s\right] - \frac{(\xi - 3)}{2}\ln\left(\frac{T}{\theta_{rot,m}}\right)$$

$$- \frac{a_R}{b_R k_B T^{3/2}}\ln\left(\frac{V + N b_R}{V}\right) - \frac{a_R N}{k_B T^{3/2}(V + N b_R)}, \tag{6.65}$$

$$U = \frac{\xi N}{2}k_B T - \frac{3a_R N}{2b_R T^{1/2}}\ln\left(\frac{V + N b_R}{V}\right), \tag{6.66}$$

$$H = \left(\frac{\xi}{2} + \frac{V}{V - N b_R}\right)N k_B T - \frac{a_R N}{b_R T^{1/2}}\left[\frac{3}{2}\ln\left(\frac{V + N b_R}{V}\right) + \frac{N b_R}{V + N b_R}\right]. \tag{6.67}$$

Using Eq. (4.41) we can also obtain the following relation for the entropy:

$$S = N k_B\left(\frac{\xi}{2} + 1\right) + \frac{3N k_B}{2}\ln\left\{\frac{2\pi M k_B T (V - N b_R)^{2/3}}{N^{2/3}h^2}\right\} + \frac{(\xi - 3)N k_B}{2}$$

$$\times \ln\left\{\frac{T}{\theta_{rot,m}}\right\} + N k_B \ln\left\{\frac{\pi^{(\xi-5)/2}}{\sigma_s}\right\} - \frac{a_R N}{2b_R T^{3/2}}\ln\left(\frac{V + N b_R}{V}\right). \tag{6.68}$$

The relations obtained for U and S reflect the fact that this is an empirical model that has a limited range of applicability. Specifically, the last terms on the right side of these relations become infinitely large as $T \to 0$, which is physically impossible. It is expected, therefore, that the accuracy of this model becomes increasingly poor as the temperature is reduced at low temperature. Nevertheless, at moderate temperatures, this model is generally substantially better than the van der Waals model. Recommended constants for the Redlich–Kwong model for some fluids are summarized in Table 6.2.

Because Eq. (6.66) cannot be solved explicitly for T, it is not possible to combine Eqs. (6.66) and (6.68) to obtain a closed-form fundamental relation for a Redlich–Kwong fluid. However, these two relations provide a parametric representation of the fundamental relation for this fluid. The resulting fundamental equation surface for nitrogen is shown in Figure 6.5.

Table 6.2 *Redlich–Kwong Constants Determined from Critical Data for Selected Fluids*[†]

Molecule	$10^{48} \times a_R$ (Pa m^{6}K$^{1/2}$/molecule^2)	$10^{29} \times b_R$ (m^3/molecule)	ξ
CO_2	18.0	4.93	5
CO	4.81	4.56	5
CF_4H_2(R134a)	54.9	11.0	6
CH_4	8.97	4.93	6
N_2	4.34	4.45	5
O_2	4.84	3.65	5
H_2O	39.7	3.51	6

[†] Adapted from Wark [5].

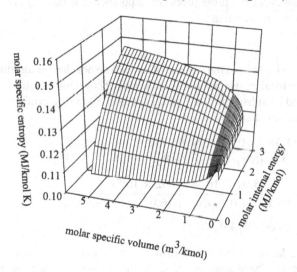

fundamental equation for nitrogen (Redlich-Kwong model)

Figure 6.5

Since the Redlich–Kwong equation was proposed, a number of other P–V–T equations of state have been developed in an effort to improve the accuracy of property prediction for dense gases and liquids. Generally these are empirically based and contain more constants that must be specified to optimize the fit to measured property values. Descriptions of these other equations of state can be found in the references by Wark [5] and Reid, Prausnitz, and Poling [20].

Example 6.3 To make popcorn kernels pop, they must be heated to approximately 204°C (400°F). Because the kernel is hard, there is little, if any, change in volume during heating, and the rise in temperature of liquid water inside the kernel is accompanied by an increase in pressure. (11–14% water content is considered optimal for popcorn kernels.) When the pressure becomes large enough, the hard shell of the kernel cracks, dropping the pressure suddenly. The superheated liquid inside the kernel then flashes to vapor explosively, causing the soft interior of the kernel to swell, producing a "popped" kernel. Estimate the pressure inside the kernel when it reaches 204°C.

Solution   Initially the kernel is presumed to be at room temperature, 20°C, and atmospheric pressure, 101 kPa. To determine the corresponding molar specific volume, we use the Redlich–Kwong equation of state (6.60). If we set $N = N_A$, then V is the molar specific volume, which we designate as $\hat{v}$. The equation of state then becomes.

$$P = \frac{N_A k_B T}{\hat{v} - N_A b_R} - \frac{a_R N_A^2}{T^{1/2} \hat{v}(\hat{v} + N_A b_R)}.$$

For water, $a_R = 3.97 \times 10^{-47}$ Pa m^{6}K$^{1/2}$/molecule^2 and $b_R = 3.51 \times 10^{-29}$ m^3/molecule. Substituting these values of the constants and setting $P = 101$ kPa and $T = 293$ K, we can determine the value of specific volume of the liquid in the kernel by iteratively solving the above equation for $\hat{v}$. (The lowest of the three roots corresponds to the liquid.) For liquid at these conditions, this process predicts that

$$\hat{v} = 0.0243 \text{ m}^3/\text{kmol}.$$

Since the amount of liquid in the kernel and the volume of its shell do not change significantly during the heating process, we idealize the process as occurring at constant specific volume. The pressure at 204°C is found by substituting the initial specific volume determined above into the Redlich–Kwong equation of state:

$$P = \frac{(6.02 \times 10^{26})(1.38 \times 10^{-23})(204 + 273)}{0.0243 - (6.02 \times 10^{26})(3.51 \times 10^{-29})}$$

$$- \frac{(3.97 \times 10^{-48})(6.02 \times 10^{26})^2}{(204 + 273)^{1/2}(0.0243)[0.0243 + (6.02 \times 10^{26})(3.51 \times 10^{-29})]}$$

$$= 6.36 \times 10^8 \text{ Pa} = 636,000 \text{ kPa}.$$

Thus, the pressure inside the kernel is estimated to reach more than 6,000 atmospheres if the kernel is heated to 204°C. The generation of such high pressures is a consequence of the fact that $(\partial P/\partial T)_{\hat{v}}$ is large for water at these conditions. By differentiating the equation of state and evaluating the derivative at 20°C and $\hat{v} = 0.0243$ m^3/kmol, it is found that $(\partial P/\partial T)_{\hat{v}} \cong 3.2 \times 10^6$ Pa/K for water. The high pressure that results from heating the kernel cracks the shell of the kernel, which is the first step in the cascade of events described above, which ultimately results in a "popped" kernel.

Example 6.4   Using the relations for the Redlich–Kwong model, determine the value of $\hat{c}_P$ for saturated liquid nitrogen and saturated nitrogen vapor at 100 K and 778 kPa. Compare the results to the recommended values indicated in Example 6.2.

Solution   As noted in Example 6.2, it can be shown by considering changes in H along a line of constant P that

$$\hat{c}_P = \left[\left(\frac{\partial H}{\partial T} \right)_{P,N} \right]_{N=N_A} = \left[\left(\frac{\partial H}{\partial T} \right)_{V,N} - \left(\frac{\partial H}{\partial V} \right)_{T,N} \frac{(\partial P/\partial T)_{V,N}}{(\partial P/\partial V)_{T,N}} \right]_{N=N_A}.$$

Differentiating Eqs. (6.67) and (6.60) as required and substituting Avogadro's number for N yields

$$\left(\frac{\partial H}{\partial T}\right)_{V,N} = \left(\frac{\xi}{2} + \frac{V}{V - N_A b_R}\right) N_A k_B + \frac{a_R N_A}{2 b_R T^{3/2}}$$
$$\times \left[\frac{3}{2}\ln\left(\frac{V + N_A b_R}{V}\right) + \frac{N_A b_R}{V + N_A b_R}\right],$$

$$\left(\frac{\partial H}{\partial V}\right)_{T,N} = \left(\frac{1}{V - N_A b_R} + \frac{V}{(V - N_A b_R)^2}\right) N_A k_B T$$
$$- \frac{a_R N_A}{b_R T^{1/2}}\left[\frac{3}{2}\left(\frac{1}{V + N_A b_R}\right) - \frac{3}{2}\left(\frac{1}{V}\right) - \frac{N_A b_R}{(V + N_A b_R)^2}\right],$$

$$\left(\frac{\partial P}{\partial T}\right)_{V,N} = \frac{N_A k_B}{V - N_A b_R} + \frac{a_R N_A^2}{2 T^{3/2} V (V + N_A b_R)},$$

$$\left(\frac{\partial P}{\partial V}\right)_{T,N} = -\frac{N_A k_B T}{(V - N_A b_R)^2} + \frac{a_R N_A^2}{T^{1/2} V^2 (V + N_A b_R)} + \frac{a_R N_A^2}{T^{1/2} V (V + N_A b_R)^2}.$$

For nitrogen, $a_R = 4.34 \times 10^{-48}$ Pa m^{6}K$^{1/2}$/molecule^2, $b_R = 4.45 \times 10^{-29}$ m^3/molecule, and $\xi = 5$. We first compute V/N_A for the liquid and vapor phases at specified temperature and pressure conditions using Eq. (6.26). (These are the lowest and highest of the three solution values for V/N_A.) Doing so, we find that $V/N_A = 9.99 \times 10^{-29}$ m^3 for the liquid phase and $V/N_A = 1.536 \times 10^{-27}$ m^3 for the vapor phase. Substituting into the above equation then yields

$$\hat{c}_P = 86.0 \text{ kJ/(kmol K) for the liquid,}$$
$$\hat{c}_P = 38.5 \text{ kJ/(kmol K) for the vapor.}$$

The liquid value is within 35% of the recommended value for liquid cited in Example 6.2. The vapor value is within 10% of the recommended value for vapor.

Example 6.5   One kmol of methane gas initially occupies a volume of 0.32 m^3 and has a temperature of 290 K. The gas expands reversibly and adiabatically until its volume is 0.64 m^3. Find the final temperature and pressure of the methane after the expansion process.

Solution   Evaluating the right side of Eq. (6.68) at the initial conditions, T_1 and V_1, and at the final conditions, T_2 and V_2, subtracting the results for the two cases, and rearranging yields

$$S_2 - S_1 = \frac{\xi N k_B}{2}\ln\left\{\frac{T_2}{T_1}\right\} + N k_B \ln\left\{\frac{V_2 - N b_R}{V_1 - N b_R}\right\}$$
$$- \frac{a_R N}{2 b_R T_2^{3/2}}\ln\left(\frac{V_2 + N b_R}{V_2}\right) + \frac{a_R N}{2 b_R T_1^{3/2}}\ln\left(\frac{V_1 + N b_R}{V_1}\right).$$

Because the process is reversible and adiabatic, the entropy of the system is unchanged and the right side of the above equation must equal zero. Since the system contains one kmol, we set N to N_A. Substituting N_A for N and the initial values for T and V, we obtain

$$\frac{\xi N_A k_B}{2} \ln\left\{\frac{T_2}{290}\right\} + N_A k_B \ln\left\{\frac{0.64 - N_A b_R}{0.32 - N_A b_R}\right\}$$

$$- \frac{a_R N_A}{2 b_R T_2^{3/2}} \ln\left(\frac{0.64 + N_A b_R}{0.64}\right) + \frac{a_R N_A}{2 b_R (290)^{3/2}} \ln\left(\frac{0.32 + N_A b_R}{0.32}\right) = 0.$$

For methane, $a_R = 8.97 \times 10^{-48}$ Pa m^{6}K$^{1/2}$/molecule^2, $b_R = 4.93 \times 10^{-29}$ m^3/molecule, and $\xi = 6$. Substituting these values and iteratively solving for T_2 yields

$$T_2 = 224 \text{ K}.$$

The Virial Expansion

In the remainder of this section we will discuss several other approaches to modeling dense gases and liquids. An alternative to the van der Waals model for dense gases is the *virial expansion*. This approach is based on the observation that the ideal gas equation of state is expected to be valid in the limit of low density $N/V \to 0$. As the density becomes larger, it is expected that the ideal gas relation will require progressively larger correction for the increasing effects of molecular interaction. The virial expansion is therefore postulated as an expansion in powers of density with the ideal gas relation as the lowest order approximation:

$$\frac{P}{k_B T} = \frac{N}{V}\left[1 + B\left(\frac{N}{V}\right) + C\left(\frac{N}{V}\right)^2 + \cdots\right]. \tag{6.69}$$

In the above relation, the $B, C, \ldots$ prefactors are referred to as the *virial coefficients*. Implementation of this approach requires that we develop a means of determining the virial coefficients. We can derive a relation for the first virial coefficient as follows. From the results of Chapter 4 we know that

$$\frac{P}{k_B T} = \left(\frac{\partial \ln Q}{\partial V}\right)_{T,N}. \tag{4.49}$$

From Eq. (6.16) we can see that the V dependence of $\ln Q$ resides only in the configuration integral Z_N. It follows that

$$\frac{P}{k_B T} = \left(\frac{\partial \ln Z_N}{\partial V}\right)_{T,N}, \tag{6.70}$$

where

$$Z_N = \int\int\int \cdots \int \exp[-\Phi(x_1, y_1, z_1, \ldots, x_N, y_N, z_N)/k_B T] \prod_{l-1}^{N} dx_l \, dy_l \, dz_l. \tag{6.11}$$

To first order, the potential energy is postulated to be equal to the sum of the potential energy contributions for each pair of molecules ϕ_{ij} in the system,

$$\Phi = \frac{1}{2} \sum_{i=1}^{N} \sum_{j=1}^{N} (\phi_{ij})_{j \neq i}. \tag{6.71}$$

Substituting Eq. (6.71) in (6.11) we get

$$Z_N = \int \int \int \cdots \int \exp\left(-\frac{1}{2} \sum_{i=1}^{N} \sum_{j=1}^{N} \left(\frac{\phi_{ij}}{k_B T}\right)_{j \neq i}\right) \prod_{l=1}^{N} dx_l \, dy_l \, dz_l. \tag{6.72}$$

For small deviations from ideal gas behavior, the potential energy interaction between molecules will be small compared to $k_B T$. Thus, each exponential factor in the integrand in Eq. (6.72) is very close to one. We therefore define

$$f_{ij} = e^{-\phi_{ij}/k_B T} - 1 \tag{6.73}$$

and rewrite the integrand in Eq. (6.72) as

$$\exp\left(-\frac{1}{2} \sum_{i=1}^{N} \sum_{j=1}^{N} \left(\frac{\phi_{ij}}{k_B T}\right)_{j \neq i}\right) = \prod_{i=1}^{N} \prod_{j=1}^{N} (1 + f_{ij})_{i \neq j}^{1/2}. \tag{6.74}$$

Because the f_{ij} terms are small compared to one we expand the product and neglect quadratic terms to obtain

$$\exp\left(-\frac{1}{2} \sum_{i=1}^{N} \sum_{j=1}^{N} \left(\frac{\phi_{ij}}{k_B T}\right)_{j \neq i}\right) = \left(1 + \frac{1}{2} \sum_{i=1}^{N} \sum_{j=1}^{N} (f_{ij})_{j \neq i}\right). \tag{6.75}$$

Inserting this result in the relation for Z_N yields

$$Z_N = \int \int \int \cdots \int \left(1 + \frac{1}{2} \sum_{i=1}^{N} \sum_{j=1}^{N} (f_{ij})_{j \neq i}\right) \prod_{l=1}^{N} dx_l \, dy_l \, dz_l. \tag{6.76}$$

Integration of the first term in the integrand gives simply V^N. Substituting the definition of f_{ij} we then obtain

$$Z_N = V^N + \int \int \int \cdots \int \left(\frac{1}{2} \sum_{i=1}^{N} \sum_{j=1}^{N} (e^{-\phi_{ij}/k_B T} - 1)_{j \neq i}\right) \prod_{l=1}^{N} dx_l \, dy_l \, dz_l. \tag{6.77}$$

Each term in the integrand summation is assumed to depend only on the separation distance between the ith and jth molecules. Integration of that term over all the $N - 2$ other sets of $dx_l, dy_l,$ and dz_l results in a prefactor of V^{N-2}:

$$Z_N = V^N + \frac{1}{2} \sum_{i=1}^{N} \sum_{j=1}^{N} \left(V^{N-2} \int \int \int \int \int \int (e^{-\phi_{ij}/k_B T} - 1)_{j \neq i} dx_i \, dy_i \, dz_i \, dx_j \, dy_j \, dz_j\right). \tag{6.78}$$

We replace the set of x_i, y_i, z_i coordinates by polar coordinates in which r is the separation distance between molecules i and j and integrate immediately over the angles since ϕ_{ij}

depends only on r. Integration over the x_j, y_j, z_j coordinates, on which ϕ_{ij} does not depend, yields another factor of V. The net result is

$$Z_N = V^N + \frac{V^{N-1}}{2} \sum_{i=1}^{N} \sum_{j=1}^{N} \left(\int_0^\infty \left(e^{-\phi_{ij}(r)/k_B T} - 1 \right)_{j \neq i} 4\pi r^2 \, dr \right), \qquad (6.79)$$

where we have set the upper integration limit to infinity since the integrand vanishes over a distance that is small compared to the extent of the system. Assuming the potential function is the same for all molecular pairs, we drop the subscript and find that the double summation is $(N^2 - N)$ times the resulting integral:

$$Z_N = V^N + V^{N-1} \left(\frac{(N^2 - N)}{2} \right) \int_0^\infty \left(e^{-\phi(r)/k_B T} - 1 \right) 4\pi r^2 \, dr \qquad (6.80)$$

Since N is much larger than one, we can further simplify this to

$$Z_N = V^N \left[1 + 2\pi (N^2/V) \int_0^\infty \left(e^{-\phi(r)/k_B T} - 1 \right) r^2 \, dr \right]. \qquad (6.81)$$

Substituting the right side of the above relation into Eq. (6.70), differentiating, and using the fact that the second term in the square brackets is small compared to one, it follows that to first order in this small term, the relation for $P/k_B T$ is

$$\frac{P}{k_B T} = \frac{N}{V} \left[1 - 2\pi \frac{N}{V} \int_0^\infty \left(e^{-\phi(r)/k_B T} - 1 \right) r^2 \, dr + \cdots \right]. \qquad (6.82)$$

Comparing this result with the virial expansion (6.69), it is clear that the second virial coefficient is given by

$$B = -2\pi \int_0^\infty \left(e^{-\phi(r)/k_B T} - 1 \right) r^2 \, dr. \qquad (6.83)$$

Equation (6.83) provides a link between the first macroscopic correction for nonideal gas effects and the intermolecular potential function. To develop accurate relations for the coefficients for higher order terms in the virial expansion, simultaneous potential interactions among more than two molecules must be considered and higher-order N/V terms must be retained throughout the analysis. Development of higher-order corrections associated with the interaction of clusters of more than two molecules is beyond the scope of this text. The interested reader can find a further discussion of this type of analysis in the texts by Hirschfelder, Curtis, and Bird [3], and McQuarrie [1] cited at the end of this chapter.

Other Models of Liquids

A variety of approaches have been proposed to develop microscopic models for liquids. As in the case of the van der Waals model described above, the usual goals of such modeling efforts are (1) to combine the model with statistical thermodynamic results to obtain macroscopic properties and (2) to facilitate an understanding of how microscopic features of the liquid affect macroscopic properties. An extensive discussion of microscale models of liquids is beyond the coverage of this text, but a sampling of other types of models will be described here.

The virial coefficient approach attempts to model a dense fluid as a very imperfect gas in which multiple particle collisions are frequent. In the analysis described above we considered only the first correction in the virial expansion, which depends only on pair interactions. By extending the analysis to consider higher-order terms and multiple particle interactions, it is possible to derive relations that are reasonable models of fluid behavior at higher density. Models of this type have been used in the study of condensation theory, but efforts to apply them at densities characteristic of liquids have yielded only limited useful information.

An alternate approach is to view the liquid as an imperfect crystal. Two versions of such models have been developed. Models of the first type are termed *cell theories*. Models of this type treat the liquid as a distorted crystal with a molecule located at or near each lattice point. The liquid nature of the system is reflected in a loss of long-range order in the system. Cell theories typically relate liquid properties to the lattice energy and the free volume in the lattice. The second category of models are *hole theories*. These attempt to account for the fact that liquids differ from crystals in that motion of the molecules would leave some of the lattice sites vacant. A discussion of cell and hole theories and equations of state that can be derived from them can be found in Hirschfelder, Curtis, and Bird [3], and Hansen and McDonald [21].

With the rapid increase in computing power over the past two decades, efforts to model liquids have increasingly focused on molecular simulation tools that computationally account for energy interactions among all the molecules in a model system. There are now very broad and active ongoing efforts to develop Monte-Carlo type stochastic simulations and deterministic simulation methods that can be used to explore the relationship between molecular characteristics and macroscopic liquid properties. The interested reader may find useful summaries of the foundations of such techniques in the texts by Hansen and McDonald [21], Allen and Tildesley [22], and Haile [23].

6.4   Analysis of Fluids with Significant Quantum Effects

In Chapter 4 we simplified the formulation of statistical thermodynamics by considering conditions in which dilute occupancy applied. We now will reconsider the more general case of a system in which quantum effects may play an important role. To explore the thermodynamics of fluids in which quantum effects are important, we will develop our analysis in terms of the grand canonical ensemble. We consider a binary mixture of particles of type a and b and define

$E_{j'}(N_a, N_b, V) =$ the energy microstates available to a system with N_a particles
of type a, N_b particles of type b, and volume V,

$\varepsilon_{l'} =$ energy of quantum state l' for particles of type a,

$n_{l'}(E_{j'}) =$ the number of type a particles in particle microstate l' when
the system is in system microstate $E_{j'}$,

$\varepsilon_{k'} =$ energy of quantum state k' for particles of type b,

$n_{k'}(E_{j'}) =$ the number of type b particles in particle microstate k' when
the system is in system microstate $E_{j'}$.

Together, the sets $\{n_{l'}\}$ and $\{n_{k'}\}$ specify the system quantum state. It follows that

$$E_{j'} = \sum_{l'=0}^{\infty} \varepsilon_{l'} n_{l'} + \sum_{k'=0}^{\infty} \varepsilon_{k'} n_{k'}, \tag{6.84}$$

$$N_a = \sum_{l'=0}^{\infty} n_{l'}, \tag{6.85a}$$

$$N_b = \sum_{k'=0}^{\infty} n_{k'}, \tag{6.85b}$$

and the canonical partition function for the system is given by

$$Q(N_a, N_b, V, T) = \sum_{j'=0}^{\infty} e^{-E_{j'}/k_{\mathrm{B}}T}, \tag{6.86}$$

which, using Eq. (6.1), we can write as

$$Q(N_a, N_b, V, T) = \sum_{\{n_{l'}\}^*} \sum_{\{n_{k'}\}^*} \exp\left[-\sum_{l'=0}^{\infty} \varepsilon_{l'} n_{l'}/k_{\mathrm{B}}T - \sum_{k'=0}^{\infty} \varepsilon_{k'} n_{k'}/k_{\mathrm{B}}T \right]. \tag{6.87}$$

The $\{n_{l'}\}^*$ and $\{n_{k'}\}^*$ notation in (6.87) implies that we sum over all sets of $n_{l'}$ and $n_{k'}$ that satisfy Eqs. (6.85a) and (6.85b). By definition, the grand canonical partition function for a binary system is given by

$$\Xi(V, T, \mu_a, \mu_b) = \sum_{N_a=0}^{\infty} \sum_{N_b=0}^{\infty} e^{(\mu_a N_a + \mu_b N_b)/k_{\mathrm{B}}T} Q(N_a, N_b, V, T). \tag{6.88}$$

We define

$$\lambda_a = e^{\mu_a/k_{\mathrm{B}}T}, \tag{6.89}$$

$$\lambda_b = e^{\mu_b/k_{\mathrm{B}}T} \tag{6.90}$$

and substitute Eq. (6.87) for Q to get

$$\begin{aligned} \Xi(V, T, \mu_a, \mu_b) &= \sum_{N_a=0}^{\infty} \sum_{N_b=0}^{\infty} \lambda_a^{N_a} \lambda_b^{N_b} \\ &\quad \times \sum_{\{n_{l'}\}^*} \sum_{\{n_{k'}\}^*} \exp\left[-\sum_{l'=0}^{\infty} \varepsilon_{l'} n_{l'}/k_{\mathrm{B}}T - \sum_{k'=0}^{\infty} \varepsilon_{k'} n_{k'}/k_{\mathrm{B}}T \right] \\ &= \sum_{N_a=0}^{\infty} \lambda_a^{N_a} \sum_{\{n_{l'}\}^*} \exp\left[-\sum_{l'=0}^{\infty} \varepsilon_{l'} n_{l'}/k_{\mathrm{B}}T \right] \\ &\quad \times \sum_{N_b=0}^{\infty} \lambda_b^{N_b} \sum_{\{n_{k'}\}^*} \exp\left[-\sum_{k'=0}^{\infty} \varepsilon_{k'} n_{k'}/k_{\mathrm{B}}T \right]. \end{aligned} \tag{6.91}$$

Using Eqs. (6.85a) and (6.85b), we replace the exponents N_a and N_b by the summations over all $n_{l'}$ and $n_{k'}$:

$$\Xi(V, T, \mu_a, \mu_b) = \sum_{N_a=0}^{\infty} \lambda_a^{\Sigma_{l'=0}^{\infty} n_{l'}} \sum_{\{n_{l'}\}^*} \exp\left[-\sum_{l'=0}^{\infty} \varepsilon_{l'} n_{l'}/k_{\mathrm{B}}T\right]$$
$$\times \sum_{N_b=0}^{\infty} \lambda_b^{\Sigma_{k'=0}^{\infty} n_{k'}} \sum_{\{n_{k'}\}^*} \exp\left[-\sum_{k'=0}^{\infty} \varepsilon_{k'} n_{k'}/k_{\mathrm{B}}T\right]. \tag{6.92}$$

We then reorganize Eq. (6.92) and write the resulting relation in the form

$$\Xi(V, T, \mu_a, \mu_b) = \xi_a \xi_b, \tag{6.93}$$

where

$$\xi_a = \sum_{N_a=0}^{\infty} \sum_{\{n_{l'}\}^*} \prod_{l'=0}^{\infty} \left(\lambda_a e^{\varepsilon_{l'} n_{l'}/k_{\mathrm{B}}T}\right)^{n_{l'}}, \tag{6.94}$$

$$\xi_b = \sum_{N_b=0}^{\infty} \sum_{\{n_{k'}\}^*} \prod_{k'=0}^{\infty} \left(\lambda_b e^{\varepsilon_{k'} n_{k'}/k_{\mathrm{B}}T}\right)^{n_{k'}}. \tag{6.95}$$

In the relation for ξ_a we are summing over all values of N_a from zero to infinity and, consequently, each $n_{l'}$ ranges over all possible values. As a result, the double summation in Eq. (6.94) can be reorganized to obtain

$$\xi_a = \sum_{n_1=0}^{n_{1,\max}} \sum_{n_2=0}^{n_{2,\max}} \sum_{n_3=0}^{n_{3,\max}} \cdots \prod_{l'=0}^{\infty} \left(\lambda_a e^{-\varepsilon_{l'} n_{l'}/k_{\mathrm{B}}T}\right)^{n_{l'}}. \tag{6.96}$$

This result can be written in the form

$$\xi_a = \sum_{n_1=0}^{n_{1,\max}} \left(\lambda_a e^{-\varepsilon_1/k_{\mathrm{B}}T}\right)^{n_1} \sum_{n_2=0}^{n_{2,\max}} \left(\lambda_a e^{-\varepsilon_2/k_{\mathrm{B}}T}\right)^{n_2} \sum_{n_3=0}^{n_{3,\max}} \left(\lambda_a e^{-\varepsilon_3/k_{\mathrm{B}}T}\right)^{n_3} \cdots \tag{6.97}$$

or in the even more compact form

$$\xi_a = \prod_{l'=0}^{\infty} \sum_{n_{l'}=0}^{n_{l',\max}} \left(\lambda_a e^{-\varepsilon_{l'}/k_{\mathrm{B}}T}\right)^{n_{l'}}. \tag{6.98}$$

Note that we have allowed for the fact that each $n_{l'}$ may have a maximum value that it may attain, $n_{l',\max}$. We can, however, let $n_{l',\max}$ be infinity, when appropriate. Some understanding of why this transformation of the relation for ξ_a is valid can be obtained by considering Example 6.6.

Example 6.6 Show that the transformation from Eq. (6.94) to (6.98) is valid for $l' = 1, 2$ with possible n_1 and n_2 values of 0, 1 and 2.

    Solution If we let

$$\zeta_{l'} = \lambda_a e^{-\varepsilon_{l'}/k_{\mathrm{B}}T}$$

Eq. (6.79) above can be written as

$$\xi_a = \sum_{N_a=0}^{\infty} \sum_{\{n_{l'}\}^*} \prod_{l'=0}^{\infty} \zeta_{l'}^{n_{l'}}.$$

For these conditions, ξ_a is determined as follows:

$$\xi_a = \left(\sum_{\{n_{l'}\}^*} \zeta_1^{n_1} \zeta_2^{n_2} \right)_{N_a=0} + \left(\sum_{\{n_{l'}\}^*} \zeta_1^{n_1} \zeta_2^{n_2} \right)_{N_a=1} + \left(\sum_{\{n_{l'}\}^*} \zeta_1^{n_1} \zeta_2^{n_2} \right)_{N_a=2}$$

$$+ \left(\sum_{\{n_{l'}\}^*} \zeta_1^{n_1} \zeta_2^{n_2} \right)_{N_a=3} + \left(\sum_{\{n_{l'}\}^*} \zeta_1^{n_1} \zeta_2^{n_2} \right)_{N_a=4} + \left(\sum_{\{n_{l'}\}^*} \zeta_1^{n_1} \zeta_2^{n_2} \right)_{N_a=5} + \cdots.$$

Note that for each term in the summation, the sets of n_1 and n_2 must be chosen so $n_1 + n_2 = N_a$. Evaluating the sums yields

$$\xi_a = \left(\zeta_1^0 \zeta_2^0 \right) + \left(\zeta_1^0 \zeta_2^1 + \zeta_1^1 \zeta_2^0 \right) + \left(\zeta_1^0 \zeta_2^2 + \zeta_1^1 \zeta_2^1 + \zeta_1^2 \zeta_2^0 \right)$$
$$+ \left(\zeta_1^2 \zeta_2^1 + \zeta_1^1 \zeta_2^2 \right) + \left(\zeta_1^2 \zeta_2^2 \right) + (0).$$
$$= 1 + \zeta_2 + \zeta_1 + \zeta_2^2 + \zeta_1 \zeta_2 + \zeta_1^2 + \zeta_1^2 \zeta_2 + \zeta_1 \zeta_2^2 + \zeta_1^2 \zeta_2^2.$$

Factoring the polynomial, we obtain

$$\xi_a = \left(1 + \zeta_1 + \zeta_1^2\right)\left(1 + \zeta_2 + \zeta_2^2\right) = \prod_{l'=1}^{2} \sum_{n_{l'}=0}^{2} (\zeta_{l'})^{n_{l'}},$$

which for $\zeta_{l'} = \lambda_a e^{-\varepsilon_{l'}/k_B T}$ has the general form of Eq. (6.98).

The rearrangement to convert Eq. (6.94) to Eq. (6.98) can also be done to the relation (6.95) for ξ_b. Doing so produces the relation

$$\xi_b = \prod_{k'=0}^{\infty} \sum_{n_{k'}=0}^{n_{k',\max}} \left(\lambda_b e^{-\varepsilon_{k'}/k_B T} \right)^{n_{k'}}. \tag{6.99}$$

Substituting Eqs. (6.98) and (6.99) into Eq. (6.93), we obtain the following relation for the partition function:

$$\Xi = \prod_{l'=0}^{\infty} \left[\sum_{n_{l'}=0}^{n_{l',\max}} \left(\lambda_a e^{-\varepsilon_{l'}/k_B T} \right)^{n_{l'}} \right] \prod_{k'=0}^{\infty} \left[\sum_{n_{k'}=0}^{n_{k',\max}} \left(\lambda_b e^{-\varepsilon_{k'}/k_B T} \right)^{n_{k'}} \right]. \tag{6.100}$$

Equation (6.100) is a general result that we can apply to fermions or bosons. Because these results can be applied to systems whose behavior is dictated by quantum effects, they are generally termed *quantum statistics*. *Fermi–Dirac statistics* are obtained by applying these results to fermions. For fermions $n_{l'}$ and $n_{k'}$ can only be either 0 or 1, since no two fermion particles may be in the same quantum state. Equation (6.100) for Ξ then reduces to

$$\Xi(V, T, \mu_a, \mu_b) = \prod_{l'=0}^{\infty} \left(1 + \lambda_a e^{-\varepsilon_{l'}/k_B T}\right) \prod_{k'=0}^{\infty} \left(1 + \lambda_b e^{-\varepsilon_{k'}/k_B T}\right) \text{(for fermions)}.$$

$$\tag{6.101}$$

Bose–Einstein statistics are based on an absence of restrictions on quantum state occupancy. For bosons, the $n_{l'}$ and $n_{k'}$ can be 0, 1, 2, ... and $n_{l',\max}$ and $n_{k',\max}$ are infinite. The relation (6.100) for Ξ therefore becomes

$$\Xi(V, T, \mu_a, \mu_b) = \prod_{l'=0}^{\infty} \left[\sum_{n_{l'}=0}^{\infty} \left(\lambda_a e^{-\varepsilon_{l'}/k_B T} \right)^{n_{l'}} \right] \prod_{k'=0}^{\infty} \left[\sum_{n_{k'}=0}^{\infty} \left(\lambda_b e^{-\varepsilon_{k'}/k_B T} \right)^{n_{k'}} \right].$$

(6.102)

We will assume that each of the $\lambda_a e^{-\varepsilon_{l'}/k_B T}$ and $\lambda_b e^{-\varepsilon_{k'}/k_B T}$ terms in the summations is less than one. Using the fact that for $y < 1$

$$\sum_{m=0}^{\infty} y^m = (1 - y)^{-1}$$

(6.103)

we can then write Eq. (6.102) as

$$\Xi(V, T, \mu_a, \mu_b) = \prod_{l'=0}^{\infty} \left(1 - \lambda_a e^{-\varepsilon_{l'}/k_B T} \right)^{-1} \prod_{k'=0}^{\infty} \left(1 - \lambda_b e^{-\varepsilon_{k'}/k_B T} \right)^{-1} \quad \text{(for bosons)}.$$

(6.104)

These two fundamental distributions of the statistical mechanics of a binary mixture of independent particles can be written in the following compact form:

$$\Xi_{f/b} = \prod_{l'=0}^{\infty} \left(1 \pm \lambda_a e^{-\varepsilon_{l'}/k_B T} \right)^{\pm 1} \prod_{k'=0}^{\infty} \left(1 \pm \lambda_b e^{-\varepsilon_{k'}/k_B T} \right)^{\pm 1},$$

(6.105)

where, as the notation suggests, the upper sign applies for fermions and the lower sign applies for bosons.

In our analysis of the grand canonical ensemble of systems containing a binary mixture, we showed that

$$N_a = k_B T \left(\frac{\partial \ln \Xi}{\partial \mu_a} \right)_{V,T},$$

(6.106)

$$N_b = k_B T \left(\frac{\partial \ln \Xi}{\partial \mu_b} \right)_{V,T},$$

(6.107)

and since

$$N_a = \langle N_a \rangle = \left\langle \sum_{l'=0}^{\infty} n_{l'} \right\rangle = \sum_{l'=0}^{\infty} \langle n_{l'} \rangle,$$

(6.108)

$$N_b = \langle N_b \rangle = \left\langle \sum_{k'=0}^{\infty} n_{k'} \right\rangle = \sum_{k'=0}^{\infty} \langle n_{k'} \rangle,$$

(6.109)

it follows that

$$\sum_{l'=0}^{\infty} \langle n_{l'} \rangle = k_B T \left(\frac{\partial \ln \Xi}{\partial \mu_a} \right)_{V,T},$$

(6.110)

$$\sum_{k'=0}^{\infty} \langle n_{k'} \rangle = k_B T \left(\frac{\partial \ln \Xi}{\partial \mu_b} \right)_{V,T}.$$

(6.111)

The definitions of λ_a and λ_b dictate that

$$k_{\mathrm{B}}T \left(\frac{\partial \ln \Xi}{\partial \mu_a} \right)_{V,T} = \lambda_a \left(\frac{\partial \ln \Xi}{\partial \lambda_a} \right)_{V,T}, \tag{6.112}$$

$$k_{\mathrm{B}}T \left(\frac{\partial \ln \Xi}{\partial \mu_b} \right)_{V,T} = \lambda_b \left(\frac{\partial \ln \Xi}{\partial \lambda_b} \right)_{V,T}. \tag{6.113}$$

Using Eqs. (6.112) and (6.113) to replace the derivatives with respect to chemical potential with derivatives with respect to λ_a and λ_b, Eqs. (6.100) and (6.101) become

$$\sum_{l'=0}^{\infty} \langle n_{l'} \rangle = \lambda_a \left(\frac{\partial \ln \Xi}{\partial \lambda_a} \right)_{V,T}, \tag{6.114}$$

$$\sum_{k'=0}^{\infty} \langle n_{k'} \rangle = \lambda_b \left(\frac{\partial \ln \Xi}{\partial \lambda_b} \right)_{V,T}. \tag{6.115}$$

Substituting the relation (6.105) into Eq. (6.114), we obtain

$$\sum_{l'=0}^{\infty} \langle n_{l'} \rangle = \lambda_a \left(\frac{\partial}{\partial \lambda_a} \ln \left[\prod_{l'=0}^{\infty} \left(1 \pm \lambda_a e^{-\varepsilon_{l'}/k_{\mathrm{B}}T} \right)^{\pm 1} \prod_{k'=0}^{\infty} \left(1 \pm \lambda_b e^{-\varepsilon_{k'}/k_{\mathrm{B}}T} \right)^{\pm 1} \right] \right)_{V,T}$$

$$= \lambda_a \left(\frac{\partial}{\partial \lambda_a} \left[\sum_{l'=0}^{\infty} \ln \left\{ \left(1 \pm \lambda_a e^{-\varepsilon_{l'}/k_{\mathrm{B}}T} \right)^{\pm 1} \right\} \right. \right.$$

$$\left. \left. + \sum_{k'=0}^{\infty} \ln \left\{ \left(1 \pm \lambda_b e^{-\varepsilon_{k'}/k_{\mathrm{B}}T} \right)^{\pm 1} \right\} \right] \right)_{V,T}$$

$$= \lambda_a \sum_{l'=0}^{\infty} (\pm)(\pm) \left(\frac{e^{-\varepsilon_{l'}/k_{\mathrm{B}}T}}{1 \pm \lambda_a e^{-\varepsilon_{l'}/k_{\mathrm{B}}T}} \right)$$

$$= \sum_{l'=0}^{\infty} \left(\frac{\lambda_a e^{-\varepsilon_{l'}/k_{\mathrm{B}}T}}{1 \pm \lambda_a e^{-\varepsilon_{l'}/k_{\mathrm{B}}T}} \right). \tag{6.116}$$

Similar substitution of Eq. (6.105) into Eq. (6.115) and evaluation of the derivative yields

$$\sum_{k'=0}^{\infty} \langle n_{k'} \rangle = \sum_{k'=0}^{\infty} \left(\frac{\lambda_b e^{-\varepsilon_{k'}/k_{\mathrm{B}}T}}{1 \pm \lambda_b e^{-\varepsilon_{k'}/k_{\mathrm{B}}T}} \right). \tag{6.117}$$

The right and left sides of Eqs. (6.116) and (6.117) are equal only if the respective terms in the sums are equal. This indicates that the average number of type a particles in quantum state l' and the average number of type b particles in quantum state k' are given by

$$\langle n_{l'} \rangle = \left(\frac{\lambda_a e^{-\varepsilon_{l'}/k_{\mathrm{B}}T}}{1 \pm \lambda_a e^{-\varepsilon_{l'}/k_{\mathrm{B}}T}} \right), \tag{6.118}$$

$$\langle n_{k'} \rangle = \left(\frac{\lambda_b e^{-\varepsilon_{k'}/k_{\mathrm{B}}T}}{1 \pm \lambda_b e^{-\varepsilon_{k'}/k_{\mathrm{B}}T}} \right). \tag{6.119}$$

As discussed in Chapter 4, in the limit of high temperature or low density, the number of available particle quantum states is much greater than the number of particles. For such conditions, the average number of particles in any state is very small (i.e., we have dilute occupancy). These arguments imply that $N/V \to 0$ for fixed T or $T \to \infty$ for fixed N/V corresponds to $\langle n_{l'} \rangle \to 0$ and $\langle n_{k'} \rangle \to 0$. From Eqs. (6.118) and (6.119) it is clear that $\langle n_{l'} \rangle \to 0$ must correspond to $\lambda_a e^{-\varepsilon_{l'}/k_\mathrm{B}T} \to 0$ and $\langle n_{k'} \rangle \to 0$ must correspond to $\lambda_b e^{-\varepsilon_{k'}/k_\mathrm{B}T} \to 0$. Hence, in the low density/high temperature dilute occupancy limit

$$\langle n_{l'} \rangle = \lambda_a e^{-\varepsilon_{l'}/k_\mathrm{B}T}, \tag{6.120}$$

$$\langle n_{k'} \rangle = \lambda_b e^{-\varepsilon_{k'}/k_\mathrm{B}T}. \tag{6.121}$$

Summing both sides of Eq. (6.120) over l' and both sides of (6.121) over k' yields

$$\sum_{l'=0}^{\infty} \langle n_{l'} \rangle = N_a = \lambda_a \sum_{l'=0}^{\infty} e^{-\varepsilon_{l'}/k_\mathrm{B}T} = \lambda_a q_a, \tag{6.122}$$

$$\sum_{k'=0}^{\infty} \langle n_{k'} \rangle = N_b = \lambda_b \sum_{k'=0}^{\infty} e^{-\varepsilon_{k'}/k_\mathrm{B}T} = \lambda_b q_b. \tag{6.123}$$

Combining Eq. (6.120) with (6.122) and Eq. (6.121) with (6.123) yields

$$\frac{\langle n_{l'} \rangle}{N_a} = \frac{e^{-\varepsilon_{l'}/k_\mathrm{B}T}}{q_a}, \tag{6.124}$$

$$\frac{\langle n_{k'} \rangle}{N_b} = \frac{e^{-\varepsilon_{k'}/k_\mathrm{B}T}}{q_b}. \tag{6.125}$$

Taking the natural log of both sides of Eq. (6.105), we obtain

$$\ln \Xi = \pm \sum_{l'=0}^{\infty} \ln \left\{ 1 \pm \lambda_a e^{-\varepsilon_{l'}/k_\mathrm{B}T} \right\} \pm \sum_{k'=0}^{\infty} \ln \left\{ 1 \pm \lambda_b e^{-\varepsilon_{k'}/k_\mathrm{B}T} \right\}. \tag{6.126}$$

Expanding the summation terms in a Taylor series for small $\lambda_a e^{-\varepsilon_{l'}/k_\mathrm{B}T}$ and small $\lambda_b e^{-\varepsilon_{k'}/k_\mathrm{B}T}$ and retaining only the first term in each, the above equation transforms to

$$\ln \Xi = \pm \left[\pm \lambda_a \sum_{l'=0}^{\infty} e^{-\varepsilon_{l'}/k_\mathrm{B}T} \pm \lambda_b \sum_{k'=0}^{\infty} e^{-\varepsilon_{k'}/k_\mathrm{B}T} \right], \tag{6.127}$$

which reduces to

$$\ln \Xi = \lambda_a q_a + \lambda_b q_b \tag{6.128}$$

or equivalently

$$\Xi = e^{\lambda_a q_a} e^{\lambda_b q_b}. \tag{6.129}$$

Replacing the exponentials with the series representation

$$e^x = \sum_{N=0}^{\infty} \frac{x^N}{N!} \tag{6.130}$$

Eq. (6.129) can be written in the form

$$\Xi = \left(\sum_{N_a=0}^{\infty} \frac{(\lambda_a q_a)^{N_a}}{N_a!} \right) \left(\sum_{N_b=0}^{\infty} \frac{(\lambda_b q_b)^{N_b}}{N_b!} \right) = \sum_{N_a=0}^{\infty} \sum_{N_b=0}^{\infty} \left(\frac{\lambda_a^{N_a} q_a^{N_a}}{N_a!} \right) \left(\frac{\lambda_b^{N_b} q_b^{N_b}}{N_b!} \right). \tag{6.131}$$

We showed previously that for a grand canonical ensemble of systems containing a binary mixture of particles

$$\Xi = \sum_{N_a=0}^{\infty} \sum_{N_b=0}^{\infty} e^{(\mu_a N_a + \mu_b N_b)/k_B T} Q = \sum_{N_a=0}^{\infty} \sum_{N_b=0}^{\infty} \lambda_a^{N_a} \lambda_b^{N_b} Q. \tag{6.132}$$

Comparing this result with that immediately above, it is clear that in the dilute occupancy limit

$$Q = \left(\frac{q_a^{N_a}}{N_a!} \right) \left(\frac{q_b^{N_b}}{N_b!} \right). \tag{6.133}$$

Equation (6.133) is identical to the relation (4.173) for a system exhibiting dilute occupancy. Thus, in the dilute occupancy limit, the Fermi–Dirac and Bose–Einstein statistics results obtained in this section are consistent with the results obtained in Chapter 4 for systems exhibiting dilute occupancy. Equations (6.105), (6.118), and (6.119) provide a starting point for the thermodynamic analysis of fluids in which quantum effects are important.

6.5   Fermion and Boson Gases

Having developed a statistical framework for fluids that exhibit quantum effects in the previous section, we now will explore how quantum effects influence the behavior of different gases. We will consider two categories of quantum gases. The first is a gas comprised of indistinguishable fermions. (Recall that fermions have half-integral spin and no two particles of this type may occupy the same quantum state.) Particles of this type include electrons and protons.

The second type of gas to be considered here is one containing indistinguishable bosons. (Recall that bosons have integral spin and have no restrictions on state occupancy.) Particles of this type that exhibit quantum effects include photons and deuterons (nuclei of the deuterium atom consisting of one neutron and one proton).

Results obtained in the previous section indicate that for either of these gas types containing one species of particle

$$\Xi_{f/b} = \prod_{k'=0}^{\infty} \left(1 \pm \lambda e^{-\varepsilon_{k'}/k_B T} \right)^{\pm 1}, \tag{6.134}$$

$$\langle n_{k'} \rangle_{f/b} = \frac{\lambda e^{-\varepsilon_{k'}/k_B T}}{1 \pm \lambda e^{-\varepsilon_{k'}/k_B T}}, \tag{6.135}$$

where

$$\lambda = e^{\mu/k_B T}. \tag{6.136}$$

Where a $\pm$ appears in Eqs. (6.134) and (6.135) the upper sign applies for fermion gases and the lower for boson gases. It follows immediately from the above results and Eq. (4.119) that for either a fermion or boson gas

$$(PV)_{f/b} = \pm k_B T \sum_{k'=0}^{\infty} \ln \left(1 \pm \lambda e^{-\varepsilon_{k'}/k_B T} \right). \tag{6.137}$$

Also, since the sum over all n'_k must equal N,

$$N_{f/b} = \sum_{k'=0}^{\infty} \left(\frac{\lambda e^{-\varepsilon_{k'}/k_B T}}{1 \pm \lambda e^{-\varepsilon_{k'}/k_B T}} \right). \qquad (6.138)$$

It was shown in the previous section that in the classical (dilute occupancy) limit, the relations for Fermi–Dirac and Bose–Einstien statistics reduce to those obtained using classical Boltzmann statistics. Systems that exhibit deviations from classical behavior are said to be degenerate. Strongly degenerate thus implies large quantum effects, whereas weakly degenerate implies weak but significant quantum effects.

A Weakly Degenerate Fermion Gas

We wish to consider the behavior of the general relations

$$N = \sum_{k'=0}^{\infty} \left(\frac{\lambda e^{-\varepsilon_{k'}/k_B T}}{1 + \lambda e^{-\varepsilon_{k'}/k_B T}} \right), \qquad (6.139)$$

$$PV = k_B T \sum_{k'=0}^{\infty} \ln \left(1 + \lambda e^{-\varepsilon_{k'}/k_B T} \right) \qquad (6.140)$$

in the limit of small but significant λ for a fermion gas of particles having translational energy storage only. From our analysis in Chapter 1, we know that for translation of a particle in a 3-D cubical box, the quantum energy levels are given by

$$\varepsilon = \frac{h^2}{8mV^{2/3}} (n_x^2 + n_y^2 + n_z^2), \qquad n_x, n_y, n_z = 1, 2, 3, \ldots. \qquad (6.141)$$

This relation indicates that the translational energy levels are close together for large V. We therefore invoke a high volume limit to allows us to treat ε as a continuous variable. As shown in Chapter 1, the degeneracy of the energy levels given by the above relation is

$$g(\varepsilon) \, d\varepsilon = \frac{\pi}{4} \left(\frac{8m}{h^2} \right)^{3/2} V \varepsilon^{1/2} \, d\varepsilon. \qquad (6.142)$$

For the large V limit, we therefore replace the summation over all quantum states with an integral over all energy values, with a multiplying factor included to account for the degeneracy of the energy distribution:

$$N = \int_0^{\infty} g(\varepsilon) \frac{\lambda e^{-\varepsilon/k_B T} \, d\varepsilon}{1 + \lambda e^{-\varepsilon/k_B T}}, \qquad (6.143)$$

$$PV = k_B T \int_0^{\infty} g(\varepsilon) \ln \left(1 + \lambda e^{-\varepsilon/k_B T} \right) d\varepsilon. \qquad (6.144)$$

Substituting for $g(\varepsilon)$ yields

$$N = \frac{\pi}{4} \left(\frac{8m}{h^2} \right)^{3/2} V \int_0^{\infty} \frac{\lambda \varepsilon^{1/2} e^{-\varepsilon/k_B T} \, d\varepsilon}{1 + \lambda e^{-\varepsilon/k_B T}}, \qquad (6.145)$$

$$PV = \frac{\pi k_B T}{4} \left(\frac{8m}{h^2} \right)^{3/2} V \int_0^{\infty} \varepsilon^{1/2} \ln \left(1 + \lambda e^{-\varepsilon/k_B T} \right) d\varepsilon. \qquad (6.146)$$

The integrals in the above expressions cannot be evaluated in closed form. However, we note from the results in the previous section that the high temperature classical limit corresponds to $\lambda e^{-\varepsilon/k_B T} \to 0$. Since high temperatures correspond to $e^{-\varepsilon/k_B T}$ near one at the lowest energy levels, this implies that the classical limit corresponds to $\lambda \to 0$. We therefore expand the integrands in the above relations as power series in λ and integrate term by term to obtain the following expansions for N/V and P:

$$\frac{N}{V} = \frac{1}{\Lambda^3} \sum_{j=1}^{\infty} \frac{(-1)^{j+1}\lambda^j}{j^{3/2}}, \tag{6.147}$$

$$\frac{P}{k_B T} = \frac{1}{\Lambda^3} \sum_{j=1}^{\infty} \frac{(-1)^{j+1}\lambda^j}{j^{5/2}}, \tag{6.148}$$

where

$$\Lambda = \left(\frac{h^2}{2\pi m k_B T} \right)^{1/2}. \tag{6.149}$$

An expansion for $P/k_B T$ as a power series in N/V is postulated as having the form

$$\frac{P}{k_B T} = \frac{N}{V} + B\left(\frac{N}{V}\right)^2 + C\left(\frac{N}{V}\right)^3 + \cdots.$$

The unknown coefficients are determined by substituting the above expansions for $P/k_B T$ and N/V and equating coefficients for each power of λ. This process yields the following series:

$$\frac{P}{k_B T} = \frac{N}{V} + \frac{\Lambda^3}{2^{5/2}}\left(\frac{N}{V}\right)^2 + \cdots. \tag{6.150}$$

The above expression is in the form of a virial expansion for pressure. Using a similar approach, the following expansion can be obtained for the internal energy:

$$U = \frac{3}{2}N k_B T \left[1 + \frac{\Lambda^3}{2^{5/2}}\left(\frac{N}{V}\right) + \cdots \right]. \tag{6.151}$$

Note that although there are no intermolecular forces, the particles experience an effective interaction through the quantum symmetry requirement of the N-body wave function. For fermions, the interaction is essentially repulsive since the first correction tends to increase the pressure. These results clearly indicate the increase in quantum effects with increasing thermal de Broglie wavelength Λ.

A Strongly Degenerate Fermion Gas

We will now consider an ideal gas of indistinguishable fermions at low temperature and/or high density. Analysis of this type of gas is useful as a model of the valence electrons in a metallic crystal. The electron gas model is a reasonable approximation because although the electrons do interact with nuclei in the metal, the interaction is due to Coulombic forces that are so long range that any individual electron experiences a nearly uniform electronic potential within the bulk metal. The success of the model as an analytical tool stems from the fact that at least some of the macroscopically observable properties of the metallic crystal are due more to quantum statistical effects than to the details of electron–nucleus and electron–electron interactions.

For the fermion case, Eq. (6.120) dictates that the average number of particles in the k'th quantum state is given by

$$\langle n_{k'} \rangle = \left(\frac{\lambda e^{-\varepsilon_{k'}/k_{\mathrm{B}}T}}{1 + \lambda e^{-\varepsilon_{k'}/k_{\mathrm{B}}T}} \right). \tag{6.152}$$

Using the definition (6.136) of λ, we can rewrite this relation as

$$\langle n_{k'} \rangle = \frac{1}{1 + e^{(\varepsilon_{k'} - \mu)/k_{\mathrm{B}}T}}. \tag{6.153}$$

The energy quantum states are presumed to be dictated by the particle in a 3-D box solution. Assuming V is large, we again treat the energy ε as a continuous variable and convert the above relation to an energy distribution

$$f(\varepsilon) = \frac{1}{1 + e^{(\varepsilon - \mu)/k_{\mathrm{B}}T}}, \tag{6.154}$$

where $f(\varepsilon)$ is the average occupancy of a microstate with energy ε. (Note that for fermions, the occupancy must be either 0 or 1.) As illustrated in Figure 6.6, in the limit of $T \to 0$, $\mu \to \mu_0$ and the distribution equals one for $\varepsilon < \mu_0$, and it equals zero for $\varepsilon > \mu_0$. This implies that at $T = 0$, all the states with $\varepsilon < \mu_0$ are occupied and those with $\varepsilon > \mu_0$ are unoccupied.

The 3-D particle-in-a-box solution dictates that the degeneracy for the particle translational energy levels is given by Eq. (6.142). For electrons, the degeneracy $g^*(\varepsilon)$ is actually a factor of two larger than that for a simple particle because the electron can be in one of two spin states ($\pm 1/2$) for each translational state. Thus

$$g^*(\varepsilon)\, d\varepsilon = 4\pi \left(\frac{2m}{h^2} \right)^{3/2} V \varepsilon^{1/2}\, d\varepsilon. \tag{6.155}$$

At $T = 0$, we can integrate the distribution over all energies to get N, the total number of particles in the system:

$$N = \int_0^\infty f(\varepsilon) g^*(\varepsilon)\, d\varepsilon = 4\pi \left(\frac{2m}{h^2} \right)^{3/2} V \int_0^{\mu_0} \varepsilon^{1/2}\, d\varepsilon = \frac{8\pi}{3} \left(\frac{2m}{h^2} \right)^{3/2} V \mu_0^{3/2}. \tag{6.156}$$

Rearranging yields the following relation for μ_0:

$$\mu_0 = \frac{h^2}{8m} \left(\frac{3}{\pi} \right)^{2/3} \left(\frac{N}{V} \right)^{2/3}. \tag{6.157}$$

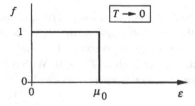

Figure 6.6

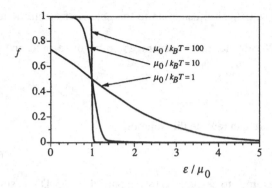

Figure 6.7

For an electron gas in a metal, μ_0 is termed the *Fermi energy*. The value of this energy is typically 1 to 5 eV or 0.2 to 0.8 fJ. It can be seen in Figure 6.7 that even for room temperature levels where μ_0/k_BT is on the order of 100, the distribution $f(\varepsilon)$ is still virtually a step function. Thus, in cases like this, where the *Fermi temperature* $\theta_F = \mu_0/k_B$ is large compared to the system temperature T, $f(\varepsilon)$ is well approximated by the step function

$$f(\varepsilon) = \begin{cases} 1 & \text{for } \varepsilon < \mu_0, \\ 0 & \text{for } \varepsilon > \mu_0. \end{cases} \tag{6.158}$$

Since the Fermi temperature for most metals is on the order of 20,000 K, the step function is a good approximation for most temperatures encountered in engineering applications. For low to moderate temperatures ($T/\theta_F \to 0$), the occupancy distribution can therefore be used to determine the internal energy and pressure of the gas as follows:

$$U_0 = \int_0^\infty f(\varepsilon)g^*(\varepsilon)\varepsilon \, d\varepsilon = 4\pi \left(\frac{2m}{h^2}\right)^{3/2} V \int_0^{\mu_0} \varepsilon^{3/2} \, d\varepsilon, \tag{6.159}$$

$$P_0 V = k_B T \int_0^\infty f(\varepsilon)g^*(\varepsilon) \ln\left(1 + \lambda e^{-\varepsilon/k_B T}\right) d\varepsilon$$

$$= 4\pi \left(\frac{2m}{h^2}\right)^{3/2} V \int_0^{\mu_0} \varepsilon^{1/2} \ln\left(1 + \lambda e^{-\varepsilon/k_B T}\right) d\varepsilon. \tag{6.160}$$

Evaluating the integrals and using the definition of the Fermi energy yields

$$U_0 = \tfrac{3}{5} N \mu_0, \tag{6.161}$$

$$P_0 = \frac{2}{5}\left(\frac{N\mu_0}{V}\right). \tag{6.162}$$

U_0 and P_0 are the zero point energy and pressure, respectively, for a fermion gas. Equation (6.161) implies that at low temperature, the electron "gas" in the metal contributes nothing to the heat capacity since $(\partial U/\partial T)_V \to 0$ there. Equation (6.162) indicates that quantum effects cause the pressure to approach a finite limit as $T \to 0$. With a bit more mathematical effort, the following expansion for U at low temperatures can be obtained:

$$U = U_0\left[1 + \frac{5\pi^2}{12}\left(\frac{k_B T}{\mu_0}\right)^2 + \cdots\right], \tag{6.163}$$

which implies that at low temperatures the molar specific heat for the electron gas varies linearly with temperature:

$$\hat{c}_V = \frac{\pi^2}{2} N_A k_B \left(\frac{k_B T}{\mu_0} \right). \tag{6.164}$$

A Weakly Degenerate Boson Gas

For a weakly degenerate ideal boson gas, the following previously obtained forms of Eqs. (6.122) and (6.123) apply:

$$PV = -k_B T \sum_{k'=0}^{\infty} \ln \left(1 - \lambda e^{\varepsilon_{k'}/k_B T} \right), \tag{6.165}$$

$$N = \sum_{k'=0}^{\infty} \left(\frac{\lambda e^{-\varepsilon_{k'}/k_B T}}{1 - \lambda e^{-\varepsilon_{k'}/k_B T}} \right). \tag{6.166}$$

Note that these relations are valid only if the summands are finite continuous functions for all values of energy ε. To avoid singularities, we must require that

$$0 \leq \lambda < e^{\varepsilon_0/k_B T},$$

where ε_0 is the ground state energy. This assures that $\lambda e^{-\varepsilon_{k'}/k_B T}$ is less than one for all states except the ground state where it may be close to one. We again invoke the idealization that for large V the energy is virtually a continuous variable, replace the summation over all quantum states with an integral over all energy values, and use the degeneracy for the 3-D particle-in-a-box solution. In doing so, we separate the ground state term from the summation so that we can avoid large values of the integrand when $\lambda e^{-\varepsilon_0/k_B T}$ is close to one. We also redefine the zero energy as corresponding to the ground state so that $\varepsilon_0 = 0$. The resulting relations for N/V and PV are

$$\frac{N}{V} = \frac{\lambda}{V(1-\lambda)} + \frac{\pi}{4} \left(\frac{8m}{h^2} \right)^{3/2} \int_{0+}^{\infty} \frac{\lambda \varepsilon^{1/2} e^{-\varepsilon/k_B T} \, d\varepsilon}{1 - \lambda e^{-\varepsilon/k_B T}}, \tag{6.167}$$

$$\frac{P}{k_B T} = -\frac{\ln(1-\lambda)}{V} - \frac{\pi}{4} \left(\frac{8m}{h^2} \right)^{3/2} \int_{0+}^{\infty} \varepsilon^{1/2} \ln \left(1 - \lambda e^{-\varepsilon/k_B T} \right) d\varepsilon. \tag{6.168}$$

For the weakly degenerate case, λ is small compared to one and for large V the first term on the right side of the above expressions can be neglected. Following the line of analysis as for the weakly degenerate fermion gas to evaluate the integrals, we can obtain the relations for U and P:

$$\frac{P}{k_B T} = \frac{N}{V} - \frac{\Lambda^3}{2^{5/2}} \left(\frac{N}{V} \right)^2 + \cdots, \tag{6.169}$$

$$U = \frac{3}{2} N k_B T \left[1 - \frac{\Lambda^3}{2^{5/2}} \left(\frac{N}{V} \right) + \cdots \right]. \tag{6.170}$$

Thus the first correction for quantum effects for a weakly degenerate boson gas is the same as that for a weakly degenerate fermion gas but with opposite sign.

A Strongly Degenerate Boson Gas

Equations (6.167) and (6.168) also apply to a strongly degenerate boson gas in the limit of large V. For convenience, we rewrite those relations in the form

$$\frac{N}{V} = \frac{I_N(\lambda)}{\Lambda^3} + \frac{\lambda}{V(1-\lambda)}, \tag{6.171}$$

$$\frac{P}{k_B T} = \frac{I_P(\lambda)}{\Lambda^3} - \frac{\ln(1-\lambda)}{V}, \tag{6.172}$$

where

$$I_N(\lambda) = \frac{\pi \Lambda^3}{4} \left(\frac{8m}{h^2}\right)^{3/2} \int_{0+}^{\infty} \frac{\lambda \varepsilon^{1/2} e^{-\varepsilon/k_B T} d\varepsilon}{1 - \lambda e^{-\varepsilon/k_B T}}, \tag{6.173}$$

$$I_P(\lambda) = -\frac{\pi \Lambda^3}{4} \left(\frac{8m}{h^2}\right)^{3/2} \int_{0+}^{\infty} \varepsilon^{1/2} \ln\left(1 - \lambda e^{-\varepsilon/k_B T}\right) d\varepsilon. \tag{6.174}$$

Equation (6.135) indicates that for a boson the average number of particles in state k' is

$$\langle n_{k'} \rangle = \frac{\lambda e^{-\varepsilon_{k'}/k_B T}}{1 - \lambda e^{-\varepsilon_{k'}/k_B T}}. \tag{6.175}$$

This implies that the average number of particles in the ground state ($\varepsilon_0 = 0$) is

$$\langle n_0 \rangle = \frac{\lambda}{1-\lambda}. \tag{6.176}$$

For the number of particles in the ground state to be finite, it is necessary that $0 \le \lambda < 1$. Note that no similar restriction on λ applies to a fermion gas. We can combine Eqs. (6.171) and (6.173) to generate an equation of state if we can solve Eq. (6.171) for λ in terms of the other variables. Over the range $0 \le \lambda < 1$, $I_N(\lambda)$ is a bounded monotonically increasing function of λ. The limit of $I_N(\lambda)$ at $\lambda = 1$ is also well defined, with $I_N(1)$ being equal to 2.612. Equation (6.171) can be rewritten as

$$\Lambda^3 \frac{N}{V} = I_N(\lambda) + \left(\frac{\Lambda^3}{V}\right) \frac{\lambda}{(1-\lambda)}. \tag{6.177}$$

Analysis of this equation reveals that the right side grows monotonically with λ for $0 \le \lambda < 1$, and hence $\Lambda^3 N/V$ increases monotonically with λ. For $\Lambda^3 N/V < 2.612$, λ is small and the second term in Eq. (6.177) can be neglected for large V. λ is then given to good accuracy by the root of

$$I_N(\lambda) - \Lambda^3 \frac{N}{V} = 0 \quad \text{for } \Lambda^3 N/V < 2.612. \tag{6.178}$$

For $\Lambda^3 N/V > 2.612$, λ is close to one and we can use the value of I_N at one to approximate the solution for λ as

$$\lambda = 1 - \frac{\Lambda^3}{V[\Lambda^3 N/V - I_N(1)]} \quad \text{for } \Lambda^3 N/V > 2.612. \tag{6.179}$$

The condition $\Lambda^3 N/V = I_N(1) = 2.612$ is a special condition for a gas of this type in the sense that it marks a transition between high-λ and low-λ behavior. By definition

$$\Lambda^3 \frac{N}{V} = \left[\frac{h^2}{2\pi m k_B T} \right]^{3/2} \frac{N}{V}$$

and $\Lambda^3 N/V$ is therefore a function of temperature for fixed density. It follows that for low temperatures, $\Lambda^3 N/V$ is large and Eq. (6.179) provides an approximate relation for λ. Substituting this relation into Eq. (6.176), we can obtain a relation for the mean occupancy of the ground state. Doing so and neglecting terms of order V^{-1} compared to one for large V, we obtain

$$\langle n_0 \rangle = \frac{V}{\Lambda^3} \left[\frac{\Lambda^3 N}{V} - I_N(1) \right]. \tag{6.180}$$

Defining a temperature T_0 as the temperature at which $\Lambda^3 N/V = I_N(1)$ it follows that

$$\frac{N}{V} \left(\frac{h^2}{2\pi m k_B T_0} \right)^{3/2} = I_N(1).$$

This relation can be used to eliminate $I_N(1)$ in Eq. (6.180). After rearranging, the resulting form of this relation becomes

$$\frac{\langle n_0 \rangle}{N} = 1 - \left(\frac{T}{T_0} \right)^{3/2}. \tag{6.181}$$

Note that the definition of T_0 implies that the transition condition $\Lambda^3 N/V = 2.612$ corresponds to $T = T_0$. Temperatures above T_0 correspond to $\Lambda^3 N/V < 2.612$ and the value of λ determined from Eq. (6.178) will be less than one and will decrease rapidly toward zero as T increases. This observation and Eq. (6.176) implies that $\langle n_0 \rangle \to 0$ as T increases above T_0. Thus when $T \gg T_0$, the number of molecules in the ground state is essentially zero, with bosons distributed smoothly over the many quantum states available to each.

Equation (6.181) implies that, as the temperature is lowered past T_0, the occupancy of the ground state abruptly increases to a significant level, with the occupancy increasing until, at $T = 0$, all the bosons are in the ground state. This trend is illustrated in Figure 6.8. The preference of the bosons for one state out of the many available to them that is exhibited at $T = T_0$ is analogous to the phase transition from a gas to a condensed phase. Because of this analogy, this abrupt enhanced occupancy of the ground state is termed *Bose–Einstein condensation*.

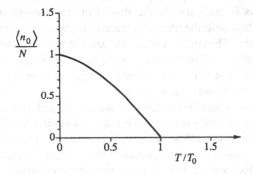

Figure 6.8

With a bit more effort, equations for $P/k_B T$ and the internal energy can be obtained for temperature ranges above and below T_0. Doing so yields the following relations for the internal energy:

$$\frac{U}{N} = \begin{cases} \left(\frac{3}{2}\right) k_B T \left(\frac{V}{N\Lambda^3}\right) I_P(\lambda) & \text{for } T > T_0, \\ \left(\frac{3}{2}\right) k_B T \left(\frac{V}{N\Lambda^3}\right) I_P(1) & \text{for } T < T_0, \end{cases} \tag{6.182}$$

where $I_P(1) = 1.342$. Differentiating these relations and taking $N = N_A$ yields the following expressions for the molar specific heat:

$$\frac{\hat{c}_V}{N_A k_B} = \begin{cases} \left(\frac{15}{4}\right)\left(\frac{\hat{v}}{N_A\Lambda^3}\right) I_P(\lambda) - \left(\frac{9}{4}\right)\frac{I_N(\lambda)}{I_P(\lambda)} & \text{for } T > T_0, \\ \left(\frac{15}{4}\right)\left(\frac{\hat{v}}{N_A\Lambda^3}\right) I_P(1) & \text{for } T < T_0. \end{cases} \tag{6.183}$$

To determine $\hat{c}_V$ for a given temperature in the low temperature range, Eq. (6.179) must first be used to determine λ for the specified temperature. The integrals $I_N(\lambda)$ and $I_P(\lambda)$ can then be evaluated and substituted into Eq. (6.183) to determine $\hat{c}_V$.

The above relations indicate a variation of $\hat{c}_V$ with T that exhibits a peak with a discontinuous slope at $T = T_0$. The variation of $\hat{c}_V$ for helium 4 at its lambda transition point is similar to the trend indicated by Eq. (6.183). Calculation of T_0 for helium yields 3.14 K, which is comparable to the lambda transition temperature of 2.18 K. The lambda transition is not a purely Bose–Einstein condensation phenomenon, but quantum statistical effects appear to be an important contributing factor in this behavior of helium at low temperatures.

Although the occurrence of Bose–Einstein condensation at low temperatures was predicted in the 1920s, experimental observation of this phenomenon in a low temperature gas has only recently been achieved. Creation of Bose–Einstein condensate by cooling rubidium gas was achieved by scientists at the Joint Institute for Laboratory Astrophysics in Colorado in 1995. These investigators used a combination of laser cooling and evaporative cooling in a magnetic trap to cool rubidium atoms in the gas to less than 100 nK. This extraordinarily low temperature must be achieved to create a collection of atoms in the ground state with the lowest allowable energy for the atoms in the magnetic trap. Atoms in the ground state are confined by the magnetic field to a region near the center of the trap. Since the energy, and hence the momentum, are fixed for atoms that drop down to the lowest level, the region they occupy must be of finite size to satisfy the uncertainty principle.

The behavior of atoms in the Bose–Einstein condensate observed in these experiments can be thought of as being analogous to that of photons in a laser beam. In a laser beam the wave behavior of all the photons is the same. In the Bose–Einstein condensate, the quantum mechanical wave functions that describe physical characteristics of the atom, such as position and velocity, are all the same. The coalescence of the particle wave function into a single wave is the essential feature of Bose–Einstein condensation. Since the experiments that created Bose–Einstein condensate in rubidium gas, other investigators have succeeded in creating Bose–Einstein condensate in lithium and sodium.

The macroscopic properties of Bose–Einstein condensates are largely unknown at this point, but it is anticipated that they may exhibit behavior that is much different from classical fluids. It may be possible that matter waves of a Bose–Einstein condensate can be reflected, focused, diffracted, and modulated in frequency and amplitude. A Bose–Einstein condensate is expected to have compressibility like a fluid, but the quantum coherence may give the fluid other unusual characteristics. There has been some speculation that the viscosity of the condensate will be extraordinarily low, making it a sort of superfluid. Exploration of the properties of the condensate is expected to be one of the main objectives of future research in this area. Further information on recent experimental investigations of Bose–Einstein condensation in low temperature gases may be found in the references by Anderson et al. [24], Andrews et al. [25], and Cornell and Wieman [26].

Exercises

6.1 For methane (CH_4) $\sigma_{LJ} = 3.82 \times 10^{-10}$ m and $\varepsilon_0 = 2.04 \times 10^{-21}$ J are recommended values for the constants in the Lennard–Jones 6-12 intermolecular potential. Use these values to predict the van der Waals constants a_v and b_v for methane. How do the resulting values compare to those in Table 6.1?

6.2 For oxygen (O_2), use the tabulated values of a_v and b_v in Table 6.1 to estimate the depth of the potential well in the interaction potential ε_0 and the approximate effective diameter of the molecules $D \cong \sigma_{LJ}$.

6.3 The chemical potential of a pure van der Waals fluid as a function of temperature, volume, and number of molecules is given by Eq. (6.27). Use this relation with Eq. (6.26) to plot the variation of $\mu / k_B T$ as a function of pressure for one kmol of water at 100°C. In your plot, show the variation for pressures between 10 kPa and 500 kPa.

6.4 Demonstrate directly using results of Section 6.2 that $G = \mu N$ for a pure van der Waals fluid.

6.5 A system containing one kmol of dense fluid a executes a change of state from V_1, T_1 to V_2, T_2 resulting in a change in internal energy $(U_2 - U_1)_a$. Another system containing one kmol of dense fluid b also executes a change of state from V_1, T_1 to V_2, T_2 resulting in a change in internal energy $(U_2 - U_1)_b$. The sizes of the species a and b molecules are the same but for the molecular interaction potentials, the depth of the potential well ε_0 is larger for species a than it is for b. Which do you expect to be larger, $(U_2 - U_1)_a$ or $(U_2 - U_1)_b$? Justify your answer.

6.6 Derive relations for the chemical potentials μ_a and μ_b as functions of T, V, N_a, and N_b for the van der Waals binary mixture considered in Section 6.2.

6.7 Derive the equation for $\hat{c}_P = [(\partial H / \partial T)_{P,N}]_{N=N_A}$ cited at the beginning of Example 6.3. (Hint: Consider changes of H along a line of constant P and assume $H = H(V, N, T)$ and $P = P(V, N, T)$ as in Eqs. (6.26) and (6.29).)

6.8 Use the van der Waals model to predict the molar specific heat $\hat{c}_P$ for saturated liquid oxygen and saturated oxygen vapor at 104 K and 352 kPa. Compare the results to the recommended values of 54.4 kJ/(kmol K) for liquid and 33.6 kJ/(kmol K) for vapor.

6.9

(a) Beginning with the differential relation

$$d\hat{h} = \left(\frac{\partial \hat{h}}{\partial T} \right)_P dT + \left(\frac{\partial \hat{h}}{\partial P} \right)_T dP,$$

use the definition of the specific heat at constant pressure and the Maxwell relation

$$\left(\frac{\partial S}{\partial P} \right)_{T,N} = -\left(\frac{\partial V}{\partial T} \right)_{P,N}$$

to derive the following general relation for the differential change in enthalpy for a system of fixed mass:

$$d\hat{h} = \hat{c}_P \, dT + \left[\hat{v} - T \left(\frac{\partial \hat{v}}{\partial T} \right)_P \right] dP.$$

(b) Use the result of part (a) to derive the following relation for the Joule–Thomson coefficient:

$$\mu_{JT} = \left(\frac{\partial T}{\partial P} \right)_h = -\frac{1}{\hat{c}_P} \left[\hat{v} - T \left(\frac{\partial \hat{v}}{\partial T} \right)_P \right].$$

(c) For many throttling flow processes, enthalpy is constant while the pressure drops. The Joule–Thomson coefficent indicates whether the corresponding temperature change is positive, negative, or zero. Consider a throttling process involving nitrogen at an initial temperature of 140 K. Use the van der Waals model results for nitrogen to determine whether the throttling process will increase or decrease the temperature for the following two initial pressures: 200 kPa and 4,000 kPa.

6.10 Oxygen gas is compressed reversibly and adiabatically from an initial condition of 20°C and 101 kPa to a final pressure of 5,000 kPa. Determine the final temperature and the work required (enthalpy change) per kmol of gas using (a) the van der Waals model and (b) relations for an ideal gas with constant specific heats. For part (b) use the equipartition prediction (assuming rotation and translation storage only) for oxygen: $\hat{c}_P = 29.1 \, kJ/(kmol \, K)$. Compare the results for the two models and comment on the reasons for the differences.

6.11 Nitrogen gas expands reversibly and adiabatically from an initial condition of 20°C and 8,000 kPa to 4,000 kPa. Determine the final temperature and the enthalpy change per kmol of gas using (a) the Redlich–Kwong model and (b) relations for an ideal gas with constant specific heats. For part (b) use the equipartition prediction (assuming rotation and translation storage only) for nitrogen: $\hat{c}_P = 29.1 \, kJ/(kmol \, K)$. Compare the results for the two models and comment on the reasons for the differences.

6.12 Use the equation for $\hat{c}_P$ cited at the beginning of Example 6.3 and the relations for a van der Waals fluid to calculate and plot the variation of $\hat{c}_P$ with temperature for nitrogen gas at 10 MPa between temperatures of 150 and 300 K. Compare your results to the predictions of equipartition theory.

6.13 Use the equation for $\hat{c}_P$ cited at the beginning of Example 6.3 and the relations for a van der Waals fluid to calculate and plot the variation of $\hat{c}_P$ with temperature for methane gas at 9 MPa between temperatures of 200 and 400 K. Compare your results to the predictions of equipartition theory.

6.14 In Chapter 3, we obtained a relation for $\hat{c}_P - \hat{c}_{\hat{v}}$ that can be written in the form

$$\hat{c}_P - \hat{c}_{\hat{v}} = \frac{\hat{v} T \beta_T^2}{\alpha_T},$$

where α_T is the isothermal compressibility and β_T is the volume coefficient of thermal expansion defined as

$$\alpha_T = -\frac{1}{\hat{v}} \left(\frac{\partial \hat{v}}{\partial P} \right)_T, \qquad \beta_T = \frac{1}{\hat{v}} \left(\frac{\partial \hat{v}}{\partial T} \right)_P.$$

For oxygen at a temperature of 90 K, use the van der Waals relations obtained in Section 6.2 to compute and plot the variations of α_T and $(\hat{c}_P - \hat{c}_{\hat{v}})/N_A k_B$ with $\hat{v}$ between $\hat{v} = 0.010$

m^3/kmol and $\hat{v} = 10.0$ m^3/kmol. Assess and comment on the physical significance of the trends in the plots.

6.15 The so-called square-well interaction potential is defined as

$$\phi(r) = \begin{cases} \infty & \text{for } r < \sigma, \\ -\varepsilon_0 & \text{for } \sigma < r < \lambda\sigma, \\ 0 & \text{for } r > \lambda\sigma. \end{cases}$$

(a) Determine a relation for the second virial coefficient as a function of temperature for molecules that obey the square-well potential defined above.

(b) For nitrogen, recommended values of the constants in the square-well potential are

$$\sigma = 3.28 \times 10^{-10} \text{ m}, \qquad \lambda = 1.58, \qquad \varepsilon_0 = 1.31 \times 10^{-21} \text{ J}$$

Using these values with the relation determined in part (a), plot the variation of B with temperature between 100 and 400 K.

6.16 The hard-sphere interaction potential is defined as

$$\phi(r) = \begin{cases} \infty & \text{for } r < \sigma, \\ 0 & \text{for } r > \sigma. \end{cases}$$

(a) Determine a relation for the second virial coefficient for molecules that obey the square-well potential defined above. How does B vary with temperature for this case?

(b) For methane a value of 3.36×10^{-10} m is recommended for σ. Use this value and the result of part (a) to determine the percent correction to PV/Nk_BT provided by the second virial coeffcient for methane at $T = 200$ K and $P = 101$ kPa.

6.17 Determine the next term in each of the series (6.150) and (6.169) for P/k_BT. Based on your results, assess the symmetry of quantum effects for fermions and bosons to order $(N/V)^3$ in these expansions.

6.18

(a) An application requires use of gaseous nitrogen (N_2) at 77 K and 76 kPa. Determine whether quantum effects will introduce a significant departure from ideal gas behavior for this system.

(b) Hydrogen is a possible alternative to the use of nitrogen in this system. Assess the importance of quantum effects for gaseous hydrogen (H_2) under the same conditions.

References

[1] McQuarrie, D. A., *Statistical Mechanics*, Chapter 12, Harper and Row, New York, 1976.

[2] Van der Waals, J. D., *On the Continuity of the Gaseous and Liquid States*, edited by J.S. Rowlinson, Studies in Statistical Mechanics, Vol. XIV, Elsevier Publishing, New York, 1988.

[3] Hirschfelder, J. O., C. F., Curtis, and R. B., Bird, *Molecular Theory of Gases and Liquids*, J. Wiley & Sons, New York, 1954.

[4] Tien, C. L. and J. H., Lienhard, *Statistical Thermodynamics*, Chapter 9, Hemisphere Publishing Corporation, New York, 1979.

[5] Wark, K., *Advanced Thermodyanmics for Engineers*, Chapter 5, McGraw-Hill, New York, 1995.

[6] Kac, M., G. E., Uhlenbeck, and P. C., Hemmer, "On the van der Waals Theory of the Vapour–Liquid Equilibrium I. Discussion of a One-Dimensional Model," *J. Math. Phys.*, 4: 216–228, 1963.

[7] Alder, B. J. and C. E., Hecht, "Studies in Molecular Dyanmics VII. Hard Sphere Distribution Functions and an Augmented van der Waals Theory," *J. Chem. Phys.*, 50: 2032–7, 1969.

[8] Rigby, M., "The van der Waals Fluid: a Renaissance," *Q. Rev. Chem. Soc.*, 24: 416–32, 1970.

[9] Hoover, W. G., G., Stell, E., Goldmark, and G. D., Degani, "Generalised van der Waals Equation of State, *J. Chem. Phys.*, 63: 5434–8, 1975.

[10] Lebowitz, J. L. and E. M., Waisman, "Statistical Mechanics of Simple Fluids: Beyond van der Waals," *Physics Today*, March, 1980, pp. 24–30.

[11] Song, Y. and E. A., Mason, "Statistical Mechanical Theory of a New Analytical Equation of State," *J. Chem. Phys.*, 91: 7840–53, 1989.

[12] Nordholm, S. and A. D. J., Haymet, "Generalized van der Waals Theory I. Basic Formulation and Application to Uniform Fields," *Aust. J. Chem.*, 33: 2013–27, 1980.

[13] Nordholm, S., M., Johnson, and B. C., Freasier, "Generalized van der Waals Theory III. The Prediction of Hard Sphere Structure," *Aust. J. Chem.*, 33: 2139–50, 1980.

[14] Nordholm, S., J., Gibson, and M. A., Hooper, "Generalized van der Waals Theory VIII. An Improved Analysis of the Liquid/Gas Interface," *J. Stat. Phys.*, 28: 391–406, 1982.

[15] Hooper, M. A. and S., Nordholm, "Generalized van der Waals Theory II. Quantum Effects on the Equation of State," *Aust. J. Chem.*, 33: 2029–35, 1980.

[16] Hooper, M. A. and S., Nordholm, "Generalized van der Waals Theory IV. Variational Determination of the Hard Sphere Diameter," *Aust. J. Chem.*, 34: 1809–18, 1981.

[17] Hooper, M. A. and S., Nordholm, "Generalized van der Waals Theory XII. Curved Interfaces in Simple Fluids," *J. Chem. Phys.*, 81: 2432–8, 1984.

[18] Penfold, R., J., Satherley, and S., Nordholm, "Generalized van der Waals Theory of Fluids. Vapour–Liquid Equilibria in Simple Binary Mixtures," *Fluid Phase Equilibria*, 109: 183–204, 1995.

[19] Redlich, O. and J., Kwong, "On the Thermodynamics of Solutions. V. An Equation of State. Fugacities of Gaseous Solutions," *Chem. Rev.*, 44: 233, 1949.

[20] Reid, R. C., J. M., Prausnitz, and B. E., Poling, *The Properties of Gases and Liquids*, 4th ed. McGraw-Hill, New York, 1986.

[21] Hansen, J. P. and I. R., McDonald, *Theory of Simple Liquids*, 2nd ed. Academic Press, Orlando, FL, 1986.

[22] Allen, M. P. and D. J., Tildesley, *Computer Simulation of Liquids*, Oxford University Press, Oxford, UK, 1989.

[23] Haile, J. M., *Molecular Dynamics Simulation: Elementary Methods*, J. Wiley & Sons, New York, 1992.

[24] Anderson, M. H., J. R., Ensher, M. R., Matthews, C. E., Wieman, and E. A., Cornell, "Observation of Bose–Einstein Condensation in a Dilute Atomic Vapor," *Science*, 269: 198–201, 1995.

[25] Andrews, M. R., C. G., Townsend, H. -J., Miesner, D. S., Durfee, D.M., Kurn, and W., Ketterle, "Observation of Interference between Two Bose Condensates," *Science*, 275: 637–41, 1997.

[26] Cornell, E. A. and C. E., Wieman, "The Bose–Einstein Condensate," *Scientific American*, 278: 40–5, 1998.

Solid Crystals

Chapter 7 demonstrates the application of statistical thermodynamics theory to crystalline solids. Because of its relevance to electron transport in metallic crystalline solids, the electron gas theory for metals is also described in this chapter. This chapter provides only an introduction to the microscale thermophysics of solids. Readers interested in more comprehensive treatments of solid state thermophysics should consult the references cited at the end of this chapter.

7.1 Monatomic Crystals

Our objective here is to use statistical thermodynamics tools to evaluate thermo-dynamic properties of solid crystals. Our first goal is to derive a relation for the partition function Q. In doing so, we will specifically consider the structure of a monatomic crystal. One approach is to model the crystal as a system of regularly spaced masses and springs as indicated schematically in Figure 7.1. The springs represent the interatomic forces that each atom experiences. The mean locations of the masses are at regularly spaced lattice points.

Actually, each atom sits in a potential well whose minimum is at a lattice point. The potential well for each atom is usually very steep. Each atom vibrates about its equilibrium position with a small amplitude, which suggests that we can work with a Taylor series representation of the potential valid near the equilibrium point.

Rather than working with the potential for a single atom, we will consider the potential for the crystal as a whole, which we will designate as Φ. The total potential energy is a function of the displacement of all N atoms in the crystal from their equilibrium position at the minimum of the potential well. The displacement of each atom has three Cartesian coordinates and the total potential is therefore a function of all $3N$ such coordinates. Designating each of the coordinates as ξ_i, it follows that

$$\Phi = \Phi(\xi_1, \xi_2, \ldots, \xi_{3N}). \tag{7.1}$$

Expanding Φ in a Taylor series for small displacements yields

$$\Phi(\xi_1, \xi_2, \ldots, \xi_{3N}) = \Phi(0, 0, \ldots, 0) + \sum_{j=1}^{3N} \left(\frac{\partial \Phi}{\partial \xi_j}\right)_0 \xi_j + \frac{1}{2} \sum_{i=1}^{3N} \sum_{j=1}^{3N} \left(\frac{\partial^2 \Phi}{\partial \xi_i \partial \xi_j}\right)_0 \xi_i \xi_j + \cdots.$$

$$\tag{7.2}$$

Since Φ is a minimum when the $\xi_j = 0$, the first derivatives are all zero, and we can write

$$\Phi(\xi_1, \xi_2, \ldots, \xi_{3N}) = \Phi(0, 0, \ldots, 0) + \frac{1}{2} \sum_{i=1}^{3N} \sum_{j=1}^{3N} k_{ij} \xi_i \xi_j + \cdots, \tag{7.3}$$

where

$$k_{ij} \equiv \left(\frac{\partial^2 \Phi}{\partial \xi_i \partial \xi_j}\right)_0 \tag{7.4}$$

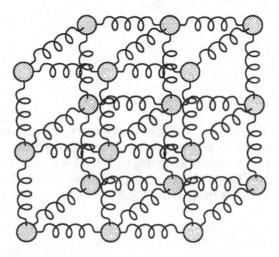

Figure 7.1

are force constants. The important results here are that:

(1) Φ is a quadratic function of the displacements.
(2) The minimum Φ is a function of the lattice spacing, which is a function of the density N/V. To emphasize this, we write this term as $\Phi(0, \ldots, 0, N/V)$.
(3) This expansion represents a system of coupled harmonic oscillators. The coupling is in the cross terms $\xi_i \xi_j$ where $i \neq j$.

Analysis of a large number of coupled harmonic oscillators is a difficult problem. However, for a polyatomic molecule, we previously argued that the complex vibrational motions of the molecule could be decomposed into a set of independent harmonic oscillators by introducing normal coordinates. Noting that a crystal containing N atoms can be considered to be just a large polyatomic molecule, this implies that we can also decompose the vibrations of the crystal into a set of independent harmonic oscillators by applying such a normal coordinate analysis.

Considering the $3N$ degrees of freedom for the crystal, three are associated with translation of the whole crystal and three more are concerned with rotation of the whole crystal. The remaining $3N - 6$ degrees of freedom are attributable to vibration. Thus, if we applied a normal coordinate analysis, we would obtain $3N - 6$ vibrational frequencies ν_j, where

$$\nu_j = \frac{1}{2\pi}\left(\frac{k_j}{m_{r,j}}\right)^{1/2}, \qquad j = 1, 2, \ldots, 3N - 6. \tag{7.5}$$

Note that k_j is an effective force constant and $m_{r,j}$ is an effective reduced mass for the jth vibrational mode. The $m_{r,j}$ and k_j are expected to depend on the atomic mass, the bond force constants, and the equilibrium lattice spacing $(\sim (V/N)^{1/3})$. For a given crystal type, it follows that the ν_js are dependent only on density (N/V).

Since the crystal does not translate or rotate, the complete partition function is given by

$$Q = e^{-\Phi(0,\ldots,0,N/V)/k_B T} \prod_{j=1}^{3N-6} q_{\text{vib},j}, \tag{7.6}$$

where $q_{vib,j}$ is the vibrational partition function associated with the jth vibrational frequency. Note that there is no factor of $N!$ in the denominator of this relation. Since each atom stays nominally at a fixed lattice point, we could, in principle, label each point and thus distinguish individual atoms. It is therefore not necessary to divide by $N!$ to avoid including contributions of identical configurations more than once.

For a harmonic oscillator with a specified characteristic frequency ν, we have already shown in our treatment of diatomic gases that

$$q_{vib} = \frac{e^{-h\nu/2k_BT}}{1 - e^{-h\nu/k_BT}}. \tag{7.7}$$

Using this result for each vibrational mode, we have

$$Q = e^{-\Phi(0,...,0,N/V)/k_BT} \prod_{j=1}^{3N-6} \frac{e^{-h\nu_j/2k_BT}}{1 - e^{-h\nu_j/k_BT}}. \tag{7.8}$$

Since there are $3N$ vibrational characteristic frequencies, it is plausible to assume that they are essentially continuously distributed. We therefore introduce a distribution function $g(\nu)$ such that $g(\nu)\,d\nu$ equals the number of normal frequencies in the interval from ν to $\nu + d\nu$. Since N is of the order of 10^{26}, the number of vibrational degrees of freedom is essentially $3N$. Taking the log of both sides of Eq. (7.8) for Q yields

$$-\ln Q = \frac{\Phi(0,...,0,N/V)}{k_BT} + \sum_{j-1}^{3N} \left[\ln\{1 - e^{-h\nu_j/k_BT}\} + \frac{h\nu_j}{2k_BT} \right]. \tag{7.9}$$

For the many closely spaced frequencies ν_j, we replace the summation by an integral by treating ν as a continuous variable and multiplying by $g(\nu)\,d\nu$:

$$-\ln Q = \frac{\Phi(0,...,0,N/V)}{k_BT} + \int_0^\infty \left[\ln\{1 - e^{-h\nu/k_BT}\} + \frac{h\nu}{2k_BT} \right] g(\nu)\,d\nu, \tag{7.10}$$

where

$$\int_0^\infty g(\nu)\,d\nu = 3N \tag{7.11}$$

since there are a total of $3N$ normal frequencies.

If we can determine $g(\nu)$, we can develop a relation for Q and subsequently calculate the thermodynamic properties of the crystal. In fact, using the relation for U derived for the canonical ensemble and the above relation for $\ln Q$, it can easily be shown that

$$U = \Phi(0,...,0,N/V) + \int_0^\infty \left[\frac{h\nu e^{-h\nu/k_BT}}{1 - e^{-h\nu/k_BT}} + \frac{h\nu}{2} \right] g(\nu)\,d\nu, \tag{7.12}$$

$$\hat{c}_V = N_A k_B \int_0^\infty (h\nu/k_BT)^2 e^{-h\nu/k_BT} (1 - e^{-h\nu/k_BT})^{-2} g(\nu)\,d\nu. \tag{7.13}$$

The main issue, then, is how to evaluate $g(\nu)$. If each of the N atoms of a crystalline solid behave as harmonic oscillators with three degrees of freedom (each having two modes of energy storage), classical (equipartition) theory predicts that each atom contributes $3 \times 2 \times (1/2)k_B = 3k_B$ to the specific heat:

$$\hat{c}_V = 3N_A k_B = 24.9 \text{ kJ/(kmol K)} = 6 \text{ cal/(mole K)}. \tag{7.14}$$

This result is known as the *law of Dulong and Petit*. It is in good agreement with the measurements for many crystalline solids at high temperatures, but it fails at low temperatures. As $T \to 0$, $\hat{c}_V \to 0$ proportional to T^3, for common metallic solids such as aluminum, gold, and silver. Einstein's model described in the next section was the first to predict a deviation from the equipartition prediction due to quantum effects as $T \to 0$.

Example 7.1   Recommended values of the specific heat for copper and aluminum at 25°C are 0.396 kJ/kg K and 0.900 kJ/kg K, respectively. Compare these values to the predictions of the law of Dulong and Petit.

Solution   The law of Dulong and Petit requires that for copper

$$c_V = \hat{c}_V / M = 3 N_A k_B / M = 3(6.02 \times 10^{26})(1.38 \times 10^{-23})/63.5$$
$$= 0.392 \text{ kJ/kg K},$$

whereas for aluminum

$$c_V = 3 N_A k_B / M = 3(6.02 \times 10^{26})(1.38 \times 10^{-23})/27.0$$
$$= 0.923 \text{ kJ/kg K}$$

The values predicted by the law of Dulong and Petit are within 3% of the recommended values.

7.2   Einstein's Model

Einstein was the first to propose a theoretical explanation for the low temperature variation of $\hat{c}_V$ for crystalline solids. His explanation was based on the quantum theoretical arguments used earlier by Planck in his treatment of electromagnetic radiation. As we have done above, Einstein argued that the crystal could be treated as a set of $3N$ independent harmonic oscillators. He further postulated that all the harmonic oscillators had the same characteristic frequency ν_E. In our formulation this is equivalent to assuming that the distribution function $g(\nu)$ is a delta function with one frequency

$$g(\nu) = 3N\delta(\nu - \nu_E). \tag{7.15}$$

Substituting this relation into the above equation for $\hat{c}_V$, after a little manipulation, we can write the result in the form

$$\hat{c}_V = 3 N_A k_B \left(\frac{\theta_E}{T}\right)^2 \frac{e^{-\theta_E/T}}{(1 - e^{-\theta_E/T})^2}, \tag{7.16}$$

where

$$\theta_E = h\nu_E / k_B. \tag{7.17}$$

Note that ν_E and hence θ_E (the Einstein temperature) are presumed to be characteristic values for the particular solid. The variation of the specific heat with temperature predicted by the Einstein model is plotted in Figure 7.2.

Agreement of Einstein's model with data is qualitatively good. However, there is a problem with its prediction of the specific heat variation at low temperatures. As $T \to 0$,

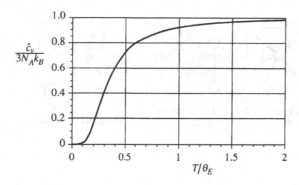

Figure 7.2

the above relation for $\hat{c}_V$ varies as

$$\hat{c}_V = 3N_A k_B \left(\frac{\theta_E}{T}\right)^2 e^{-\theta_E/T}. \tag{7.18}$$

This is not consistent with data for monatomic crystals, which imply that $\hat{c}_V \to 0$ proportional to T^3 as $T \to 0$. To avoid this deficiency, a better treatment of energy storage in the crystal at low temperatures must be included in the model. We will examine the features of energy storage in a crystal in detail in the next section.

7.3 Lattice Vibrations in Crystalline Solids

The Einstein model in the previous section grossly idealizes the nature of vibrational motions in the solid in that it assumes that all oscillations have the same frequency. In real crystals, the frequencies of lattice vibrations can span a wide range. To place the energy storage of the crystal in proper perspective, in this section we will examine some of the basic features of lattice vibrations in a crystal. This will provide a foundation for development of an improved relation for the specific heat in the following section.

The structure of the crystal is specified in terms of its lattice and its basis. The *basis* is one or more atoms with specific spacing. The *lattice* is a periodic array of points in space. The structure of the crystal is generated by repeatedly hanging the basis at each lattice point. The smallest basis that will generate the crystal structure is termed the *primitive basis*. A cell defined by the position vectors of atoms in the primitive basis is a *primitive cell* for the crystal. Possible types of primitive cells include simple cubic, body-centered cubic, face-centered cubic, and hexagonal. These are shown schematically in Figure 7.3. The body-centered cubic structure is characteristic of chromium, molybdenum, and tungsten. Crystalline forms of carbon (diamond) silicon, tin, gold, silver, copper, and sodium chloride have a face-centered cubic structure. In a compound such as sodium chloride, the basis contains one atom of each species. For pure materials such as diamond and silicon, the basis for the face-centered cubic structure also contains two atoms. Crystalline forms of zinc, magnesium, and titanium exhibit a hexagonal close packing structure.

There are many interesting and important features of the crystal structures described above. However, the main point of interest to us here is that the structure is spatially periodic. We therefore seek to determine how such a periodic structure stores energy as

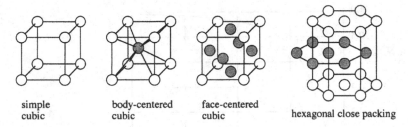

simple
cubic

body-centered
cubic

face-centered
cubic

hexagonal close packing

Figure 7.3

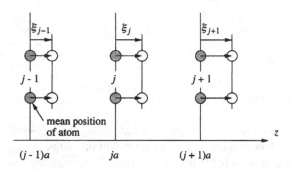

Figure 7.4

oscillatory motion of the atoms about their mean lattice locations. To explore the basic features of such lattice vibrations we will specifically consider a crystal with a monatomic basis and focus on vibrational behavior in one dimension. The situation of interest is shown schematically in Figure 7.4. Here we assume that the forces exerted on each atom are linear in the displacement of the atom relative to its neighbors. Thus, for the 1-D motion represented in Figure 7.4, the force on atoms in plane j is given by

$$f_j = C_f(\xi_{j+1} - \xi_j) + C_f(\xi_{j-1} - \xi_j), \tag{7.19}$$

where C_f is the force constant between adjacent atoms. It follows that the equation of motion for atoms in plane j is

$$m\frac{d^2\xi_j}{dt^2} = C_f(\xi_{j+1} + \xi_{j-1} - 2\xi_j), \tag{7.20}$$

where m is the mass of the atom. We expect solutions of this equation to have sinusoidal spatial and time dependence, implying that the solution is of the form

$$\xi_j = \xi_0 e^{-i(jKa+\omega t)}. \tag{7.21}$$

In the above relation, ω and K are the angular frequency and wavenumber of the wave, respectively. These are related to the frequency ν and wavelength λ of the wave as

$$\omega = 2\pi\nu, \tag{7.22}$$

$$K = 2\pi/\lambda. \tag{7.23}$$

Substituting the displacement relation (7.21) for the terms in Eq. (7.20), we obtain, after a little rearranging, the dispersion relation relating the frequency and wavenumber for the

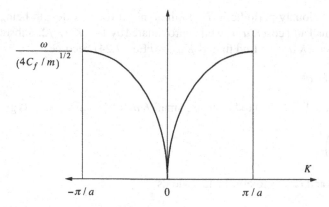

Figure 7.5

waves:

$$\omega^2 = (2C_f/m)(1 - \cos Ka). \tag{7.24}$$

A plot of ω verses K for $-\pi/a \leq K \leq \pi/a$ is shown in Figure 7.5. For K values beyond the range shown in this figure, the variation of ω with K repeats with period $2\pi/a$. Although the wave solution of the displacement amplitude mathematically varies in a continuous way with z, only its magnitude at the planes corresponding to integer multiples of a is of physical relevance. Using the solution (7.21), it is easily shown that the ratio of displacement amplitudes at adjacent planes is

$$\frac{\xi_{j+1}}{\xi_j} = \frac{\xi_0 e^{-i[(j+1)Ka+\omega t]}}{\xi_0 e^{-i[jKa+\omega t]}} = e^{-iKa}. \tag{7.25}$$

Thus, the displacement ratio for adjacent atoms is periodic in K with period $2\pi/a$, and all physically distinct variations of the displacement differences are represented by wavenumbers in the range $-\pi/a \leq K \leq \pi/a$. This range of K, which represents the entire range of possible wave motions for the atoms in the crystal, is referred to as the *first Brillouin zone*. Note that if K is outside the first Brillouin zone, it is always possible to convert it to a physically equivalent value K inside the zone by subtracting an appropriate integer multiple of $2\pi/a$. A value of K outside the first Brillouin zone reproduces the lattice wave motion of the corresponding K within the zone. This periodic dependence of frequency on wavenumber for the lattice wave motion is distinctly different from wave motion in an elastic continuum.

Another noteworthy feature of the lattice wave motion represented by Eq. (7.21) is the group velocity obtained by differentiating the dispersion relation

$$v_g \equiv \frac{d\omega}{dK}. \tag{7.26}$$

Using (7.24) and (7.26), we obtain

$$v_g = \left(\frac{C_f a^2}{m}\right)^{1/2} \cos(Ka/2). \tag{7.27}$$

The group velocity is obviously periodic in K. In the limit of the wavelength being long compared to a, Ka is small and $\cos(Ka)$ is well approximated by $1 - (Ka)^2/2$. Substituting this approximation for $\cos(Ka)$, we find that as $Ka \to 0$ Eq. (7.24) becomes

$$\omega^2 = (C_f/m)K^2a^2. \tag{7.28}$$

Taking the cosine factor in Eq. (7.27) to be one for small Ka, the group velocity is given by

$$v_g = \left(\frac{C_f a^2}{m}\right)^{1/2}. \tag{7.29}$$

Combining Eqs. (7.28) and (7.29) to eliminate a yields

$$v_g = \frac{\omega}{K} \tag{7.30}$$

for the small Ka limit, which is consistent with the theory of elastic waves in a continuum.

It should be noted that in a three-dimensional lattice, waves of three different polarizations are possible: oscillation in one of two orthogonal directions normal to the direction of propagation (transverse waves) and oscillation in the direction of propagation (longitudinal waves). In general there will be a separate dispersion relation for transverse and longitudinal wave modes in the Brillouin zone.

If the basis for the crystal lattice contains two or more atoms, multiple dispersion relations linking frequency and wavenumber in the Brillouin zone can exist for each polarization. When there is ionic bonding of different atoms in the lattice, as in a NaCl crystal, the atoms carry opposite charges and vibrations of the lattice may be induced by the electric field of a light wave. A branch of the dispersion relation that represents lattice vibrations that respond to this type of excitation is termed an *optical branch*. A branch that corresponds to waves generated by purely mechanical interaction is termed an *acoustic branch*.

The discussion above provides only a brief introduction to the physics of lattice vibrations. Several important features are evident, however. First, as postulated in the Einstein model in the previous section, the lattice waves are quantized. The energy of a given mode of vibration with angular frequency ω is taken to be that for a simple harmonic oscillator

$$\varepsilon = \left(n + \frac{1}{2}\right)h\nu = \left(n + \frac{1}{2}\right)\frac{h\omega}{2\pi}. \tag{7.31}$$

When a mode is occupied to quantum number n, the mode is interpreted as being occupied by n pseudoparticles of vibrational energy that are termed *phonons*.

Given the dispersion relation behavior described above, it is also clear that lattice vibrations will occur over a range of frequencies dictated by the dispersion relations within the first Brillouin zone. They will not have a single frequency, as postulated in the Einstein model. In general, prediction of the distribution of frequencies $g(\nu)$ discussed in the previous section is not a simple matter. To do so we must consider the features of oscillations in the lattice in detail. However, we have also demonstrated that in the limit of the wavelength being long compared to the interatomic spacing a, the wave behavior for the acoustic branch approaches that for waves in a continuum. In the next section we will use this information to construct the Debye model for energy storage in a crystal, which improves upon the Einstein model described in the previous section.

7.4   The Debye Model

In quantum theory, the energy level of a harmonic oscillator is directly proportional to its frequency. Because the lower energies are most populated at very low temperatures, the inadequacy of Einstein's model at low temperatures is likely due to inaccurate treatment of low frequency (or long-wavelength) modes. The theory of the heat capacity of crystals, later developed by Debye, treats the long-wavelength frequencies of the crystal in an exact manner and consequently predicts the correct low temperature variation of $\hat{c}_V$.

Debye reasoned that modes having wavelengths that are long compared to the atomic spacing do not depend strongly on the detailed atomic-level structure of the solid. He further argued that as a consequence, long-wavelength modes can be calculated by assuming that the crystal is a continuous elastic body. This is a viable approach because, as we noted in the previous section, the lattice vibrations approach the behavior of waves in a continuum in the long-wavelength limit. The Debye theory treatment of the distribution of mode frequencies includes calculation of the modes as standing waves within the extent of the crystal. We therefore let the imaginary part of

$$\Lambda = A e^{i(\mathbf{k} \cdot \mathbf{r} - \omega t)} \tag{7.32}$$

represent a wave of amplitude A traveling through the crystal in vector direction $\mathbf{k}$ with frequency $\omega = 2\pi \nu$. Here $\mathbf{k}$ is called the wave vector, and its magnitude is $2\pi/\lambda$. The velocity of this wave is given by $\upsilon = \omega/|\mathbf{k}| = \nu \lambda$. A standing wave can be obtained by superimposing two waves traveling in opposite directions. Doing so yields

$$\Lambda = 2 A e^{i \mathbf{k} \cdot \mathbf{r}} \cos \omega t. \tag{7.33}$$

To make this a confined standing wave, we require that the imaginary part of the above relation vanish at the edges of the crystal. If we take the crystal to be a cube of length L, this requirement becomes

$$k_x L = n_x \pi, \tag{7.34}$$

$$k_y L = n_y \pi, \tag{7.35}$$

$$k_z L = n_z \pi, \tag{7.36}$$

where

$$n_x, n_y, n_z = 1, 2, 3, \ldots. \tag{7.37}$$

The magnitude of the wave vector for any combination of ns is therefore given by

$$|\mathbf{k}|^2 = \left(\frac{\pi}{L}\right)^2 \left(n_x^2 + n_x^2 + n_x^2\right). \tag{7.38}$$

Since we expect the ns to typically be large, the number of standing waves with a wave vector of magnitude less than $|\mathbf{k}|$ is essentially equal to one eighth of the volume of the sphere defined by the above equation (see our earlier quantum analysis of energy level degeneracy for a particle in a 3-D box in Chapter 1):

$$N_{<|\mathbf{k}|} = \frac{\pi}{6} \left(\frac{L|\mathbf{k}|}{\pi}\right)^3 = \frac{L^3 |\mathbf{k}|^3}{6\pi^2} = \frac{V |\mathbf{k}|^3}{6\pi^2}. \tag{7.39}$$

The number between $|\mathbf{k}|$ and $|\mathbf{k}| + d|\mathbf{k}|$ is obtained by differentiating this relation:

$$\hat{g}(|\mathbf{k}|)d|\mathbf{k}| = \frac{dN_{<|\mathbf{k}|}}{d|\mathbf{k}|}d|\mathbf{k}| = \frac{V|\mathbf{k}|^2 d|\mathbf{k}|}{2\pi^2}. \tag{7.40}$$

We can convert this result to a frequency distribution $\hat{g}(\nu)\,d\nu$ by using the relation

$$\nu = \nu_w/\lambda = \nu_w|\mathbf{k}|/2\pi \tag{7.41}$$

and the equivalent differential form

$$d|\mathbf{k}| = \frac{2\pi}{\nu_w}\,d\nu, \tag{7.42}$$

where ν_w is the wave velocity. Substituting (7.41) and (7.42) into (7.40) yields

$$\hat{g}(\nu)\,d\nu = \frac{4\pi V \nu^2}{\nu_w^3}\,d\nu. \tag{7.43}$$

This relation accounts for one type of wave. As we noted in the previous section, two types of waves are possible: (1) transverse waves in which the medium vibrates perpendicular to the direction of propagation (i.e., the direction of $|\mathbf{k}|$) and (2) longitudinal waves in which the medium vibrates in the same direction that the wave propagates.

Since transverse waves can vibrate in two normal directions, there are three wave polarizations possible for the medium: two transverse waves and one longitudinal wave. All three of these waves make a contribution equal to $\hat{g}(\nu)$ to the overall distribution function $g(\nu)$. Combining the three contributions yields the complete expression for the Debye approximation to $g(\nu)\,d\nu$:

$$g(\nu)\,d\nu = \left(\frac{2}{\nu_t^3} + \frac{1}{\nu_l^3}\right)4\pi V \nu^2\,d\nu, \tag{7.44}$$

where ν_t and ν_l are the transverse and longitudinal wave velocities, respectively. Defining a mean wave velocity as

$$\nu_m = 3^{1/3}\left(\frac{2}{\nu_t^3} + \frac{1}{\nu_l^3}\right)^{-1/3}, \tag{7.45}$$

we can write the relation for $g(\nu)\,d\nu$ in the form

$$g(\nu)\,d\nu = \frac{12\pi V}{\nu_m^3}\nu^2\,d\nu. \tag{7.46}$$

This relation is exact in the limit of low frequencies where we expect that crystal behaves like a continuous elastic body. Although this relation is not expected to be accurate at high frequencies, the Debye theory uses this distribution for all the normal frequencies. However, the total number of frequencies must equal $3N$. Debye therefore defined a cutoff frequency ν_D so that the integral of $g(\nu)\,d\nu$ from 0 to ν_D equals $3N$:

$$\int_0^{\nu_D} g(\nu)\,d\nu = \int_0^{\nu_D} \frac{12\pi V}{\nu_m^3}\nu^2\,d\nu = 3N. \tag{7.47}$$

Executing the integral for the above $g(\nu)$ relation yields

$$\nu_D = \left(\frac{3N}{4\pi V} \right)^{1/3} \nu_m. \tag{7.48}$$

The Debye theory frequency distribution can then be stated as

$$g(\nu)\,d\nu = \begin{cases} \frac{9N}{\nu_D^3}\nu^2\,d\nu & \text{for } 0 \le \nu \le \nu_D, \\ 0 & \text{for } \nu > \nu_D. \end{cases} \tag{7.49}$$

This distribution can be inserted into the previously derived relations for thermodynamic properties of the crystal. Doing so yields the following relation for the specific heat:

$$\hat{c}_V = 9N_A k_B \left(\frac{\theta_D}{T} \right)^{-3} \int_0^{\theta_D/T} \frac{\eta^4 e^\eta}{(e^\eta - 1)^2}\,d\eta, \tag{7.50}$$

where θ_D is the Debye temperature defined as

$$\theta_D = \frac{h\nu_D}{k_B}. \tag{7.51}$$

The integral cannot be evaluated in terms of simple functions and must be evaluated numerically. It is traditional to define the Debye function $D(\xi)$ as

$$D(\xi) = 3\xi^3 \int_0^{\xi^{-1}} \frac{\eta^4 e^\eta}{(e^\eta - 1)^2}\,d\eta, \tag{7.52}$$

so that the heat capacity is given by

$$\hat{c}_V = 3N_A k_B D(T/\theta_D). \tag{7.53}$$

Tabulations of the Debye function are available and it can be evaluated using any number of numerical methods. Values of θ_D that give the best overall fit to experimental $\hat{c}_V$ data are listed in Table 7.1 for a variety of monatomic solids.

The limiting behavior of the Debye $\hat{c}_V$ relation can be evaluated as follows. At high temperatures, the upper limit of the integral in the $\hat{c}_V$ relation becomes very small. Since we then need to evaluate the integrand for small η, we can use an expansion of it:

$$\int_0^{\theta_D/T} \frac{\eta^4 e^\eta}{(e^\eta - 1)^2}\,d\eta = \int_0^{\theta_D/T} \frac{\eta^4(1 + \eta + \cdots)}{(1 + \eta + \cdots - 1)^2}\,d\eta = \int_0^{\theta_D/T} \eta^2\,d\eta = \frac{1}{3}\left(\frac{\theta_D}{T} \right)^3. \tag{7.54}$$

Substituting into the $\hat{c}_V$ relation yields

$$\hat{c}_V = 3N_A k_B \quad (\text{for } T \to \infty), \tag{7.55}$$

which is in agreement with the law of Dulong and Petit. The low temperature limit can be obtained by letting the upper limit of the integral go to infinity:

$$\int_0^{\theta_D/T} \frac{\eta^4 e^\eta}{(e^\eta - 1)^2}\,d\eta = \int_0^\infty \frac{\eta^4 e^\eta}{(e^\eta - 1)^2}\,d\eta = \frac{4\pi^4}{15}. \tag{7.56}$$

Table 7.1 *Representative Values of Debye and Fermi Temperatures for Various Monatomic Crystalline Solids*

Solid	θ_D (K)	θ_F (K)	Number of Valence Electrons
Na	158	37,500	1
K	91	24,600	1
Cu	343	81,200	1
Ag	225	63,600	1
Au	165	63,900	1
Mg	400	82,700	2
Zn	327	109,000	2
Cd	209	86,600	2
Al	428	134,900	3
Pb	105	108,700	4

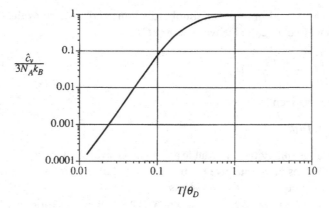

Figure 7.6

Substituting, we find that

$$\hat{c}_V = \frac{12\pi^4}{5} N_A k_B \left(\frac{T}{\theta_D}\right)^3 \quad \text{(as } T \to 0\text{)}, \tag{7.57}$$

which agrees with the T^3 variation observed experimentally. The variation of the specific heat with temperature predicted by the Debye model is plotted in Figure 7.6. The proportionality of $\hat{c}_V$ to T^3 at low temperature can be seen in this plot.

The Debye model is based on a very idealized treatment of the distribution of lattice vibration frequencies. It is a successful model because the resulting distribution is accurate at very low temperatures and because the specific heat and other properties depend on integrals of the distribution function, which weakens their sensitivity to inaccuracies in the distribution at higher frequencies. In general, the Debye temperature varies with temperature

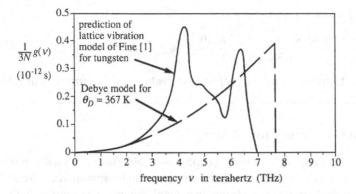

Figure 7.7 Frequency distribution of lattice vibrations predicted by the lattice vibration analysis of Fine [1] for tungsten. Also shown is the frequency distribution predicted by the Debye model for $\theta_D = 367$ K, which is typical of tungsten.

for a given material. The variation is usually weak, however, usually no more than about 10% between 0 and 100 K, and assuming a constant value of the Debye temperature is often a reasonable approximation for engineering applications.

The discussion in the previous section indicates that the frequency distribution for real crystals may be much more complex. For real crystals, a more accurate representation of the frequency distribution can be obtained either from detailed analysis of the vibrational characteristics of the lattice or from experiments. Figure 7.7 shows a comparison of the Debye distribution and the variation predicted by a lattice vibration analysis for tungsten. In principle, a more realistic distribution could be incorporated into the above analysis to get an improved prediction of the specific heat variation with temperature. The interested reader can find more information on frequency distributions for real crystal lattices in references devoted to modern treatments of solid state physics (see, for example, the text by Kittel [2]).

Example 7.2 Solid pure silver has a density of 10,500 kg/m^3. Reported propagation velocities for longitudinal and transverse waves in silver are 3,650 and 1,610 m/s, respectively. Use these values to estimate the Debye cutoff frequency and the Debye temperature for silver.

Solution The Debye mean wave velocity is computed using Eq. (7.45),

$$v_m = 3^{1/3} \left(\frac{2}{v_t^3} + \frac{1}{v_l^3} \right)^{-1/3} = 3^{1/3} \left(\frac{2}{(1,610)^3} + \frac{1}{(3,650)^3} \right)^{-1/3} = 1,817 \text{ m/s}.$$

The cutoff frequency is obtained by using Eq. (7.48). Using the fact that the mass density ρ is $(N/V)(M/N_A)$, we find

$$\nu_D = \left(\frac{3N}{4\pi V} \right)^{1/3} v_m = \left(\frac{\rho N_A / M}{4\pi / 3} \right)^{1/3} v_m$$

$$= \left(\frac{(10,500)6.02 \times 10^{26}/107.9}{4\pi/3} \right)^{1/3} 1,817 = 4.37 \times 10^{12} \text{ Hz}.$$

From Eq. (7.51) we then obtain

$$\theta_D = \frac{h\nu_D}{k_B} = \frac{6.63 \times 10^{-34}(4.37 \times 10^{12})}{1.38 \times 10^{-23}} = 210 \text{ K}.$$

This value of the Debye temperature is close to the value $\theta_D = 225$ for silver in Table 7.1.

7.5 Electron Gas Theory for Metals

Many pure metals, such as iron, copper, and gold, have a crystalline structure and the energy storage associated with lattice vibrations can be appropriately treated with the model described in the previous sections. In a metal, the valence electrons in the outer shells are only weakly bound to the nucleus of the atom. These become conduction electrons, which move about freely in the volume of the metal. Because the valence electrons have escaped, the atoms at the lattice points in the metal effectively have a net positive charge. Conduction electrons do interact with these ions as they move about in the metal. To analyze the contribution of these valence electrons to the thermodynamics properties of the metal, we model them as a fermion gas occupying the volume of the crystal. To use the results of the fermion gas analysis presented in the last chapter we must determine the quantum energy states and degeneracy for the electrons.

For a cubic crystal with characteristics dimension $V^{1/3}$, energy levels of electrons are modeled as a particle in a cubic 3-D box with side $V^{1/3}$. In Chapter 1, we found that for this model the energy levels are given by

$$\varepsilon = \frac{h^2}{8mV^{2/3}}\left(n_x^2 + n_y^2 + n_z^2\right), \tag{7.58}$$

where n_x, n_y, n_z are contained in $\{1, 2, 3, \ldots\}$ and the number density of the energy states between ε and $\varepsilon + d\varepsilon$ (degeneracy) is given by

$$g(\varepsilon)d\varepsilon = \frac{\pi}{4}\left(\frac{8mV^{2/3}}{h^2}\right)^{3/2} \varepsilon^{1/2}\, d\varepsilon. \tag{7.59}$$

Here we will designate the number density of electron quantum states as $g^*(\varepsilon)d\varepsilon$. For fermions, no two particles can be in the same quantum state. Two electrons can occupy each energy state in the particle-in-a-box solution, one with spin up and the other with spin down. Consequently, the density of states for the electrons is twice that for the particle-in-a-box solution:

$$g^*(\varepsilon)d\varepsilon = \frac{\pi}{2}\left(\frac{8mV^{2/3}}{h^2}\right)^{3/2} \varepsilon^{1/2}\, d\varepsilon. \tag{7.60}$$

If electrons were classical particles we would attribute to them a quadratic component of kinetic energy for translation in three orthogonal directions, and the equipartition theorem would dictate that the mean kinetic energy of each would be $(3/2)k_B T$. It follows that for a kmol (containing N_A atoms), with each atom having one electron in the valence shell, the contribution to the molar heat capacity would be $(3/2)N_A k_B$. This is not consistent with experimental measurements, which indicate that the electronic contribution to the heat capacity varies linearly with temperature. Equipartition predictions are not valid because the electrons in our model gas are not classical particles.

The internal energy in one kmol of the electron gas is given by

$$\hat{u}_{el} = \int_0^\infty \varepsilon g^*(\varepsilon) f(\varepsilon) \, d\varepsilon, \tag{7.61}$$

where $f(\varepsilon)$ is the equilibrium occupancy of energies between ε and $\varepsilon + d\varepsilon$ for a fermion gas. In the previous chapter we showed that this distribution is

$$f(\varepsilon) = \frac{1}{1 + e^{(\varepsilon - \mu)/k_B T}}. \tag{7.62}$$

At low to moderate temperatures, the chemical potential is virtually the same as that attained in the limit of $T \to 0$. Designating this value as the Fermi energy ε_F, the distribution $f(\varepsilon)$ can be written

$$f(\varepsilon) = \frac{1}{1 + e^{(\varepsilon - \varepsilon_F)/k_B T}}. \tag{7.63}$$

The product $g^*(\varepsilon) f(\varepsilon)$ represents the density of accessible states. Its variation with ε is shown in Figure 7.8.

By definition, the electron gas molar specific heat is given by

$$\hat{c}_{V,el} = \frac{d\hat{u}_{el}}{dT} = \int_0^\infty \varepsilon g^*(\varepsilon) \left(\frac{df}{dT} \right) d\varepsilon. \tag{7.64}$$

We further note that the definitions of g^* and f require that the integral of their product must equal the total number of electrons in the system. Thus, for a system containing one kmol of electrons

$$N_A = \int_0^\infty g^*(\varepsilon) f(\varepsilon) \, d\varepsilon. \tag{7.65}$$

Differentiating both sides of the above relation with respect to temperature yields

$$0 = \int_0^\infty g^*(\varepsilon) \left(\frac{df}{dT} \right) d\varepsilon. \tag{7.66}$$

For convenience, we multiply both sides of Eq. (7.66) by ε_F and combine the result with

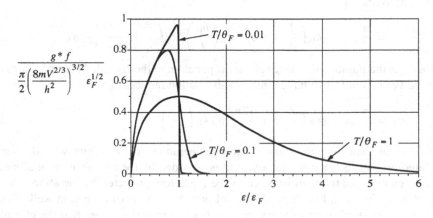

Figure 7.8

Eq. (7.64) to obtain

$$\hat{c}_{V,\text{el}} = \int_0^\infty (\varepsilon - \varepsilon_\text{F}) g^*(\varepsilon) \left(\frac{df}{dT} \right) d\varepsilon. \tag{7.67}$$

If $k_\text{B}T \ll \varepsilon_\text{F}$ ($T \ll \theta_\text{F} = \varepsilon_\text{F}/k_\text{B}$), which is often the case, df/dT is nonzero only when $\varepsilon \cong \varepsilon_\text{F}$. Hence, the relation (7.67) is well approximated as

$$\hat{c}_{V,\text{el}} \cong g^*(\varepsilon_\text{F}) \int_0^\infty (\varepsilon - \varepsilon_\text{F}) \left(\frac{df}{dT} \right) d\varepsilon. \tag{7.68}$$

Differentiating Eq. (7.63), we obtain

$$\left(\frac{df}{dT} \right) = \left(\frac{\varepsilon - \varepsilon_\text{F}}{k_\text{B}T^2} \right) \frac{e^{(\varepsilon - \varepsilon_\text{F})/k_\text{B}T}}{[1 + e^{(\varepsilon - \varepsilon_\text{F})/k_\text{B}T}]^2}. \tag{7.69}$$

Substituting the right side of Eq. (7.69) for df/dT and changing the integration variable to $\eta = (\varepsilon - \varepsilon_\text{F})/k_\text{B}T$ transforms Eq. (7.68) into

$$\hat{c}_{V,\text{el}} \cong k_\text{B}^2 T g^*(\varepsilon_\text{F}) \int_{-\varepsilon_\text{F}/k_\text{B}T}^\infty \frac{\eta^2 e^\eta}{(1 + e^\eta)^2} \, d\eta. \tag{7.70}$$

For $k_\text{B}T \ll \varepsilon_\text{F}$ the lower limit of the integral may be taken to be $-\infty$. The resulting definite integral equals $\pi^2/3$, and the relation for the specific heat reduces to

$$\hat{c}_{V,\text{el}} \cong (\pi^2/3) k_\text{B}^2 T g^*(\varepsilon_\text{F}). \tag{7.71}$$

To evaluate $g^*(\varepsilon_\text{F})$ we use Eq. (7.60) together with the relation (6.157) for the Fermi energy derived in Chapter 6. Substituting the result into Eq. (7.71) and rearranging, the relation for the specific heat becomes

$$\hat{c}_{V,\text{el}} = \frac{\pi^2}{2} N_\text{A} k_\text{B} \left(\frac{T}{\theta_\text{F}} \right), \tag{7.72}$$

where $\theta_\text{F} = \varepsilon_\text{F}/k_\text{B}$ is the Fermi temperature.

For metallic crystals, contributions of both lattice vibrations (phonons) and electrons must be included. For temperatures below the Fermi temperature, we combine Eqs. (7.50) and (7.72) to obtain

$$\hat{c}_V = \frac{\pi^2}{2} r_\text{ve} N_\text{A} k_\text{B} \left(\frac{T}{\theta_\text{F}} \right) + 9 N_\text{A} k_\text{B} \left(\frac{\theta_\text{D}}{T} \right)^{-3} \int_0^{\theta_\text{D}/T} \frac{\eta^4 e^\eta}{(e^\eta - 1)^2} \, d\eta, \tag{7.73}$$

where r_ve is the number of valence electrons per atom in the crystal. At temperatures below both the Fermi and Debye temperatures, this relation reduces to

$$\hat{c}_V = \frac{\pi^2}{2} r_\text{ve} N_\text{A} k_\text{B} \left(\frac{T}{\theta_\text{F}} \right) + \frac{12\pi^4}{5} N_\text{A} k_\text{B} \left(\frac{T}{\theta_\text{D}} \right)^3. \tag{7.74}$$

Experimental data for pure metals at low temperature are consistent with the variation indicated by this equation in that a best fit is provided by the combination of a linear and a cubic temperature term. In some cases the prefactors predicted by the above model are a good fit to the data. In others, the model prefactors do not fit the data well. The main cause of the disagreement in many cases is that the model assumes that the electrons do not interact with the lattice. In fact they may do so, and phonon–electron interactions may have a significant effect on energy storage in some metals. In real crystals, the potential field imposed by the lattice is spatially periodic. It can be shown that this spatial periodicity changes the accessible energy levels for the electrons, resulting in bands of energy that they can occupy and bands that are inaccessible. A full discussion of these issues is beyond the scope of this text. Further discussion of electron interactions with the lattice and electronic energy bands in solids can be found in the references by Kittel [2] and Hook and Hall [3].

Example 7.3 Assess the importance of the electronic contribution to the specific heat of aluminum at 70 K.

Solution The specified temperature is well below the Debye and Fermi temperatures for aluminum. We therefore use Eq. (7.74), which can be written in the form

$$\frac{\hat{c}_v}{N_A k_B} = r_{ve} \frac{\pi^2}{2} \left(\frac{T}{\theta_F} \right) + \frac{12\pi^4}{5} \left(\frac{T}{\theta_D} \right)^3.$$

Substituting r_{ve}, θ_F, and θ_D values from Table 7.1 yields

$$\frac{\hat{c}_v}{N_A k_B} = 3 \frac{\pi^2}{2} \left(\frac{70}{134,900} \right) + \frac{12\pi^4}{5} \left(\frac{70}{428} \right)^3$$
$$= 0.0077 + 1.023 = 1.030.$$

In this case the electronic contribution is slightly less than 1% of the specific heat value. More than 99% is due to phonon energy storage.

7.6   Entropy and the Third Law

Now that we have examined thermodynamic characteristics of gas, liquid, and solid phases at high and low temperatures, we are in a position to draw some general conclusions about how the entropy of substances varies at low temperatures. First, for a Debye crystal, we substitute

$$U = k_B T \left(\frac{\partial \ln Q}{\partial T} \right)_{N,V}$$

from Chapter 4 and the relations

$$g(\nu)\,d\nu = \begin{cases} \frac{9N}{\nu_D^3} \nu^2 \, d\nu & \text{for } 0 \le \nu \le \nu_D, \\ 0 & \text{for } \nu > \nu_D, \end{cases} \tag{7.49}$$

$$-\ln Q = \frac{\Phi(0,\ldots,0,N/V)}{k_B T} + \int_0^\infty \left[\ln\{1 - e^{-h\nu/k_B T}\} + \frac{h\nu}{2k_B T} \right] g(\nu)\,d\nu \tag{7.10}$$

for a Debye crystal into the relation for entropy obtained in Chapter 4,

$$\frac{S}{k_B} = \frac{U}{k_B T} + \ln Q. \tag{4.41}$$

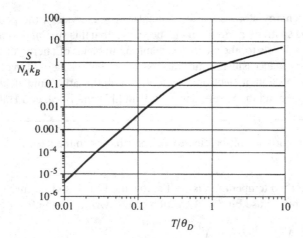

Figure 7.9 Entropy versus T/θ_D for a solid Debye crystal.

Doing so yields the following relation for the entropy of a monatomic crystal:

$$S = 9Nk_B \left(\frac{T}{\theta_D}\right)^3 \left[\int_0^{\theta_D/T} \frac{\eta^2 e^{-\eta}}{1-e^{-\eta}}\, d\eta - \int_0^{\theta_D/T} \eta^2 \ln(1-e^{-\eta})\, d\eta\right]. \qquad (4.75)$$

This relation is plotted in Figure 7.9. As $T \to 0$, the integrals in the above relation approach constants, with the net result that $S \to 0$ as $T \to 0$ proportional to T^3. This is fully consistent with the following statement of the third law of thermodynamics presented without justification in Chapter 3:

♦   *The entropy of a pure crystalline solid vanishes as the absolute temperature approaches zero:* $(\lim_{T\to 0} S = 0)_{\text{pure solid crystal}}$.

Note that this suggests that there is an absolute scale of entropy for a pure crystalline solid that is referenced to absolute zero temperature. The work of Walther Nernst in the early part of the twentieth century implied this limiting behavior. Since that time, the work of other investigators indicates that only pure crystalline solids have zero entropy at absolute zero temperature. If we substitute the single-species form of relation (6.127) for the partition function for fermion or boson gases into the entropy relation obtained in Chapter 4,

$$S = k_B \ln \Xi + \left(\frac{\partial \ln \Xi}{\partial T}\right)_{V,\mu}, \qquad (4.113)$$

we obtain the following relation for the entropy of a fermion or boson gas:

$$S_{f/b} = k_B \sum_{l'=0}^{\infty} \left[\pm\ln\left\{1 \pm e^{(\mu-\varepsilon_{l'})/k_B T}\right\} + \frac{(\varepsilon_{l'}/k_B T)e^{(\mu-\varepsilon_{l'})/k_B T}}{1 + e^{(\mu-\varepsilon_{l'})/k_B T}}\right]. \qquad (4.76)$$

Analysis of the limiting behavior of the right of this relation would allow us to determine the limiting value of entropy as $T \to 0$ for either a fermion or boson gas. At first glance this appears to be a significant result. However, virtually all substances undergo transition to a liquid and/or a solid phase as the temperature approaches zero. The fermion/boson gas model thus becomes inapplicable to virtually all substances at temperatures near zero. Determination of the limiting entropy value for an arbitrary substance at zero temperature requires very detailed analysis of the substance's behavior near zero. Analyses of this type indicate that the entropy approaches a finite value for substances other than a pure crystalline solid. Further discussion of the third law can be found in the references by Zemansky and Dittman [4], Lay [5], and Goldstein and Goldstein [6].

Exercises

7.1 For a monatomic crystal of the type considered in Section 7.1, show that the crystal obeys the law of Dulong and Petit in the high temperature limit for any $g(\nu)$ that has the following characteristics:
   (1) $g(\nu) = 0$ for ν greater than some cutoff frequency ν_{max}.
   (2) $g(\nu)$ is finite for $0 \leq \nu \leq \nu_{max}$.

7.2 Show that in the limit $T \to 0$ that the entropy of a Debye crystal is given by $S = (4/5)\pi^4 N k_B (T/\theta_D)^3$.

7.3 Derive a relation for the entropy of an Einstein crystal and compare its predictions to that of the relation for the Debye crystal as $T \to 0$.

7.4 We argued in Section 7.4 that in the limit $T \to 0$ the integral in Eq. (7.50) approached $4\pi^2/15$ and consequently that the specific heat approached the cubic variation specified by Eq. (7.57) as $T \to 0$. Rewrite the integral in two parts:

$$\int_0^{\theta_D/T} \frac{\eta^4 e^\eta}{(e^\eta - 1)^2}\, d\eta = \int_0^\infty \frac{\eta^4 e^\eta}{(e^\eta - 1)^2}\, d\eta - \int_{\theta_D/T}^\infty \frac{\eta^4 e^\eta}{(e^\eta - 1)^2}\, d\eta.$$

Evaluate the second integral approximately for large η and derive a next correction term for the specific heat in the limit of small but finite temperature T.

7.5 Solid pure zinc has a density of 7,100 kg/m^3. In most metals, the propagation velocity for longitudinal waves is about twice that for transverse waves. Use this fact to estimate the velocity of longitudinal lattice waves in zinc.

7.6 For diamond, the Debye temperature is 2,230 K. Will the specific heat agree with the law of Dulong and Petit at room temperature? Quantitatively justify your answer.

7.7 Reported propagation velocities for longitudinal and transverse waves in titanium are 3,320 and 1,670 m/s, respectively. Solid titanium has a density of 7,300 kg/m^3. Use these data to estimate the Debye cutoff frequency and the Debye temperature for titanium.

7.8 You are designing a mechanical component and you want the material for the component to have the lowest possible average acoustic velocity. If you only had density and specific heat data available, would you select a metal for the component with high or low values of density and specific heat? Briefly justify your answer.

7.9 A block of pure copper has a mass of 30 kg. Write a computer program to numerically determine the energy required to raise the block's temperature from 20 to 400 K.

7.10 A tank with aluminum walls contains 6.0 kg of nitrogen gas at 800 kPa and 290 K. The walls of the tank have a total mass of 20 kg. The tank is dipped into and out of liquid nitrogen, which suddenly cools the tank's walls to 77 K, while the gas inside is still at 290 K. Heat exchange then occurs between the tank walls and the interior gas. Assuming that no exchange occurs between the walls and the air outside the tank, determine the equilibrium temperature in the gas and in the tank wall. For the composite system, consisting of the nitrogen gas and the tank walls, write a computer program to determine the net change in entropy between the conditions after chilling the walls and the final equilibrium.

7.11 Estimate the specific heat of lead at 40 K. Assess the importance of the electronic contribution to the specific heat at this temperature and compare the overall value at 40 K to the value predicted by the law of Dulong and Petit.

7.12 Derive a relation for the fraction of the specific heat that is due to phonon energy storage in a pure metallic crystal. Plot the variation of this fraction with temperature between 0 and 300 K for aluminum. Based on your plot, how low must the temperature be before electron contributions are significant?

7.13 Based on the theory presented in this chapter, experimental measurements of the heat capacity of metals at very low temperature are often fit with a curve of the form

$$\hat{c}_v = \gamma_c T + \beta_c T^3.$$

Measured data indicate that $\gamma_c = 2.08$ J/(kmol K^2) for potassium and $\gamma_c = 0.69$ J/(kmol K^2) for copper. Compare each of these values to the corresponding value predicted by simple electron gas theory. What does the comparison suggest about the accuracy of the electron gas model?

7.14 Write a computer program to compute the specific heat of gold as a function of temperature from 0 to 200 K and plot the results. At what temperature are electronic and phonon energy storage about equally important?

7.15 Initially a 0.6 kg block of pure solid gold is at 30 K and a block of pure aluminum with a mass of 1.0 kg is at 90 K. The two masses are brought into contact with one another. Assuming the blocks do not thermally interact with the surroundings, determine the temperature in each block after equilibrium is reached.

7.16 Initially a 2.5 kg block of pure solid gold is at 20 K, another 1.0 kg block of gold is at 40 K, and a block of pure aluminum with a mass of 1.0 kg is at 90 K. The three masses are brought into contact with one another. Assuming the blocks do not thermally interact with the surroundings, determine the temperature in each block after equilibrium is reached.

References

[1] Fine, P. C., "The Normal Modes of Vibration of a Body-Centered Cubic Lattice," *Physical Review*, 56:355–9, 1939.

[2] Kittel, C., *Introduction to Solid State Physics*, 6th ed., John Wiley & Sons, New York, 1986.

[3] Hook, J. R. and H. E. Hall, *Solid State Physics*, 2nd ed., John Wiley & Sons, Chichester, England, 1991.

[4] Zemanksy, M. W. and R. Dittman, *Heat and Thermodynamics*, Chapter 19, McGraw-Hill, New York, 1981.

[5] Lay, J. E., *Statistical Mechanics and Thermodynamics of Matter*, Chapter 5, Harper and Row, New York, 1990.

[6] Goldstein, M. and I. F. Goldstein, *The Refrigerator and the Universe, Understanding the Laws of Energy*, Chapter 14, Harvard University Press, Cambridge, MA, 1993.

Phase Transitions and Phase Equilibrium

Chapter 8 approaches the topics of phase equilibrium and phase transitions from a microscale perspective. Specifically, the roles of fluctuations and system stability in the onset of phase transitions are examined in detail. Using aspects of statistical thermodynamics theory developed in earlier chapters, the van der Waals model is used to demonstrate how fluctuations and system instability give rise to phase transitions in fluid systems. Binary fluid systems are considered, with pure fluid results being recovered when the mass of one species is set to zero. It is shown that critical exponents and the law of corresponding states can be deduced from the van der Waals model for pure fluids. Microscale aspects of solid–liquid transitions are also considered.

8.1 Fluctuations and Phase Stability

In the development of thermodynamics presented in the preceding chapters, we have identified different categories of substances according to the density of the substance and the nature of molecular interactions in the substance. Gases have low density and the molecules spend most of the time traveling through space with momentum and energy being exchanged between molecules only through brief collisions. In liquids, the molecules are free to roam about within the system but the density is much higher than in gases, with the mean distance between adjacent molecules being only one to two molecular diameters. Because the molecules are close to their neighbors, they continuously are subject to force interactions with nearby molecules, resulting in continuous exchange of momentum and energy.

In solids the atoms have fixed average locations in space and they interact with adjacent atoms by stretching or compressing the bonds with the adjacent atoms. Lattice vibrations facilitate storage and transport of energy within a crystalline solid. The solid, liquid, and gas categories are termed *phases* of a material. A *phase* is defined as a quantity of matter that is macroscopically homogeneous throughout. Thus far, we have said nothing about how transitions from one phase to another can occur, or how the microscale characteristics of the system affect such transitions. In this chapter we will explore these issues in detail. In particular, we want to explore the roles of microscale fluctuations and system stability in the onset of phase transitions. In doing so we will use the van der Waals model to examine features of phase stability and phase transitions. Although this is a very simplistic model, it exhibits the main features of phase transitions observed in most real systems. We will also examine ways that we can generalize the thermodynamic framework to treat phase transitions in all real fluids.

Fluctuations

In Section 4.4 we examined the predictions of statistical thermodynamics theory regarding microscale fluctuations in a system at macroscale equilibrium. We specifically

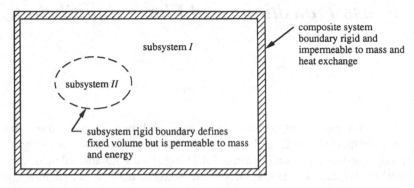

Figure 8.1

considered the variance of properties, defined for an arbitrary property Y as

$$\sigma_Y^2 = \langle Y - \langle Y \rangle \rangle^2. \tag{8.1}$$

The square root of the variance of σ_Y is the standard deviation of Y from its mean value.

Here we will consider fluctuations in a permeable control volume within a larger system at equilibrium. The region within the control volume we will designate as subsystem II, whereas the region outside the control volume but within the larger system is designated as system I. The walls of the composite system are rigid and permit no heat or mass exchange, effectively isolating it from the rest of the universe. The system contains two species a and b. This situation is shown schematically in Figure 8.1. The subsystem II is in equilibrium with the surrounding system I, and hence it is at the same pressure, temperature, and chemical potential values as system I. For a system held at fixed P, T, μ_a, and μ_b, the grand canonical ensemble provides the model for statistical analysis. For a system of this type we demonstrated in Section 4.4 that the variance in the number of type a particles N_a in the system is given by

$$\sigma_{N_a}^2 = k_B T \left(\frac{\partial \langle N_a \rangle}{\partial \mu_a} \right)_{V,T,\mu_b}. \tag{4.125}$$

We similarly noted that the variance of the total number of particles $N = N_a + N_b$ and the variance of the internal energy within the system are given by

$$\sigma_N^2 = k_B T \left[\left(\frac{\partial \langle N_a \rangle}{\partial \mu_a} \right)_{V,T,\mu_b} + \left(\frac{\partial \langle N_b \rangle}{\partial \mu_b} \right)_{V,T,\mu_a} + \left(\frac{\partial \langle N_a \rangle}{\partial \mu_b} \right)_{V,T,\mu_a} + \left(\frac{\partial \langle N_b \rangle}{\partial \mu_a} \right)_{V,T,\mu_b} \right], \tag{4.126}$$

$$\sigma_U^2 = k_B T^2 \left(\frac{\partial U}{\partial T} \right)_{V,N_a,N_b} + \Phi \sigma_{N_a}^2 \left(\frac{\partial U}{\partial N_a} \right)_{V,T,\mu_b}^2 + \Phi \sigma_{N_b}^2 \left(\frac{\partial U}{\partial N_b} \right)_{V,T,\mu_a}^2, \tag{4.127a}$$

where

$$\Phi = \frac{\left(\frac{\partial \langle N_a \rangle}{\partial \mu_a} \right)_{V,T,\mu_b} \left(\frac{\partial \langle N_b \rangle}{\partial \mu_b} \right)_{V,T,\mu_a}}{\left(\frac{\partial \langle N_a \rangle}{\partial \mu_a} \right)_{V,T,\mu_b} \left(\frac{\partial \langle N_b \rangle}{\partial \mu_b} \right)_{V,T,\mu_a} - \left(\frac{\partial \langle N_a \rangle}{\partial \mu_b} \right)_{V,T,\mu_a} \left(\frac{\partial \langle N_b \rangle}{\partial \mu_a} \right)_{V,T,\mu_b}}. \tag{4.127b}$$

Given these relations, we can draw some conclusions regarding possible levels of fluctuations in the system within the control volume in Figure 8.1. First, we note that in Eq. (4.127a) we can equivalently write the relation for $(\partial U/\partial T)_{V,N_a,N_b}$ as

$$
\left(\frac{\partial U}{\partial T}\right)_{V,N_a,N_b} = \left(\frac{\partial(N\hat{u}/N_A)}{\partial T}\right)_{V,N_a,N_b} = \frac{N}{N_A}\left(\frac{\partial\hat{u}}{\partial T}\right)_{V,N_a,N_b} = \frac{N\hat{c}_V}{N_A}, \tag{8.2}
$$

and hence the relation for σ_U^2 can be written

$$
\sigma_U^2 = \frac{Nk_BT^2\hat{c}_V}{N_A} + \Phi\sigma_{N_a}^2\left(\frac{\partial U}{\partial N_a}\right)_{V,T,\mu_b}^2 + \Phi\sigma_{N_b}^2\left(\frac{\partial U}{\partial N_b}\right)_{V,T,\mu_a}^2 . \tag{8.3}
$$

The first term on the right side of Eq. (8.3) is well behaved as long as the molar specific heat is positive and finite. The derivatives $(\partial U/\partial N_a)_{V,T,\mu_b}$ and $(\partial U/\partial N_b)_{V,T,\mu_a}$ represent the amount by which the system internal energy increases when one particle of type a or b are added to the system. For a system with a large number of particles, this is expected to be a finite number comparable to the mean energy per molecule in the system. The effect of the second and third terms therefore depends primarily on $\sigma_{N_a}^2$ and $\sigma_{N_b}^2$, respectively. Equation (4.125) indicates that $\sigma_{N_a}^2$ is finite and positive as long as $(\partial\mu_a/\partial N_a)_{V,T,\mu_b}$ is positive. If $(\partial\mu_a/\partial N_a)_{V,T,\mu_b}$ approaches zero, $\sigma_{N_a}^2$ and σ_U^2 will become very large, implying that there will be very large fluctuations in the molecular density and internal energy within the control volume. The derivatives that appear in Φ and σ_N^2 also clearly may affect the level of fluctuations in the system. Note that one of the Maxwell relations is

$$
\left(\frac{\partial\mu_a}{\partial N_b}\right)_{V,T,\mu_b} = \left(\frac{\partial\mu_b}{\partial N_a}\right)_{V,T,\mu_a}
$$

and so there are actually three derivatives that dictate the magnitude of Φ and σ_N^2: $(\partial\mu_a/\partial N_a)_{V,T,\mu_b}$, $(\partial\mu_b/\partial N_b)_{V,T,\mu_a}$, and $(\partial\mu_a/\partial N_b)_{V,T,\mu_b}$. If these are all finite and the molar specific heat is finite, the magnitude of fluctuations in N, N_a, and U are finite and well defined. However, as any of these approaches zero, the system mass and energy fluctuations become very large. When the fluctuations become very large, the system no longer has a well-defined equilibrium state in the usual thermodynamic sense. One of two things may happen in such circumstances. Either the system remains in a choatic state with high fluctuation levels or one of the large fluctuations may cause the system to change to a new state in which fluctuation levels are low enough that the state is stable and well defined in the usual thermodynamic sense.

The above relations apply to a binary mixture. For a system containing a pure substance, these relations reduce to simpler forms. To apply them to a single-species system we set N_b to zero, and drop μ_b as a parameter. For a single-species system, here we will also drop the a subscript and designate the number of particles and chemical potential simply as N and μ, respectively. It follows that for a single-species system, the relation for the variance in the number of particles N is given by

$$
\sigma_N^2 = \frac{k_BT}{(\partial\mu/\partial N)_{V,T}}. \tag{8.4}
$$

In Section 4.4 we showed that this relation can be converted to the form

$$
\sigma_N^2 = \frac{N^2k_BT}{-(\partial P/\partial V)_{N,T}V^2}. \tag{8.5}
$$

For a single-species system Φ reduces to one and the relation for the variance in the internal energy becomes

$$\sigma_U^2 = \frac{N k_B T^2 \hat{c}_v}{N_A} + \sigma_N^2 \left(\frac{\partial U}{\partial N} \right)_{V,T}^2. \tag{8.6}$$

These relations indicate that for a pure system the subsystems within the control volume in Figure 8.1 will have well-defined mean properties with a low level of fluctuations about the mean values only if $\hat{c}_V$ and $-(\partial P / \partial V)_{N,T}$ are positive and finite. For any physically realistic system, N and U are always positive and they have positive mean values within the control volume. For such systems, the definition of the variance makes it mathematically impossible for σ_N^2 or σ_U^2 to have a negative value. This implies that $-(\partial P / \partial V)_{N,T}$ cannot be negative for physically realistic systems. Although Eq. (8.6) indicates that $\hat{c}_V$ could be negative and result in a positive σ_U^2, this also seems unrealistic since it would imply that adding internal energy to a system would decrease its temperature.

For virtually all real systems, $\hat{c}_V$ is finite and greater than zero and its value does not contribute to high or unrealistic fluctuations in the system. However, as we shall see, in some real systems $-(\partial P / \partial V)_{N,T}$ can become small enough to produce high fluctuation levels in subregions of a system like that in Figure 8.1.

The above results indicate that for a pure or binary system, it is possible to encounter circumstances in which the necessary conditions for equilibrium are met but fluctuations in energy or particle number are large. Although we have assumed that the system is in equilibrium, we have not explored the stability of the equilibrium within the system. The question of stability is especially relevant to systems that exhibit high levels of density or energy fluctuations. These observations suggest that if the system is not stable, such fluctuations could cause the system to undergo a transition to some new configuration. To examine this issue we will now consider the stability of a thermodynamic system in more detail.

Intrinsic Stability

We now will examine the stability of the overall system shown in Figure 8.1. Because the system is isolated and its wall are rigid, its mass, volume, and internal energy are fixed. In Chapter 3 we showed that the entropy maximization principle leads to a specification of the necessary conditions for equilibrium between two subsystems like subsystems I and II in Figure 8.1. This result was obtained by requiring $dS = 0$ for the composite system at equilibrium. The stability of the system is actually dictated by additional requirements that result from a more complete specification of the entropy maximum condition.

We know from our analysis of thermodynamic systems in Chapter 4 that there will be fluctuations in the system in Figure 8.1. If the system is perfectly isolated, its total mass, volume, and energy are fixed. However, internal fluctuations could cause changes in the number of particles or energy in subsystems I and II, provided that there are no net changes in these parameters for the system as a whole. In our analysis here, the fundamental equation is assumed to apply to each of the subsystems. The control volume defining subsystem II is assumed to be fixed, and since the volume of the total system is fixed, it follows that the volume of subsystem I is also fixed.

For perturbations δU^I, δN_a^I, and δN_b^I in system I, the resulting change in entropy for system I is obtained by expanding the fundamental equation in a Taylor series

$$\Delta S^I = \delta S^I + \left(\frac{1}{2!}\right)\delta^2 S^I + \left(\frac{1}{3!}\right)\delta^3 S^I + \cdots, \tag{8.7}$$

where the δS^I, $\delta^n S^I$ terms are a shorthand way of representing the terms of successive order:

$$\delta S^I = \sum_{\eta=U^I,N_a^I,N_b^I} \left(\frac{\partial S}{\partial \eta}\right)^I \delta\eta, \tag{8.8}$$

$$\delta^2 S^I = \sum_{\eta=U^I,N_a^I,N_b^I}\ \sum_{\xi=U^I,N_a^I,N_b^I} \left(\frac{\partial^2 S}{\partial\eta\partial\xi}\right)^I \delta\eta\,\delta\xi, \tag{8.9}$$

$$\delta^3 S^I = \sum_{\eta=U^I,N_a^I,N_b^I}\ \sum_{\xi=U^I,N_a^I,N_b^I}\ \sum_{\zeta=U^I,N_a^I,N_b^I} \left(\frac{\partial^3 S}{\partial\eta\partial\xi\partial\zeta}\right)^I \delta\eta\delta\xi\delta\zeta. \tag{8.10}$$

Identical reasoning results in corresponding relations for subsystem II:

$$\Delta S^{II} = \delta S^{II} + \left(\frac{1}{2!}\right)\delta^2 S^{II} + \left(\frac{1}{3!}\right)\delta^3 S^{II} + \cdots, \tag{8.11}$$

$$\delta S^{II} = \sum_{\eta=U^{II},N_a^{II},N_b^{II}} \left(\frac{\partial S}{\partial \eta}\right)^{II} \delta\eta, \tag{8.12}$$

$$\delta^2 S^{II} = \sum_{\eta=U^{II},N_a^{II},N_b^{II}}\ \sum_{\xi=U^{II},N_a^{II},N_b^{II}} \left(\frac{\delta^2 S}{\partial\eta\partial\xi}\right)^{II} \delta\eta\delta\xi, \tag{8.13}$$

$$\delta^3 S^{II} = \sum_{\eta=U^{II},N_a^{II},N_b^{II}}\ \sum_{\xi=U^{II},N_a^{II},N_b^{II}}\ \sum_{\zeta=U^{II},N_a^{II},N_b^{II}} \left(\frac{\partial^3 S}{\partial\eta\partial\xi\partial\zeta}\right)^{II} \delta\eta\delta\xi\delta\zeta. \tag{8.14}$$

The additivity property of entropy dictates that the change in the entropy of the total system equals the sum of the changes in the subsystems. We can therefore write the total change in entropy for the composite system ΔS as

$$\Delta S = \delta S + \delta^2 S + \delta^3 S + \cdots, \tag{8.15}$$

where

$$\delta S = \delta S^I + \delta S^{II}, \tag{8.16}$$

$$\delta^n S = \delta^n S^I + \delta^n S^{II}, \qquad n = 2, 3, 4, \ldots. \tag{8.17}$$

For the subsystems, since the mass and energy of overall system is fixed, fluctuations in these quantities for the subsystems must satisfy

$$\delta U^I = -\delta U^{II}, \tag{8.18}$$

$$\delta N_a^I = -\delta N_a^{II}, \tag{8.19}$$

$$\delta N_b^I = -\delta N_b^{II}. \tag{8.20}$$

Substituting Eqs. (8.8) and (8.12) into (8.16) and using Eqs. (8.18), (8.19), and (8.20) yields

$$
\delta S = \left[\left(\frac{\partial S}{\partial U} \right)^{I}_{N_a, N_b, V} - \left(\frac{\partial S}{\partial U} \right)^{II}_{N_a, N_b, V} \right] \delta U^{I}
$$

$$
+ \left[\left(\frac{\partial S}{\partial N_a} \right)^{I}_{U, N_b, V} - \left(\frac{\partial S}{\partial N_a} \right)^{II}_{U, N_b, V} \right] \delta N^{I}_{a}
$$

$$
+ \left[\left(\frac{\partial S}{\partial N_b} \right)^{I}_{U, N_a, V} - \left(\frac{\partial S}{\partial N_b} \right)^{II}_{U, N_a, V} \right] \delta N^{I}_{b}. \tag{8.21}
$$

Since by definition

$$
(\partial S / \partial U)_{N_a, N_b, V} = T,
$$

$$
(\partial S / \partial N_a)_{U, N_b, V} = -\mu_a / T,
$$

$$
(\partial S / \partial N_b)_{U, N_a, V} = -\mu_b / T,
$$

it follows from the equilibrium requirements that the chemical potential values and temperature be equal in systems I and II that the first variation of S is zero,

$$
\delta S = 0 \text{ (at equilibrium).} \tag{8.22}
$$

The condition $\delta S = 0$ reflects a zero slope condition at equilibrium. Basic calculus dictates that a necessary condition for a maximum in S is that $(\delta^n S)_{min} < 0$, where $(\delta^n S)_{min}$ is the lowest-order nonvanishing variation of S for $n = 2, 3, 4, \ldots$. The maximum in S assures stability because for equilibrium at the maximum, fluctuations away from equilibrium result in a decrease of the system entropy. Spontaneous subsequent fluctuations that increase the entropy back toward the maximum are then most likely to occur, tending to restore the original equilibrium. To determine the necessary conditions for a stable equilibrium, we therefore seek to determine the conditions that result in a maximum value of entropy at the equilibrium point. Thus, for the composite system we are considering here, the criteria for equilibrium and stability may be stated as

$$
\delta S = 0 \qquad \text{criterion of equilibrium,} \tag{8.23}
$$

$$
\left. \begin{array}{l} \delta^n S < 0 \quad \text{for the smallest} \\ n = 2, 3, 4 \ldots \quad \text{at which } \delta^n S \neq 0 \end{array} \right\} \text{criterion of stability.} \tag{8.24}
$$

The above criteria indicate that we must first consider $\delta^2 S$ to determine the necessary conditions for the system to be stable. Summing the rights sides of Eqs. (8.9) and (8.13), we obtain

$$
\delta^2 S = \left(\frac{\partial^2 S}{\partial N_a^2} \right)^{I} (\partial N_a^I)^2 + \left(\frac{\partial^2 S}{\partial N_b^2} \right)^{I} (\delta N_b^I)^2 + \left(\frac{\partial^2 S}{\partial U^2} \right)^{I} (\delta U^I)^2
$$

$$
+ 2 \left(\frac{\partial^2 S}{\partial N_a \partial N_b} \right)^{I} \delta N_a^I \delta N_b^I + 2 \left(\frac{\partial^2 S}{\partial U^I \partial N_a^I} \right)^{I} \delta U^I \delta N_a^I
$$

$$
+ 2 \left(\frac{\partial^2 S}{\partial U^I \partial N_b^I} \right)^{I} \delta U^I \delta N_b^I + \left(\frac{\partial^2 S}{\partial N_a^2} \right)^{II} (\delta N_a^{II})^2 + \left(\frac{\partial^2 S}{\partial N_b^2} \right)^{II} (\delta N_b^{II})^2
$$

$$+ \left(\frac{\partial^2 S}{\partial U^2}\right)^{\mathrm{II}} (\delta U^{\mathrm{II}})^2 + 2 \left(\frac{\partial^2 S}{\partial N_a \partial N_b}\right)^{\mathrm{II}} \delta N_a^{\mathrm{II}} \delta N_b^{\mathrm{II}}$$

$$+ 2 \left(\frac{\partial^2 S}{\partial U \partial N_a}\right)^{\mathrm{II}} \delta U^{\mathrm{II}} \delta N_a^{\mathrm{II}} + 2 \left(\frac{\partial^2 S}{\partial U \partial N_b}\right)^{\mathrm{II}} \partial U^{\mathrm{II}} \partial N_b^{\mathrm{II}}. \qquad (8.25)$$

Because the terms in the above equation are part of a Taylor series, the partial derivatives are evaluated at the preperturbation state, which is the same in both subsystems. For a given state, however, the derivatives of extensive properties scale with the size of the system $N^{\mathrm{I}} = N_a^{\mathrm{I}} + N_b^{\mathrm{I}}$ because at equilibrium S, U, N_a, and N_b do.

As an example, consider

$$\left(\frac{\partial^2 S}{\partial U \partial N_a}\right)^{\mathrm{I}} \sim \frac{N^{\mathrm{I}}}{N^{\mathrm{I}} \times N^{\mathrm{I}}} \sim \frac{I}{N^{\mathrm{I}}}.$$

Thus, for the two subsystems of different size at the same equilibrium state, it follows that

$$\left(\frac{\partial^2 S}{\partial U \partial N_a}\right)^{\mathrm{I}} \bigg/ \left(\frac{\partial^2 S}{\partial U \partial N_a}\right)^{\mathrm{II}} = \frac{N^{\mathrm{II}}}{N^{\mathrm{I}}}, \qquad (8.26)$$

or equivalently

$$\left(\frac{\partial^2 S}{\partial U \partial N_a}\right)^{\mathrm{II}} = \frac{N^{\mathrm{I}}}{N^{\mathrm{II}}} \left(\frac{\partial^2 S}{\partial U \partial N_a}\right)^{\mathrm{I}}, \qquad (8.27)$$

where $N^{\mathrm{II}} = N_a^{\mathrm{II}} + N_b^{\mathrm{II}}$. Similarly, it can be shown that

$$\left(\frac{\partial^2 S}{\partial U^2}\right)^{\mathrm{II}} = \frac{N^{\mathrm{I}}}{N^{\mathrm{II}}} \left(\frac{\partial^2 S}{\partial U^2}\right)^{\mathrm{I}}, \qquad \left(\frac{\partial^2 S}{\partial N_a^2}\right)^{\mathrm{II}} = \frac{N^{\mathrm{I}}}{N^{\mathrm{II}}} \left(\frac{\partial^2 S}{\partial N_a^2}\right)^{\mathrm{I}}, \qquad (8.28)$$

$$\left(\frac{\partial^2 S}{\partial N_b^2}\right)^{\mathrm{II}} = \frac{N^{\mathrm{I}}}{N^{\mathrm{II}}} \left(\frac{\partial^2 S}{\partial N_b^2}\right)^{\mathrm{I}}, \qquad \left(\frac{\partial^2 S}{\partial U \partial N_b}\right)^{\mathrm{II}} = \frac{N^{\mathrm{I}}}{N^{\mathrm{II}}} \left(\frac{\partial^2 S}{\partial U \partial N_b}\right)^{\mathrm{I}}, \qquad (8.29)$$

$$\left(\frac{\partial^2 S}{\partial N_a \partial N_b}\right)^{\mathrm{II}} = \frac{N^{\mathrm{I}}}{N^{\mathrm{II}}} \left(\frac{\partial^2 S}{\partial N_a \partial N_b}\right)^{\mathrm{I}}. \qquad (8.30)$$

We next modify Eq. (8.25) by using Eqs. (8.18)–(8.20) and (8.27)–(8.30) to eliminate terms with II superscripts. Equation (8.25) then becomes

$$\delta^2 S = \left(1 + \frac{N_{\mathrm{I}}}{N_{\mathrm{II}}}\right) \left[\left(\frac{\partial^2 S_{\mathrm{I}}}{\partial U_{\mathrm{I}}^2}\right) (\delta U_{\mathrm{I}})^2 + \left(\frac{\partial^2 S_{\mathrm{I}}}{\partial N_{a,\mathrm{I}}^2}\right) (\delta N_{a,\mathrm{I}})^2 \right.$$

$$+ \left(\frac{\partial^2 S_{\mathrm{I}}}{\partial N_{b,\mathrm{I}}^2}\right) (\delta N_{b,\mathrm{I}})^2 + 2 \left(\frac{\partial^2 S_{\mathrm{I}}}{\partial U_{\mathrm{I}} \partial N_{a,\mathrm{I}}}\right) \delta U_{\mathrm{I}} \delta N_{a,\mathrm{I}}$$

$$\left. + 2 \left(\frac{\partial^2 S_{\mathrm{I}}}{\partial U_{\mathrm{I}} \partial N_{b,\mathrm{I}}}\right) \delta U_{\mathrm{I}} \delta N_{b,\mathrm{I}} + 2 \left(\frac{\partial^2 S_{\mathrm{I}}}{\partial N_{a,\mathrm{I}} \partial N_{b,\mathrm{I}}}\right) \delta N_{a,\mathrm{I}} \delta N_{b,\mathrm{I}} \right]. \qquad (8.31)$$

Since N^{I} and N^{II} are both positive definite numbers, the sign of $\delta^2 S$ is dictated by the term in square brackets in Eq. (8.31). Inside the square brackets all the terms are evaluated for subsystem I. We will therefore drop the I superscript with the understanding that these terms are evaluated in subsystem I. We can say that the equilibrium will be stable if $\delta^2 S^*$ is less than zero, where $\delta^2 S^*$ is defined as

$$
\delta^2 S^* = \frac{\delta^2 S}{1 + N^{\mathrm{I}}/N^{\mathrm{II}}} = \left[\left(\frac{\partial^2 S}{\partial U^2} \right) (\delta U)^2 + \left(\frac{\partial^2 S}{\partial N_a^2} \right) (\delta N_a)^2 + \left(\frac{\partial^2 S}{\partial N_b^2} \right) (\delta N_b)^2 \right.
$$

$$
+ 2 \left(\frac{\partial^2 S}{\partial U \, \partial N_a} \right) \delta U \delta N_a + 2 \left(\frac{\partial^2 S}{\partial U \, \partial N_b} \right) \delta U \delta N_b
$$

$$
\left. + 2 \left(\frac{\partial^2 S}{\partial N_a \partial N_b} \right) \delta N_a \delta N_b \right]. \tag{8.32}
$$

Rearranging the terms on the right side of the relation for $\delta^2 S^*$, we obtain

$$
\delta^2 S^* = \xi_1 \left\{ \delta U + \frac{(\partial^2 S/\partial U \, \partial N_a)}{(\partial^2 S/\partial U^2)} \delta N_a + \frac{(\partial^2 S/\partial U \, \partial N_b)}{(\partial^2 S/\partial U^2)} \delta N_b \right\}^2
$$

$$
+ \xi_2 \left\{ \delta N_a + \frac{\frac{\partial^2 S}{\partial N_a \partial N_b} - \frac{(\partial^2 S/\partial U \partial N_a)(\partial^2 S/\partial U \partial N_b)}{(\partial^2 S/\partial U^2)}}{\frac{\partial^2 S}{\partial N_a^2} - \frac{(\partial^2 S/\partial U \partial N_a)^2}{(\partial^2 S/\partial U^2)}} \delta N_b \right\}^2 + \xi_3 \{\delta N_b\}^2, \tag{8.33}
$$

where

$$
\xi_1 = \frac{\partial^2 S}{\partial U^2}, \tag{8.34}
$$

$$
\xi_2 = \frac{\partial^2 S}{\partial N_a^2} - \frac{(\partial^2 S/\partial U \, \partial N_a)^2}{(\partial^2 S/\partial U^2)}, \tag{8.35}
$$

$$
\xi_3 = \frac{\partial^2 S}{\partial N_b^2} - \frac{(\partial^2 S/\partial U \, \partial N_b)^2}{(\partial^2 S/\partial U^2)} - \frac{\left[\frac{\partial^2 S}{\partial N_a \partial N_b} - \frac{(\partial^2 S/\partial U \partial N_a)(\partial^2 S/\partial U \partial N_b)}{(\partial^2 S/\partial U^2)} \right]^2}{\frac{\partial^2 S}{\partial N_a^2} - \frac{(\partial^2 S/\partial U \partial N_a)^2}{(\partial^2 S/\partial U^2)}} \tag{8.36}
$$

In Eq. (8.33) the squared terms in curly brackets are positive definite. The conditions under which $\delta^2 S^*$ is negative are dictated by the prefactors of these terms. To determine the conditions for which $\delta^2 S^*$ is negative, we must therefore evaluate the partial derivatives in the prefactors of the squared terms. Using results from Chapters 2 and 4, we evaluate the derivatives as follows. Since by definition $(\partial S/\partial U)_{N_a, N_b, V} = 1/T$,

$$
\xi_1 = \frac{\partial^2 S}{\partial U^2} = \left(\frac{\partial}{\partial U} \left(\frac{\partial S}{\partial U} \right)_{N_a, N_b, V} \right)_{N_a, N_b, V} = -\frac{1}{T^2} \left(\frac{\partial T}{\partial U} \right)_{N_a, N_b, V}. \tag{8.37}
$$

Using Eq. (8.2) to evaluate the derivative in (8.37) we obtain

$$
\xi_1 = -\frac{N_A}{N \hat{c}_V T^2}. \tag{8.38}
$$

Using the relations $(\partial S/\partial N_a)_{N_b, U, V} = -\mu_a/T$, $(\partial S/\partial N_b)_{N_a, U, V} = -\mu_b/T$, it can similarly be shown that

$$\xi_2 = \frac{-\frac{1}{T}\left(\frac{\partial N_b}{\partial \mu_b}\right)_{V,T,\mu_a}}{\left(\frac{\partial N_a}{\partial \mu_a}\right)_{V,T,\mu_b}\left(\frac{\partial N_b}{\partial \mu_b}\right)_{V,T,\mu_a} - \left(\frac{\partial N_a}{\partial \mu_b}\right)_{V,T,\mu_a}\left(\frac{\partial N_b}{\partial \mu_a}\right)_{V,T,\mu_b}}, \tag{8.39}$$

$$\xi_3 = -\frac{1}{T}\left(\frac{\partial \mu_b}{\partial N_b}\right)_{T,V,\mu_a}. \tag{8.40}$$

The stability criterion is satisfied if ξ_1, ξ_2, and ξ_3 are all negative. Examination of Eqs. (8.38)–(8.40) indicates that the parameters that dictate stability for a binary system are $\hat{c}_V$, $(\partial \mu_a/\partial N_a)_{V,T,\mu_b}$, $(\partial \mu_b/\partial N_b)_{V,T,\mu_a}$, and $(\partial \mu_a/\partial N_b)_{V,T,\mu_b}$. Since the Maxwell relations require that $(\partial \mu_a/\partial N_b)_{V,T,\mu_b} = (\partial \mu_b/\partial N_a)_{V,T,\mu_a}$, we need not consider both cross derivatives. Reexamination of the relations derived above for $\sigma_{N_a}^2$, σ_N^2, and σ_U^2 reveals that the parameters that dictate stability are exactly the same as those that dictate the magnitude of fluctuations in the system. It is readily seen that $-k_B/\xi_1$ is the first term in Eq. (8.3) for σ_U^2 and that

$$\xi_2 = -\frac{\Phi}{T}\left(\frac{\partial \mu_a}{\partial N_a}\right)_{V,T,\mu_b} = -\frac{\Phi k_B}{\sigma_{N_a}^2}, \tag{8.41}$$

$$\xi_3 = -\frac{k_B}{\sigma_{N_b}^2}. \tag{8.42}$$

This is a particularly noteworthy result because it provides insight as to the physical causes of system instability. For example, our analysis above indicates that stability is assured if ξ_1, ξ_2, and ξ_3 are negative. Negative ξ_3 requires that $(\partial \mu_b/\partial N_b)_{V,T,\mu_a}$ be positive. If the value of this derivative were decreased until it passed through zero and became negative, the stability of the system would be in doubt. Our analysis of fluctuations in the system indicates that the fluctuations in N_b and U would be infinite if the value of this derivative reached zero. A system with infinite fluctuations is certainly not stable in the usual thermodynamic sense. This strongly implies a link between the stability of the binary system and the magnitude of the fluctuations in the system.

Since the choice of a and b species designations is arbitrary, the requirement that ξ_3 be less than zero is equivalent to the requirement that $(\partial \mu_a/\partial N_a)_{V,T,\mu_b}$ and $(\partial \mu_b/\partial N_b)_{V,T,\mu_a}$ be greater than zero. Using the Maxwell relation for the cross derivative and doing a bit of rearranging, we can state the requirement that ξ_2 be less than zero as

$$\frac{\left(\frac{\partial \mu_a}{\partial N_a}\right)_{V,T,\mu_b}}{1 - \left(\frac{\partial N_a}{\partial \mu_b}\right)_{V,T,\mu_a}^2 \left(\frac{\partial \mu_a}{\partial N_a}\right)_{V,T,\mu_b}\left(\frac{\partial \mu_b}{\partial N_b}\right)_{V,T,\mu_a}} > 0. \tag{8.43}$$

Our analysis implies that for a stable system, $(\partial \mu_a/\partial N_a)_{V,T,\mu_b}$ and $(\partial \mu_b/\partial N_b)_{V,T,\mu_a}$ are positive. For a stable system, since the above inequality is also satisfied, the second term in the denominator is less than one. If conditions change so that $(\partial \mu_a/\partial N_a)_{V,T,\mu_b}$ or $(\partial \mu_b/\partial N_b)_{V,T,\mu_a}$ decreases toward zero, this second term can only become smaller, and the inequality (8.43) will be violated only when $(\partial \mu_a/\partial N_a)_{V,T,\mu_b}$ becomes negative. Stability is therefore assured if $(\partial \mu_a/\partial N_a)_{V,T,\mu_b}$ and $(\partial \mu_b/\partial N_b)_{V,T,\mu_a}$ are both positive, and the additional requirement (8.43) is redundant. For the binary system considered here, we can therefore state that stability is assured if the following conditions are met:

$$\hat{c}_V > 0, \tag{8.44}$$

$$\left(\frac{\partial \mu_a}{\partial N_a}\right)_{V,T,\mu_b} > 0, \qquad \left(\frac{\partial \mu_b}{\partial N_b}\right)_{V,T,\mu_a} > 0. \tag{8.45}$$

These are both necessary and sufficient conditions for stability of a binary mixture system. If these conditions are met, the system is said to be *intrinsically stable*.

The analysis developed in this section for binary mixtures can be extended to systems containing more than two species. In principle, such an extension is straightforward, but the mathematics becomes increasingly complicated as the number of components increases. Further information on analysis of thermodynamic stability for systems containing more than two species can be found in the text by Modell and Reid [1].

The analysis described above must apply to a single-species system in the limit of $N_b \to 0$. As we noted in connection with our analysis of fluctuations, to apply the binary mixture results to a single-species system we set N_b to zero and drop μ_b as a parameter. We also drop the a subscript on μ and N. Equation (8.44) still applies to a single-species system, but the second requirement becomes

$$\left(\frac{\partial \mu}{\partial N}\right)_{V,T} > 0. \tag{8.46}$$

In Chapter 4 we showed that for a binary mixture

$$\left(\frac{\partial N_a}{\partial \mu_a}\right)_{V,T,\mu_b} = -\frac{N_a^2}{V^2}\left(\frac{\partial V}{\partial P}\right)_{N_a,T,\mu_b}, \tag{4.140}$$

which implies that for a single-species system

$$\left(\frac{\partial \mu}{\partial N}\right)_{V,T} = -\frac{V^2}{N^2}\left(\frac{\partial P}{\partial V}\right)_{N,T}. \tag{8.47}$$

Since V and N are positive definite, the condition (8.46) is equivalent to $(\partial P/\partial V)_{N,T} < 0$. We can therefore state the necessary and sufficient conditions for intrinsic stability of a pure system as

$$\hat{c}_V > 0, \tag{8.48}$$

$$\left(\frac{\partial P}{\partial V}\right)_{N,T} < 0. \tag{8.49}$$

Because it relates to thermal energy storage, the first condition is sometimes referred to as the *criterion of thermal stability*. Similarly, because the second condition is associated with the mechanical compressibility of the system, it is sometimes referred to as the *criterion of mechanical stability*.

Implications for Dense Gases and Liquids

As discussed in Chapter 6, the van der Waals equation provides a prototype model of the behavior of pure dense gases and liquids. At low temperatures, the van der Waals model predicts isotherms that look like those in Figure 8.2. At low values of the molar volume $\hat{v} = V N_A / N$, points along the isotherms correspond to a liquid phase, whereas at high $\hat{v}$, the fluid is in a gaseous phase.

To explore the behavior of fluids at low specific volume, we consider the fluid inside the piston and cylinder apparatus shown in Figure 8.3. This apparatus initially contains vapor phase fluid with high specific volume at temperature T_1. The state point for the system is on isotherm T_1 at the far right in Figure 8.2. We move the piston to slowly decrease the volume of the system exchanging heat with an isothermal reservoir so as to keep the system temperature constant at T_1. As this process occurs, the state point follows the isotherm to the left. At first this results in an increase in pressure. If the system follows the isotherm, the van der Waals model predicts that the pressure eventually would pass through a maximum and then a minimum before rising steadily at low $\hat{v}$. These extrema in the pressure occur only for temperatures below a certain value referred to as the *critical temperature*, T_c. We note that at the maximum and minimum $(\partial P / \partial \hat{v})_T = 0$, which implies that $(\partial P / \partial V)_{N,T} = 0$ and that $\sigma_N^2 \to \infty$ at these points. This implies that the system would exhibit very large fluctuations near these conditions. Furthermore, in the region between the pressure minimum and maximum, $(\partial P / \partial \hat{v})_T > 0$, which implies that $(\partial P / \partial V)_{N,T} > 0$.

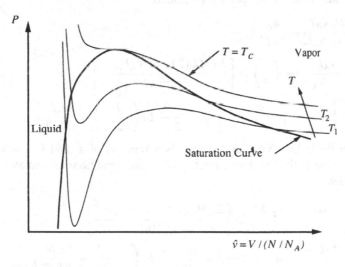

Figure 8.2

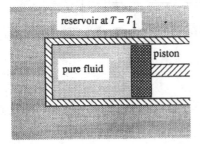

Figure 8.3

It follows from our analysis above that the requirements for intrinsic stability are not satisfied in this region.

The above observations lead to the conclusion that stable thermodynamic states cannot exist in the range of conditions between the pressure minimum and maximum on the isotherm. As $\hat{v}$ is decreased isothermally if the system passes through a sequence of equilibrium states, the observed behavior of a real system does, in fact, deviate from the sequence of states predicted by the van der Waals isotherm. At some point before reaching the maximum pressure on the isotherm, the pressure levels off and stays constant as $\hat{v}$ drops. The system condition bifurcates with a portion remaining at the vapor state at which the pressure first leveled off and the remaining portion exhibiting a higher density liquid state. The pressure at which this bifurcation occurs is the *saturation pressure* P_{sat}, which varies with temperature for a pure fluid: $P_{sat} = P_{sat}(T)$.

At the saturation pressure the gas and liquid phase can coexist, with the former referred to as *saturated vapor* and latter as *saturated liquid*. For the vapor and liquid phase subsystems to be in equilibrium within the overall system, we know that the temperature, pressure, and chemical potential must be the same in both phases. In Chapter 6 we derived Eqs. (6.26) and (6.27) for the pressure and chemical potential of a pure van der Waals fluid. These relations can be written in terms of the molar specific volume as

$$P = \frac{N_A k_B T}{\hat{v} - N_A b_v} - \frac{a_v N_A^2}{\hat{v}^2}, \tag{8.50}$$

$$\frac{\mu}{k_B T} = \frac{N_A b_v}{\hat{v} - N_A b_v} - \left(\frac{3}{2}\right) \ln \left[\frac{2\pi M k_B T (\hat{v} - N_A b_v)^{2/3}}{N_A^{2/3} h^2}\right]$$
$$- \left[\frac{(\xi - 5)}{2} \ln \pi - \ln \sigma_s\right] - \frac{(\xi - 3)}{2} \ln \left(\frac{T}{\theta_{rot,m}}\right) - \frac{2a_v N_A}{\hat{v} k_B T}. \tag{8.51}$$

Clearly for a given fluid, the chemical potential is a function of T and $\hat{v}$. For a given temperature T below the critical temperature, the saturation conditions are dictated by the following five relations:

$$\frac{\mu_l}{k_B T} = \frac{N_A b_v}{\hat{v}_l - N_A b_v} - \left(\frac{3}{2}\right) \ln \left[\frac{2\pi M k_B T (\hat{v}_l - N_A b_v)^{2/3}}{N_A^{2/3} h^2}\right]$$
$$- \left[\frac{(\xi - 5)}{2} \ln \pi - \ln \sigma_s\right] - \frac{(\xi - 3)}{2} \ln \left(\frac{T}{\theta_{rot,m}}\right) - \frac{2a_v N_A}{\hat{v}_l k_B T}, \tag{8.52}$$

$$\frac{\mu_v}{k_B T} = \frac{N_A b_v}{\hat{v}_v - N_A b_v} - \left(\frac{3}{2}\right) \ln \left[\frac{2\pi M k_B T (\hat{v}_v - N_A b_v)^{2/3}}{N_A^{2/3} h^2}\right]$$
$$- \left[\frac{(\xi - 5)}{2} \ln \pi - \ln \sigma_s\right] - \frac{(\xi - 3)}{2} \ln \left(\frac{T}{\theta_{rot,m}}\right) - \frac{2a_v N_A}{\hat{v}_v k_B T}, \tag{8.53}$$

$$\mu_v = \mu_l, \tag{8.54}$$

$$P = \frac{N_A k_B T}{\hat{v}_l - N_A b_v} - \frac{a_v N_A^2}{\hat{v}_l^2}, \tag{8.55}$$

$$P = \frac{N_A k_B T}{\hat{v}_v - N_A b_v} - \frac{a_v N_A^2}{\hat{v}_v^2}, \tag{8.56}$$

where for $\hat{v}$ and μ the subscript l denotes a saturated liquid property and the subscript v denotes a saturated vapor property. These equations embody the requirements that the vapor and liquid satisfy the van der Waals equation of state and chemical potential relations and that the pressure and chemical potential are the same in both phases at equilibrium. There are five equations for the five unknowns μ_v, μ_l, $\hat{v}_v$, $\hat{v}_l$, and P. Although the equations are nonlinear, they can be solved to determine the five unknown saturation properties for a specified value of temperature. The saturation conditions dictated by these relations are indicated as the saturated vapor and liquid curves in Figure 8.2 and Figure 8.4.

Figure 8.5 shows the variation of the saturation pressure as a function of temperature predicted by solving Eqs. (8.52)–(8.56) using the van der Waals constants for nitrogen from Table 6.1. Also shown is the variation indicated by standard thermodynamic tables based

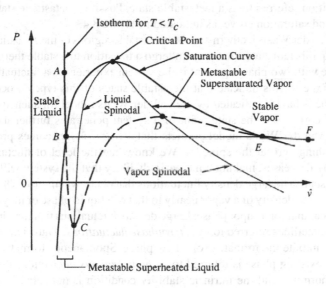

Figure 8.4

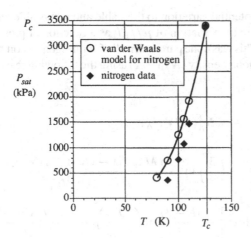

Figure 8.5

on experimental measurements. Qualitatively, the model prediction is consistent with the variation in the measured data, but the numerical agreement becomes increasingly worse as the temperature decreases. This is not surprising since the constants in Table 6.1 were obtained by matching nitrogen properties at the critial point.

The curves in Figure 8.4 denoting the limit of intrinsic stability for the vapor and liquid are termed the *vapor spinodal limit* and the *liquid spinodal limit*, respectively. It has been widely demonstrated that liquid can be superheated and vapor can be supercooled (supersaturated) beyond their equilibrium saturation conditions without a phase transformation occurring, even though the system equilibrium state would lie within the saturation "dome" in Figure 8.4 at equilibrium. Vapor that is supercooled below its equilibrium saturation temperature and liquid that is superheated above its equilibrium saturation temperature exist in a nonequilibrium condition referred to as a metastable state. Possible metastable states lie between the spinodal and saturation curve, as indicated in Figure 8.4.

The portions of the van der Waals isotherms in the metastable regions of the $P-\hat{v}$ diagram represent metastable equilibrium states that may undergo a transition to a stable thermodynamic equilibrium state with two phases present if the system experiences a fluctuation of sufficient magnitude. Experiments indicate that metastable states of this type do occur in real systems and that the behavior indicated by the van der Waals model for such systems is at least qualitatively correct. As the state point of a system penetrates further into the metastable region, the van der Waals model predicts that $-(\partial P/\partial \hat{v})_T$ becomes progressively smaller, approaching zero at the spinodal. We know that the level of fluctuations increases as $-(\partial P/\partial \hat{v})_T$ decreases, making it increasingly likely that the system will jump to a stable equilibrium state. The large density fluctuations that occur when $-(\partial P/\partial V)_{N,T}$ becomes small may raise the density of a vapor nearly to that of a liquid phase or may lower the density of a liquid near that for a vapor phase. Large density fluctuations that may initiate a change of phase are sometimes referred to as *heterophase fluctuations*. Thus, jumping to a new equilibrium may initiate the formation of a new phase. Spontaneous formation of a new phase within a preexisting phase is referred to as *homogeneous nucleation*. Because fluctuations become enormous and the intrinsic stability condition is not satisfied at the spinodal curve, the spinodal condition is interpreted as a limiting condition beyond which a phase transition must occur.

The relationship of the chemical potential variation to the stable and metastable regimes can be seen in Figure 8.6. In this figure the variation of $\mu/k_B T$ as a function of pressure is plotted for three different isotherms. This was done parametrically for nitrogen by computing $\mu/k_B T$ and pressure over a range of molar density ($N/V N_A$) using the following rearranged versions of Eqs. (6.26) and (6.27):

$$P = \frac{(N/V N_A)N_A k_B T}{1 - (N/V N_A)N_A b_v} - a_v N_A^2 (N/V N_A)^2, \tag{8.57}$$

$$\frac{\mu}{k_B T} = \frac{(N/V N_A)N_A b_v}{1 - (N/V N_A)N_A b_v} - \left(\frac{3}{2}\right) \ln \left[\frac{2\pi M k_B T [1 - (N/V N_A)N_A b_v]^{2/3}}{(N/V N_A)^{2/3} N_A^{2/3} h^2} \right]$$

$$- \left[\frac{(\xi - 5)}{2} \ln \pi - \ln \sigma_s \right] - \frac{(\xi - 3)}{2} \ln \left(\frac{T}{\theta_{\text{rot},m}} \right) - \frac{2a_v (N/V N_A)N_A}{k_B T}. \tag{8.58}$$

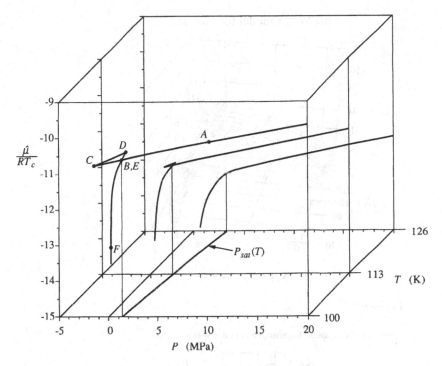

Figure 8.6 Chemical potential variation for nitrogen (van der Waals model).

For a specified temperature, the computed value of $\mu / k_B T = \hat{\mu}/RT$ at each value of (N/VN_A) was plotted as a function of the corresponding pressure. The resulting curves shown in Figure 8.6 thus indicate the variation of $\hat{\mu}/RT$ with P along an isotherm predicted by the van der Waals model for nitrogen. In Figure 8.6, the letters on the 100 K isotherm indicate states that are the same as the corresponding locations in Figure 8.4. Segment AB is subcooled liquid, BC is superheated liquid, DE is supersaturated vapor, and EF is superheated vapor. The segment CD is inaccessible because of stability considerations. As shown in Figure 8.6, as the temperature approaches the critical point (126 K), the cusps in the chemical potential curve disappear, leaving a smooth variation of μ with P at temperatures above T_c.

Another useful perspective can be obtained by considering the variation of $\mu / k_B T$ with molar density (N/VN_A) and temperature as predicted by Eq. (8.58). This variation for nitrogen is shown in Figure 8.7. We showed above that when $(\partial \mu / \partial N)_{V,T} \to 0$ the level of fluctuations in the system becomes enormous and intrinsic stability is no longer assured. In Figure 8.7, this derivative will approach zero at locations where a line of constant temperataure on the $\mu / k_B T$ surface has zero slope. It can be seen in this figure that at high temperature (>140 K) the slope of the surface along a line of constant temperature is steep at high densities, becoming less steep at low densities, but never going to zero. However, at lower temperatures (<60 K) there clearly exist two locations where the slope of the surface along an isotherm goes to zero. These correspond to the two points where the isotherm intersects the liquid and vapor spinodals. The wavy nature of this surface at low temperatures and densities thus indicates a region where the substance lacks intrinsic stability. This region is inaccessible to the system and attempting to traverse it generally will result in a phase transition.

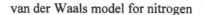

van der Waals model for nitrogen

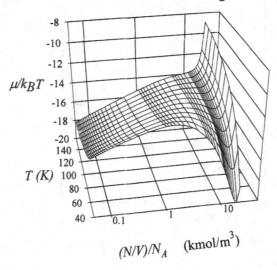

μ/k_BT

T (K)

$(N/V)/N_A$ (kmol/m^3)

Figure 8.7

fundamental equation for nitrogen (van der Waals model)

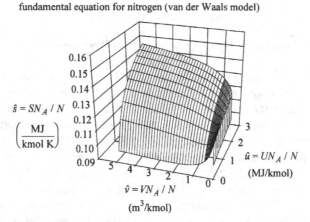

$\hat{s} = SN_A / N$

$\left(\dfrac{MJ}{kmol\ K}\right)$

$\hat{u} = UN_A / N$

(MJ/kmol)

$\hat{v} = VN_A / N$

(m^3/kmol)

Figure 8.8

Useful insight can also be obtained by examining the fundamental relation. In our analysis above, we considered a closed system with a fixed number of molecules. For such a system, at equilibrium, the entropy must be at a maximum, which mathematically implies that $\delta^2 S < 0$. This mathematical constraint on the second variance of the entropy is equivalent to the geometric requirement that the fundamental relation have negative curvature at equilibrium. Figure 8.8 shows the fundamental equation surface for nitrogen obtained using the van der Waals model in Chapter 6. Note that the surface in this plot represents the relation among $\hat{s}$, $\hat{u}$, and $\hat{v}$ obtained when the fundamental Eq. (6.31) for a van der Waals fluid is rearranged to a form containing only molar specific properties. Curvature trends in the specific properties are equivalent to those for the corresponding extensive properties S, U, and V for a system with fixed N.

In Figure 8.8, stable portions of the fundamental surface with negative curvature are concave downward. It can be seen that in Figure 8.8, the surface is concave downward everywhere, except for a small region at low $\hat{v}$ and $\hat{u}$. This corresponds to the range of conditions where we have shown that intrinsic stability is not assured. This, of course, corresponds exactly to the region between the spinodal curves in Figure 8.4. Stable equilibrium cannot exist in regions of upward curvature on the fundamental surface because fluctuations become very large as they are approached, and at the location where the curvature changes from positive to negative, intrinsic stability is no longer assured. The phase transition that occurs when the system state attempts to penetrate this region can be viewed as the system bifurcating, with a portion jumping across this region of positive curvature to a stable equilibrium on the other side with negative curvature.

It should be noted that although we have frequently used the van der Waals model in examples, the conclusions reached in this section are applicable to any real fluid. This is based on the expectation that relations among the properties like those derived for the van der Waals fluids exist for all real fluids, although we may not be able to easily write them down in closed form.

Example 8.1 Use the results for the van der Waals model for a pure substance to predict the temperature at the limit of intrinsic stability for liquid nitrogen at atmospheric pressure. Also determine the level of density fluctuations in one cubic micron of the liquid at 1 K below the temperature at the limit of intrinsic stability.

Solution The necessary condition for intrinsic stability of a pure substance, Eq. (8.49),

$$\left(\frac{\partial P}{\partial V}\right)_{N,T} < 0$$

can be expressed in terms of the molar specific volume as

$$\left(\frac{\partial P}{\partial \hat{v}}\right)_T < 0.$$

The limit of intrinsic stability corresponds to $(\partial P/\partial \hat{v})_T = 0$. Differentiating Eq. (8.50), we find that the specific volume at the stability limit must satisfy

$$\left(\frac{\partial P}{\partial \hat{v}}\right)_T = -\frac{N_A k_B T}{(\hat{v} - N_A b_v)^2} + \frac{2a_v N_A^2}{\hat{v}^3} = 0.$$

Combining this equation with the van der Waals equation of state to eliminate T, we obtain

$$P = \frac{a_v N_A^2 (\hat{v} - 2N_A b_v)}{\hat{v}^3}.$$

For the specified value of pressure, this cubic equation has three solutions for $\hat{v}$. Substituting $a_v = 3.76 \times 10^{-49}$ Pa m^6/molecule^2 and $b_v = 6.40 \times 10^{-29}$ m^3/molecule for nitrogen from Table 6.1, we iteratively compute solutions of $\hat{v} = 0.07941$, 1.120, and -1.20 m^3/kmol. The negative value is not physically possible and is ignored. The larger positive value is the spinodal limit for supersaturated vapor. The value of interest here for superheated liquid is the lower positive value $\hat{v} = 0.07941$ m^3/kmol. Substituting this value into the above equation for the spinodal condition and solving for T yields a value of 107 K for the temperature at the limit of intrinsic stability. Note that this is well above the saturation temperature of 77 K for nitrogen at atmospheric pressure.

To evaluate the level of density fluctuations at 106 K, we first use the van der Waals equation of state (8.50) to iteratively determine the specific volume for liquid at 101 kPa and 106 K. There are three solutions to the resulting cubic equation. The lowest, $\hat{v} = 0.07065$ m^3/kmol, corresponds to superheated metastable liquid. Noting that $(\partial P/\partial V)_{N,T} = (N_A/N)$ $(\partial P/\partial \hat{v})_T$ we can rewrite Eq. (8.5) as

$$\frac{\sigma_N}{N} = \left[\frac{RT}{-(\partial P/\partial \hat{v})_T \hat{v}^2 N} \right]^{1/2}.$$

In this equation, N is the number of molecules in one cubic micron, determined as

$$N = \frac{\left(10^{-6}\right)^3}{\hat{v}/N_A} = \frac{10^{-18}}{0.07065/6.02 \times 10^{26}} = 8.52 \times 10^9 \text{ molecules.}$$

Using the equation noted above for $(\partial P/\partial \hat{v})_T$,

$$\left(\frac{\partial P}{\partial \hat{v}} \right)_T = -\frac{N_A k_B T}{(\hat{v} - N_A b_v)^2} + \frac{2 a_v N_A^2}{\hat{v}^3},$$

we substitute to obtain

$$\left(\frac{\partial P}{\partial \hat{v}} \right)_T = -\frac{6.02 \times 10^{26} (1.38 \times 10^{-36}) 106}{[0.07065 - (6.02 \times 10^{26}) 6.40 \times 10^{-29}]^2}$$

$$+ \frac{2(3.76 \times 10^{-49})(6.02 \times 10^{26})^2}{(0.07065)^3} = -8.00 \times 10^7 \text{ Pa kmol/m}^3.$$

Substituting the above results and the value of N into the relation for σ_N/N, we find that

$$\frac{\sigma_N}{N} = \left[\frac{8308(106)}{8.00 \times 10^7 (0.07065)^2 8.52 \times 10^9} \right]^{1/2} = 1.6 \times 10^{-5}.$$

Thus the fractional standard deviation in the number of molecules in a cubic micron is very small, even within 1 K of the intrinsic limit of stability. The system must get very close to this limit before the level of fluctuations becomes large.

Solid–Liquid Phase Transitions

The discussion above focused on fluctuations within the system and their impact on phase stability and phase transitions. The nature of such fluctuations can be analyzed in a straightforward manner for gases and liquids. Fluctuations are clearly different for solids, however, and to understand how microscale features affect solid–liquid phase transitions, we must consider the system from a different perspective. In this section we will specifically consider the possibility of a transition from a liquid to a crystalline solid structure, or vice versa. The nature of processes that take place when a solid melts continues to be the subject of ongoing research. A simplistic but useful perspective on such processes can be developed by considering the basic characteristics of energy storage in solid crystal. We argued in Chapter 7 that energy is stored by lattice vibrations in a crystal, which are basically concerted vibrations of atoms about their lattice locations in the crystal. Considering these to be classical harmonic oscillators, we would expect that such vibrations could be sustained in a stable manner as long as their amplitude does not become too large. For a cubic lattice structure, if $\hat{v}$ is the molar specific volume, the mean spacing δ_l between lattice sites occupied by atoms must be

$$\delta_l = (\hat{v}/N_A)^{1/3}. \tag{8.59}$$

The energy in each classical oscillator with a maximum displacement of z_{max} is $(1/2)k_s z_{max}^2$. We expect that the crystal lattice will begin to shake itself apart when z_{max} becomes close to δ_l. If we interpret this as the onset of melting, we expect that the melting point will correspond to conditions where the mean energy of the most energetic oscillations is close to $(1/2)k_s(\hat{v}/N_A)^{2/3}$. Since the mean energy per oscillator scales with $(1/2)k_B T$, it follows that the melting point temperature T_m should approximately be given by a relation of the form.

$$(1/2)k_B T_m = C_0(1/2)k_s(\hat{v}/N_A)^{2/3}, \tag{8.60}$$

where C_0 is an unknown constant. Our analysis of a harmonic oscillator in Chapter 1 also indicated that the effective spring constant is related to the mass per atom m_a and the frequency of the oscillator as

$$k_s = m_a(2\pi \nu)^2. \tag{8.61}$$

Picking the Debye cutoff frequency ν_D as the characteristic frequency of the most energetic oscillations, the above relation becomes

$$k_s = m_a(2\pi \nu_D)^2. \tag{8.62}$$

Substituting the right side of (8.62) for k_s in (8.60) and using the definition of the Debye temperature $\theta_D = h\nu_D/k_B$, we can rewrite Eq. (8.60) as

$$T_m = C_0(1/2)\frac{m_a k_B(2\pi \theta_D)^2(\hat{v}/N_A)^{2/3}}{h^2}. \tag{8.63}$$

For a species with atomic weight $\tilde{M}$, the mass per atom is $\tilde{M}/N_A$. We can therefore replace m_a with $\tilde{M}/N_A$ to obtain

$$T_m = C_0 \frac{\tilde{M}}{N_A k_B}\left(\frac{2\pi k_B \theta_D}{h}\right)^2 \left(\frac{\hat{v}}{N_A}\right)^{2/3}. \tag{8.64}$$

By absorbing the numerical factors into the unknown constant, which we now designate as C_m, we can write the above relation for the meting point temperature as

$$T_m = C_m\left[\frac{\tilde{M}}{N_A k_B}\left(\frac{k_B \theta_D}{h}\right)^2 \left(\frac{\hat{v}}{N_A}\right)^{2/3}\right]. \tag{8.65}$$

This heuristic model provides a prediction of how the melting temperature of a crystalline solid varies with molecular weight, Debye temperature, and molar specific volume of the crystal. In Figure 8.9, the melting temperature for several metallic crystals is plotted as a function of the term in square brackets in Eq. (8.65). Most of the data do lie close to a straight

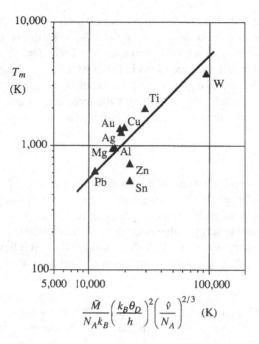

$$\frac{\tilde{M}}{N_A k_B}\left(\frac{k_B \theta_D}{h}\right)^2\left(\frac{\hat{v}}{N_A}\right)^{2/3} \text{ (K)}$$

Figure 8.9

line, which corresponds approximately to $C_m = 0.06$. Some metals, such as zinc and tin, deviate significantly from the line. Part of the reason for the scatter is that the cited Debye temperature is usually a best fit to energy storage in the crystal at low temperatures, and it may not be representative of the crystal's behavior at higher temperatures near the melting point.

Another interpretation of the above analysis can be obtained by rearranging Eq. (8.65) to the form

$$\frac{N_A k_B T_m}{\tilde{M}}\left(\frac{h}{k_B \theta_D}\right)^2\left(\frac{N_A}{\hat{v}}\right)^{2/3} = C_m. \tag{8.66}$$

Since this relation represents the conditions for incipient melting, it implies that the onset of melting will occur when the dimensionless group on the left side of Eq. (8.66) reaches the threshold value of C_m. This, in a sense, is a stability limit for the crystal, beyond which it will vibrate so severely that its crystalline structure will begin to break down. The plot of data for various metals in Figure 8.9 indicates that breakdown begins to occur when the value of this dimensionless group approaches 0.06. The scatter in the data also indicates that the exact threshold value may vary significantly from one material to another.

8.2   Phase Transitions and Saturation Conditions

First- and Second-Order Phase Transitions

The discussion in the previous section indicates that for a pure substance, a phase transition from liquid to vapor or vapor to liquid will occur at constant pressure and temperature but will result in a change in the molar specific volume. This is also generally true for solid–liquid phase transitions in a pure substance. For a van der Waals fluid in particular, Eq. (6.30) can be rearranged to obtain the following relation for the molar entropy of the fluid:

$$\hat{s} = N_A k_B \left(\frac{\xi}{2} + 1 \right) + \frac{3 N_A k_B}{2} \ln \left\{ \frac{2\pi M k_B T \, (\hat{v} - N_A b_v)^{2/3}}{N_A^{2/3} h^2} \right\}$$

$$+ \frac{(\xi - 3) N_A k_B}{2} \ln \left\{ \frac{T}{\theta_{\text{rot},m}} \right\} + N_A k_B \ln \left\{ \frac{\pi^{(\xi-5)/2}}{\sigma_s} \right\}. \qquad (8.67)$$

This relation implies that, for fixed T, a change in molar specific volume will produce a change in entropy for a van der Waals fluid. In general, a change of phase for a pure substance is accompanied by a change in specific volume and specific entropy. In Chapter 3 we demonstrated that for a pure substance the molar Gibbs free energy $\hat{g}$ is equal to the molar chemical potential $\hat{\mu} = N_A \mu$. The Gibbs–Duhem equation for a pure substance, Eq. (3.50), can therefore be written in the form

$$d\hat{g} = \hat{v} \, dP - \hat{s} \, dT. \qquad (8.68)$$

Because P and T are constant during a solid–liquid or liquid–vapor phase change of a pure substance, $\hat{g}$ is unchanged during the process. The differential relation (8.68) implies that $\hat{g}$ can be considered a function of P and T. It follows mathematically that

$$d\hat{g} = \left(\frac{\partial \hat{g}}{\partial P} \right)_T dP + \left(\frac{\partial \hat{g}}{\partial T} \right)_P dT. \qquad (8.69)$$

Since both Eq. (8.68) and Eq. (8.69) are valid, we find that

$$\left(\frac{\partial \hat{g}}{\partial P} \right)_T = \hat{v}, \qquad \left(\frac{\partial \hat{g}}{\partial T} \right)_P = -\hat{s}. \qquad (8.70)$$

The conclusion that $\hat{v}$ and $\hat{s}$ change discontinuously during the phase change implies that these first-order derivatives of the Gibbs function change discontinuously. A phase change in which the first-order derivatives of the Gibbs function change discontinuously is known as a *phase change of first order*. A phase change in which second-order derivatives of the Gibbs function change discontinuously is categorized as a *phase change of second order*.

If we consider a system containing N_A molecules or atoms, the definition of enthalpy $H = U + PV$ implies the following relation among molar specific properties:

$$\hat{h} = \hat{u} + P\hat{v}. \qquad (8.71)$$

Differentiating this relation we obtain

$$d\hat{h} = d\hat{u} + P \, d\hat{v} + \hat{v} \, dP.$$

Equation (3.32), $dU = T \, dS - P \, dV$, when applied to a system containing N_A molecules, can be written as

$$d\hat{u} = T \, d\hat{s} - P \, d\hat{v}. \qquad (8.72)$$

In Eqs. (8.71) and (8.72) $\hat{u}$ and $\hat{h}$ are the molar specific internal energy and enthalpy, respectively. Combining Eqs. (8.71) and (8.72) to eliminate $d\hat{u}$ yields

$$d\hat{h} = T \, d\hat{s} + \hat{v} \, dP. \qquad (8.73)$$

Dividing both sides of this relation by dT and considering a process at constant pressure, in which the ordinary differentials become partial differentials, we obtain

$$\left(\frac{\partial \hat{h}}{\partial T}\right)_P = T \left(\frac{\partial \hat{s}}{\partial T}\right)_P. \tag{8.74}$$

With this result we can write the definition (3.164) for the specific heat at constant pressure for a pure system as

$$\hat{c}_P = T \left(\frac{\partial \hat{s}}{\partial T}\right)_P. \tag{8.75}$$

Because $\hat{s}$ changes by a finite amount with no change in temperature, $(\partial \hat{s}/\partial T)_P$ is infinite and it follows that the molar specific heat $\hat{c}_P$ is infinite when both phases are present in the system. Note that $\hat{c}_P$ is finite for a system containing only one of the saturated phases.

The Clapeyron Equation

We now will consider what happens when a system containing a pure substance undergoes a change from an initial saturated condition to another saturated condition for which the pressure and temperature differ by dP and dT, respectively. Because the initial and final states are both saturation states, the molar Gibbs function for the two phases must be the same before and after the change. The changes in $\hat{g}$ for phase I and for phase II during the state change must therefore be the same:

$$d\hat{g}^{\mathrm{I}} = d\hat{g}^{\mathrm{II}}. \tag{8.76}$$

Using the Gibbs–Duhem equation to replace the differential Gibbs function changes, we obtain

$$\hat{v}^{\mathrm{I}} dP - \hat{s}^{\mathrm{I}} dT = \hat{v}^{\mathrm{II}} dP - \hat{s}^{\mathrm{II}} dT. \tag{8.77}$$

Noting that dT and dP correspond to changes along the saturation curve, we rearrange this relation to the form

$$\left(\frac{dP}{dT}\right)_{\mathrm{sat}} = \frac{\hat{s}^{\mathrm{II}} - \hat{s}^{\mathrm{I}}}{\hat{v}^{\mathrm{II}} - \hat{v}^{\mathrm{I}}}. \tag{8.78}$$

Equation (8.78) is the general form of the *Clapeyron equation*, which relates changes in volume and entropy to the slope of the saturation P–T curve for a pure substance. We showed above that at constant pressure $d\hat{h} = T d\hat{s}$, which implies that for the phase change at constant T and P, $\hat{s}^{\mathrm{II}} - \hat{s}^{\mathrm{I}} = (\hat{h}^{\mathrm{II}} - \hat{h}^{\mathrm{I}})/T$. The Clapeyron Eq. (8.78) can therefore be written in the equivalent form

$$\left(\frac{dP}{dT}\right)_{\mathrm{sat}} = \frac{\hat{h}^{\mathrm{II}} - \hat{h}^{\mathrm{I}}}{(\hat{v}^{\mathrm{II}} - \hat{v}^{\mathrm{I}})T}. \tag{8.79}$$

Using the shorthand notation

$$\Delta \hat{h} = \hat{h}^{\mathrm{II}} - \hat{h}^{\mathrm{I}}, \qquad \Delta \hat{v} = \hat{v}^{\mathrm{II}} - \hat{v}^{\mathrm{I}}, \tag{8.80}$$

we can write the Clapeyron equation in the form

$$\left(\frac{dP}{dT}\right)_{\mathrm{sat}} = \frac{\Delta \hat{h}}{T \Delta \hat{v}}. \tag{8.81}$$

This form of the Clapeyron equation is particularly useful because it relates the slope of the saturation curve to measurable quantities. The volume change can be measured directly and the enthalpy change associated with the phase change can be measured using standard calorimetry techniques.

For the case of a liquid–vapor phase change away from the critical point, we can invoke two idealizations: (1) at saturation, the liquid volume is negligible compared to the vapor volume and (2) the vapor behaves as an ideal gas. This allows us to replace $\Delta \hat{v}$ in the Clapeyron equation with $N_A k_B T / P$:

$$\left(\frac{dP}{dT} \right)_{\text{sat}} = \frac{P \Delta \hat{h}}{N_A k_B T^2}. \tag{8.82}$$

This relation is known as the *Clausius–Clapeyron equation*. Over a limited temperature range, the enthalpy of vaporization may be approximated as being constant. The Clausius–Clapeyron equation can then be integrated to obtain

$$\ln P_{\text{sat}} = -\frac{\Delta \hat{h}}{N_A k_B T} + C_0, \tag{8.83}$$

where C_0 is a constant that must be determined from measured data or a model of the fluid. This relation suggests that a plot of $\ln P_{\text{sat}}$ versus $1/T$ would result in a straight line over a limited range of temperature. In fact, a plot of data for many substances produces very nearly a straight line from the lowest temperature where a liquid and vapor coexist (the triple point) all the way to the critical temperature. This makes this relation particularly useful in fitting vapor pressure data for real substances. The reasons for the better-than-expected accuracy of Eq. (8.83) will be examined further in Section 8.4.

Example 8.2   If Eq. (8.82) is integrated assuming $\Delta \hat{h}$ is a linear function of temperature, a relation of the form

$$\ln P_{\text{sat}} = A - \frac{B}{T} + C \ln T$$

is obtained for the vapor pressure curve. This form has been suggested as an improved relation for fitting vapor pressure data. Show that the van der Waals model for a pure substance discussed in Section 8.1 predicts a relation of this form if the following idealizations are invoked: (1) the liquid specific volume is constant and (2) the vapor behaves as an ideal gas.

Solution   The saturation conditions are dictated by Eqs. (8.52)–(8.56) with the additional idealizations indicated above. Since the vapor is to be treated as an ideal gas, the a_v and b_v terms in Eqs. (8.53) and (8.56) are set to zero, which reduces them to the forms

$$\frac{\mu_v}{k_B T} = -\left(\frac{3}{2} \right) \ln \left[\frac{2\pi M k_B T \hat{v}_v^{2/3}}{N_A^{2/3} h^2} \right] - \left[\frac{(\xi - 5)}{2} \ln \pi - \ln \sigma_s \right]$$

$$- \frac{(\xi - 3)}{2} \ln \left(\frac{T}{\theta_{\text{rom},m}} \right),$$

$$P = N_A k_B T \hat{v}_v.$$

Setting the right side of the above equation for μ_v/k_BT equal to the right side of Eq. (8.52), substituting $N_A k_B T/P$ for $\hat{v}_v$, simplifying, and rearranging, we obtain

$$\ln\left(\frac{\hat{v}_l - N_A b_v}{N_A k_B T/P}\right) = \frac{N_A b_v}{\hat{v}_l - N_A b_v} - \frac{2a_v N_A}{\hat{v}_l k_B T}.$$

This relation can be further rearranged to the form

$$\ln P = \left[\frac{N_A b_v}{\hat{v}_l - N_A b_v} - \ln(\hat{v}_l - N_A b_v) + \ln(N_A k_B)\right] - \frac{2a_v N_A/(\hat{v}_l k_B)}{T} + \ln T.$$

If the molar specific volume of the liquid is taken to be constant, this relation clearly has the form $\ln P_{\text{sat}} = A - (B/T) + C \ln T$.

The Gibbs Phase Rule

We now will consider the conditions at equilibrium in a general system that contains m phases and r species. In doing so we consider each phase that is present to be a subsystem. It follows from results obtained in Chapter 3 that, at equilibrium, the temperature, pressure, and chemical potential of each species must be the same in each phase. Denoting phases with roman numerals and species with number 1 to r, we can state these requirements as

$$T_{\text{I}} = T_{\text{II}} = \cdots = T_m, \qquad (8.84a)$$

$$P_{\text{I}} = P_{\text{II}} = \cdots = P_m, \qquad (8.84b)$$

$$\left.\begin{aligned}\mu_{1,\text{I}} &= \mu_{1,\text{II}} = \cdots = \mu_{1,m}, \\ \mu_{2,\text{I}} &= \mu_{2,\text{II}} = \cdots = \mu_{2,m}, \\ &\ \ \vdots \\ \mu_{r,\text{I}} &= \mu_{r,\text{II}} = \cdots = \mu_{r,m}.\end{aligned}\right\} \qquad (8.85)$$

At equilibrium, the fundamental equation must apply to the system as a whole and any subsystems we define. As discussed in Chapter 3, we can represent the fundamental equation in several different forms. Beginning with the energy form $U = U(S, V, N_1, \ldots, N_r)$, we can execute the partial Legendre transform that replaces S with T and V with P to obtain the Gibbs function representation $G = G(T, P, N_1, \ldots, N_r)$, where by definition

$$G = U - TS + PV. \qquad (8.86)$$

Since U, S, and V are additive over subsystems and T and P must be same in all subsystems, the above relation implies that G is additive over the subsystems.

We define the mole fraction of species i as

$$X_i = N_i/N, \qquad (8.87)$$

where N is the total number of particles of all species,

$$N = \sum_{i=1}^{r} N_i. \qquad (8.88)$$

Equation (8.87) implies that in the Gibbs function representation of the fundamental equation, we can replace N_i with $X_i N$:

$$G = G(T, P, X_1 N, X_2 N, \ldots, X_r N). \tag{8.89}$$

But we know that G is additive over any subsystems. If we consider four subsystems with N particles each, the value of G for the total system of $4N$ particles must be four times that of each subsystem having N particles. This line of reasoning can be applied to any number of subsystems, which implies that G must be linearly proportional to the total number of particles in the system. It follows directly that

$$G(T, P, X_1 N, X_2 N, \ldots, X_r N) = N G(T, P, X_1, X_2, \ldots, X_r), \tag{8.90}$$

and hence the Gibbs function for any system can be determined as

$$G = N G(T, P, X_1, X_2, \ldots, X_r). \tag{8.91}$$

This result implies that the ratio G/N is a function of T, P, and the mole fraction of each species in the system,

$$G/N = f(T, P, X_1, X_2, \ldots, X_r). \tag{8.92}$$

Differentiating G/N with respect to the number of species i particles and using the chain rule from basic calculus yields

$$\left(\frac{\partial (G/N)}{\partial N_i} \right)_{T,P,N_{j\neq i}} = \left(\frac{1}{N} \right) \left(\frac{\partial G}{\partial N_i} \right)_{T,P,N_{j\neq i}} - \left(\frac{G}{N^2} \right) \left(\frac{\partial N}{\partial N_i} \right)_{T,P,N_{j\neq i}}. \tag{8.93}$$

Since $N = N_1 + N_2 + \cdots + N_r$, $\partial N / \partial N_i = 1$. Using this result and the fact that by definition $\mu_i = (\partial G / \partial N_i)_{T,P,N_{j\neq i}}$, we can rearrange the above relation to the form

$$\mu_i = N \left(\frac{\partial (G/N)}{\partial N_i} \right)_{T,P,N_{j\neq i}} + \left(\frac{G}{N} \right). \tag{8.94}$$

Since for any species j, $X_j = N_j/N$ and N is the sum of all the species numbers, each mole fraction X_j is a function of all the species numbers. Accounting for the N_i dependence of all the mole fractions, the derivative in Eq. (8.94) can be rewritten using the chain rule as

$$\mu_i = N \sum_{j=1}^{r} \left(\frac{\partial (G/N)}{\partial X_j} \right)_{T,P,N_{j\neq i}} \left(\frac{\partial X_j}{\partial N_i} \right)_{T,P,N_{j\neq i}} + \left(\frac{G}{N} \right). \tag{8.95}$$

From the definitions of X_i and N, it can be easily shown that

$$\left(\frac{\partial X_j}{\partial N_i} \right)_{T,P,N_{k\neq i}} = \frac{\delta_{ij} - X_j}{N}, \tag{8.96}$$

where δ_{ij} is the Kronecker delta. Substituting this result into Eq. (8.95) we obtain

$$\mu_i = \sum_{j=1}^{r} (\delta_{ij} - X_j) \left(\frac{\partial (G/N)}{\partial X_j} \right)_{T,P,N_{j\neq i}} + \left(\frac{G}{N} \right). \tag{8.97}$$

This rather lengthy manipulation has led us to an important conclusion. Since G/N is a function of T, P, and the mole fractions, the right side of Eq. (8.97) is a function only of these parameters. We have therefore demonstrated that the chemical potential for each species can, in principle, be written as a function of T, P, and the mole fraction of each species in the system:

$$\mu_i = \mu_i(T, P, X_1, X_2, \ldots, X_r). \tag{8.98}$$

Note that this is a relation among intensive parameters only. A relation of this type exists for each species i in each phase in the system. Thus, in a system with m phases and r species, there are $m \times r$ relations of this type that must be satisfied at equilibrium. The equality of chemical potential in each phase for each species embodied in Eq. (8.85) results in an additional $(m - 1) \times r$ equations that must be satisfied. For each phase it must also be true that the sum of the mole fractions must equal one, which results in m additional equations. The total number of equations to be satisfied is therefore

$$\text{total number of equations} = m \times r + (m - 1) \times r + m = 2mr - r + m.$$

The intensive properties involved in these equations are P, T, r chemical potentials for each of m phases, and r mole fractions in each of m phases. The total number of variables is given by

$$\text{total number of variables} = 2 + m \times r + m \times r = 2mr + 2.$$

The number of independently variable intensive properties in the overall system is the difference between the total number of variables and the number of equations they must satisfy. Designating the number of independently variable intensive properties as η, it follows from our above analysis that

$$\eta = 2mr + 2 - [2mr - r + m],$$

which simplifies to

$$\eta = r - m + 2. \tag{8.99}$$

Equation (8.99) is known as the *Gibbs phase rule*. It permits us to determine the minimum number of independent thermodynamic properties from the set $\{T, P, X_1, X_2, \ldots, X_r\}$ that must be specified to specify the equilibrium state of a system containing r species and m phases. For a pure single-phase system ($r = 1$, $m = 1$) it indicates that two independent intensive properties must be specified to establish the state of the system. For a pure ideal gas, specifying pressure and temperature is sufficient to determine the state of the system, as we already know from the ideal gas equation of state. The Gibbs phase rule further stipulates that, for a binary ideal gas mixture, it is necessary to specify three independent properties from the set $\{T, P, X_1, X_2\}$.

   For a single-species system with two phases present, only one intensive property can be independently varied. The presence of two phases is a saturation condition. As we discussed in the previous section, the requirements for equilibrium lead to a relation between temperature and pressure for a van der Waals fluid at saturation. The same is true for any real pure substance, with the result that the pressure and temperature cannot be chosen independently. The relation between pressure and temperature for saturated mixtures of liquid and vapor is referred to as the *vapor-pressure curve* for the substance because it indicates the equilibrium pressure of vapor above a liquid surface at a specified temperature. It is also possible for three phases of a pure substance to coexist in a system at equilibrium. The Gibbs phase rule then indicates that the number of independently variable intensive properties from the set

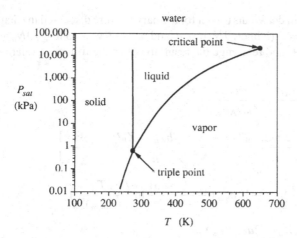

Figure 8.10

$\{T, P\}$ is zero. This implies that there is a unique combination of the intensive properties from this set that correspond to the equilibrium state with three coexisting phases. Such a state is termed a *triple point* for the substance. Triple points for substances are often specified as reference points for properties or instrument calibration because they are fixed states that can be uniquely reproduced. Figure 8.10 shows the primary saturation curves for water (it does not show possible solid–solid phase transitions) and indicates that single combination of pressure and temperature corresponding to the triple point.

It is noteworthy that the solid–liquid coexistence curve for water in Figure 8.10 has a negative slope. For most common solids, the slope of this curve is positive. This behavior is associated with the fact that, unlike most substances, water expands when it freezes, with the result that solid ice at saturation has a higher specific volume than saturated water at the same pressure. The Clapeyron Eq. (8.79) indicates that $(dP/dT)_{sat}$ must be negative if solidification reduces the enthalpy and increases the specific volume of the substance. For most substances, freezing decreases both the enthalpy and specific volume, resulting in a positive slope to the solid–liquid saturation curve. For water the negative slope implies that if the pressure on ice is increased at constant temperature, the ice will melt. This behavior facilitates ice skating by causing a liquid film to form under the skate where the pressure is elevated by the weight of the skater acting on the skate blade. This liquid film provides lubrication, allowing the skater to glide smoothly over the ice surface. Refreezing of the liquid film after the pressure is relieved is sometimes referred to as *regelation*.

8.3 Phase Equilibria in Binary Mixtures

Systems containing a liquid and vapor phase and more than one species are commonly encountered in engineering practice. The simplest system of this type is one containing a binary mixture with coexisting liquid and vapor phases. In this section we will explore the features of such a system in some detail.

The Gibbs phase rule indicates that for a binary mixture (of species a and b) with two phases present, specification of two properties from the set $\{T, P, X_a, X_b\}$ will establish the equilibrium state for the system. We can examine how specification of two properties dictates the state by considering the van der Waals model for a binary mixture discussed in Chapter 6. Using the relations $\mu_a / k_B T = -(\partial \ln Q / \partial N_a)_{V,T,N_b}$ and $\mu_b / k_B T = -(\partial \ln Q / \partial N_b)_{V,T,N_a}$, we can differentiate Eq. (6.49) and rearrange the result to obtain the following relations for the chemical potentials:

$$\frac{\mu_a}{k_B T} = \frac{N_A b_{v,a}}{\hat{v} - X_a N_A b_{v,a} - X_b N_A b_{v,b}}$$

$$- \left(\frac{3}{2}\right) \ln \left[\frac{2\pi M_a k_B T (\hat{v} - X_a N_A b_{v,a} - X_b N_A b_{v,b})^{2/3}}{N_A^{2/3} X_a^{2/3} h^2}\right]$$

$$- \left[\frac{(\xi_a - 5)}{2} \ln \pi - \ln \sigma_{s,a}\right] - \frac{(\xi_a - 3)}{2} \ln \left(\frac{T}{(\theta_{rot,m})_a}\right)$$

$$- \frac{2a_{v,aa} N_A X_a + 2a_{v,ab} N_A X_b}{\hat{v} k_B T}, \tag{8.100}$$

$$\frac{\mu_b}{k_B T} = \frac{N_A b_{v,b}}{\hat{v} - X_a N_A b_{v,a} - X_b N_A b_{v,b}}$$

$$- \left(\frac{3}{2}\right) \ln \left[\frac{2\pi M_b k_B T (\hat{v} - X_a N_A b_{v,a} - X_b N_A b_{v,b})^{2/3}}{N_A^{2/3} X_b^{2/3} h^2}\right]$$

$$- \left[\frac{(\xi_b - 5)}{2} \ln \pi - \ln \sigma_{s,b}\right] - \frac{(\xi_b - 3)}{2} \ln \left(\frac{T}{(\theta_{rot,m})_b}\right)$$

$$- \frac{2a_{v,bb} N_A X_b + 2a_{v,ab} N_A X_b}{\hat{v} k_B T}, \tag{8.101}$$

where $\hat{v}$ is the molar specific volume and X_a and X_b are the mole fractions defined as

$$\hat{v} = V N_A / (N_a + N_b), \tag{8.102}$$

$$X_a = N_a / (N_a + N_b), \tag{8.103}$$

$$X_b = N_b / (N_a + N_b). \tag{8.104}$$

If we specify the temperature and pressure in the system, as required by the Gibbs phase rule, we can determine other properties for the equilibrium state by enforcing the necessary conditions for equilibrium. We therefore take the temperature, pressure, and chemical potential for each species to be the same in both phases. In addition, in each phase the sum of the mole fractions must equal one, which allows us to replace X_b with $1 - X_a$ for this mixture. We obtain two equations by evaluating the chemical potential for each species in each phase and setting them equal, and we obtain two more by requiring that the pressure equation of state be satisfied in both phases. After a bit of simplifying, these relations can be written in the forms

$$P = \frac{N_A k_B T}{\hat{v}_v - X_{a,v} N_A b_{v,a} - X_{b,v} N_A b_{v,b}}$$

$$- \frac{a_{v,aa} N_A^2 X_{a,v}^2 + 2a_{v,ab} N_A^2 X_{a,v} X_{b,v} + a_{v,bb} N_A^2 X_{b,v}^2}{\hat{v}_v^2}, \tag{8.105}$$

$$P = \frac{N_A k_B T}{\hat{v}_l - X_{a,l} N_A b_{v,a} - X_{b,l} N_A b_{v,b}}$$

$$- \frac{a_{v,aa} N_A^2 X_{a,l}^2 + 2a_{v,ab} N_A^2 X_{a,l} X_{b,l} + a_{v,bb} N_A^2 X_{b,l}^2}{\hat{v}_l^2}, \tag{8.106}$$

$$\frac{N_A b_{v,a}}{\hat{v}_l - X_{a,l} N_A b_{v,a} - X_{b,l} N_A b_{v,b}} - \frac{N_A b_{v,a}}{\hat{v}_v - X_{a,v} N_A b_{v,a} - X_{b,v} N_A b_{v,b}}$$

$$+ \frac{2a_{v,aa} N_A X_{a,v} + 2a_{v,ab} N_A X_{b,v}}{\hat{v}_v k_B T} - \frac{2a_{v,aa} N_A X_{a,l} + 2a_{v,ab} N_A X_{b,l}}{\hat{v}_l k_B T}$$

$$= \ln\left[\frac{(\hat{v}_l - X_{a,l} N_A b_{v,a} - X_{b,l} N_A b_{v,b}) X_{a,v}}{(\hat{v}_v - X_{a,v} N_A b_{v,a} - X_{b,v} N_A b_{v,b}) X_{a,l}}\right], \tag{8.107}$$

$$\frac{N_A b_{v,b}}{\hat{v}_l - X_{a,l} N_A b_{v,a} - X_{b,l} N_A b_{v,b}} - \frac{N_A b_{v,b}}{\hat{v}_v - X_{a,v} N_A b_{v,a} - X_{b,v} N_A b_{v,b}}$$

$$+ \frac{2a_{v,bb} N_A X_{b,v} + 2a_{v,ab} N_A X_{b,v}}{\hat{v}_v k_B T} - \frac{2a_{v,bb} N_A X_{b,l} + 2a_{v,ab} N_A X_{b,l}}{\hat{v}_l k_B T}$$

$$= \ln\left[\frac{(\hat{v}_l - X_{a,l} N_A b_{v,a} - X_{b,l} N_A b_{v,b}) X_{b,v}}{(\hat{v}_v - X_{a,v} N_A b_{v,a} - X_{b,v} N_A b_{v,b}) X_{b,l}}\right]. \tag{8.108}$$

With the additional two constraints

$$X_{a,v} + X_{b,v} = 1, \tag{8.109}$$

$$X_{a,l} + X_{b,l} = 1 \tag{8.110}$$

we have a complete set of equations that can be used to determine the equilibrium conditions. If T and P are specified, Eqs. (8.105)–(8.110) can, in principle, be solved for the six unknowns $\hat{v}_l$, $\hat{v}_v$, $X_{a,l}$, $X_{b,l}$, $X_{a,v}$, and $X_{b,v}$. Because the equations are nonlinear, this is not a simple task. However, if we idealize the vapor phase as having ideal gas behavior, we can simplify the solution process by neglecting the van der Waals terms for the vapor phase in the equations for the first iteration. A solution can then be generated with the following algorithm:

(1) For the specified P and T, determine $\hat{v}_v$ using the ideal gas law: $\hat{v}_v = N_A k_B T / P$.
(2) Guess $X_{a,l}$.
(3) Compute $X_{b,l}$ using Eq. (8.110).
(4) Solve Eq. (8.106) iteratively to determine $\hat{v}_l$.
(5) Solve Eqs. (8.107) and (8.108) simultaneously for $X_{a,v}$ and $X_{b,v}$.
(6) If the values of $X_{a,v}$ and $X_{b,v}$ do not satisfy Eq. (8.109), the guessed $X_{a,l}$ value is corrected and steps (3)–(5) are repeated. When Eq. (8.109) is satisfied $\hat{v}_v$ is recomputed by iteratively solving Eq. (8.105) using the most recent values of $X_{a,l}$, $X_{b,l}$, $X_{a,v}$, and $X_{b,v}$ and steps (2)–(6) are repeated. When successive values of the mole fractions and specific volumes do not change significantly, the algorithm terminates.

One additional aspect of this model must be considered. The parameters $b_{v,a}$, $b_{v,b}$, $a_{v,aa}$, and $a_{v,bb}$ are taken to be the pure component van der Waals constants for species a and b.

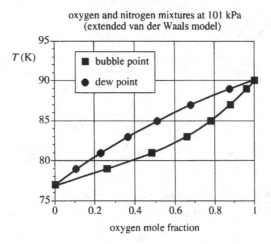

Figure 8.11

The constant $a_{v,ab}$, however, represents interactions of a type molecules with b type molecules. Accurate specification of this parameter requires knowledge of the potential force interaction between a type and b type molecules (see Chapter 6).

To demonstrate the use of the analysis described above, we will specifically consider a binary mixture of oxygen and nitrogen at low temperatures. Figure 8.11 shows the phase equilibrium diagram for such a mixture at atmospheric pressure predicted by the extended van der Waals model and computational algorithm outlined above. Species a and b were chosen to be nitrogen and oxygen, respectively, and the pure component van der Waals constants were taken to be

$$a_{v,aa} = 3.76 \times 10^{-49} \text{ Pa m}^6/\text{molecule}^2, \qquad b_{v,a} = 4.90 \times 10^{-29} \text{ m}^3/\text{molecule},$$
$$(8.111)$$

$$a_{v,bb} = 3.78 \times 10^{-49} \text{ Pa m}^6/\text{molecule}^2, \qquad b_{v,b} = 3.98 \times 10^{-29} \text{ m}^3/\text{molecule}.$$
$$(8.112)$$

The $a_{v,aa}$ and $a_{v,bb}$ values are the same as those in Table 6.1. The $b_{v,a}$ and $b_{v,b}$ values were altered from those listed in Table 6.1. The values used here were chosen so that the pure fluid equilibrium saturation temperature at 101 kPa matched established values of 77 K and 90 K for nitrogen and oxygen, respectively. To evaluate the interspecies interaction constant, the following plausible, if ad hoc, relation was used here:

$$a_{v,ab} = \frac{a_{v,aa} + a_{v,bb}}{2}. \qquad (8.113)$$

Although this extended van der Waals model is crude in many respects, the binary phase diagram in Figure 8.11 is remarkably close to that indicated by measured data for nitrogen–oxygen mixtures.

The upper curve in Figure 8.11 is the *dew point curve*, where condensation first occurs when a vapor mixture of fixed concentration is slowly reduced in temperature. The lower curve is termed the *bubble point curve* because it represents conditions where vapor first forms when a liquid mixture of fixed concentration is slowly heated. Combinations of T and oxygen mole fraction below this line correspond to a liquid phase only. Combinations above the dew point line correspond to vapor phase only. At points in the lens-shaped region between the bubble point and dew point lines, a liquid and vapor phase coexist. In such a two-phase system, the liquid concentration is the value on the bubble point line at the specified temperature, and the vapor concentration is the value on the dew point line at the specified temperature. Thus, for this mixture, the concentrations in the vapor and liquid phases are different at equilibrium, except at the pure species endpoints.

Example 8.3   Use the extended van der Waals model discussed above to estimate the limits of intrinsic stability for a liquid binary mixture of oxygen and nitrogen at atmospheric pressure.

Solution   As discussed in Section 8.1, the necessary and sufficient conditions for intrinsic stability for a binary mixture are that $(\partial \mu_a / \partial N_a)_{V,T,\mu_b} > 0$ with either species taken as species a. It can be shown from basic mathematical considerations that

$$\left(\frac{\partial \mu_a}{\partial N_a}\right)_{V,T,\mu_b} = \left(\frac{\partial \mu_a}{\partial N_a}\right)_{V,T,N_b} - \left(\frac{\partial \mu_a}{\partial N_b}\right)_{V,T,N_a} \left(\frac{\partial \mu_b}{\partial N_a}\right)_{V,T,N_b} \Big/ \left(\frac{\partial \mu_b}{\partial N_b}\right)_{V,T,N_a}.$$

Also, one of the Maxwell relations requires that

$$\left(\frac{\partial \mu_a}{\partial N_b}\right)_{V,T,N_a} = \left(\frac{\partial \mu_b}{\partial N_a}\right)_{V,T,N_b}.$$

The necessary and sufficient conditions for intrinsic stability can therefore be stated as the requirement that for either fluid taken as a or b

$$\left(\frac{\partial \mu_a}{\partial N_a}\right)_{V,T,N_b} - \frac{(\partial \mu_a / \partial N_b)^2_{V,T,N_a}}{(\partial \mu_b / \partial N_b)_{V,T,N_a}} > 0.$$

For the oxygen and nitrogen mixture considered here, we designate oxygen terms with an o subscript and nitrogen terms with an n subscript. The necessary and sufficient conditions for stability can then be stated as

$$\left(\frac{\partial \mu_o}{\partial N_o}\right)_{V,T,N_n} - \frac{(\partial \mu_o / \partial N_n)^2_{V,T,N_o}}{(\partial \mu_n / \partial N_n)_{V,T,N_o}} > 0,$$

$$\left(\frac{\partial \mu_n}{\partial N_n}\right)_{V,T,N_o} - \frac{(\partial \mu_o / \partial N_n)^2_{V,T,N_o}}{(\partial \mu_o / \partial N_o)_{V,T,N_n}} > 0.$$

The three partial derivatives in the above relations are evaluated as follows. First, the right side of Eqs. (8.103) and (8.104) are substituted into Eqs. (8.100) and (8.101) to replace the mole fractions with N_a and N_b. The resulting equations for μ_a and μ_b are then differentiated with respect to N_a and N_b. Arbitrarily taking a to be oxygen and b to be nitrogen, we obtain the following relations for the derivatives in the stability criteria relations:

$$\left(\frac{\partial \mu_o}{\partial N_o}\right)_{V,T,N_n} = \frac{k_B T}{N_o + N_n}\left[\frac{2N_A b_{v,o}}{\hat{v} - X_o N_A b_{v,o} - X_n N_A b_{v,n}}\right.$$

$$\left. + \frac{(N_A b_{v,o})^2}{(\hat{v} - X_o N_A b_{v,o} - X_n N_A b_{v,n})^2} + \frac{1}{X_o} - \frac{2a_{v,oo}N_A^2}{\hat{v}N_A k_B T}\right],$$

$$\left(\frac{\partial \mu_o}{\partial N_n}\right)_{V,T,N_o} = \frac{k_B T}{N_o + N_n}\left[\frac{N_A(b_{v,o} + b_{v,n})}{\hat{v} - X_o N_A b_{v,o} - X_n N_A b_{v,n}}\right.$$

$$\left. + \frac{N_A^2 b_{v,o} b_{v,n}}{(\hat{v} - X_o N_A b_{v,o} - X_n N_A b_{v,n})^2} - \frac{2a_{v,on}N_A^2}{\hat{v}N_A k_B T}\right],$$

$$\left(\frac{\partial \mu_n}{\partial N_n}\right)_{V,T,N_o} = \frac{k_B T}{N_o + N_n}\left[\frac{2N_A b_{v,n}}{\hat{v} - X_o N_A b_{v,o} - X_n N_A b_{v,n}}\right.$$

$$\left. + \frac{(N_A b_{v,n})^2}{(\hat{v} - X_o N_A b_{v,o} - X_n N_A b_{v,n})^2} + \frac{1}{X_n} - \frac{2a_{v,nn}N_A^2}{\hat{v}N_A k_B T}\right].$$

To determine the stability limits, the following procedure is used. The pressure is fixed at one atmosphere, and the mole fraction of oxygen is specified. The mole fraction of nitrogen is calculate simply as $X_n = 1 - X_o$. Beginning at a very low value of specific volume, the above relations for the derivatives are used to compute the left side of the stability condition inequalities for progressively larger values of specific volume $\hat{v}$. For each $\hat{v}$ value, the extended van der Waals equation of state is used to compute the temperature,

$$P = \frac{N_A k_B T}{\hat{v} - X_o N_A b_{v,o} - X_n N_A b_{v,n}} - \frac{a_{v,oo}N_A^2 X_o^2 + 2a_{v,on}N_A^2 X_o X_n + a_{v,nn}N_A^2 X_n^2}{\hat{v}^2}.$$

Beginning well below the bubble point for the mixture, as the specific volume increases, T increases and the computed values for the left sides of the stability criterion inequalities decrease, eventually reaching zero. The point where one of these expressions equals zero is interpreted as the limit of intrinsic stability. In these example calculations, the same van der Waals constants were used as in the calculations represented in Figure 8.11. The resulting limits of stability are plotted for several mixture concentrations in Figure 8.12. Note that these results imply that the mixtures may be superheated substantially above the equilibrium bubble point without a loss of intrinsic stability. The fluctuation relations (4.125) and (4.126) imply, however, that the level of concentration and density fluctuation in the liquid will become very large as the stability limit is approached.

In some binary systems, the interactions among the two molecular species are such that the concentrations in the vapor and liquid phases are the same for a mixture at a specific temperature and pressure. The phase diagram then looks like the plot in Figure 8.13. A two-phase binary mixture in which the concentrations in the two phases are the same is termed an *azeotrope* or an *azeotropic mixture*. In general, the conditions corresponding to

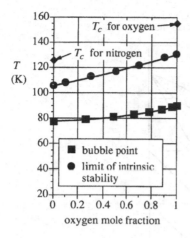

Figure 8.12

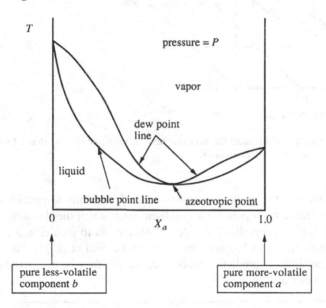

Figure 8.13

the two-phase region in the binary phase diagram vary with pressure, as illustrated in Figure 8.14 for nitrogen–oxygen mixtures.

Figure 8.14 was generated using the extended van der Waals model described above to predict the phase equilibria at three pressure levels for nitrogen–oxygen mixtures. The shaded lens-shaped regions are the cross section of a three-dimensional region in which a liquid and vapor phases coexist. Combinations of T, P, and system oxygen mole fraction above this region in Figure 8.14 correspond to a single-phase system containing only a vapor phase. For combinations below the two-phase region, the system contains only a liquid phase. Although the van der Waals model is very idealized, the variation in the binary phase diagram indicated by Figure 8.14 is close to that observed for real nitrogen–oxygen mixtures.

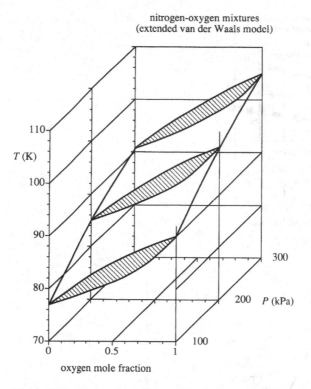

nitrogen-oxygen mixtures
(extended van der Waals model)

Figure 8.14 Binary phase diagrams for three pressures as predicted by the extended van der Waals model for nitrogen–oxygen mixtures.

As discussed in Chapter 6, the model analysis used in this section to predict vapor–liquid equilibria in binary mixtures represents a simplistic extension of the classical van der Waals theory. Use of extended generalized van der Waals models to predict vapor–liquid equilibria in simple binary mixtures has been explored by Penfold et al. [2]. For simple fluids their results indicate that, as predictive tools, such models may be competitive with fully empirical equations of state.

Example 8.4   In a cryogenic system, an uninsulated portion of a pipe carrying liquid nitrogen at atmospheric pressure is in contact with dry air at atmospheric pressure and 20°C. The liquid nitrogen inside is at 77 K and because of heat exchange with the surroundings, the outside wall of the pipe is at the slightly higher temperature of 85 K. A layer of liquid condensate forms on the outside of the bared pipe. If local thermodynamic equilibrium exists at the surface of the liquid layer, estimate the oxygen concentration in the liquid.

   Solution   In this analysis we idealize the surrounding air as a binary mixture of oxygen and nitrogen with an oxygen mole fraction of 0.21. At atmospheric pressure, the phase equilibrium diagram in Figure 8.11 indicates that for local thermodynamic equilibrium to exist at the liquid–gas interface at 85 K, the local oxygen concentration in the vapor $X_{O_2,v}$

(on the dew point line) and the liquid $X_{O_2,l}$ (on the bubble point line) must be

$$X_{O_2,v} = 0.51, \qquad X_{O_2,l} = 0.78.$$

This implies that the concentrations in the vapor and liquid phases tend toward these values if the pipe, and the fluid adjacent to it, are held at 85 K. Thus, the liquid condensate on the pipe is expected to have an oxygen concentration of approximately 0.78. If the liquid drips off of the pipe and lands in a warmer location, it will evaporate. Because of the liquid's higher O_2 concentration, this may result in a local oxygen concentration that is much higher than the normal atmospheric concentration of 0.21 at the location where the liquid evaporates. Since the flammability of many substances increases as O_2 concentration increases, condensation on the liquid nitrogen lines can be a fire hazard if the liquid condensate drips off of the lines into locations where flammable materials exist.

Empirical Models

An alternative approach to constructing a model of binary phase equilibria is to make use of empirical models. If the vapor is idealized as being a mixture of independent ideal gases and the mixture therefore obeys Dalton's law, the partial pressure of each component P_a is equal to the total pressure multiplied by the vapor mole fraction:

$$P_a = P X_{a,v}. \tag{8.114}$$

Two idealized models that relate the vapor partial pressure of a component to its concentration in the liquid phase are *Henry's law* and *Raoult's law*. Henry's law states that the partial pressure P_a is proportional to the mole fraction of component a in the liquid,

$$P_a = C_H X_{a,l}. \tag{8.115}$$

In the above relation, C_H designates the Henry's law proportionality constant. Henry's law is usually a good approximation for components having low liquid concentrations $X_{a,l}$.

Raoult's law states that the partial pressure for component a is given by

$$P_a = P_{sat,a} X_{a,l}, \tag{8.116}$$

where $P_{sat,a}$ is the saturation pressure for pure component a at the specified system temperature T, and $X_{a,l}$ is the liquid mole fraction of species a. Raoult's law is generally a good approximation for values of $X_{a,l}$ near one.

If Raoult's law is assumed to apply to a binary system, it follows that

$$P_a = P_{sat,a} X_{a,l}, \qquad P_b = P_{sat,b} X_{b,l}. \tag{8.117}$$

Since Dalton's law requires that the sum of the partial pressures must equal the system pressure P,

$$P = P_{sat,a} X_{a,l} + P_{sat,b} X_{b,l}. \tag{8.118}$$

Using the fact that $X_{a,l} + X_{b,l} = 1$, the above relation can be rearranged to obtain

$$X_{a,l} = \frac{P - P_{sat,b}(T)}{P_{sat,a}(T) - P_{sat,b}(T)}. \tag{8.119}$$

In Eq. (8.119) the functional dependence of the pure component vapor pressures on temperature has been explicitly indicated. Thus, if the binary mixture conforms to Raoult's law, the bubble point curve $X_{a,1}(T)$ on the phase diagram for a specified pressure can be predicted from the pure component vapor pressure curves using Eq. (8.119). If, in addition, Dalton's law applies to the vapor mixture at equilibrium, then the partial pressure indicated by Dalton's law and that indicated by Raoult's law must be equal for each species. It follows that

$$P_{\text{sat},a} X_{a,1} = P X_{a,v}. \tag{8.120}$$

Rearranging the above relation yields

$$X_{a,v} = \frac{P_{\text{sat},a}(T)}{P} X_{a,1}(T). \tag{8.121}$$

Thus, having determined $X_{a,1}(T)$ as described above, Eq. (8.121) can be used to determine the dew point curve $X_{a,v}(T)$ for a specified pressure, completing the phase diagram. The phase diagram shown in Figure 8.15 was constructed by applying this idealized method to a mixture of refrigerants R-11 and R-113 at atmospheric pressure. The predictions of this model are actually fairly close to the actual observed mixture behavior for these fluids at atmospheric pressure. However, some mixtures deviate strongly from the predictions of Raoult's law and Dalton's law, and this type of model will not be accurate for such mixtures.

Phase Equilibria in Binary Mixtures – General Considerations

Systems containing two phases and two components arise commonly in applications and many of these do not conform to the extended van der Waals model or the ideal mixture model described above. To deal with such systems, we must consider them from a more basic perspective. The temperature, pressure, and chemical potential for each component must be the same in both phases. The main difficulty in applying these requirements is that we need to relate the chemical potential constraints to other, more measurable

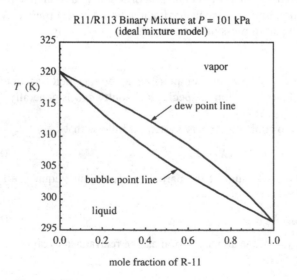

Figure 8.15

properties. The extended van der Waals model discussed above accomplishes this, but some mixtures are not accurately treated by this model. For pure substances, the Gibbs–Duhem equation can be used to relate molar chemical potential to changes in temperature and pressure:

$$d\hat{\mu} = -\hat{s}\,dT + \hat{v}\,dP. \tag{8.122}$$

Integrating this relation between a reference state (designated with a 0 subscript) and an arbitrary state, we obtain

$$\hat{\mu}(T, P) = \hat{\mu}_0 - \int_{T_0}^{T} \hat{s}\,dT + \int_{P_0}^{P} \hat{v}\,dP. \tag{8.123}$$

Use of this relation requires that we have a means of evaluating the chemical potential at the reference temperature and pressure. There is no fundamental thermodynamic requirement that allows us to specify the absolute value of the chemical potential at some particular point. As we have shown, statistical thermodynamics allows us to predict an absolute value of chemical potential if we have an appropriate model of the fluid, such as the van der Waals model discussed above. However, this is only approximate for many substances. Because we only desire the ability to compute changes in chemical potential, we elect to evaluate the reference condition in an arbitrary, but consistent manner for the substances we consider.

If we consider an ideal gas, we can use the ideal gas equation of state to evaluate $\hat{v}$ in Eq. (8.123) and integrate along a line of constant temperature to obtain

$$\hat{\mu}(T, P) = \hat{\mu}_0 + RT \ln(P/P_0), \tag{8.124}$$

where $R = N_A k_B$ is the universal gas constant. This result implies that, for an ideal gas, the change in chemical potential along a line of constant pressure is simply related to the log of the pressure ratio. To generalize this relation so it applies to substances other than ideal gases, G. N. Lewis defined the *fugacity* f such that for any substance

$$\hat{\mu}(T, P) = \hat{\mu}_0 + RT \ln(f/f_0). \tag{8.125}$$

In this defining relationship either $\hat{\mu}_0$ of f_0 may be chosen arbitrarily, but once one is chosen, the other must be fixed to satisfy Eq. (8.125) at the reference state. In the limit of low pressure, if the real substance approaches ideal gas behavior, we expect that f/f_0 approaches P/P_0. The ratio f/f_0 is termed the *activity* and is often denoted by a separate symbol $\tilde{a}$. This term arises from the fact that $\tilde{a} = f/f_0$ indicates the difference between the chemical potential in an arbitrary state and that in the standard state, when both states are at the same temperature. The associated difference in the chemical potential is presumed to be an indicator of how "active" the substance will be with respect to mass transfer.

It follows from the definition of fugacity that, at fixed T,

$$d\hat{\mu}(T, P) = RT\,d(\ln f). \tag{8.126}$$

Substituting into Eq. (8.122), we obtain (for T fixed)

$$d(\ln f) = \hat{v}\,dP/RT. \tag{8.127}$$

For fluids in general, we define the *compressibility factor* Z as

$$Z = P\hat{v}/RT. \tag{8.128}$$

The compressibility factor quantifies the deviation of the fluid from ideal gas behavior. If Z is close to 1, the behavior is essentially ideal. In general Z differs significantly from one for dense gases and liquids. The value of Z can be predicted using the virial expansion or van der Waals models discussed in Chapter 6. Using the definition of Z, we can rewrite (8.127) as

$$d(\ln f) = \frac{Z\,dP}{P}. \tag{8.128}$$

Subtracting $d(\ln P)$ from both sides of Eq. (8.128), we obtain

$$d[\ln(f/P)] = (Z - 1)\frac{dP}{P}. \tag{8.129}$$

We integrate both sides from zero pressure, where $f = P$, to an arbitrary state along a line of constant temperature:

$$\ln(f/P) = \left[\int_0^P (Z - 1)\frac{dP}{P} \right]_{T=\text{constant}}. \tag{8.130}$$

This relation can be solved explicitly for the fugacity to give

$$f = P \exp\left\{ \int_0^P (Z - 1)\frac{dP}{P} \right\}_{T=\text{constant}}. \tag{8.131}$$

Thus, for a pure substance, if we know $Z(P, T)$ we can generate $f(P, T)$ by executing the integration indicated in Eq. (8.131) with T held constant.

For each component in a two-phase two-component system, a different relation between chemical potential and fugacity exists for each of the two phases:

$$\hat{\mu}_a^{\text{I}} = \hat{\mu}_{a,0}^{\text{I}} + RT \ln\left(f_a^{\text{I}}/f_{a,0}^{\text{I}}\right), \tag{8.132}$$

$$\hat{\mu}_a^{\text{II}} = \hat{\mu}_{a,0}^{\text{II}} + RT \ln\left(f_a^{\text{II}}/f_{a,0}^{\text{II}}\right). \tag{8.133}$$

Identical relations apply for species b. Equating the right sides of Eqs. (8.132) and (8.133) yields

$$\hat{\mu}_{a,0}^{\text{I}} + RT \ln\left(f_a^{\text{I}}/f_{a,0}^{\text{I}}\right) = \hat{\mu}_{a,0}^{\text{II}} = RT \ln\left(f_a^{\text{II}}/f_{a,0}^{\text{II}}\right). \tag{8.134}$$

Since Eq. (8.125) applies to any two states for a specific substance at the same temperature, it can be used to relate reference states for the two phases if they are at the same temperature but have different pressures and concentrations:

$$\hat{\mu}_{a,0}^{\text{I}} = \hat{\mu}_{a,0}^{\text{II}} + RT \ln\left(f_{a,0}^{\text{I}}/f_{a,0}^{\text{II}}\right). \tag{8.135}$$

Substituting the right side of this relation for $\hat{\mu}_{a,0}^{\text{I}}$ in Eq. (8.134) and simplifying leads to

$$f_a^{\text{I}} = f_a^{\text{II}}. \tag{8.136}$$

Using identical arguments for species b converts the equality of chemical potential to the condition

$$f_b^{\text{I}} = f_b^{\text{II}}. \tag{8.137}$$

Thus if we define the chemical potential reference state in all phases at the same temperature, the equality of chemical potential in all phases can be replaced by the equality of fugacity for each component at equilibrium. In some applications, statement of the equilibrium conditions in terms of fugacities is advantageous because of their close link to pressure.

Regardless of whether equilibrium is specified in terms of chemical potential or fugacity, to generate a binary mixture phase diagram like Figure 8.11 we must develop a relation for fugacity or chemical potential as a function of temperature, pressure, and concentration. We have shown above that, for a pure substance, f can be determined from the dependence of the compressibility factor on T and P. Unfortunately, extending this approach to predict the fugacity of a component in a mixture is generally a very difficult task. For gas mixtures, it is common to define the *fugacity coefficient* for component a as

$$\phi_a \equiv \frac{f_a}{X_{a,\mathrm{v}} P}, \tag{8.138}$$

where $X_{a,\mathrm{v}}$ is the mole fraction of species a in the vapor phase. For liquid mixtures, it is conventional to define the *activity coefficient* for component a as

$$\gamma_a \equiv \frac{f_a}{X_{a,\mathrm{l}} f_{a,0}}, \tag{8.139}$$

where $X_{a,\mathrm{l}}$ is the liquid mole fraction of species a and $f_{a,0}$ is the fugacity of component a at some convenient reference condition. With these definitions, the equality of fugacity for species a in the liquid and vapor phases can be written as

$$\phi_a X_{a,\mathrm{v}} P = \gamma_a X_{a,\mathrm{l}} f_{a,0}. \tag{8.140}$$

With this formulation, we can establish relations among $X_{a,\mathrm{v}}$, $X_{a,\mathrm{l}}$, P, and T at equilibrium if we can develop relations for ϕ_a and γ_a as functions of these variables:

$$\phi_a = \phi_a(T, P, X_{a,\mathrm{v}}, X_{b,\mathrm{v}}), \tag{8.141}$$

$$\gamma_a = \gamma_a(T, P, X_{a,\mathrm{l}}, X_{b,\mathrm{l}}). \tag{8.142}$$

If the vapor is a mixture of ideal gases, the definition of f_a dictates that $\phi_a = 1$. An ideal liquid mixture is taken to correspond to $\gamma_a = 1$ with $f_{a,0}$ equal to the saturation pressure of pure liquid species a at the given system temperature. If the gas and liquid phases are both ideal mixtures, Eq. (8.140) reduces to

$$X_{a,\mathrm{v}} P = X_{a,\mathrm{l}} P_{\mathrm{sat},a}. \tag{8.143}$$

Since the left side of (8.143) is just the partial pressure of vapor component a, this relation is equivalent to Raoult's law. Note that identical reasoning will generate relations equivalent to Eqs. (8.138)–(8.143) for species b. The value of this formulation is that it provides a framework to account for the deviation of real mixtures from the ideal behavior embodied in Dalton's law and Raoult's law. The deviation of ϕ_a, ϕ_b, γ_a, and γ_b from one indicate the degree to which these mixtures deviate from the ideal behavior.

It is possible to obtain phase equilibrium data experimentally and use it directly to establish a phase equilibrium diagram or to develop empirical relations for the fugacity and activity coefficients as functions of temperature, pressure, and concentrations. Analytical determination of relations for ϕ_a, ϕ_b, γ_a, and γ_b for nonideal mixtures requires that we incorporate methods to accurately model the force interactions between molecules in the vapor and liquid phases. Often this is a challenging task because the nonideal behavior is a consequence of complex force interactions among the molecules resulting from factors such as the nonspherically symmetric nature of the molecules and the polar nature of the molecules. The effects of these molecular characteristics on fugacity and activity coefficients are beyond the scope of this text. The interested reader may find a detailed discussion of these issues in the text by Prausnitz et al. [3].

8.4   Thermodynamic Similitude and the Principle of Corresponding States

Models such as the van der Waals model and Redlich–Kwong model discussed in Chapter 6 can be used to predict fluid properties in either the vapor or liquid phase. An interesting feature of these models is that they suggest that values of thermodynamic properties of different pure materials can be treated in a unified way when the property is normalized with its value at the critical point. Beginning with the general form of the van der Waals equation of state,

$$P = \frac{N_A k_B T}{\hat{v} - N_A b_v} - \frac{a_v N_A^2}{\hat{v}^2}, \tag{8.144}$$

we can evaluate the constants a_v and b_v by requiring that the slope of the van der Waals isotherm and its curvature be zero at the critical point. The latter requirement establishes an inflection point there. Invoking these requirements results in the following relations for the van der Waals constants:

$$b_v = \frac{\hat{v}_c}{3N_A} = \frac{k_B T_c}{8 P_c}, \tag{8.145}$$

$$a_v = \frac{27}{64} \left(\frac{k_B^2 T_c^2}{P_c} \right). \tag{8.146}$$

We define reduced properties, denoted with an r subscript, by normalizing each with its value at the critical point:

$$T_r = T/T_c, \qquad P_r = P/P_c, \qquad \hat{v}_r = \hat{v}/\hat{v}_c. \tag{8.147}$$

Replacing the physical properties in Eq. (8.144) with the product of the reduced property and its value at the critical point, we can write the van der Waals equation of state in terms of reduced properties as

$$P_r = \frac{8T_r}{3\hat{v}_r - 1} - \frac{3}{\hat{v}_r^2}. \tag{8.148}$$

In addition, it is also easily shown that the compressibility factor $Z = P\hat{v}/RT$ can be expressed in terms of reduced properties as

$$Z = Z_c \frac{P_r \hat{v}_r}{T_r}, \tag{8.149}$$

where for a van der Waals fluid, the critical compressibility factor $Z_c = P_c \hat{v}_c / RT_c$ is 0.375. Since in principle, the above two relations (8.148) and (8.149) can be combined to eliminate $\hat{v}_r$ as a variable, they imply that the compressibility factor can be considered to be only a function of reduced pressure and temperature $Z = Z(P_r, T_r)$. For a van der Waals fluid this functional dependence can be determined by specifying P_r and T_r, solving Eq. (8.148) iteratively for physically possible $\hat{v}_r$ values, and then computing Z

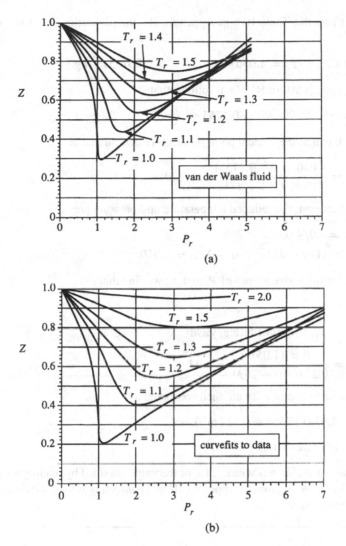

Figure 8.16

from (8.149) with $Z_c = 0.375$. The variation of $Z = Z(P_r, T_r)$ computed in this manner for a van der Waals fluid is shown in Figure 8.16a. The compressibility data for some real fluids can, in fact, be collapsed onto a single curve in this way. The prediction for a van der Waals fluid is qualitatively the same as for many real fluids, but the quantitative accuracy varies considerably. Figure 8.16b shows the variation of $Z = Z(P_r, T_r)$, which closely fits data for water, CO_2, nitrogen, and a variety of hydrocarbons presented by Su [4].

Example 8.5   Use the compressibility charts to determine the density of R134a at 120°C and 4.0 Mpa.

Solution From the Table III.s in Appendix III, the properties of R134a at the critical point are

$$T_c = 101.0°C, \qquad P_c = 4.056 \, kPa, \qquad v_c = 0.00195 \, m^3/kg.$$

Since the molecular mass is 90 for R134a, it follows that

$$\hat{v}_c = M v_c = (90)(0.00195) = 0.1755 \, m^3/kmol.$$

The compressibility factor at the critical point is therefore determined as

$$Z_c = \frac{P_c \hat{v}_c}{RT_c} = \frac{4.056 \times 10^6 (0.1755)}{(8308)(101 + 273)} = 0.229.$$

For the conditions of interest, the reduced temperature and pressure are

$$P_r = P/P_c = 4.0/4.056 = 0.986,$$
$$T_r = T/T_c = (120 + 273)/(101 + 273) = 1.050.$$

Using Figure 8.16b, for the above values of P_r and T_r, we find that

$$Z = 0.64.$$

Solving Eq. (8.149) for $\hat{v}_r$, and substituting, yields

$$\hat{v}_r = \frac{ZT_r}{Z_c P_r} = \frac{0.64(1.05)}{0.229(101 + 273)} = 2.97.$$

The molar specific volume and density are then computed as

$$\hat{v} = \hat{v}_r \hat{v}_c = 2.97(0.1755) = 0.521 \, m^3/kmol,$$
$$\hat{\rho} = 1/\hat{v} = 1/0.521 = 1.92 \, kmol/m^3.$$

The molar density indicated above is within 14% of the value obtained by reading the mass density from the R134a $P–h$ diagram in Appendix III and converting it to molar density by dividing by the molecular mass.

For a van der Waals fluid, the difference between $\hat{\mu}/T$ and its value at the critical point $\hat{\mu}_c/T_c$ can also be expressed as a function only of reduced temperature and volume:

$$\frac{\hat{\mu}}{T} - \frac{\hat{\mu}_c}{T_c} = N_A k_B \left[\frac{1}{3\hat{v}_r - 1} - \frac{1}{2} - \frac{\xi}{2} \ln T_r - \ln \left(\frac{3\hat{v}_r - 1}{2} \right) - \frac{9}{4} \left(\frac{1}{\hat{v}_r T_r} - 1 \right) \right],$$

$$(8.150)$$

where

$$\frac{\hat{\mu}_c}{N_A k_B T_c} = \frac{N_A b_v}{\hat{v}_c - N_A b_v} - \left(\frac{3}{2} \right) \ln \left[\frac{2\pi M k_B T_c (\hat{v}_c - N_A b_v)^{2/3}}{N_A^{2/3} h^2} \right]$$
$$- \left[\frac{\xi - 5}{2} \ln \pi - \ln \sigma_s \right] - \frac{\xi - 3}{2} \ln \left(\frac{T_c}{\theta_{rot,m}} \right) - \frac{2a_v N_A}{k_B \hat{v}_c T_c}. \qquad (8.151)$$

This implies that when we invoke the necessary conditions for equilibrium in a two-phase system, the requirement of equality of chemical potential can be expressed in terms of reduced properties. For a van der Waals fluid, the deviation of the molar specific entropy from its value at the critical point can also be expressed as a function of reduced properties:

$$\frac{\hat{s} - \hat{s}_c}{N_A k_B} = \frac{\xi}{2} \ln T_r + \ln\left(\frac{3\hat{v}_r - 1}{2}\right), \tag{8.152}$$

where

$$\frac{\hat{s}_c}{N_A k_B} = \left(\frac{\xi}{2} + 1\right) + \left(\frac{3}{2}\right) \ln\left[\frac{2\pi M k_B T_c (\hat{v}_c - N_A b_v)^{2/3}}{N_A^{2/3} h^2}\right]$$
$$+ \left[\frac{\xi - 5}{2} \ln \pi - \ln \sigma_s\right] + \frac{\xi - 3}{2} \ln\left(\frac{T_c}{\theta_{rot,m}}\right). \tag{8.153}$$

For a system containing the liquid and vapor phase of a pure van der Waals fluid Eqs. (8.148) and (8.150) can be used to state the necessary conditions for equilibrium in terms of reduced properties. It follows from these equations and the definitions of the reduced properties that at equilibrium the reduced pressure and temperature must be the same in both phases. Using Eq. (8.150) to evaluate the chemical potential in each phase and setting the resulting expressions for the liquid and vapor chemical potentials equal to each other and using the fact that the reduced temperature is the same in both phases yields

$$\frac{1}{3\hat{v}_{r,l} - 1} - \ln(3\hat{v}_{r,l} - 1) - \frac{9}{4\hat{v}_{r,l} T_{r,sat}} = \frac{1}{3\hat{v}_{r,v} - 1} - \ln(3\hat{v}_{r,v} - 1) - \frac{9}{4\hat{v}_{r,v} T_{r,sat}}, \tag{8.154}$$

where $\hat{v}_{r,l}$ and $\hat{v}_{r,v}$ are the reduced molar specific volumes in the liquid and vapor phases, respectively. The van der Waals equation of state must also be satisfied in both phases:

$$P_{r,sat} = \frac{8T_{r,sat}}{3\hat{v}_{r,l} - 1} - \frac{3}{\hat{v}_{r,l}^2}, \tag{8.155}$$

$$P_{r,sat} = \frac{8T_{r,sat}}{3\hat{v}_{r,v} - 1} - \frac{3}{\hat{v}_{r,v}^2}. \tag{8.156}$$

For specified $T_{r,sat}$, the above three nonlinear equations can be solved for $P_{r,sat}$, $\hat{v}_{r,l}$, and $\hat{v}_{r,v}$. The variations of $P_{r,sat}$ with $T_{r,sat}$ and $\hat{v}_{r,l}$ and $\hat{v}_{r,v}$ with $T_{r,sat}$ obtained by iteratively solving these equations are shown in Figures 8.17 and 8.18. In Figure 8.17, $P_{r,sat}$ is plotted as a function of $1/T_r$ on a semilog plot to demonstrate that the van der Waals model predicts a nearly linear variation of $\ln(P_{r,sat})$ with $1/T_r$, which was suggested by our consideration of the Clapeyron equation in Section 8.2. In Figure 8.18, the vapor values of reduced specific volume correspond to $\hat{v}_r > 1$ and liquid values correspond to $\hat{v}_r < 1$.

It is also noteworthy that the relation for the enthalpy of a van der Waals fluid derived in Chapter 6 can be manipulated to obtain the following relation for the change in molar specific enthalpy for conversion of saturated liquid to saturated vapor:

$$\frac{\Delta\hat{h}_{lv}}{N_A k_B T_c} = T_r \left[\frac{3\hat{v}_{r,v}}{3\hat{v}_{r,v} - 1} - \frac{3\hat{v}_{r,l}}{3\hat{v}_{r,l} - 1}\right] - \frac{9}{4}\left[\frac{1}{\hat{v}_{r,v}} - \frac{1}{\hat{v}_{r,l}}\right]. \tag{8.157}$$

The pairs of liquid and vapor reduced specific volumes on the right side of the above relation are functions of only T_r. This implies that $\Delta\hat{h}_{lv}/N_A k_B T_c$ is a function only of

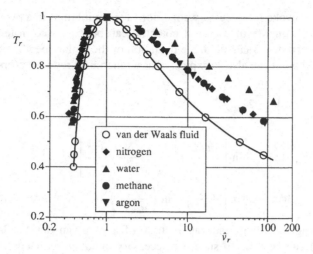

Figure 8.17

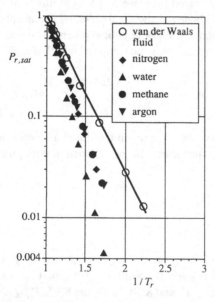

Figure 8.18

reduced temperature for a van der Waals fluid. For the values of T_r indicated in Figure 8.18, corresponding pairs of $\hat{v}_r$ values computed using the van der Waals model can be used to determine $\Delta\hat{h}_{lv}/N_A k_B T_c$. The resulting variation of $\Delta\hat{h}_{lv}/N_A k_B T_c$ with T_r is shown in Figure 8.18.

In addition to the computed results for the van der Waals model, points representing established properties for several real pure fluids are plotted in reduced parameter form in Figures 8.17–8.19. It can be seen that, although the qualitative behavior of the van der Waals model is consistent with that of the real fluids, the predictions of the van der Waals model differ substantially from the data for real fluids, particularly at low temperatures. The trends in the data for real fluids in these figures is linked to the nature of the molecules. For argon,

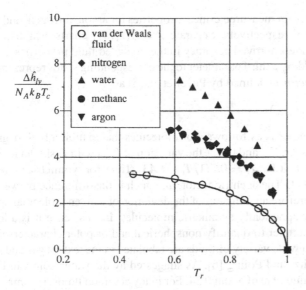

Figure 8.19

nitrogen, and methane, the molecules are nearly spherically symmetric geometrically and in terms of their potential interactions with neighboring molecules. The points for these fluids lie essentially on the same curve in these figures. The points for water differ significantly from those for argon, nitrogen, and methane. Water molecules have a nonspherical structure and their polar nature results in nonspherically symmetric potential force interactions with neighboring molecules. The trends in these data suggest that the corresponding states framework implied by the van der Waals model works well when the molecules are close to spherically symmetric, but deviations from such symmetry cause the thermodynamic similarity to break down.

In Chapter 6, we derived the van der Waals equation of state from an interaction potential model that contained two parameters: the depth of the potential well and a distance parameter σ_{LJ}. It is perhaps not surprising then that we can characterize the behavior of a van der Waals fluid in terms of two macroscopic parameters P_c and T_c (or equivalently a_v and b_v). The behavior of some real fluids is well characterized with two parameters. Generally, this is true for molecules that nominally exhibit spherical symmetry in their shape and in their force interactions with other molecules. For many substances, two parameters is not enough to provide useful correlation accuracy. Molecules that show the strongest deviation from the two-parameter corresponding states behavior are most commonly those having geometry and interaction potentials that are extremely nonspherical and those with strong dipole moments. Intermediate molecules that are slightly nonspherical and/or have weak to moderate dipole moments behave in a manner similar to that predicted by the two-parameter corresponding states model, but they exhibit significant deviations from its predictions for some states.

An obvious means to patch up the corresponding states model so as to account for nonsymmetric and polar behavior is to include an additional characterizing factor. One way of doing so is to include the critical compressibility $Z_c = P_c \hat{v}_c / RT_c$ as a third parameter. Values of Z_c vary among different fluids. For a van der Waals fluid, we noted that $Z_c = 0.375$.

The values of Z_c computed from measured critical properties for argon, nitrogen, and water are 0.291, 0.290, and 0.236, respectively. Separate $Z = Z(P_r, T_r)$ charts established for different Z_c values do provide improved accuracy in the predicted fluid properties.

An alternate means of adding a third parameter to the corresponding states representation is to include the *acentric factor* ω defined by Pitzer et al. [5] as

$$\omega \equiv -1.0 - \log_{10}(P_{r,\text{sat}})_{T_r=0.7}. \tag{8.158}$$

The rationale for this parameter is as follows. For molecules that exhibit a high degree of spherical symmetry (such as Ar, N_2, and CH_4), the variation of $P_{r,\text{sat}}$ with T_r shown in Figure 8.17 indicates that $P_{r,\text{sat}}$ equals 0.1 at $T_r = 0.7(1/T_r = 1.43)$. Thus, for symmetric molecules the base ten log of $P_{r,\text{sat}}$ is -1. The acentric factor, ω, which is the difference between -1 and $\log_{10}(P_{r,\text{sat}})_{T_r=0.7}$, is therefore an indicator of the deviation of a given molecular species from behavior exhibited by spherically symmetric molecules. In this sense it is a logical choice as an additional parameter to quantify nonspherical and/or polar characteristics of molecules. Values of ω for a wide variety of fluids are tabulated in the text on gas and liquid properties by Reid, Prausnitz, and Poling [6]. As suggested by the water points in Figure 8.17, water has a positive value of ω of about 0.34. For many common fluids, recommended values of ω are positive and less than 0.5. A few, such as neon, have recommended values that are negative. Pitzer et al. [5] proposed a relation for the compressibility factor having the form

$$Z = Z_0(T_r, P_r) + \omega Z_1(T_r, P_r), \tag{8.159}$$

where the Z_0 function applies to molecules exhibiting spherical symmetry and ωZ_1 is a correction for nonsymmetric behavior. Pitzer et al. [5] tabulated values of Z_0 and Z_1 as functions of T_r and P_r. Updated and extended versions of these tables are provided by Reid et al. [6]. Further discussion of extensions of the corresponding states framework may also be found in the text by Reid et al. [6].

Near Critical State Behavior

In our examination of phase equilibria for a pure system we found that the molar specific volumes of a coexisting liquid and vapor phase change with saturation temperature. As the temperature is increased in a pure two-phase system of fixed volume, the specific volume of the vapor decreases and the specific volume of the liquid increases. As the temperature approaches the critical temperature, the molar specific volume of both phases approaches the specific volume at the critical point. The critical state is of special importance to the thermodynamic characteristics of the substance for two reasons. First, it defines the ranges of temperature and pressure where a liquid–vapor phase transition is possible. In addition, the critical state is of special interest because the thermodynamic characteristics of a pure fluid system become quite extraordinary near the critical point. If a system molar specific volume is equal to $\hat{v}_c$ and the temperature is higher than the critical temperature, the state of the substance is said to be *supercritical*. If the temperature of the system is gradually lowered toward the critical temperature, properties behave in an anomalous way as the critical point is approached. The critical isotherm has zero slope and zero curvature (an inflection point) at the critical point. Equations (8.5) and (8.6) imply that density and energy fluctuations become infinite as $(\partial P/\partial \hat{v})_T \to 0$. At the critical point the system thus does not have a well-defined equilibrium state in the usual thermodynamic sense. The resulting density fluctuations cause light passing through the fluid to be scattered so that the fluid intermittently looks opaque in localized regions within the system. This phenomena is referred to as *critical opalescence*.

In addition, because $(\partial P/\partial \hat{v})_T$ approaches zero as the critical point is approached, the isothermal compressibility $\kappa_T = -(1/V)(\partial V/\partial P)_{T,N} = -(1/\hat{v})(\partial \hat{v}/\partial P)_T$ approaches infinity there. This implies that the small hydrostatic pressure variation in the system will produce a very large density gradient from the top to the bottom of the system. Any systems on the earth that are close to the critical state strongly depart from our usual definition of an equilibrium state, and our usual definitions of thermodynamic properties break down in this range of conditions. In addition to the theoretical complexities associated with this behavior, the high levels of property fluctuations and the gravity-induced variation of density makes experimental determination of physical properties very difficult near the critical point.

A widely used means of quantifying property behavior near the critical point is to examine how properties deviate from their value at the critical point in the near-critical region. The usual approach is to postulate a power law variation and quantify the behavior in terms of the exponent of the power law dependence. For a fluid system of the type considered here, it is common to define pressure, temperature, and volume parameters as

$$\tilde{p} \equiv P_r - 1 = (P - P_c)/P_c, \tag{8.160}$$

$$\tilde{t} \equiv T_r - 1 = (T - T_c)/T_c, \tag{8.161}$$

$$\tilde{v} \equiv \hat{v}_r - 1 = (\hat{v} - \hat{v}_c)/\hat{v}_c. \tag{8.162}$$

These parameters are obviously defined so that their deviation from zero indicates the fractional deviation of the associated physical property from the critical value. When $\tilde{t}$ is positive $(T > T_c)$, only a single value of $\tilde{v}$ is possible, whereas when $\tilde{t}$ is negative $(T < T_c)$, $\tilde{v}$ can take on two values: $\tilde{v}_l$ for the liquid phase and $\tilde{v}_v$ for the vapor phase. To characterize the behavior near the critical point, we are interested in the power law variations:

$$\tilde{p} \sim \tilde{v}^\delta, \tag{8.163}$$

$$\kappa_T \sim \begin{cases} \tilde{t}^{-\gamma} & (\tilde{t} > 0), \\ (-\tilde{t})^{-\gamma'} & (\tilde{t} < 0), \end{cases} \tag{8.164}$$

$$\tilde{v}_v - \tilde{v}_l \sim (-\tilde{t})^\beta, \tag{8.165}$$

$$\hat{c}_v \sim \begin{cases} \tilde{t}^{-\alpha} & (\tilde{t} > 0), \\ (-\tilde{t})^{-\alpha'} & (\tilde{t} < 0). \end{cases} \tag{8.166}$$

The exponents in the power law variations above are termed *critical exponents* for the system. They can be determined explicitly for a van der Waals fluid. Using Eqs. (8.160)–(8.162) to replace P_r, T_r, and $\hat{v}_r$ in the van der Waals equation of state (8.148), we obtain

$$\tilde{p} = \frac{4(1 + \tilde{t})}{1 + 3\tilde{v}/2} - \frac{3}{(1 + \tilde{v})^2} - 1. \tag{8.167}$$

Expanding the right side to third order in the small variables $\tilde{t}$ and $\tilde{v}$ we find

$$\tilde{p} = -\frac{3}{2}\tilde{v}^3 + \tilde{t}(4 - 6\tilde{v} + 9\tilde{v}^2) + \cdots. \tag{8.168}$$

This implies that $\delta = 3$ and the critical isotherm ($\tilde{t} = 0$) varies as $\tilde{p} = -(3/2)\tilde{v}^3$ near the critical point. Differentiating this relation, we can also show that

$$\kappa_T = \frac{1}{6}P_c\tilde{t}^{-1},\tag{8.169}$$

which implies that $\gamma = \gamma' = 1$.

As discussed above, the values of molar specific volume for saturated liquid and vapor are dictated by the requirements that temperature, pressure, and chemical potential must be the same in both phases. In the near-critical region we invoke the requirement of equal pressure and temperature by evaluating the right side of Eq. (8.168) for the saturated liquid at $\tilde{t}$ and $\tilde{v}_l$ and setting it equal to the right side of Eq. (8.168) evaluated for saturated vapor at $\tilde{t}$ and $\tilde{v}_v$. Doing so and retaining terms to third order yields, after a little rearranging we get

$$\left(\tilde{v}_v^3 - \tilde{v}_l^3\right) + 4\tilde{t}(\tilde{v}_v - \tilde{v}_l) - 6\tilde{t}\left(\tilde{v}_v^2 - \tilde{v}_l^2\right) = 0.\tag{8.170}$$

Dividing through by $(\tilde{v}_v - \tilde{v}_l)$, we can reduce this to

$$\left(\tilde{v}_v^2 + \tilde{v}_l\tilde{v}_v + \tilde{v}_l^2\right) + 4\tilde{t} - 6\tilde{t}(\tilde{v}_v + \tilde{v}_l) = 0.\tag{8.171}$$

To invoke the requirement of equal chemical potential and temperature, we use Eqs. (8.160)–(8.162) to replace the P_r, T_r, and $\hat{v}_r$ terms in Eq. (8.154) and expand the resulting terms for small $\tilde{t}$, $\tilde{v}_l$ and $\tilde{v}_v$. Rearranging the resulting equation, we obtain

$$\left(\tilde{v}_v^2 + \tilde{v}_l\tilde{v}_v + \tilde{v}_l^2\right) + 4\tilde{t} - 4\tilde{t}^2 - 4\tilde{t}\left(\tilde{v}_v + \tilde{v}_l\right) = 0.\tag{8.172}$$

In principle, Eqs. (8.171) and (8.172) solved simultaneously dictate $\tilde{v}_l$ and $\tilde{v}_v$ for specified $\tilde{t}$ near the critical point. These relations can be rearranged to the following equivalent forms:

$$3(\tilde{v}_v + \tilde{v}_l)^2 + (\tilde{v}_v - \tilde{v}_l)^2 + 16\tilde{t} - 24\tilde{t}(\tilde{v}_v + \tilde{v}_l) = 0,\tag{8.173}$$

$$3(\tilde{v}_v + \tilde{v}_l)^2 + (\tilde{v}_v - \tilde{v}_l)^2 + 16\tilde{t} - 16\tilde{t}^2 - 16\tilde{t}(\tilde{v}_v + \tilde{v}_l) = 0.\tag{8.174}$$

When two phases are present, $\hat{v}_v$ is larger than $\hat{v}_c$ and $\hat{v}_l$ is smaller than $\hat{v}_c$. It follows directly that $\tilde{v}_v$ must be positive and $\tilde{v}_l$ must be negative. As a result $\tilde{v}_v + \tilde{v}_l$ is much smaller than $\tilde{v}_v - \tilde{v}_l$. We therefore neglect the $3(\tilde{v}_v + \tilde{v}_l)^2$ term in the above equations as well as the $\tilde{t}^2$ term in (8.174) (compared to the $\tilde{t}$ term) and combine Eqs. (8.173) and (8.174) to eliminate $\tilde{v}_v + \tilde{v}_l$. After simplification, the resulting relation becomes

$$\tilde{v}_v - \tilde{v}_l = 4(-\tilde{t})^{1/2},\tag{8.175}$$

which implies that the critical exponent β is $1/2$. Using similar reasoning together with the other property relations for a van der Waals fluid it can be shown that the critical constants α and α' are both zero for a van der Waals fluid.

For real pure fluids, the critical exponents generally span a range of values. The exponents α and α' generally lie between -0.2 and 0.3. For real fluids, β is typically between 0.3 and 0.5 and δ is typically between 4 and 5. For the isothermal compressibility of pure fluids the critical exponents generally lie in the ranges $1.2 < \gamma < 1.4$ and $1.0 < \gamma' < 1.3$. Figure 8.20 shows the computed variation of $\tilde{v}_v - \tilde{v}_l$ with $\tilde{t}$ indicated by the iteratively computed saturation values of molar specific volume for a van der Waals fluid plotted in the figure. Also shown are the established data for nitrogen and the near-critical variation for a van der Waals fluid indicated by Eq. (8.175). The near-critical relation agrees well with computed

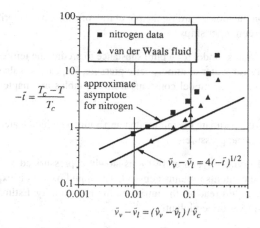

Figure 8.20

results for the van der Waals fluid even at large deviation from the critical temperature. The data for nitrogen exhibit a slope that implies a β value of 0.45.

In addition to the pure fluid system examined here, critical phenomena arise in a number of other physical systems, including multicomponent systems and systems in which electromagnetic effects are important. The limited discussion here only considers a small portion of the larger statistical physics framework that has been developed to analyze characteristics of near-critical systems. The interested reader can find further information on this topic in the texts by Callen [7], Stanley [8], and Yeomans [9].

Exercises

8.1 Derive Eqs. (8.39) and (8.40) from Eqs. (8.35) and (8.36).

8.2 Use the van der Waals model to determine the variation of the chemical potential with pressure for methane at $T = 160\,\text{K}$. Use a computer program to determine $\mu/k_B T$ as a function of $\hat{v}$ and simultaneously compute P for each value of $\hat{v}$ along the isotherm. Plot your results and interpret the nature of each segment of the isotherm. Also compare the equilibrium saturation pressure indicated by your model results to the established value of 1,592 kPa at this temperature.

8.3 Use the van der Waals model to determine the variation of the chemical potential with pressure for argon at $T = 130\,\text{K}$. Use a computer program to determine $\mu/k_B T$ as a function of $\hat{v}$ and simultaneously compute P for each value of $\hat{v}$ along the isotherm. Plot your results and interpret the nature of each segment of the isotherm. Also compare the equilibrium saturation pressure indicated by your model results to the established value of 2,027 kPa at this temperature.

8.4 If liquified natural gas (methane) leaks from containers on a ship it may spill onto the surface of water near the ship, which is at a temperature of 15°C. The liquid methane at atmospheric pressure that contacts the water may be heated close to 15°C. Liquid methane is known to vaporize explosively when heated close to the intrinsic limit of stablity. Use the results for the van der Waals model for a pure substance to predict the temperature at the limit of intrinsic stability for liquid methane at atmospheric pressure. Repeat the calculation using the Redlich–Kwong model. Use model constants from the tables in Chapter 6 for these calculations. What are your conclusions regarding the possibility of explosive vaporization of methane when it leaks from container ships?

8.5   Use the results for the van der Waals model for a pure substance to predict the temperature at the limit of intrinsic stability for water at atmospheric pressure. Repeat the calculation using the Redlich–Kwong model. Use model constants from the tables in Chapter 6 for these calculations.

8.6   Using the Redlich–Kwong model, estimate the minimum temperature at which water vapor is intrinsically stable at atmospheric pressure.

8.7   A vessel containing saturated liquid oxygen at 140 K is slowly depressurized in such a way that the temperature of the contents remains constant. The liquid oxygen is expected to vaporize violently when the liquid reaches the intrinsinc limit of stability. Estimate the pressure at which the liquid will vaporize violently.

8.8   A vessel containing saturated nitrogen vapor at 80 K is slowly compressed in such a way that the temperature of the contents remains constant. The nitrogen vapor is expected to suddenly initiate formation of liquid droplets when the vapor reaches the intrinsic limit of stability. Estimate the pressure at which droplet formation is expected to be observed.

8.9   The equation of state (6.60) and chemical potential relation (6.65) for the Redlich–Kwong model discussed in Chapter 6 can be written in terms of molar specific volume $\hat{v}$ and molar chemical potential ($\hat{\mu}$) as follows:

$$P = \frac{N_A k_B T}{\hat{v} - N_A b_R} - \frac{a_R N_A^2}{T^{1/2} \hat{v}(\hat{v} + N_A b_R)}, \tag{i}$$

$$\frac{\hat{\mu}}{N_A k_B T} = \frac{N_A b_R}{\hat{v} - N_A b_R} - \left(\frac{3}{2}\right) \ln \left[\frac{2\pi M k_B T (\hat{v} - N_A b_R)^{2/3}}{N_A^{2/3} h^2}\right] - \left[\frac{(\xi - 5)}{2} \ln \pi - \ln \sigma_s\right]$$

$$- \frac{(\xi - 3)}{2} \ln \left(\frac{T}{\theta_{rot,m}}\right) - \frac{a_R N_A^2}{(N_A b_R) N_A k_B T^{3/2}} \ln \left(\frac{\hat{v} + N_A b_R}{\hat{v}}\right)$$

$$- \frac{a_R N_A^2}{N_A k_B T^{3/2}(\hat{v} + N_A b_R)}. \tag{ii}$$

These equations can be used to predict the saturation conditions for a fluid if the constants for the Redlich–Kwong model are known. At equilibrium, the pressure, temperature, and chemical potential must have equal values in the liquid and vapor phase. The saturation values of P, $\hat{v}_l$, and $\hat{v}_v$ for a specified T can be calculated as follows:

1.   A value of pressure is guessed to initiate the scheme.
2.   For the value of pressure and temperature Eq. (i) is iteratively solved for $\hat{v}_l$ and $\hat{v}_v$.
3.   The values of $\hat{v}_l$ and T are inserted into Eq. (ii) to compute $\hat{\mu}_l$, and the values of $\hat{v}_v$ and T are inserted into Eq. (ii) to compute $\hat{\mu}_v$.
4.   If $\hat{\mu}_v$ equals $\hat{\mu}_l$, the guessed value of P is correct, as are the values of $\hat{v}_l$ and $\hat{v}_v$ computed in step 1. If $\hat{\mu}_v$ does not equal $\hat{\mu}_l$, a correction to P is generated using a Newton–Raphson scheme with $\hat{\mu}_l - \hat{\mu}_v$ as the error function. The calculation scheme then returns to step 2.

Write a computer program to use the algorithm above to determine $T_{sat}(P)$ for $R134a$ at atmospheric pressure and five other points between the critical pressure and one atmosphere. Use values of a_R and b_R for R134a from Table 6.2 with $\xi = 6$ and $\sigma_s = 1$. The value of $\theta_{rot,m}$ for R134a is unknown, but a value of 5 K can be used here without loss of accuracy. Run the program and tabulate and plot your results. Compare the predictions of this model with saturation data for R134a tabulated in Appendix III.

8.10 Air in the atmosphere at 101 kPa and 20°C has a water vapor mole fraction of 0.02. Consistent with Dalton's law, the water vapor behaves as if it alone occupied the volume accessible to the gas at its partial pressure and the mixture temperature. This implies that when the system is in equilibrium, condensation of water vapor may occur when the partial pressure of the vapor equals the saturation pressure of water at the system temperature. Similarly, for a metastable mixture, the limit of intrinsic stability is expected to correspond to the partial pressure of water vapor being equal to the pressure at the intrinsic stability limit for pure water vapor at the system temperature. If the mixture is cooled at constant (overall) pressure, at what temperature will the intrinsic limit of stability be reached? (Hint: Use the van der Waals model to predict the intrinsic limit.)

8.11 Derive Eq. (8.100) from Eq. (6.49) from Chapter 6.

8.12 Use the extended van der Waals model discussed above to estimate the limit of intrinsic stability for a binary mixture of oxygen and nitrogen vapor at atmospheric pressure with a nitrogen mole fraction of 0.3.

8.13 Use the extended van der Waals model to estimate the temperature limit of intrinsic stability for pure oxygen vapor at atmospheric pressure. To reduce the possibility of condensation when the gas is cooled at constant pressure, it is suggested that nitrogen vapor be added to make a mixture with a nitrogen mole fraction of 0.1. Evaluate whether this will have the desired effect.

8.14 Derive Eq. (8.150) from Eq. (8.51).

8.15 Derive Eq. (8.154) from Eq. (8.150).

8.16 Derive Eq. (8.157) from Eq. (6.29).

8.17 (a) Derive the reduced equation of state $P_r = P_r(T_r, \hat{v}_r)$ for the Redlich–Kwong model. (b) For the Redlich–Kwong model, derive the relation that embodies equivalence of chemical potential in coexisting liquid and vapor phases at equilibrium in terms of reduced properties. (This should look similar to Eq. (8.154) derived for the van der Waals model.)

8.18 Derive Eq. (8.168) from Eq. (8.167) for small $\tilde{t}$ and $\tilde{v}$.

8.19 Derive Eq. (8.169) from Eq. (8.168) for small $\tilde{t}$ and $\tilde{v}$.

8.20 Plot the saturation data for methane from Appendix III near the critical point and estimate the value of β in Eq. (8.165).

References

[1] Modell, M. and R. C., Reid. *Thermodynamics and Its Applications,* 2nd ed., Prentice-Hall, Englewood Cliffs, NJ, 1983.

[2] Penfold, R., J. Satherley, and S. Nordholm. "Generalized van der Waals Theory of Fluids. Vapour–Liquid Equilibria in Simple Binary Mixtures," *Fluid Phase Equilibria,* 109: 183–204, 1995.

[3] Prausnitz, J. M., R. N. Lichtenthaler, and E. Gomes de Azevedo. *Molecular Thermodynamics of Fluid-Phase Equilibria,* Prentice-Hall, Englewood Cliffs, NJ, 1986.

[4] Su, G-J. "Molecular Law of Corresponding Sates for Real Gases," *Industrial and Engineering Chemistry,* 38:803, 1946.

[5] Pitzer, K. S., D. Z. Lippmann, R. F. Curl, C. M. Huggins, and D. E. Petersen. "The Volumetric and Thermodynamic Properties of Fluids. II. Compressibility Factor, Vapor Pressure and Enthalpy of Vaporization," *J. Am. Chem. Soc.,* 77:3433, 1955.

[6] Reid, R. C., J. M. Prausnitz, and B. E. Poling. *The Properties of Gases and Liquids,* 4th ed. McGraw-Hill, New York, 1987.

[7] Callen, H. B., *Thermodynamics and an Introduction to Thermostatistics,* 2nd ed. John Wiley & Sons, New York, 1985.

[8] Stanley, H. E., *Introduction to Phase Transitions and Critical Phenomena,* Oxford University Press, New York, 1971.

[9] Yeomans, J. M., *Statistical Mechanics of Phase Transitions,* Oxford University Press, New York, 1992.

Nonequilibrium Thermodynamics

The statistical and classical thermodynamics framework developed in Chapters 1–8 of this text is based on analysis of systems at equilibrium. In Chapter 9 we explore the extension of this framework to systems that are not in equilibrium. This chapter focuses on systems that exhibit steady spatial variations of properties. Systems of this type are modeled as having local thermodynamic equilibrium and obeying a linear relation between fluxes and affinities. Analysis of microscale features of such linear systems is shown to link correlation moments and kinetic coefficients. The Onsager reciprocity relations are subsequently derived. Thermoelectric effects are examined as an example application of the nonequilibrium linear theory developed in this chapter.

9.1 Properties in Nonequilibrium Systems

The thermodynamic theoretical framework developed in previous chapters of this text is limited to analysis of equilibrium states. Often, however, it is the process that takes the system from one state to another that is of primary interest. Overall changes accomplished during the process can be determined by analyzing the initial and final states using equilibrium thermodynamics. If the process is very slow, it may be well approximated by a sequence of equilibrium states, and a quasistatic model may adequately predict the outcome of the process.

In many real processes, the departure from equilibrium is so severe that the quasistatic model is too inaccurate to be useful. The objective of this chapter is to develop thermodynamic tools that can be applied to irreversible processes in nonequilibrium systems. In doing so, we also want to examine how such processes relate to the microscale features of nonequilibrium systems.

In the equilibrium theory developed thus far, the system considered for analysis is postulated to be macroscopically (spatially) homogeneous. Real systems often are not in equilibrium and are not spatially homogeneous. An example of a system in which a steady nonequilibrium condition exists is shown in Figure 9.1. In this system, the opposite walls of a container of gas are held at different temperatures. The separation distance between the walls is 0.1 m and the pressure in the gas is one atmosphere.

If a small temperature sensor is used to measure the temperature at several locations between the opposing walls, the readings would indicate a linear variation of temperature with distance between the walls having different temperatures. Measurements of this type are meaningful if the sensor is small enough that molecules that interact with it have energy distributions close to the Boltzmann distributions characterizing an equilibrium system at the temperature indicated by the sensor. Three conditions must be satisfied for this to be true:

(1) The region containing molecules interacting with the sensor must be small enough that variation of the mean properties of the molecules in the region is small.

297

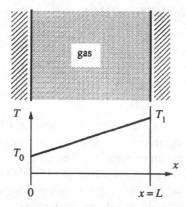

Figure 9.1

(2) The region interacting with the sensor must contain enough molecules so that a large number of molecules interact with the sensor within a time τ_m that is short compared to macroscopic time scales associated with the process. This ensures that the sensor reading accurately reflects the mean properties of the molecules in the region and that a well-defined property exists within the time available to measure it.

(3) The region interacting with the sensor must be large compared with the mean free path of the molecules so that the sensor is responding to collisions with molecules that underwent their last collision within the region of interest.

In the next chapter, we will further explore departures from thermodynamic equilibrium. For now it is enough to acknowledge that if all these conditions are met, measurement of local properties will be meaningful within subregions of the size specified.

These observations suggest that in general we may expect that local thermodynamic properties can be defined for regions satisfying the three conditions above. Continuum models of systems in which conditions are not spatially uniform are generally based on the assumption that differential regions within the system satisfy the above three conditions. Along with this assumption of *local thermodynamic equilibrium*, it is also generally assumed that the functional relations among thermodynamic properties that hold for equilibrium systems also hold among local thermodynamic properties. This allows us to use all the analytical machinery developed in previous sections to relate local properties in our analytical treatment of nonequilibrium systems. While these idealizations are not fully valid in all cases, they provide a useful analytical framework for many real systems.

9.2   Entropy Production, Affinities, and Fluxes

We begin our development of nonequilibrium thermodynamics by considering a differential volume of a continuous system that is not in equilibrium. The system contains a binary mixture of two particle types a and b. We adopt the idealization of local thermodynamic equilibrium discussed in the previous section. It follows directly that the fundamental relation relates local properties in the differential volume:

$$S = S(U, V, N_a, N_b).\tag{9.1}$$

For convenience, we elect to work in terms of properties per unit volume. Since the fundamental equation is first order, we can divide both sides by the volume of the local differential system to obtain a relation of the form

$$\tilde{S} = \tilde{S}(\tilde{U}, \tilde{N}_a, \tilde{N}_b), \tag{9.2}$$

where the $\tilde{S}$, $\tilde{U}$, $\tilde{N}_a$, and $\tilde{N}_b$ are the entropy, internal energy, and number of molecules per unit volume. Expanding Eq. (9.2), retaining first-order terms, and using the definitions of temperature and chemical potential yields

$$d\tilde{S} = \frac{d\tilde{U}}{T} - \frac{\mu_a}{T}d\tilde{N}_a - \frac{\mu_b}{T}d\tilde{N}_b. \tag{9.3}$$

The significance of Eq. (9.3) is that it implies that transport of energy or mass into the volume changes $\tilde{U}$, $\tilde{N}_a$, and $\tilde{N}_b$ and therefore changes the entropy of the system. Thus, the rate of change of entropy in the system can be linked to the rates of change of energy and mass in the system. We therefore treat entropy as a transportable quantity and define $\mathbf{J}_S$, $\mathbf{J}_U$, $\mathbf{J}_a$, and $\mathbf{J}_b$ as vector fluxes of entropy, energy, species a, and species b in the continuous system, respectively. Note that these are vectors which have units of entropy, energy or number of molecules per unit area per unit time. For the differential volume with total differential surface area A, it follows directly from these definitions that

$$d\tilde{S} = \int_A \mathbf{J}_S \cdot d\mathbf{A}, \tag{9.4}$$

$$d\tilde{U} = \int_A \mathbf{J}_U \cdot d\mathbf{A}, \tag{9.5}$$

$$d\tilde{N}_a = \int_A \mathbf{J}_a \cdot d\mathbf{A}, \tag{9.6}$$

$$d\tilde{N}_b = \int_A \mathbf{J}_b \cdot d\mathbf{A}. \tag{9.7}$$

In the above relation $d\mathbf{A}$ is a vector with magnitude equal to the differential area element in the integration and direction normal to the surface element. The integral is carried out by summing the integral contributions of the six faces of the volume element in Figure 9.2.

Substituting Eqs. (9.4)–(9.7) into Eq. (9.3) yields, after combining integral terms,

$$\int_A \left(\mathbf{J}_S - \frac{1}{T}\mathbf{J}_U + \frac{\mu_a}{T}\mathbf{J}_a + \frac{\mu_b}{T}\mathbf{J}_b \right) \cdot d\mathbf{A} = 0. \tag{9.8}$$

This relation is valid for an arbitrary differential volume only if the integrand is zero. It follows that

$$\mathbf{J}_S = \frac{1}{T}\mathbf{J}_U - \frac{\mu_a}{T}\mathbf{J}_a - \frac{\mu_b}{T}\mathbf{J}_b. \tag{9.9}$$

Equation (9.9), which relates the entropy flux at an arbitrary location to fluxes of energy and mass, provides a starting point for analysis of transport in systems that are out of equilibrium.

Treating entropy as a transported quantity, the total rate of accumulation (substantive derivative) of entropy within the differential volume (fixed in the laboratory reference frame) is given by

$$\frac{D\tilde{S}}{Dt} = \frac{\partial \tilde{S}}{\partial t} + \nabla \cdot \mathbf{J}_S. \tag{9.10}$$

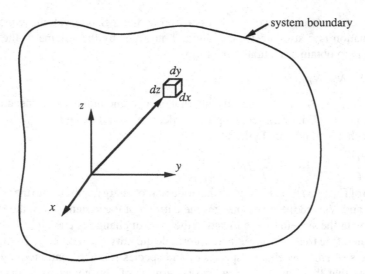

Figure 9.2

Conservation of energy and of each species similarly require that

$$\frac{D\tilde{U}}{Dt} = \frac{\partial \tilde{U}}{\partial t} + \nabla \cdot \mathbf{J}_U = 0, \tag{9.11}$$

$$\frac{D\tilde{N}_a}{Dt} = \frac{\partial \tilde{N}_a}{\partial t} + \nabla \cdot \mathbf{J}_a = 0, \tag{9.12}$$

$$\frac{D\tilde{N}_b}{Dt} = \frac{\partial \tilde{N}_b}{\partial t} + \nabla \cdot \mathbf{J}_b = 0. \tag{9.13}$$

Taking the divergence of Eq. (9.9), substituting it into (9.10), and using Eqs. (9.11)–(9.13) yields

$$\frac{D\tilde{S}}{Dt} = \nabla\left(\frac{1}{T}\right) \cdot \mathbf{J}_U - \nabla\left(\frac{\mu_a}{T}\right) \cdot \mathbf{J}_a - \nabla\left(\frac{\mu_b}{T}\right) \cdot \mathbf{J}_b. \tag{9.14}$$

Expanding the vector notation in Cartesian coordinates, we can write Eq. (9.14) as

$$\frac{D\tilde{S}}{Dt} = \frac{\partial}{\partial x}\left(\frac{1}{T}\right) J_{U,x} + \frac{\partial}{\partial y}\left(\frac{1}{T}\right) J_{U,y} + \frac{\partial}{\partial z}\left(\frac{1}{T}\right) J_{U,z}$$

$$- \frac{\partial}{\partial x}\left(\frac{\mu_a}{T}\right) J_{a,x} - \frac{\partial}{\partial y}\left(\frac{\mu_a}{T}\right) J_{a,y} - \frac{\partial}{\partial z}\left(\frac{\mu_a}{T}\right) J_{a,z}$$

$$- \frac{\partial}{\partial x}\left(\frac{\mu_b}{T}\right) J_{b,x} - \frac{\partial}{\partial y}\left(\frac{\mu_b}{T}\right) J_{b,y} - \frac{\partial}{\partial z}\left(\frac{\mu_b}{T}\right) J_{b,z}. \tag{9.15}$$

In Eq. (9.15), each term on the right side is the product of a gradient term and a corresponding flux term for one of the Cartesian coordinate directions. The prefactor for each flux term is referred to as an *affinity*. Each affinity can be viewed as a driving force for the corresponding flux. The reasons for this follow from equilibrium thermodynamics. If adjacent regions of the system are at equal temperature, there is no temperature gradient and the affinity terms for temperature are zero. Since the temperatures in adjacent regions are the same, they are thermally in equilibrium and no net flux of internal energy will occur. Hence, the flux of internal energy is directly associated with nonzero values of the affinity factors. Similar reasoning applies to the terms involving chemical potential and mass flux. The total rate of entropy generation is thus seen to be the sum of terms, each of which is the product of an affinity and a flux. This can be written more generically for our binary mixture system as

$$\frac{D\tilde{S}}{Dt} = \sum_{i=1}^{9} a_i J_i,$$
(9.16)

where the J_i are three energy fluxes and six mass fluxes associated with the orthogonal Cartesian coordinate and a_i is the corresponding affinity for flux J_i.

Evaluation of the rate of entropy generation within the differential volume obviously requires that we determine the fluxes of energy and each species at the point of interest. The relation for the flux depends on the nature of the molecules and their state in the system. From a macroscopic viewpoint, each flux may, in general, be expected to be a function of the system state variables P, T, μ_a, and μ_b and the affinities a_i. In fact, in most real systems the fluxes are, to a good level of accuracy, functions of the affinities alone:

$$J_i = J_i(a_1, a_2, \ldots, a_9).$$
(9.17)

Since each affinity is viewed as a driving force for the corresponding flux, we expect that each flux vanishes if the affinities go to zero. One approach to generating a functional form for the fluxes would be to develop a Taylor series expansion for each flux J_i:

$$J_i = \sum_{j=1}^{9} L_{ij} a_j + \frac{1}{2} \sum_{j=1}^{9} \sum_{k=1}^{9} L_{ijk} a_j a_k + \cdots,$$
(9.18)

where

$$L_{ij} = \frac{\partial J_i}{\partial a_j},$$
(9.19a)

$$L_{ijk} = \frac{\partial^2 J_i}{\partial a_j \partial a_k}.$$
(9.19b)

Fortunately, in many systems of practical and scientific interest, the system behavior is well approximated by a linear relation between fluxes and affinities. Linear systems are considered in more detail in the next section.

9.3 Analysis of Linear Systems

Treating the binary mixture system considered in the previous section as a linear system, we retain only the linear terms in Eq. (9.18):

$$J_i = \sum_{j=1}^{9} L_{ij} a_j.$$
(9.20)

Although the above linear relation may be applicable to a variety of different system types, here we will specifically consider a system in which energy and each species are transported only in the x direction. Replacing each affinity term a_j with the equivalent gradient term implied by the equivalence of Eqs. (9.15) and (9.16), and taking J_i as $-J_{a,x}$ and $-J_{b,x}$ for species a and b, respectively, we can reduce the linear flux relations to

$$J_{U,x} = L_{UU}\frac{\partial}{\partial x}\left(\frac{1}{T}\right) + L_{Ua}\frac{\partial}{\partial x}\left(\frac{\mu_a}{T}\right) + L_{Ub}\frac{\partial}{\partial x}\left(\frac{\mu_b}{T}\right), \tag{9.21}$$

$$-J_{a,x} = L_{aU}\frac{\partial}{\partial x}\left(\frac{1}{T}\right) + L_{aa}\frac{\partial}{\partial x}\left(\frac{\mu_a}{T}\right) + L_{ab}\frac{\partial}{\partial x}\left(\frac{\mu_b}{T}\right), \tag{9.22}$$

$$-J_{b,x} = L_{bU}\frac{\partial}{\partial x}\left(\frac{1}{T}\right) + L_{ba}\frac{\partial}{\partial x}\left(\frac{\mu_a}{T}\right) + L_{bb}\frac{\partial}{\partial x}\left(\frac{\mu_b}{T}\right). \tag{9.23}$$

In Eq. (9.21), $J_{U,x}$ is the flux of total internal energy in the material. However, we are often more interested in heat flow. By analogy with Eq. (3.86) for quasisteady heat exchange in a system of fixed volume,

$$\delta Q = T dS, \tag{3.86}$$

we define a heat flux $J_{Q,x}$ by the relation

$$J_{Q,x} = T J_{S,x}. \tag{9.24}$$

The x-direction component of the vector Eq. (9.9) indicates that

$$J_{S,x} = \frac{1}{T}J_{U,x} - \frac{\mu_a}{T}J_{a,x} - \frac{\mu_b}{T}J_{b,x}. \tag{9.25}$$

Combining Eqs. (9.24) and (9.25) yields

$$J_{Q,x} = J_{U,x} - \mu_a J_{a,x} - \mu_b J_{b,x}. \tag{9.26}$$

Substituting the right sides of Eqs. (9.21), (9.22), and (9.23) for $J_{U,x}$, $J_{a,x}$, and $J_{b,x}$ in Eq. (9.26), the resulting relation and Eqs. (9.22) and (9.23) can be written as

$$J_{Q,x} = L_{QQ}\frac{\partial}{\partial x}\left(\frac{1}{T}\right) + L_{Qa}\left(\frac{1}{T}\right)\frac{\partial\mu_a}{\partial x} + L_{Qb}\left(\frac{1}{T}\right)\frac{\partial\mu_b}{\partial x}, \tag{9.27}$$

$$-J_{a,x} = L_{aQ}\frac{\partial}{\partial x}\left(\frac{1}{T}\right) + L_{aa}\left(\frac{1}{T}\right)\frac{\partial\mu_a}{\partial x} + L_{ab}\left(\frac{1}{T}\right)\frac{\partial\mu_b}{\partial x}, \tag{9.28}$$

$$-J_{b,x} = L_{bQ}\frac{\partial}{\partial x}\left(\frac{1}{T}\right) + L_{ba}\left(\frac{1}{T}\right)\frac{\partial\mu_a}{\partial x} + L_{bb}\left(\frac{1}{T}\right)\frac{\partial\mu_b}{\partial x}, \tag{9.29}$$

where

$$\begin{aligned} L_{QQ} &= L_{UU} + \mu_a(L_{aU} + L_{Ua}) + \mu_b(L_{bU} + L_{Ub}) \\ &\quad + \mu_a\mu_b(L_{ba} + L_{ab}) + \mu_a^2 L_{aa} + \mu_b^2 L_{bb}, \end{aligned} \tag{9.30}$$

$$L_{Qa} = L_{Ua} + \mu_a L_{aa} + \mu_b L_{ba}, \tag{9.31}$$

$$L_{aQ} = L_{aU} + \mu_a L_{aa} + \mu_b L_{ab}, \tag{9.32}$$

$$L_{Qb} = L_{Ub} + \mu_b L_{bb} + \mu_a L_{ab}, \tag{9.33}$$

$$L_{bQ} = L_{bU} + \mu_b L_{bb} + \mu_a L_{ba}. \tag{9.34}$$

Note that the affinities for the fluxes $J_{Q,x}$, $-J_{a,x}$, and $-J_{b,x}$ are now different than for the original internal energy and mass fluxes. In this formulation of the analysis they are $\nabla(1/T)$, $(1/T)\nabla\mu_a$, and $(1/T)\nabla\mu_b$.

Although linear, the above flux relations (9.27)–(9.29) are awkwardly cast in terms of chemical potential gradients. Generally, however, it is often possible to express the chemical potential as a function of other properties, making it possible to recast the relation in a more useful form. For example, if we consider a binary mixture of van der Waal fluids, the relations for the chemical can be written in the forms

$$
\frac{\mu_a}{k_B T} = \frac{(\rho_a + \rho_b)b_{v,a}}{1 - \rho_a b_{v,a} - \rho_b b_{v,b}} - \left(\frac{3}{2}\right) \ln\left[\frac{2\pi M_a k_B T (1 - \rho_a b_{v,a} - \rho_b b_{v,b})^{2/3}}{\rho_a^{2/3} h^2}\right]
$$
$$
- \left[\frac{(\xi_a - 5)}{2} \ln\pi - \ln\sigma_{s,a}\right] - \frac{(\xi_a - 3)}{2} \ln\left(\frac{T}{(\theta_{rot}, m)_a}\right)
$$
$$
- \frac{2a_{v,aa}\rho_a + 2a_{v,ab}\rho_b}{k_B T}, \tag{9.35}
$$

$$
\frac{\mu_b}{k_B T} = \frac{(\rho_a + \rho_b)b_{v,b}}{1 - \rho_a b_{v,a} - \rho_b b_{v,b}} - \left(\frac{3}{2}\right) \ln\left[\frac{2\pi M_b k_B T (1 - \rho_a b_{v,a} - \rho_b b_{v,b})^{2/3}}{\rho_b^{2/3} h^2}\right]
$$
$$
- \left[\frac{(\xi_b - 5)}{2} \ln\pi - \ln\sigma_{s,b}\right] - \frac{(\xi_b - 3)}{2} \ln\left(\frac{T}{(\theta_{rot}, m)_b}\right)
$$
$$
- \frac{2a_{v,bb}\rho_b + 2a_{v,ab}\rho_a}{k_B T}, \tag{9.36}
$$

where ρ_a and ρ_b are molecular number densities of each species defined as

$$
\rho_a = N_a/V, \qquad \rho_b = N_b/V. \tag{9.37}
$$

These relations indicate that μ_a and μ_b are functions of ρ_a, ρ_b, and T. We can therefore use the chain rule to reorganize the flux relations to the forms

$$
J_{Q,x} = \left[-\frac{L_{QQ}}{T^2} + \frac{L_{Qa}}{T}\frac{\partial\mu_a}{\partial T} + \frac{L_{Qb}}{T}\frac{\partial\mu_b}{\partial T}\right]\frac{\partial T}{\partial x} + \left[\frac{L_{Qa}}{T}\frac{\partial\mu_a}{\partial\rho_a} + \frac{L_{Qb}}{T}\frac{\partial\mu_b}{\partial\rho_a}\right]\frac{\partial\rho_a}{\partial x}
$$
$$
+ \left[\frac{L_{Qa}}{T}\frac{\partial\mu_a}{\partial\rho_b} + \frac{L_{Qb}}{T}\frac{\partial\mu_b}{\partial\rho_b}\right]\frac{\partial\rho_b}{\partial x}, \tag{9.38}
$$

$$
-J_{a,x} = \left[-\frac{L_{aQ}}{T^2} + \frac{L_{aa}}{T}\frac{\partial\mu_a}{\partial T} + \frac{L_{ab}}{T}\frac{\partial\mu_b}{\partial T}\right]\frac{\partial T}{\partial x} + \left[\frac{L_{aa}}{T}\frac{\partial\mu_a}{\partial\rho_a} + \frac{L_{ab}}{T}\frac{\partial\mu_b}{\partial\rho_a}\right]\frac{\partial\rho_a}{\partial x}
$$
$$
+ \left[\frac{L_{aa}}{T}\frac{\partial\mu_a}{\partial\rho_b} + \frac{L_{ab}}{T}\frac{\partial\mu_b}{\partial\rho_b}\right]\frac{\partial\rho_b}{\partial x}, \tag{9.39}
$$

$$
-J_{b,x} = \left[-\frac{L_{bQ}}{T^2} + \frac{L_{bb}}{T}\frac{\partial\mu_b}{\partial T} + \frac{L_{ba}}{T}\frac{\partial\mu_a}{\partial T}\right]\frac{\partial T}{\partial x} + \left[\frac{L_{bb}}{T}\frac{\partial\mu_b}{\partial\rho_a} + \frac{L_{ba}}{T}\frac{\partial\mu_a}{\partial\rho_a}\right]\frac{\partial\rho_a}{\partial x}
$$
$$
+ \left[\frac{L_{bb}}{T}\frac{\partial\mu_b}{\partial\rho_b} + \frac{L_{ba}}{T}\frac{\partial\mu_a}{\partial\rho_b}\right]\frac{\partial\rho_b}{\partial x}. \tag{9.40}
$$

The form of these relations is consistent with those associated with Fourier heat conduction, which links heat flux to temperature gradient, and Fick's law, which links mass diffusion flux to concentration or density gradients. They indicate, however, that contributions to the heat flux may result from a concentration gradient as well as a temperature gradient. Transport of thermal energy associated with mass concentration gradients is sometimes referred to as the *diffusion thermo effect* or the *Dufour effect*. The above relations also indicate that contributions to the mass flux can result from a temperature gradient as well as a concentration gradient. Transport of mass induced by a temperature gradient is sometimes termed the *thermal diffusion effect* or the *Soret effect*.

For linear systems, the parameters L_{ij} in Eq. (9.20) play a central role in the linkage between property gradients and fluxes of mass and energy in a system. These parameters are called *kinetic coefficients*. In the next two sections we will explore the nature of these coefficients in more detail.

9.4   Fluctuations and Correlation Moments

An important characteristic of the kinetic coefficients can be understood by considering the behavior of fluctuations in the system. Specifically, we are interested in the microscale fluctuations in a system containing two species a and b that is in contact with a reservoir that holds the temperature and chemical potential of each species in the system constant. These intensive properties are fixed for the system, while the extensive variables U, N_a, and N_b may fluctuate because of exchanges of mass and energy with the reservoir in contact with the system. In our analysis, the volume of the system is assumed to be fixed (although the analysis can be extended to also consider changes in volume). The system considered here is equivalent to a member system in a grand canonical ensemble (see Chapter 4). More importantly, this system is equivalent to a control volume within a much larger system. The fluctuations within this type of system are therefore equivalent to the local fluctuations in a control volume in a large system at equilibrium. For an arbitrary thermodynamic property Y, in our analysis here we will denote the instantaneous values as Y'. The mean value of the property, which we take to be equivalent to the thermodynamic property at equilibrium, we will designate simply as Y.

For the type of system of interest here, at equilibrium, the fundamental equation relates the entropy to the volume and the mean values of U, N_a, and N_b:

$$S = S(V, U, N_a, N_a). \tag{9.41}$$

When U, N_a, and N_b fluctuate, their instantaneous values will generally differ from their respective mean values. The value of entropy computed for the system with the fundamental equation using instantaneous values of U, N_a, and N_b we will interpret as the instantaneous entropy of the system S':

$$S' = S\left(V, U', N_a', N_b'\right) \tag{9.42}$$

By definition (see Chapter 2), the entropy is related to the number of microstates W for the system as

$$S = k_B \ln W. \tag{9.43}$$

This relation holds at equilibrium, and we can also use it to determine the number of microstates W' that correspond to the instantaneous entropy for a particular set of fluctuating U', N_a', and N_b' values:

$$S' = k_B \ln W'. \tag{9.44}$$

Combining Eqs. (9.43) and (9.44), we can obtain the following relation:

$$W' = W e^{\Delta S / k_B}, \tag{9.45}$$

where W is the number of microstates for the equilibrium condition at the mean values of U, N_a, and N_b and

$$\Delta S = S' - S(V, U, N_a, N_b). \tag{9.46}$$

Since we are interested in systems that behave classically, the properties U', N_a', and N_b' are taken to be continuous and we use Eq. (9.45) to construct a probability density function $\tilde{p}$ defined such that $\tilde{p} \, dU' dN_a' dN_b'$ is the probability of a fluctuation with combinations of these variables that lie within intervals U' to $U' + dU'$, N_a' to $N_a' + dN_a'$, and N_b' to $N_b' + dN_b'$. This probability must be proportional to the number of microstates W' and must be properly normalized. We therefore define the probability density function as

$$\tilde{p} = \frac{W'}{\int_0^\infty \int_0^\infty \int_0^\infty W' dU' dN_a' dN_b'} = \frac{W e^{\Delta S / k_B}}{\int_0^\infty \int_0^\infty \int_0^\infty W e^{\Delta S / k_B} dU' dN_a' dN_b'}. \tag{9.47}$$

Since W is a constant for the equilibrium condition about which the fluctuations occur, this can be simplified to

$$\tilde{p} = \tilde{p}_0 e^{\Delta S / k_B}, \tag{9.48}$$

where $\tilde{p}_0$ is given by

$$\tilde{p}_0 = \frac{1}{\int_0^\infty \int_0^\infty \int_0^\infty e^{\Delta S / k_B} dU' dN_a' dN_b'}. \tag{9.49}$$

Note that if $\tilde{p}$ is integrated over all possible U', N_a', and N_b', the result is unity, as is necessary for a normalized distribution function.

We now will use Eq. (9.48) to explore correlations among fluctuations. For small fluctuations in the extensive properties U, N_a, and N_b about the equilibrium (mean) values, the resulting fluctuating entropy can be represented as a Taylor series expansion about the equilibrium state values. For simplicity in notation, we set $Y_1 = U$, $Y_2 = N_a$, and $Y_3 = N_b$. The Taylor series expansion can then be written as

$$S' = S(V, Y_1, Y_2, Y_3) + \sum_{i=1}^{3} \left(\frac{\partial S'}{\partial Y_i'} \right)_0 (Y_i' - Y_i)$$

$$+ \frac{1}{2} \sum_{i=1}^{3} \sum_{j=1}^{3} \left(\frac{\partial^2 S'}{\partial Y_i' \partial Y_j'} \right)_0 (Y_i' - Y_i)(Y_j' - Y_j) + \cdots, \tag{9.50}$$

where the zero subscript implies that the derivatives are evaluated where $Y_i' - Y_i = 0$ for all i. The fluctuating entropy must be a maximum at equilibrium to ensure that fluctuations away from equilibrium tend to be restored by spontaneous processes that tend to increase entropy. This is a necessary condition for the equilibrium to be stable. Because the fluctuating entropy exhibits a maximum at equilibrium, it follows directly that the first derivatives in the expansion above must be zero. The expansion therefore simplifies to

$$S' = S(V, Y_1, Y_2, Y_3) + \frac{1}{2}\sum_{i=1}^{3}\sum_{j=1}^{3}\left(\frac{\partial^2 S'}{\partial Y_i' \partial Y_j'}\right)_0 (Y_i' - Y_i)(Y_j' - Y_j) + \cdots.$$

(9.51)

Since the second derivatives in the double summation above are evaluated at the equilibrium condition and are nonzero, we take them to be equal to the corresponding second derivative of S with respect to equilibrium values of extensive properties. The above relation then becomes

$$S' = S(V, Y_1, Y_2, Y_3) + \frac{1}{2}\sum_{i=1}^{3}\sum_{j=1}^{3}\left(\frac{\partial^2 S}{\partial Y_i \partial Y_j}\right)(Y_i' - Y_i)(Y_j' - Y_j) + \cdots.$$

(9.52)

For convenience, we next introduce the following variables:

$$\Delta S = S' - S(V, U, N_a, N_b),$$

(9.53)

$$\varepsilon_i = Y_i' - Y_i.$$

(9.54)

With some simple rearrangement, Eq. (9.52) can be expressed in terms of these variables as

$$\Delta S = \frac{1}{2}\sum_{i=1}^{3}\sum_{j=1}^{3}\left(\frac{\partial^2 S}{\partial Y_i \partial Y_j}\right)\varepsilon_i\varepsilon_j + \cdots.$$

(9.55)

If we now truncate the above series and insert the right side into Eq. (9.48) for ΔS, we obtain the following relation for the probability density function for fluctuations with extensive property deviations $Y_1' - Y_1 = \varepsilon_1$, $Y_2' - Y_2 = \varepsilon_2$, and $Y_3' - Y_3 = \varepsilon_3$:

$$\tilde{p} = \tilde{p}_0 \exp\left\{\frac{1}{2}\sum_{i=1}^{3}\sum_{j=1}^{3}\left(\frac{\partial^2 S}{\partial Y_i \partial Y_j}\right)\frac{\varepsilon_i\varepsilon_j}{k_B}\right\},$$

(9.56)

where

$$\tilde{p}_0 = \frac{1}{\int_{-\infty}^{\infty}\int_{-\infty}^{\infty}\int_{-\infty}^{\infty} e^{\Delta S/k_B}\, d\varepsilon_1 d\varepsilon_2 d\varepsilon_3}.$$

(9.57)

Note that by changing the variables of integration in Eq. (9.57), the limits of integration are changed to span all possible fluctuations. Equation (9.56) is sometimes called the *Einstein fluctuation formula*. This relation reflects the fact that, to first order, ΔS is a quadratic function of the deviation of extensive properties from their equilibrium values. Furthermore, because the fluctuating entropy is a maximum for zero deviation from the equilibrium state, the second derivatives in the double sum must be negative. It follows that ΔS is zero at $\varepsilon_1 = \varepsilon_2 = \varepsilon_3 = 0$ and becomes increasingly negative as ε_1, ε_2, and ε_3 deviate from zero. As ε_1, ε_2, and ε_3 approach $\pm\infty$, $\Delta S \to -\infty$. Einstein's fluctuation formula indicates that the probability of fluctuations approaches zero as the absolute values of ε_1, ε_2, and ε_3 become large.

For each of the extensive variables associated with the equilibrium state there is a driving potential associated with changes in that variable: $1/T$ for U, μ_a/T, for N_a and μ_b/T for N_b. For each external property Y_i the corresponding potential is $(\partial S/\partial Y_i)_{Y_{j\neq i}}$. We now will extend this definition to the fluctuating properties we have defined here as follows: We define a driving potential ϕ_i' for fluctuating external property Y_i' as

$$\phi_i' = \left(\partial S'/\partial Y_i'\right)_{Y_{j\neq i}'}. \tag{9.58}$$

Since $(\partial S'/\partial Y_i')_{Y_{j\neq i}'}$ is a function of the external properties Y_i, we construct a Taylor series expansion for ϕ_i' about the equilibrium state:

$$\phi_i' = (\phi_i')_0 + \sum_{j=1}^{3}\left(\frac{\partial \phi_i'}{\partial Y_j'}\right)_0 (Y_j' - Y_j) + \cdots, \tag{9.59}$$

where the zero subscript again indicates that the derivatives are evaluated at the equilibrium state. Defining α_i' as the deviation of ϕ_i' from its value at the equilibrium state,

$$\alpha_i' = \phi_i' - (\phi_i')_0, \tag{9.60}$$

it follows that

$$\alpha_i' = \left(\frac{\partial S'}{\partial Y_i'}\right) - \left(\frac{\partial S'}{\partial Y_i'}\right)_0, \tag{9.61}$$

where the zero subscript denotes the value of the derivative in the limit $Y_i' - Y_i \to 0$. Since $(\partial S'/\partial Y_i')_0 = 0$, this reduces to

$$\alpha_i' = \left(\frac{\partial S'}{\partial Y_i'}\right). \tag{9.62}$$

Also, since S is a constant for the equilibrium state about which the fluctuations occur, Eq. (9.53) implies that

$$\left(\frac{\partial S'}{\partial Y_i'}\right) = \left(\frac{\partial \Delta S}{\partial Y_i'}\right), \tag{9.63}$$

and since Eq. (9.53) indicates that $\varepsilon_i = Y_i' - Y_i$, the above relation can be written

$$\left(\frac{\partial S'}{\partial Y_i'}\right) = \left(\frac{\partial \Delta S}{\partial \varepsilon_i}\right). \tag{9.64}$$

Combining Eqs. (9.62) and (9.64) we find that

$$\alpha_i' = \frac{\partial \Delta S}{\partial \varepsilon_i}. \tag{9.65}$$

Note that the definition of α_i' is such that

$$\text{if } Y_i = U \quad \text{then } \alpha_i' = \frac{1}{T'} - \frac{1}{T}, \tag{9.66}$$

$$\text{if } Y_i = N_a \quad \text{then } \alpha_i' = \left(\frac{\mu_a}{T}\right)' - \frac{\mu_a}{T}, \tag{9.67}$$

and

$$\text{if } Y_i = N_b \quad \text{then } \alpha_i' = \left(\frac{\mu_b}{T}\right)' - \frac{\mu_b}{T}. \tag{9.68}$$

Before proceeding further a comment regarding fluctuation of intensive and extensive properties is in order. In our development of statistical thermodynamics we considered the grand canonical ensemble in which each member system is in contact with reservoirs (or other member systems) that exchange mass and energy in such a way that the temperature and chemical potential for each species are the same for all systems. In this ensemble, the extensive variables U, N_a, and N_b fluctuate (vary among the ensemble members) about the mean value, but within this statistical model the conjugate driving potentials $1/T$, μ_a/T, and μ_b/T cannot vary. Real systems are not subject to the restrictions of this statistical model, however. For example, sensors that measure temperatures operate by reaching thermal equilibrium with the surrounding portion of a system. If the energy of the system fluctuates, the sensor output will usually exhibit a corresponding fluctuation in temperature.

We therefore interpret fluctuations of the conjugate driving potentials $1/T$, μ_a/T, and μ_b/T in the following way: In the control volume of fixed volume considered here, fluctuations in U, N_a, and N_b with interacting reservoirs cause the state to fluctuate and the state variation for the system can be parameterized either in terms of fluctuations in the external properties U, N_a, and N_b or in terms of the changes in the driving potentials that result from fluctuations in U, N_a, and N_b.

The correlation moment of an extensive variable fluctuation ε_i and a driving potential fluctuation α'_j can be computed as

$$\langle \varepsilon_i \alpha'_j \rangle = \int_{-\infty}^{\infty} \int_{-\infty}^{\infty} \int_{-\infty}^{\infty} \varepsilon_i \alpha'_j \, \tilde{p} \, d\varepsilon_1 d\varepsilon_2 d\varepsilon_3, \tag{9.69}$$

where $\tilde{p}$ is the probability density function defined by Eq. (9.48). Substituting for α_j and $\tilde{p}$ using Eqs. (9.65) and (9.48) gives

$$\langle \varepsilon_i \alpha'_j \rangle = \int_{-\infty}^{\infty} \int_{-\infty}^{\infty} \int_{-\infty}^{\infty} \varepsilon_i \left(\frac{\partial \Delta S}{\partial \varepsilon_j} \right) \tilde{p}_0 e^{\Delta S/k_B} \, d\varepsilon_1 d\varepsilon_2 d\varepsilon_3. \tag{9.70}$$

With some straightforward manipulation, this relation can be rearranged to the form

$$\langle \varepsilon_i \alpha'_j \rangle = k_B \tilde{p}_0 \int_{-\infty}^{\infty} \int_{-\infty}^{\infty} \int_{-\infty}^{\infty} \varepsilon_i \frac{\partial}{\partial \varepsilon_j} (e^{\Delta S/k_B}) \, d\varepsilon_1 d\varepsilon_2 d\varepsilon_3. \tag{9.71}$$

We now consider two possibilities: $i = j$ and $i \neq j$. First, for $i = j$ we can reorganize Eq. (9.71) in the form

$$\langle \varepsilon_i \alpha'_j \rangle = k_B \tilde{p}_0 \int_{-\infty}^{\infty} \int_{-\infty}^{\infty} \int_{-\infty}^{\infty} \frac{\partial}{\partial \varepsilon_i} (\varepsilon_i e^{\Delta S/k_B}) \, d\varepsilon_1 d\varepsilon_2 d\varepsilon_3$$

$$- k_B \tilde{p}_0 \int_{-\infty}^{\infty} \int_{-\infty}^{\infty} \int_{-\infty}^{\infty} e^{\Delta S/k_B} \, d\varepsilon_1 d\varepsilon_2 d\varepsilon_3. \tag{9.72}$$

From Eq. (9.57) it follows that the second term in Eq. (9.72) equals $-k_B$. In the first term, i must equal 1, 2, or 3, and the integration with respect to ε_i can be done first to obtain

$$\langle \varepsilon_i \alpha'_j \rangle = k_B \tilde{p}_0 \int_{-\infty}^{\infty} \int_{-\infty}^{\infty} [\varepsilon_i e^{\Delta S/k_B}]_{-\infty}^{\infty} \frac{d\varepsilon_1 d\varepsilon_2 d\varepsilon_3}{d\varepsilon_i} - k_B. \tag{9.73}$$

As discussed above, ΔS approaches negative infinity as ε_i approaches $\pm\infty$, and therefore the term in square brackets is zero at both integration limits. The first term in Eq. (9.73) is therefore zero and for $i = j$, the correlation moment reduces to the simple result

$$\langle \varepsilon_i \alpha_j' \rangle = -k_B. \tag{9.74}$$

For $i \neq j$, the integration on the right side of Eq. (9.71) can be done first with respect to ε_j to obtain

$$\langle \varepsilon_i \alpha_j' \rangle = k_B \tilde{p}_0 \int_{-\infty}^{\infty} \int_{-\infty}^{\infty} \varepsilon_i [e^{\Delta S / k_B}]_{-\infty}^{\infty} \frac{d\varepsilon_1 d\varepsilon_2 d\varepsilon_3}{d\varepsilon_j}. \tag{9.75}$$

Here again, ΔS approaches negative infinity as ε_j approaches $\pm\infty$, and therefore the term in square brackets is zero at both integration limits. Thus, for $i \neq j$, $\langle \varepsilon_i \alpha_j' \rangle = 0$. We can summarize the results for both cases as

$$\langle \varepsilon_i \alpha_j' \rangle = \begin{cases} -k_B & \text{for } i = j, \\ 0 & \text{for } i \neq j. \end{cases} \tag{9.76}$$

These results imply that fluctuations in extensive variables are correlated with fluctuations in the corresponding driving potential, but not with fluctuations in other driving potentials. As will be shown in the next section, this impacts the interrelationship among kinetic coefficients.

9.5   Onsager Reciprocity of Kinetic Coefficients

In classical thermodynamics, the extensive properties for a system in contact with a reservoir for that property are taken to be fixed constants. As noted above, this is not strictly true. If energy or mass is permitted to flow freely between a system and a reservoir, at a microscopic level the exchange will produce fluctuations in the property. In large systems, these fluctuations are small and rapid and the value of the property stays close to the average for the system.

On rare occasions a large fluctuation may occur, which, for example, might alter the energy in the system by a nonnegligible amount. If the system were decoupled from the reservoir just after such a fluctuation, the energy and temperature of the system would be at least locally different. If the system were not decoupled, the fluctuation would decay by the spontaneous flow of energy between the reservoir and the system. Onsager [1] realized that this spontaneous flow is in some sense driven by the temperature difference generated by the original fluctuation. He connected macroscopic processes to nonequilibrium thermodynamics by assuming that the decay of a spontaneous fluctuation is identical to the macroscopic flow of energy or mass between a reservoir and a system that are not in equilibrium.

The usefulness of Onsager's assumption can be seen by considering the following analysis. We consider again the constant-volume control volume system discussed in the previous section. The system is in contact with a reservoir with which it can exchange energy and particles of species a and b. For convenience, we will continue to use the notation defined in the previous section. In the previous section we discussed correlation moments. For two extensive properties Y_i and Y_j, the correlation moment is defined as

$$\langle \varepsilon_i \varepsilon_j \rangle = \int_{-\infty}^{\infty} \int_{-\infty}^{\infty} \int_{-\infty}^{\infty} \varepsilon_i \varepsilon_j \tilde{p} \, d\varepsilon_1 d\varepsilon_2 d\varepsilon_3, \tag{9.77}$$

where $\tilde{p}$ is the probability density function defined by Eq. (9.48). An extension of this concept is the delayed correlation moment $\langle \varepsilon_i \varepsilon_j(\tau) \rangle$, which is the average product of the deviations ε_i and ε_j, with ε_j being observed a time τ after ε_i is observed. In the absence of magnetic field effects, the physical laws governing the motion of the particles are assumed to be symmetric with respect to the reversal of time. This implies that in the absence of a magnetic field, the delayed correlation moment must be unchanged by the replacement of τ by $-\tau$:

$$\langle \varepsilon_i \varepsilon_j(\tau) \rangle = \langle \varepsilon_i \varepsilon_j(-\tau) \rangle. \tag{9.78}$$

Because only the relative time interval between the samples is important, the right side is equivalent to $\langle \varepsilon_i(\tau) \varepsilon_j \rangle$. It follows that

$$\langle \varepsilon_i \varepsilon_j(\tau) \rangle = \langle \varepsilon_i(\tau) \varepsilon_j \rangle. \tag{9.79}$$

Subtracting $\langle \varepsilon_i \varepsilon_j \rangle$ from both sides of Eq. (9.79) and dividing by τ, we obtain

$$\left\langle \varepsilon_i \frac{\varepsilon_j(\tau) - \varepsilon_j}{\tau} \right\rangle = \left\langle \frac{\varepsilon_j(\tau) - \varepsilon_i}{\tau} \varepsilon_j \right\rangle. \tag{9.80}$$

This relation must hold for all τ, and we therefore expect it to be valid in the limit $\tau \to 0$. Taking this limit, we can write Eq. (9.80) in terms of time derivatives as

$$\langle \varepsilon_i \dot{\varepsilon}_j \rangle = \langle \dot{\varepsilon}_i \varepsilon_j \rangle. \tag{9.81}$$

Consistent with the discussion above, we assume that the decay of a fluctuation $\dot{\varepsilon}_i$ is governed by the same linear laws as macroscopic processes. We can then relate $\dot{\varepsilon}_i$ and $\dot{\varepsilon}_j$ to fluctuations of thermodynamic potentials α_l' as

$$\dot{\varepsilon}_i = \sum_{l=1}^{3} L_{li} \alpha_l', \tag{9.82}$$

$$\dot{\varepsilon}_j = \sum_{l=1}^{3} L_{lj} \alpha_l'. \tag{9.83}$$

Using Eqs. (9.82) and (9.83) to replace $\dot{\varepsilon}_i$ and $\dot{\varepsilon}_j$ in (9.81) and interchanging the average and summation operations yields

$$\sum_{l=1}^{3} L_{lj} \langle \varepsilon_i \alpha_l' \rangle = \sum_{l=1}^{3} L_{li} \langle \alpha_l' \varepsilon_j \rangle. \tag{9.84}$$

Equation (9.76) indicates that the bracketed terms in the above equation are nonzero only if the indices are the same ($l = i$ or $l = j$). Using Eq. (9.76) to evaluate the terms in Eq. (9.84) leads to

$$L_{ij}(-k_B) = L_{ji}(-k_B),$$

which simplifies to

$$L_{ij} = L_{ji}. \tag{9.85}$$

The above relationship among the kinetic coefficients is known as *Onsager reciprocity*. In our analysis of this relationship we have considered a system with no electromagnetic effects. If a system with such effects in considered, it can be shown that a broader statement of this relation is that the value of kinetic coefficient L_{ij} measured in external magnetic field $\mathbf{B}_{\text{ext}}$ is equal to the value of L_{ji} measured in external magnetic field $-\mathbf{B}_{\text{ext}}$:

$$L_{ij}(\mathbf{B}_{\text{ext}}) = L_{ji}(-\mathbf{B}_{\text{ext}}). \tag{9.86}$$

(See, for example, Callen [2].)

Clearly, if there is no magnetic field, the statements of Onsager reciprocity (9.85) and (9.86) are equivalent.

These results imply that for the kinetic coefficients in Eqs. (9.21)–(9.23), $L_{Ua} = L_{aU}$, $L_{Ub} = L_{bU}$, and $L_{ab} = L_{ba}$. Inspection of Eqs. (9.30)–(9.39) reveals that the validity of these reciprocity relations for the mass and internal energy fluxes assures that reciprocity holds for the kinetic coefficients associated with the heat and mass fluxes in relations (9.27)–(9.29): $L_{Qa} = L_{aQ}$, $L_{Qb} = L_{bQ}$, and $L_{ab} = L_{ba}$. The analytical framework for nonequilibrium thermodynamics developed in this chapter is particularly useful for analyzing thermoelectric effects. We consider such effects in the next section.

9.6 Thermoelectric Effects

In this section we will consider a conductor in which one-dimensional flow of heat and electric current occur. The electric current is assumed to be a result of electron motion in the conductor. We therefore apply Eqs. (9.27) and (9.28), taking species a to be the electrons and neglecting the species b terms:

$$J_{Q,x} = L_{QQ}\frac{\partial}{\partial x}\left(\frac{1}{T}\right) + L_{Qa}\frac{\partial}{\partial x}\left(\frac{\mu_a}{T}\right), \tag{9.87}$$

$$-J_{a,x} = L_{aU}\frac{\partial}{\partial x}\left(\frac{1}{T}\right) + L_{aa}\frac{\partial}{\partial x}\left(\frac{\mu_a}{T}\right), \tag{9.88}$$

where

$$L_{QQ} = L_{UU} + \mu_a(L_{Ua} + L_{aU}) + \mu_a^2 L_{aa}, \tag{9.89}$$

$$L_{Qa} = L_{Ua} + \mu_a L_{aa}, \tag{9.90}$$

$$L_{aQ} = L_{aU} + \mu_a L_{aa}. \tag{9.91}$$

There are positive ions in the conductor, but they are assumed to be immobile and therefore do not contribute to the energy or current flow.

Electrical and Thermal Conductivities

In considering flow of electrons we break the overall chemical potential for the electrons into two parts: a chemical portion μ_{ch} and an electrical portion μ_e:

$$\mu_a = \mu_{\text{ch}} + \mu_e. \tag{9.92}$$

In the absence of a magnetic field, if the charge on an electron is e_e, then

$$\mu_e = e_e\phi_e, \tag{9.93}$$

where ϕ_e is the electrostatic potential. The chemical portion of the chemical potential is a function of the electron concentration and the temperature. The electrochemical potential per unit charge is μ_a/e_e. It follows that the gradient of the electrochemical potential per unit charge is the sum of two components

$$\nabla(\mu_a/e_e) = (1/e_e)\nabla\mu_e + (1/e_e)\nabla\mu_{ch}. \tag{9.94}$$

The first term on the right side of Eq. (9.94) is the gradient of the electrostatic potential, which is the electric field. The second term is a driving force arising from a concentration gradient.

The electrical conductivity σ_e is defined as the ratio of the electric current density $-e_e J_{a,x}$ to the potential gradient $\nabla(\mu_a/e_e)$ in a isothermal system:

$$\sigma_e \equiv \frac{-e_e J_{a,x}}{(1/e_e)(\partial\mu_a/\partial x)} \quad \text{for} \quad \frac{\partial T}{\partial x} = 0. \tag{9.95}$$

Substituting the right side of (9.88) for $-J_{a,x}$ for zero temperature gradient converts the above relation to

$$\sigma_e = \frac{e_e^2 L_{aa}}{T}. \tag{9.96}$$

The thermal conductivity is similarly defined as the heat flux per unit temperature gradient for zero electric current:

$$k_t \equiv \frac{-J_{Q,x}}{\partial T/\partial x} \quad \text{for} \quad J_{a,x} = 0. \tag{9.97}$$

Setting the right side of Eq. (9.88) to zero we obtain

$$\left(\frac{1}{T}\right)\frac{\partial\mu_a}{\partial x} = -\frac{L_{aQ}}{L_{aa}}\frac{\partial}{\partial x}\left(\frac{1}{T}\right). \tag{9.98}$$

Combining this equation with Eq. (9.87) to eliminate $(1/T)(\partial\mu_a/\partial x)$ and substituting the expression for $-J_{a,x}$ into Eq. (9.97) yields

$$k_t = \frac{L_{QQ}L_{aa} - L_{Qa}L_{aQ}}{L_{aa}T^2}. \tag{9.99}$$

The Seebeck Effect

The circuit composed of two different conducting materials shown schematically in Figure 9.3 is known as a thermocouple. Thermocouples constructed with two different metals are commonly used for measuring temperature. The junctions where the two dissimilar metals are connected are at different temperatures T_1 and T_2, and the impedance of the voltmeter is high enough that negligible current flows in the circuit. The terminals of the voltmeter are presumed to be at the same temperature T_m. The production of an electromotive force in a thermocouple under conditions of zero current flow is termed the *Seebeck effect*. Here we designate the two materials in the circuit as A and B.

Assuming that the chemical potential and temperature are only a function of distance along the conductors, the zero current requirement (9.97) can be rearranged to the form

$$\frac{d\mu_a}{dx} = \frac{L_{aQ}}{TL_{aa}}\frac{dT}{dx}. \tag{9.100}$$

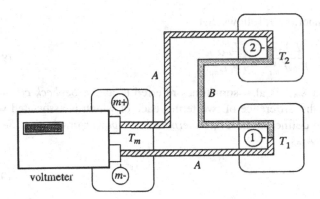

Figure 9.3

We can multiply through by dx and integrate this relation for each of the conductor segments in Figure 9.3 to obtain

$$\mu_{a,2} - \mu_{a,1} = \int_{T_1}^{T_2} \frac{L_{aQ}^A}{TL_{aa}^A} \, dT, \tag{9.101}$$

$$\mu_{a,2} - \mu_{a,m+} = \int_{T_m}^{T_2} \frac{L_{aQ}^B}{TL_{aa}^B} \, dT, \tag{9.102}$$

$$\mu_{a,1} - \mu_{a,m-} = \int_{T_m}^{T_1} \frac{L_{aQ}^B}{TL_{aa}^B} \, dT. \tag{9.103}$$

Combining these relation to eliminate $\mu_{a,1}$ and $\mu_{a,2}$, we obtain

$$\mu_{a,m+} + -\mu_{a,m-} = \int_{T_1}^{T_2} \left(\frac{L_{aQ}^A}{TL_{aa}^A} - \frac{L_{aQ}^B}{TL_{aa}^B} \right) dT. \tag{9.104}$$

Since the temperature is the same at both terminals of the voltmeter, the voltage difference measured by the meter is $\Delta\phi_m = (\mu_{a,m+} - \mu_{a,m-})/e_e$. Thus the voltage difference at the meter is given by

$$\Delta\phi_m = \int_{T_1}^{T_2} \left(\frac{L_{aQ}^A}{Te_e L_{aa}^A} - \frac{L_{aQ}^B}{Te_e L_{aa}^B} \right) dT. \tag{9.105}$$

This relation relates the voltage measured by the meter to the temperatures at each of the thermocouple junctions. If one junction is held at a reference condition, the second junction temperature can be inferred from the measured voltage. A common way of establishing a reference condition for one junction is to immerse it in an ice-water bath, which has a saturation temperature of $0°C$ at atmospheric pressure.

It is conventional to define the thermoelectric power of the thermocouple $\hat{\varepsilon}_{AB}$ as the change in voltage $\Delta\phi_m$ per unit change of temperature difference $T_2 - T_1$. The sign of $\hat{\varepsilon}_{AB}$ is chosen so that it is positive if the voltage change tends to drive current from material A

to material B at the hot junction. It follows that

$$\hat{\varepsilon}_{AB} = \frac{d\Delta\phi_m}{dT_2} = \frac{L_{aQ}^A}{Te_eL_{aa}^A} - \frac{L_{aQ}^B}{Te_eL_{aa}^B}.$$ (9.106)

The thermoelectric power $\hat{\varepsilon}_{AB}$ is also sometimes referred to as the *Seebeck coefficient*. Because $\hat{\varepsilon}_{AB}$ is equal to the difference of two terms, each of which is associated with a specific material, we also define the *absolute thermoelectric power* or *absolute Seebeck coefficient* of a substance A as

$$\hat{\varepsilon}_A = \frac{-L_{aQ}^A}{Te_eL_{aa}^A}.$$ (9.107)

The thermoelectric power for the thermocouple is then given by

$$\hat{\varepsilon}_{AB} = \hat{\varepsilon}_B - \hat{\varepsilon}_A.$$ (9.108)

Note that if the variation of the thermoelectric absolute powers with temperature is known and T_1 is fixed, the variation of thermocouple voltage with temperature can be computed as

$$\Delta\phi_m = \int_{T_1}^{T_2} (\hat{\varepsilon}_B - \hat{\varepsilon}_A)\, dT.$$ (9.109)

This is the usual way that the voltage output is determined for thermocouples having different combinations of materials.

The original transport Eqs. (9.87) and (9.88) contained three kinetic coefficients L_{QQ}, L_{aa}, and $L_{Qa} = L_{aQ}$. If we know these three parameters we can calculate the electrical conductivity, the thermal conductivity, and the thermoelectric absolute power using Eqs. (9.96), (9.99), and (9.107). Moreover, the reverse is also true. We can use these equations to calculate the three kinetic coefficients if we determine σ_e, k_t, and $\hat{\varepsilon}_A$ by measurement.

Example 9.1   You are told that for copper at 300 K, the absolute thermoelectric power is 1.83 μV/K, the thermal conductivity is 400 W/mK, and the electrical conductivity is $5.88 \times 10^7\ \Omega^{-1}\mathrm{m}^{-1}$. Use these data to determine the kinetic coefficients L_{QQ}, L_{aa}, and L_{aQ} for copper at this temperature.

Solution   Solving Eq. (9.86) for L_{aa} yields

$$L_{aa} = \frac{e_e^2}{\sigma_e T}.$$

Substituting $e_e = -1.602 \times 10^{-19}$ coulombs for the charge on one electron and the given values of σ_e and temperature, we obtain

$$L_{aa} = \frac{(-1.602 \times 10^{-19})^2}{5.88 \times 10^7 (300)} = 6.87 \times 10^{47}\ \frac{\mathrm{K}}{\mathrm{mVAs}^2}\ \text{or}\ \frac{\mathrm{K}}{\mathrm{mJs}}.$$

We can also rearrange Eq. (9.107) to solve for L_{aQ}:

$$L_{aQ} = -\hat{\varepsilon}_A Te_eL_{aa}.$$

Substituting, we find that

$$L_{aQ} = -1.83 \times 10^{-6}(300)(-1.602 \times 10^{-19})6.87 \times 10^{47} = 6.05 \times 10^{25} \frac{K}{ms}.$$

Finally, we invert Eq. (9.99) and substitute the previously determined values of L_{aa} and L_{aQ} to determine L_{QQ}:

$$L_{QQ} = k_t T^2 + \frac{L_{aQ}^2}{L_{aa}} = 400(300)^2 + \frac{(6.05 \times 10^{25})^2}{6.87 \times 10^{47}} = 3.60 \times 10^7 \frac{WK}{m}.$$

Example 9.2 In real thermocouple temperature measurement systems, it is common to connect the leads of the potentiometer to the thermocouple using two copper lead wires as shown schematically in Figure 9.4. Show that this provides an accurate measurement of the voltage difference $\phi_{e,5} - \phi_{e,2}$ if $T_1 = T_6$ and $T_5 = T_2$.

Solution The chemical potential difference between locations 1 and 6 can be written

$$\mu_{a,6} - \mu_{a,1} = (\mu_{a,6} - \mu_{a,5}) + (\mu_{a,5} - \mu_{a,2}) + (\mu_{a,2} - \mu_{a,1}).$$

Multiplying both sides of Eq. (9.100) by dx, using (9.107) to replace the kinetic coefficients, and integrating from 1 to 2 yields

$$\mu_{a,2} - \mu_{a,1} = -\int_{T_1}^{T_2} e_e \hat{\varepsilon}_{Cu} \, dT.$$

Executing a similar integration from 5 to 6, we obtain

$$\mu_{a,6} - \mu_{a,5} = -\int_{T_5}^{T_6} e_e \hat{\varepsilon}_{Cu} \, dT.$$

Substituting the right side of the relations for $\mu_{a,2} - \mu_{a,1}$ and $\mu_{a,6} - \mu_{a,5}$ into the above relation for $\mu_{a,6} - \mu_{a,1}$ we can write

$$\mu_{a,6} - \mu_{a,1} = -\int_{T_5}^{T_6} e_e \hat{\varepsilon}_{Cu} \, dT + (\mu_{a,5} - \mu_{a,2}) - \int_{T_1}^{T_2} e_e \hat{\varepsilon}_{Cu} \, dT.$$

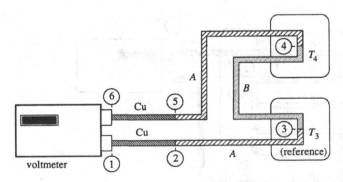

Figure 9.4

We can reorganize the integrals in this equation into the form

$$\mu_{a,6} - \mu_{a,1} = (\mu_{a,5} - \mu_{a,2}) - \int_{T_1}^{T_6} e_e \hat{\varepsilon}_{Cu} \, dT + \int_{T_2}^{T_5} e_e \hat{\varepsilon}_{Cu} \, dT.$$

Finally, since in each conductor the chemical component of the chemical potential is uniform, the change in chemical potential within the conductor is equal to the change in electrostatic potential $e_e \phi_e$, where ϕ_e is the voltage. We can therefore write the above relation as

$$e_e \phi_{e,6} - e_e \phi_{e,1} = (e_e \phi_{e,5} - e_e \phi_{e,2}) - \int_{T_1}^{T_6} e_e \hat{\varepsilon}_{Cu} \, dT + \int_{T_2}^{T_5} e_e \hat{\varepsilon}_{Cu} \, dT.$$

Canceling e_e throughout, we can reduce this to

$$\phi_{e,6} - \phi_{e,1} = (\phi_{e,5} - \phi_{e,2}) - \int_{T_1}^{T_6} \hat{\varepsilon}_{Cu} \, dT + \int_{T_2}^{T_5} \hat{\varepsilon}_{Cu} \, dT.$$

Clearly, if $T_1 = T_6$ and $T_5 = T_2$, the contribution of the integrals vanishes and the measured voltage difference at the potentiometer is identical to that between points 2 and 5. If these equalities are not true, the measured voltage will be in error by an amount dictated by the integral terms.

Example 9.3   In the arrangement of thermocouples shown schematically in Figure 9.5, junctions 2, 4, and 6 are held at a reference temperature T_{ref} and junctions 3, 5, and 7 simultaneously sense the same temperature T_p. This circuit configuration is known as a *thermopile*. Show that for this arrangement, the voltage difference $\phi_{e,8} - \phi_{e,2}$ is three times that for a single thermocouple of the same type under the same conditions.

Solution   Multiplying both sides of Eq. (9.100) by dx, using (9.107) to replace the kinetic coefficients, integrating across each segment of the circuit from 2 to 9, and combining the resulting equations yields

$$\mu_{a,8} - \mu_{a,2} = -\int_{T_2}^{T_3} e_e \hat{\varepsilon}_A \, dT - \int_{T_3}^{T_4} e_e \hat{\varepsilon}_B \, dT - \int_{T_4}^{T_5} e_e \hat{\varepsilon}_A \, dT - \int_{T_5}^{T_6} e_e \hat{\varepsilon}_B \, dT$$

$$- \int_{T_6}^{T_7} e_e \hat{\varepsilon}_A \, dT - \int_{T_7}^{T_8} e_e \hat{\varepsilon}_B \, dT.$$

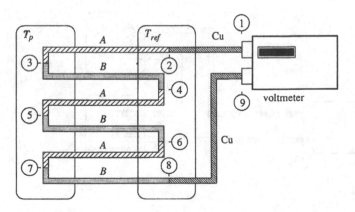

Figure 9.5

Using the facts that $T_2 = T_4 = T_6 = T_{ref}$ and $T_3 = T_5 = T_7 = T_p$, we can write the above relation as

$$\mu_{a,8} - \mu_{a,2} = -\int_{T_{ref}}^{T_p} e_e \hat{\varepsilon}_A \, dT - \int_{T_p}^{T_{ref}} e_e \hat{\varepsilon}_B \, dT - \int_{T_{ref}}^{T_p} e_e \hat{\varepsilon}_A \, dT - \int_{T_p}^{T_{ref}} e_e \hat{\varepsilon}_B \, dT$$

$$- \int_{T_{ref}}^{T_p} e_e \hat{\varepsilon}_A \, dT - \int_{T_p}^{T_{ref}} e_e \hat{\varepsilon}_B \, dT,$$

which reduces to

$$\mu_{a,8} - \mu_{a,2} = 3 \int_{T_{ref}}^{T_p} e_e (\hat{\varepsilon}_A - \hat{\varepsilon}_B) \, dT.$$

Since in each conductor the chemical component of the chemical potential is uniform, the change in chemical potential within the conductor is equal to the change in electrostatic potential $e_e \phi_e$, where ϕ_e is the voltage. Substituting $e_e \phi_e$ for μ_a in the above equation and canceling e_e throughout, we obtain

$$\phi_{e,8} - \phi_{e,2} = 3 \int_{T_{ref}}^{T_p} (\hat{\varepsilon}_A - \hat{\varepsilon}_B) \, dT.$$

The integral in the above equation is equal to the voltage difference across a single thermocouple of the same type having one junction at T_p and the other at T_{ref}. Thus the voltage difference across the entire thermopile is three times that amount.

The Peltier Effect

Figure 9.6 shows the junction between two different materials that conduct electric current and heat. The absorption or evolution of heat when electric current flows steadily across an isothermal junction of two materials is referred to as the *Peltier effect*.

As current flows across the junction, conservation of electrons requires that the flux of electrons must be the same in both materials. From Eq. (9.26) we know that for either material (neglecting b terms since we are only considering electron motion)

$$J_{U,x} = J_{Q,x} + \mu_a J_{a,x}. \tag{9.110}$$

Since at the junction μ_a is the same in both materials and $J_{a,x}$ is continuous across the junction, $\mu_a J_{a,x}$ is the same in both materials at the interface. It follows, then, from

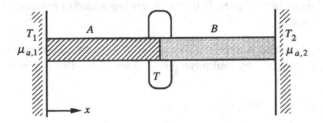

Figure 9.6

Eq. (9.110) that

$$J_{U,x}^A - J_{Q,x}^A = J_{U,x}^B - J_{Q,x}^B, \tag{9.111}$$

or equivalently

$$J_{U,x}^A - J_{U,x}^B = J_{Q,x}^A - J_{Q,x}^B. \tag{9.112}$$

Thus, the energy flux may change by an amount equal to the change in heat flux across the junction. For an isothermal junction, setting the temperature gradient terms to zero and combining Eqs. (9.87) and (9.88) for material A yields

$$J_{Q,x}^A = -\frac{L_{Qa}^A}{L_{aa}^A} J_{a,x}^A. \tag{9.113}$$

Using Eq. (9.107), we can rewrite (9.113) as

$$J_{Q,x}^A = T e_e \hat{\varepsilon}_A J_{a,x}^A. \tag{9.114}$$

Identical reasoning for material B yields

$$J_{Q,x}^B = T e_e \hat{\varepsilon}_B J_{a,x}^B. \tag{9.115}$$

Subtracting the right and left sides of Eqs. (9.113) and (9.115) and using the fact that $J_{a,x}^A = J_{a,x}^B = J_{a,x}$, we obtain

$$J_{Q,x}^B - J_{Q,x}^A = T(\hat{\varepsilon}_B - \hat{\varepsilon}_A) e_e J_{a,x}. \tag{9.116}$$

Thus the difference in the heat fluxes, which equals the heat absorbed or rejected per unit junction area, is related to the electric current flux and to the difference in the values of the absolute thermoelectric power for the materials. In connection with this process, the *Peltier coefficient* $\hat{\pi}_{AB}$ is defined as the heat that must be supplied to the junction per unit of electric current passing from material A to material B:

$$\hat{\pi}_{AB} = \frac{J_{Q,x}^B - J_{Q,x}^A}{e_e J_{a,x}} = T(\hat{\varepsilon}_B - \hat{\varepsilon}_A). \tag{9.117}$$

Thus the Peltier coefficient can be determined for a junction between two materials if the junction temperature and the absolute thermoelectric power for each material at that temperature are known.

Example 9.4   A junction between a copper and a nickel wire is held at a temperature of 300 K. The diameter of the wire is one millimeter. If a current of 20 A flows from the copper to the nickel, determine the rate at which the Peltier heat must be added or removed to hold the junction temperature constant.

Solution   Taking material A to be the copper and B to be the nickel, we find the total Peltier heat for the junction $\dot{Q}_P$ by multiplying both sides of Eq. (9.116) by the wire cross-sectional area A_c:

$$\dot{Q}_P = \left(J_{Q,x}^{Ni} - J_{Q,x}^{Cu}\right) A_c = T(\hat{\varepsilon}_{Ni} - \hat{\varepsilon}_{Cu}) e_e J_{a,x} A_c.$$

At 300 K, $\hat{\varepsilon}$ for copper is 1.83 μV/K and $\hat{\varepsilon}$ for nickel is -19.5 μV/K. Noting that $e_e J_{a,x} A_c$ is the total current flow, which equals 20 A here, we substitute to compute $\dot{Q}_P$:

$$\dot{Q}_P = 300(-19.5 \times 10^{-6} - 1.83 \times 10^{-6})20 = -0.128 \text{ W}.$$

The negative value of $\dot{Q}_P$ indicates that the heat must be removed from the junction to maintain a constant temperature.

The Thomson Effect

An additional thermoelectric effect, known as the Thomson effect, arises when electric current flows through a temperature gradient. To explore this effect, we consider the situation shown in Figure 9.7 in which heat flows through a material from a hot reservoir to a cold reservoir. In the absence of electrical current flow, a temperature gradient is established in the material in accordance with the temperature dependence of the kinetic coefficients for the material. At each point along the conducting material, we now bring a thermal reservoir into contact with the material having a temperature equal to the local temperature in the material. Since each reservoir is in thermal equilibrium with the material at each location, no heat is exchanged with the reservoirs.

Having brought these reservoirs into contact with the material, we now impose an electrical potential difference between points 1 and 2 to drive electrical current through the conducting material. At any location along the conducting material, the energy flow will differ from that established initially by a temperature gradient alone. The first law requires that any change in the total energy flow in the conductor must be accompanied by an energy exchange with the reservoir in contact with the material at that point. The heat exchange per unit volume of material q''' with the reservoir must equal the divergence of the energy flux $\nabla \cdot J_U$. Using Eq. (9.103) for this one-dimensional case, we obtain

$$q''' = \frac{\partial J_{U,x}}{\partial x} = \frac{\partial}{\partial x}(J_{Q,x} + \mu_a J_{a,x}) = \frac{\partial J_{Q,x}}{\partial x} + \frac{\partial \mu_a}{\partial x} J_{a,x}. \tag{9.118}$$

Combining (9.80) and (9.81) yields the following relation for $J_{Q,x}$:

$$J_{Q,x} = \left(\frac{L_{QQ}L_{aa} - L_{Qa}L_{aQ}}{L_{aa}} \right) \frac{\partial}{\partial x}\left(\frac{1}{T} \right) - \frac{L_{Qa}}{L_{aa}} J_{a,x}. \tag{9.119}$$

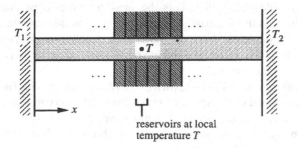

reservoirs at local temperature T

Figure 9.7

We then substitute the right side of Eq. (9.119) for $J_{Q,x}$ and use (9.88) to eliminate $\partial \mu_a / \partial x$ in Eq. (9.118) to obtain

$$q''' = \frac{\partial}{\partial x} \left[\left(\frac{L_{QQ}L_{aa} - L_{Qa}L_{aQ}}{L_{aa}} \right) \frac{\partial}{\partial x} \left(\frac{1}{T} \right) - \frac{L_{Qa}}{L_{aa}} J_{a,x} \right]$$
$$+ \left[-\frac{T}{L_{aa}} J_{a,x} - \frac{L_{aQ}T}{L_{aa}} \frac{\partial}{\partial x} \left(\frac{1}{T} \right) \right] J_{a,x}. \qquad (9.120)$$

Using Eqs. (9.96), (9.97), and (9.107) we next replace the kinetic coefficients with the conductivities and the absolute thermoelectric power:

$$q''' = \frac{\partial}{\partial x} \left[k_t T^2 \frac{\partial}{\partial x} \left(\frac{1}{T} \right) \right] + \frac{\partial}{\partial x} [T e_e \hat{\varepsilon} J_{a,x}] + \left[-\frac{e_e^2}{\sigma_e} J_{a,x} + T^2 e_e \hat{\varepsilon} \frac{\partial}{\partial x} \left(\frac{1}{T} \right) \right] J_{a,x}. \qquad (9.121)$$

The temperature variation, which is now held fixed by the reservoirs, was determined for the case of no heat exchange and no electric current. Setting q''' and $J_{a,x}$ to zero in the above relation reduces it to the following relation which dictates the temperature variation in the material:

$$0 = \frac{\partial}{\partial x} \left[k_t T^2 \frac{\partial}{\partial x} \left(\frac{1}{T} \right) \right]. \qquad (9.122)$$

The term on the right side of the above relation appears on the right side of Eq. (9.121). In accordance with Eq. (9.122), we set this term to zero in Eq. (9.121) to obtain, after a bit of rearranging,

$$q''' = T e_e J_{a,x} \frac{\partial \hat{\varepsilon}}{\partial x} - \frac{1}{T} (e_e J_{a,x})^2. \qquad (9.123)$$

Since the absolute thermoelectric power is a function only of local temperature, we can write this result in the form

$$q''' = e_e J_{a,x} T \left(\frac{d\hat{\varepsilon}}{dT} \right) \frac{\partial T}{\partial x} - \frac{1}{\sigma_e} (e_e J_{a,x})^2. \qquad (9.124)$$

The second term on the right side of Eq. (9.124) is the *Joule heat* that is produced by the flow of electric current through the material. The negative sign on this term indicates that heat must be removed volumetrically from the material to maintain a constant local temperature. Note that this well-known resistance heating of the material occurs even when no temperature gradient exists in the material.

The first term on the right side of Eq. (9.124) is known as the *Thomson heat*. The Thomson heat is the amount of heat exchanged (volumetrically) with the local reservoir to maintain a steady temperature when current $e_e J_{a,x}$ traverses temperature gradient $\partial T / \partial x$ in the material. From classical thermodynamic reasoning, Sir William Thomson, who later became Lord Kelvin, predicted this result in 1855. Thomson's deduction was experimentally verified for a number of materials long before the microscale aspects of thermoelectric effects were understood. It is conventional to define the *Thomson coefficient* $\hat{\tau}$ as the Thomson heat per unit current and per unit temperature gradient:

$$\hat{\tau} = \frac{e_e J_{a,x} T \left(\frac{d\hat{\varepsilon}}{dT}\right) \frac{\partial T}{\partial x}}{e_e J_{a,x} \left(\frac{\partial T}{\partial x}\right)} = T \left(\frac{d\hat{\varepsilon}}{dT}\right). \tag{9.125}$$

The Thomson coefficient is therefore directly related to the temperature derivative of the absolute thermoelectric power. By convention it is defined so that, if it is positive, the Thomson heat must flow into the material to keep the temperature constant. If we differentiate both sides of Eq. (9.117) with respect to temperature, we obtain

$$\frac{d\hat{\pi}_{AB}}{dT} = \hat{\varepsilon}_B - \hat{\varepsilon}_A + T \left(\frac{d\hat{\varepsilon}_B}{dT} - \frac{d\hat{\varepsilon}_A}{dT}\right). \tag{9.126}$$

Using the definition of the Thomson coefficient (9.125) we can write Eq. (9.126) as

$$\frac{d\hat{\pi}_{AB}}{dT} = \hat{\varepsilon}_B - \hat{\varepsilon}_A + \hat{\tau}_B - \hat{\tau}_A. \tag{9.127}$$

Note that Eq. (9.127) relates coefficients associated with the Peltier, Seebeck, and Thomson effects. In acknowledgment of Thomson's pioneering work on thermoelectric phenomena, Eqs. (9.127) and (9.117) are known as *Kelvin's first relation* and *Kelvin's second relation*, respectively.

Both the Seebeck coefficient $\hat{\varepsilon}_{AB}$ and the Peltier coefficient $\hat{\pi}_{AB}$ are defined for a pair of materials connected at a junction, but the Thomson coefficient and the absolute thermoelectric power are properties of a single material. For a superconducting material, it is impossible to sustain an electrostatic potential within the material because infinite current flow would result. Hence the absolute thermoelectric power must be zero since a temperature gradient cannot induce a finite electrostatic potential. If we pick a superconductor as material A, then $\hat{\varepsilon}_{AB} = \hat{\varepsilon}_B - \hat{\varepsilon}_A = \hat{\varepsilon}_B$ and the absolute thermoelectric power of material B can be determined from measurements of the rate of change of thermocouple voltage with temperature (at 2 in Figure 9.4) using Eq. (9.106):

$$\hat{\varepsilon}_B = \frac{dV_{\mathrm{m}}}{dT_2}. \tag{9.128}$$

Experiments have, in fact, been done in which a junction between lead and a superconductor was formed and the voltage versus temperature difference was determined for temperatures up to 18 K. It can be argued based on the third law of thermodynamics that the Seebeck coefficient must be zero when the junction temperature reaches absolute zero, and low temperature measurements are consistent with this line of reasoning.

We can also integrate Eq. (9.125) for an arbitrary material to compute the absolute thermoelectric power as

$$\hat{\varepsilon} = \int_{T_1}^{T} \frac{\hat{\tau}}{T} dT + \hat{\varepsilon}_1. \tag{9.129}$$

The Thomson coefficient has been determined experimentally for lead from 20 K to room temperature. The constant $\hat{\varepsilon}_1$ at T_1 can be determined from the low temperature $\hat{\varepsilon}$ data mentioned above and the $\hat{\tau}$ data can be used to numerically evaluate the integral in Eq. (9.129) to determine values of the absolute thermoelectric power for lead from 20 K to room temperature. With this procedure it is possible to establish values of the absolute thermoelectric power for lead over a broad range of temperatures. The absolute thermoelectric power of any other conductor can then be determined simply by pairing it with lead in a thermocouple and using Eq. (9.106) to determine $\hat{\varepsilon} - \hat{\varepsilon}_{\text{lead}}$ from the slope of the measured variation of thermocouple voltage with temperature. This approach makes it possible to determine absolute thermoelectric power of any conductor of interest. Once determined, values of the absolute thermoelectric power can be used to analyze the thermoelectric phenomena described in this section.

In closing this section, it should be noted that we have considered only the simplest of the interactions among the generalized fluxes that may arise in physical systems. Other thermoelectric and thermomagnetic effects can be defined and analyzed using methods that are basically the same as those developed in this section. The interested reader may find further discussion of thermoelectric effects in the texts by Zemanksy and Dittman [3] and Callen [2], the paper by Callen [4], and the handbook edited by Rowe [5].

Exercises

9.1 Derive Eqs. (9.27)–(9.29) from Eqs. (9.21)–(9.23) and (9.26).

9.2 Using the fact that μ_a and μ_b are functions of ρ_a, ρ_b, and T, derive Eqs. (9.38)–(9.40) from Eqs. (9.27)–(9.29).

9.3 You are told that for nickel at 300 K, the absolute thermoelectric power is $-19.5\ \mu$V/K, the thermal conductivity is 91 W/mK, and the electrical conductivity is $1.43 \times 10^7\ \Omega^{-1}\text{m}^{-1}$. Use these data to determine the kinetic coefficients L_{QQ}, L_{aa}, and L_{aQ} for nickel at this temperature.

9.4 For chromium at 300 K, the absolute thermoelectric power, the thermal conductivity, and the electrical conductivity are estimated to be 21.8 μV/K, 94 W/mK, and $7.80 \times 10^6\ \Omega^{-1}\text{m}^{-1}$ respectively. Use these data to determine the kinetic coefficients L_{QQ}, L_{aa}, and L_{aQ} for chromium at this temperature.

9.5 A copper–nickel thermocouple is to operate between 273 and 310 K. The variations of the absolute thermoelectric power for these materials over this range are given approximately by

$$\hat{\varepsilon}_{\text{Ni}} = -19.5 + 0.055(300 - T), \qquad \hat{\varepsilon}_{\text{Cu}} = 1.83 - 0.0032(300 - T),$$

where T is in K and $\hat{\varepsilon}_{\text{Ni}}$ and $\hat{\varepsilon}_{\text{Cu}}$ are in μV/K. If the reference junction is held at 0°C, compute and plot the variation of the voltage measured at the potentiometer with temperature T for temperatures between 290 and 310 K.

9.6 A thermocouple is to be used to measure temperatures between 273 and 320 K. For nickel, copper, and chromium, the variations of absolute thermoelectric power for these materials over this temperature range are given approximately by

$$\hat{\varepsilon}_{\text{Ni}} = -19.5 + 0.055(300 - T),$$
$$\hat{\varepsilon}_{\text{Cu}} = 1.83 - 0.0032(300 - T),$$
$$\hat{\varepsilon}_{\text{Cr}} = 21.8 - 0.084(300 - T),$$

where T is in K and $\hat{\varepsilon}_{\text{Ni}}$ and $\hat{\varepsilon}_{\text{Cu}}$ are in μV/K. To get the highest sensitivity for the resolution of the potentiometer (most voltage difference per degree kelvin), which two of these materials should be used for the thermocouple?

9.7 Copper and chromium are to be used in a thermopile like that shown in Example 9.3 with the reference junctions held at $0°C$. For the purposes of this analysis, assume that over the temperature range of interest

$$\hat{\varepsilon}_{Cu} = 1.83 - 0.0032(300 - T), \qquad \hat{\varepsilon}_{Cr} = 21.8 - 0.084(300 - T),$$

where T is in K and $\hat{\varepsilon}_{Cu}$ and $\hat{\varepsilon}_{Cr}$ are in $\mu V/K$. Determine the variation of output voltage difference with temperature for the thermopile for temperatures between 280 and 320 K. Plot your results.

9.8 Two wires, one of nickel and the other of chromium, meet at a junction immersed in an ice–water mixture, essentially holding the junction at $0°C$. Between 200 and 300 K the variations of absolute thermoelectric power for these elements are given approximately by

$$\hat{\varepsilon}_{Ni} = -19.5 + 0.055(300 - T), \qquad \hat{\varepsilon}_{Cr} = 21.8 - 0.084(300 - T),$$

where T is in K and $\hat{\varepsilon}_{Ni}$ and $\hat{\varepsilon}_{Cr}$ are in $\mu V/K$. If a current of 50 A flows from the nickel to the chromium, determine the rate at which the Peltier heat is transferred to or from the ice–water mixture.

9.9 The junction between a chromium wire and an aluminum wire is held at $0°C$ by immersing it in an ice–water mixture at atmospheric pressure. Between 270 and 300 K, the variations of the absolute thermoelectric power for these materials are given approximately by

$$\hat{\varepsilon}_{Cr} = 21.8 - 0.084(300 - T), \qquad \hat{\varepsilon}_{Al} = -1.66 - 0.0027(300 - T),$$

where T is in K and $\hat{\varepsilon}_{Cr}$ and $\hat{\varepsilon}_{Al}$ are in $\mu V/K$. If electric current flows through the junction, how large must it be and which way must it flow for the Peltier heat delivered to the ice water to equal 0.1 W?

9.10 Electric current is to be driven through a junction formed by two of the following materials: copper, aluminum, chromium, nickel, or iron. The junction is held at $0°C$ by immersing it in an ice–water mixture. Between 270 and 300 K, the variations of the absolute thermoelectric power for these materials are given approximately by

$$\hat{\varepsilon}_{Cu} = 1.83 - 0.0032(300 - T), \qquad \hat{\varepsilon}_{Al} = -1.66 - 0.0027(300 - T),$$
$$\hat{\varepsilon}_{Cr} = 21.8 - 0.084(300 - T), \qquad \hat{\varepsilon}_{Ni} = -19.5 + 0.055(300 - T),$$
$$\hat{\varepsilon}_{Fe} = 15.0 - 0.017(300 - T).$$

where T is in K and the absolute thermoelectric power values are in $\mu V/K$. Which combination of materials will maximize the Peltier coefficient for the junction?

9.11 The table below lists data indicating the variation of the absolute thermoelectric power with temperature for nickel and chromium. Fit each set of data with a second-order polynomial in temperature. Use the resulting relations to determine and plot the variation of the Thomson coefficient for each material between 200 and 400 K.

T (K)	$\hat{\varepsilon}_{Ni}$ ($\mu V/K$)	$\hat{\varepsilon}_{Cr}$ ($\mu V/K$)
100	−8.5	5.0
300	−19.5	21.8
500	−25.8	16.6

9.12 The table below lists data indicating the variation of the absolute thermoelectric power with temperature for aluminum and iron. Fit each set of data with a second-order polynomial in temperature. Use the resulting relations to determine and plot the variation of the Thomson coefficient for each material between 200 and 400 K.

T (K)	$\hat{\varepsilon}_{Al}$ (μV/K)	$\hat{\varepsilon}_{Fe}$ (μV/K)
100	-2.2	11.6
300	-1.66	15.0
500	-1.96	3.0

References

[1] Onsager, L., "Reciprocal Relations in Irreversible Processes," *Phys. Rev.*, 37: 407, 1931.

[2] Callen, H. B., *Thermodynamics and an Introduction to Thermostatistics*, 2nd ed., Chapter 14, John Wiley & Sons, New York, 1985.

[3] Zemanksy, M. W. and Dittman, R., *Heat and Thermodynamics*, McGraw-Hill, New York, 1981.

[4] Callen, H. B., "Application of Onsager's Reciprocal Relations to Thermoelectric, Thermomagnetic and Galvanomagnetic Effects," *Phys. Rev.*, 73: 1349, 1948.

[5] Rowe, D. M., ed., *CRC Handbook of Thermoelectrics*, CRC Press, Inc., Boca Raton, FL, 1995.

CHAPTER 10

Nonequilibrium and Noncontinuum Elements of Microscale Systems

There are many important systems that exhibit nonequilibrium or noncontinuum behavior. This final chapter examines some important examples of such systems. In doing so, we have two objectives. The first is to understand how, and under what conditions, the system behavior may deviate from the idealizations embodied in equilibrium theory or continuum theory. The second is to demonstrate theories and methods that are commonly used to model nonequilibrium and noncontinuum systems. Because they are commonly used to analyze such systems, kinetic theory and the Boltzmann transport equation are introduced. Nonequilibrium and noncontinuum phenomena associated with multiphase systems and electron transport in solids are examined in detail. The final section of Chapter 10 uses results from previous chapters to examine length scales and time scales at which classical and continuum theories become suspect. Doing so defines the range of conditions for which we expect classical and continuum theories to be accurate models of real physical systems. Although limited in its coverage, this chapter provides an introduction to microscale aspects of nonequilibrium and noncontinuum phenomena and serves to illustrate how they relate to the theoretical framework developed in the preceding chapters.

10.1 Basic Kinetic Theory

With increasing frequency engineers are dealing with microscale systems in which the applicability of classical macroscopic equilibrium thermodynamics becomes questionable. Generally, the applicability of classical equilibrium theory breaks down because the system is far from equilibrium and/or the system behavior deviates from a continuum model. The departure from equilibrium may be due to spatial nonuniformity of properties in the system or it may be because the system is in a metastable state or because the system has not had sufficient time to relax to equilibrium. A departure from continuum behavior may be due to dimensions in the system being comparable to the mean spatial separation of the molecules, giving the system an intrinsic granularity, or it may be due to the presence of phases separated by an interface.

As we noted in the previous chapter, the assumption of local thermodynamic equilibrium is widely used to extend classical equilibrium thermodynamic analysis to nonequilibrium systems. In some cases, however, even this approach does not provide an accurate model of the thermophysics of the system. In this final chapter of this text we will explore examples of microscale systems in which nonequilibrium and/or noncontinuum effects cause important deviations from classical thermodynamic predictions. In doing so our goals will be (1) to define the limits of applicability of classical theory and (2) to develop additional analysis tools from statistical thermodynamics and kinetic theory that can be applied to such systems. As a first step in developing additional analysis tools for nonequilibrium systems, in the first section of this chapter we will establish the basic framework of kinetic theory.

The Maxwell Distribution

We now consider a fluid composed of molecules that are free to move about in space. The total energy of a microstate occupied by a molecule is given by

$$\varepsilon = \varepsilon_{trx} + \varepsilon_{try} + \varepsilon_{trz} + \varepsilon_{other}. \tag{10.1}$$

The first three terms on the right side of Eq. (10.1) are the kinetic energies associated with translation in the x, y, and z directions. The fourth term represents the energy associated with all other energy storage modes. It follows directly that the molecular partition function is given by

$$q = q_{trx} q_{try} q_{trz} q_{other}. \tag{10.2}$$

The translational contributions are given by relations of the type

$$q_{trx} = \sum_{n'=0}^{\infty} e^{-\varepsilon_{trx,n'}/k_B T}. \tag{10.3}$$

The summation is over the quantum energy states as dictated by the 1-D particle-in-a-box quantum solution,

$$\varepsilon_{trx,n'} = \frac{h^2}{8mL^2}\left(n'_x\right)^2, \qquad n'_x = 1, 2, 3, \ldots. \tag{10.4}$$

Substituting yields

$$q_{trx} = \sum_{n'_x=0}^{\infty} e^{-h^2(n'_x)^2/8mL^2 k_B T}. \tag{10.5}$$

Considering the translational storage in the x direction, independent of other storage modes, the probability that a molecule is in a translational microstate between n'_x and $n'_x + \Delta n'_x$ is

$$\tilde{P}_{n'_x} = \frac{\sum_{n'=n'_x}^{n'_x+\Delta n'_x} e^{-h^2(n')^2/8mL^2 k_B T}}{\sum_{n'_x=0}^{\infty} e^{-h^2(n'_x)^2/8mL^2 k_B T}}. \tag{10.6}$$

Assuming we are at high enough temperature that the n'_x values are large, the spacing of the quantum levels is relatively close and we can replace the summations with integrals:

$$\tilde{P}_{n'_x} = \frac{\int_{n'=n'_x}^{n'_x+\Delta n'_x} e^{-h^2(n')^2/8mL^2 k_B T} \, dn'}{\int_{n'_x=0}^{\infty} e^{-h^2(n'_x)^2/8mL^2 k_B T} \, dn'_x}. \tag{10.7}$$

Letting $\Delta n'_x = dn'$ in the limit and evaluating the integral in the denominator yields

$$\tilde{P}_{n'_x} = \sqrt{\frac{h^2}{2\pi mL^2 k_B T}} e^{-h^2(n'_x)^2/8mL^2 k_B T} \, dn'_x. \tag{10.8}$$

Solving (10.4) for n'_x and differentiating gives

$$dn'_x = \sqrt{\frac{2mL^2}{h^2}} \varepsilon_{trx}^{-1/2} \, d\varepsilon_{trx}. \tag{10.9}$$

Substituting (10.4) and (10.9) into Eq. (10.8), we obtain a relation for the probability that the x direction translational energy is between $\varepsilon_{\mathrm{trx}}$ and $\varepsilon_{\mathrm{trx}} + d\varepsilon_{\mathrm{trx}}$:

$$\tilde{P}\left(\varepsilon_{\mathrm{trx}} \text{ to } \varepsilon_{\mathrm{trx}} + d\varepsilon_{\mathrm{trx}}\right) = \sqrt{\frac{1}{\pi k_{\mathrm{B}} T}} \varepsilon_{\mathrm{trx}}^{-1/2} e^{-\varepsilon_{\mathrm{trx}}/k_{\mathrm{B}} T} \, d\varepsilon_{\mathrm{trx}}. \tag{10.10}$$

Since $\varepsilon_{\mathrm{trx}} = mu^2/2$, the probability that $\varepsilon_{\mathrm{trx}}$ lies between $\varepsilon_{\mathrm{trx}}$ and $\varepsilon_{\mathrm{trx}} + d\varepsilon_{\mathrm{trx}}$ equals the probability that $|u|$ is between $|u|$ and $|u| + d|u|$. It follows that $d\varepsilon_{\mathrm{trx}} = d(mu^2)/2$ and

$$\tilde{P}(|u| \text{ to } |u| + d|u|) = \sqrt{\frac{2}{\pi m |u|^2 k_{\mathrm{B}} T}} e^{-m|u|^2/2k_{\mathrm{B}} T} (m/2) d(|u|^2). \tag{10.11}$$

By symmetry, the probability that u is between u and $u + du$ must be one half the probability that $|u|$ is between $|u|$ and $|u| + d|u|$. Thus

$$\tilde{P}(u \text{ to } u + du) = \frac{1}{2} \sqrt{\frac{2}{\pi m u^2 k_{\mathrm{B}} T}} e^{-mu^2/2k_{\mathrm{B}} T} (m/2) d(u^2). \tag{10.12}$$

Interpreting $\tilde{P}(u \text{ to } u + du)$ as being equivalent to the fraction of molecules with u velocities between u and $u + du$, dN_u/N, we can write Eq. (10.12) in the form

$$\frac{dN_u}{N} = \sqrt{\frac{m}{2\pi k_{\mathrm{B}} T}} e^{-mu^2/2k_{\mathrm{B}} T} \, du. \tag{10.13}$$

An identical analysis for the v and w velocity components yields

$$\frac{dN_v}{N} = \sqrt{\frac{m}{2\pi k_{\mathrm{B}} T}} e^{-mv^2/2k_{\mathrm{B}} T} \, dv, \tag{10.14}$$

$$\frac{dN_w}{N} = \sqrt{\frac{m}{2\pi k_{\mathrm{B}} T}} e^{-mw^2/2k_{\mathrm{B}} T} \, dw. \tag{10.15}$$

Interpreting the relations (10.13)–(10.15) as the probabilities that a molecule has a velocity within a differential amount of u, v, or w, it follows that the probability that a molecule simultaneously is within differential amounts of specific u, v, and w values is,

$$\frac{dN_{uvw}}{N} = \left(\frac{dN_u}{N}\right)\left(\frac{dN_v}{N}\right)\left(\frac{dN_w}{N}\right)$$

$$= \left(\frac{m}{2\pi k_{\mathrm{B}} T}\right)^{3/2} e^{-m(u^2 + v^2 + w^2)/2k_{\mathrm{B}} T} \, du \, dv \, dw. \tag{10.16}$$

Equations (10.13)–(10.16) are various forms of the Maxwell velocity distribution. We are also interested in the number density of molecules with speed c in the interval c to $c + dc$ regardless of direction. We therefore convert the above relation to spherical coordinates by using

$$u = c \sin\theta \cos\phi, \tag{10.17}$$
$$v = c \sin\theta \sin\phi, \tag{10.18}$$
$$w = c \cos\theta, \tag{10.19}$$
$$c = [u^2 + v^2 + w^2]^{1/2}. \tag{10.20}$$

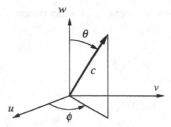

Figure 10.1   Relation between polar and Cartesian coordinates.

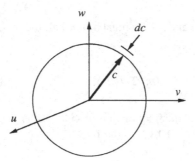

Figure 10.2   A differential change in c as a spherical shell in Cartesian velocity space.

and the corresponding differential forms of these relations. Executing the change of variables yields

$$\frac{dN_{c\theta\phi}}{N} = \left(\frac{m}{2\pi k_B T}\right)^{3/2} c^2 e^{-mc^2/2k_B T} \sin\theta \, d\theta \, d\phi \, dc. \tag{10.21}$$

In this relation the subscript on dN has been changed to $c\theta\phi$ to reflect the fact that this is now the number of molecules with speeds between c and $c + dc$ and directional angles between θ and $\theta + d\theta$ and ϕ and $\phi + d\phi$ (see Figures 10.1 and 10.2).

The number of molecules with speeds in the range c to $c + dc$ irrespective of direction can be found by integrating the above equation over all values of θ and ϕ:

$$dN_c = \int_{\phi=0}^{\phi=2\pi} \int_{\theta=0}^{\theta=\pi} dN_{c\theta\phi}. \tag{10.22}$$

Substituting the number distribution relation (10.21), we obtain

$$dN_c = 4\pi N \left(\frac{m}{2\pi k_B T}\right)^{3/2} c^2 e^{-mc^2/2k_B T} dc. \tag{10.23}$$

This is the Maxwell distribution of molecular speeds obtained for monatomic gases in Chapter 2. This distribution is plotted in Figure 10.3. Associated with the Maxwell distribution we can define the most probable speed, c_{mp}, the root mean square (rms) speed, $\langle c^2 \rangle^{1/2}$, and the average speed, $\langle c \rangle$. The most probable speed c_{mp} is the speed that maximizes dN_c/dc. Solving the Maxwell distribution relation for this quantity, differentiating,

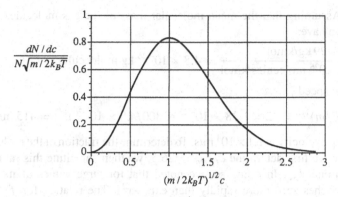

Figure 10.3 The Maxwell speed distribution.

and setting the result equal to zero yields

$$c_{mp} = \left(\frac{2k_BT}{m}\right)^{1/2}.$$
(10.24)

The mean speed $\langle c \rangle$ and the rms speed $\langle c^2 \rangle^{1/2}$ are computed by weighting the appropriate parameter with the fraction of the molecules in each speed interval and integrating over all possible speeds:

$$\langle c \rangle = \int_0^\infty c \left(\frac{dN_c}{Ndc}\right) dc = \left(\frac{8k_BT}{\pi m}\right)^{1/2},$$
(10.25)

$$\langle c^2 \rangle^{1/2} = \left[\int_0^\infty c^2 \left(\frac{dN_c}{Ndc}\right) dc\right]^{1/2} = \left(\frac{3k_BT}{m}\right)^{1/2}.$$
(10.26)

Note that these three representative molecular velocities are all proportional to $(k_BT/m)^{1/2}$ with the proportionality constant ranging from $2^{1/2}$ to $3^{1/2}$.

The fraction of molecules with speeds greater than a given value c is determined by integrating dN_c/N from c to infinity. Designating the fraction as $f_{>c}$, we have

$$f_{>c} = \int_c^\infty \left(\frac{dN_c}{N}\right) = 4\pi \left(\frac{m}{2\pi k_BT}\right)^{3/2} \int_c^\infty c^2 e^{-mc^2/2k_BT} dc.$$
(10.27)

Upon evaluating the integral, the resulting relation can be written as

$$f_{>c} = 1 + \frac{2c}{\sqrt{\pi}c_{mp}} \exp\{-(c/c_{mp})^2\} - \text{erf}\{c/c_{mp}\},$$
(10.28)

where $c_{mp} = (2k_BT/m)^{1/2}$.

Example 10.1 Determine the fraction of air molecules that would have velocities greater than the escape velocity at 300 K on the earth and the moon. What do the results imply about the tendency for these bodies to hold a gaseous atmosphere near their surface?

Solution   Assuming that the mean molecular mass of the gas molecules is the same as that for air we have:

$$m = \frac{29 \text{ kg/kmol}}{6.02 \times 10^{26} \text{ molecules/kmol}} = 4.8 \times 10^{-26} \text{ kg/moloecule.}$$

For the most probable speed

$$c_{mp} = (2k_B T/m)^{1/2} = (2 \times 1.38 \times 10^{-23} \times 300/4.8 \times 10^{-26})^{1/2} = 415 \text{ m/s.}$$

For the earth, the escape velocity is 1.1×10^4 m/s. To determine the fraction of the molecules that exceed this value we first determine $c/c_{mp} = 26.5$; we then substitute this ratio into Eq. (10.28) to determine $f_{>c}$. In doing so, we noted that for large values of its argument $1-$ erf x approaches zero more rapidly than $\exp(-x^2)$. The relation for $f_{>c}$ then becomes

$$f_{>c} \cong \frac{2c}{\sqrt{\pi} c_{mp}} \exp\{-(c/c_{mp})^2\} = 30e^{-700},$$

which is an extraordinarily small number ($\sim 10^{-300}$). On the moon, the escape velocity is only 2,000 m/s. Thus $c/c_{mp} = 4.82$ and it follows that

$$f_{>c} \cong \frac{2(4.82)}{\sqrt{\pi}} \exp\{-(4.82)^2\} = 4.4 \times 10^{-10}.$$

This fraction is significant given that the number of molecules might be $\sim 10^{40}$. Thus, on the moon there would be a steady loss of molecules escaping gravity into space as a result of their thermal motion. Eventually the entire atmosphere would escape into space.

We now turn our attention to the following question: What is the flux of molecules through (or striking) a flat surface in a gas? Considering the region within the box shown in Figure 10.4, we want to know the number of molecules in the box that strike the shaded surface Σ_x^* per unit area, per unit time. Regardless of v and w, a molecule with a given u must be within $u\,\Delta t$ at the beginning of the time interval Δt to pass through the surface. We know from previous results that the fraction of molecules with u velocities in the interval u to $u + du$ is

$$\frac{dN_u}{N} = \sqrt{\frac{m}{2\pi k_B T}} e^{-mu^2/2k_B T} \, du. \qquad (10.29)$$

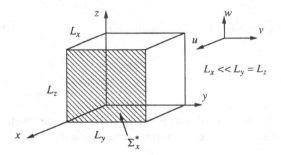

Figure 10.4

We also know that of the molecules in the box with velocity u, only those within a fraction of the total box volume equal to $u\Delta t/L_x$ will pass through the surface in time Δt. We define the differential flux dj_u as the number of molecules with velocity u passing through surface Σ_x^* per unit time, per unit area. It follows from the above discussion that dj_u must be given by

$$dj_u = \frac{dN_u}{N}(N)\left(\frac{u\Delta t}{L_x}\right)(L_yL_z)^{-1}(\Delta t)^{-1} = \frac{u}{V}dN_u, \tag{10.30}$$

where $V = L_xL_yL_z$. Integrating to get contributions to the total flux from all possible u velocity values, we obtain the following relation for the total flux j of molecules across surface Σ_x^*:

$$j = \int_u dj_u = \int_0^\infty \left(\frac{N}{V}\right)\sqrt{\frac{m}{2\pi k_B T}}ue^{-mu^2/2k_B T}\,du, \tag{10.31}$$

which reduces to

$$j = \frac{1}{4}\left(\frac{N}{V}\right)\sqrt{\frac{8k_B T}{\pi m}}. \tag{10.32}$$

Thus j is proportional to the density of molecules (N/V) and the mean speed $\langle c \rangle$.

We consider now the escape of molecules of a gas through an opening in a wall of a container, as indicated in Figure 10.5. If the size of the opening is large compared with the mean spacing between molecules, a bulk flow of the gas results that can be analyzed with continuum fluid dynamics theory. If the region outside the container is evacuated, choked sonic flow occurs through the opening. The mass flux, from compressible flow theory, is given by

$$j_{m,\text{large}} = \left(\frac{mN}{V}\right)\sqrt{\frac{\gamma k_B T}{m}}, \tag{10.33}$$

where $\gamma = \hat{c}_P/\hat{c}_V$ is the ratio of specific heats. If, however, the opening size is smaller than the mean spacing of the molecules, the mass flux is j multiplied by the mass per molecule:

$$j_{m,\text{small}} = \frac{1}{4}\left(\frac{mN}{V}\right)\sqrt{\frac{8k_B T}{\pi m}}. \tag{10.34}$$

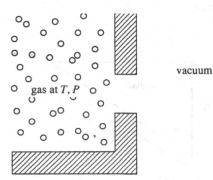

vacuum

gas at T, P

Figure 10.5

By taking the ratio of these two relations, we find

$$\frac{j_{m,\text{large}}}{j_{m,\text{small}}} = \sqrt{2\pi\gamma}.$$

(10.35)

Thus, the mass flux for the small hole is about a factor of three lower than the macroscopic compressible flow prediction.

Collision Processes

In this section we will develop a simplistic model of molecular collisions processes. In doing so we will consider a system containing a single kind of molecule. The average number density of molecules $\langle N \rangle / V$ will be designated as ρ_N. In this simple model, all molecules move with the same speed c_m, although the direction of the velocity vector is uniformly distributed in all directions. We begin by considering one specific target molecule in the system and define $\bar{g}$ as the mean relative speed of other molecules in the system with respect to this one molecule. In our simple model, the molecules are presumed to be spherically symmetric with diameter D. Molecules collide if their motion brings their centers to within a distance less than their diameter D (see Figure 10.6). This is essentially a hard sphere model. The *collision cross section* σ_c presented by the target molecule is given by

$$\sigma_c = \pi D^2.$$

(10.36)

Molecules that will collide with the target molecule in time dt have their centers within a circular cylinder of length $\bar{g}\,dt$ and volume $\sigma_c \bar{g}\,dt$. The number of molecules in this volume is $\rho_N \sigma_c \bar{g}\,dt$. It follows directly that the mean frequency of collision with the target molecule $\bar{\nu}_c$ is given by

$$\bar{\nu}_c = \rho_N \sigma_c \bar{g}.$$

(10.37)

To determine the mean relative speed, we consider the velocity vectors of two arbitrarily chosen molecules 1 and 2, as indicated in Figure 10.7. From the geometry, the magnitude of the relative velocity vector $|\mathbf{g}_{12}|$ is given by

$$|\mathbf{g}_{12}| = [|\mathbf{v}_2|^2 \sin^2 \theta + (|\mathbf{v}_1| - |\mathbf{v}_2| \cos \theta)^2]^{1/2}.$$

(10.38)

Equation (10.38) can be rearranged to obtain

$$|\mathbf{g}_{12}|^2 = |\mathbf{v}_2|^2 + |\mathbf{v}_1|^2 - 2|\mathbf{v}_1||\mathbf{v}_2| \cos \theta.$$

(10.39)

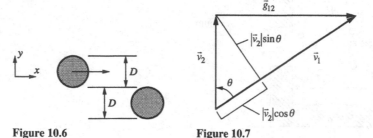

Figure 10.6   Figure 10.7

Taking $\bar{g}$ as being the root mean square value of $|\mathbf{g}_{12}|$, we have

$$\bar{g} = \langle|\mathbf{g}_{12}|^2\rangle^{1/2} = \langle|\mathbf{v}_2|^2 + |\mathbf{v}_1|^2 - 2|\mathbf{v}_1||\mathbf{v}_2|\cos\theta\rangle^{1/2}. \tag{10.40}$$

In our model the magnitude of the velocity vector for any molecule is c_m. Hence $|\mathbf{v}_1| = c_m$ and $|\mathbf{v}_2| = c_m$ and the relation for $\bar{g}$ reduces to

$$\bar{g} = \langle|\mathbf{g}_{12}|^2\rangle^{1/2} = \langle 2c_m^2 - 2c_m^2\cos\theta\rangle^{1/2}. \tag{10.41}$$

If we take the average over all values of θ from 0 to 2π, the cosine term in Eq. (10.41) averages to zero, and the relation for $\bar{g}$ becomes

$$\bar{g} = \sqrt{2}c_m. \tag{10.42}$$

Substituting this result into Eq. (10.37) for the collision frequency yields

$$\bar{\nu}_c = \sqrt{2}\rho_N \sigma_c c_m. \tag{10.43}$$

The mean time between collisions $\langle\tau_c\rangle$ is just the reciprocal of the collision frequency:

$$\langle\tau_c\rangle = \frac{1}{\bar{\nu}_c} = \frac{1}{\sqrt{2}\rho_N \sigma_c c_m}. \tag{10.44}$$

The mean free path λ_c is defined as the mean distance traveled by a given molecule between collisions. The mean free path is taken to be equal to the product of the mean time between collisions $\langle\tau_c\rangle$ and the mean speed for a given particle $\langle c\rangle$:

$$\lambda_c = \langle\tau_c\rangle\langle c\rangle. \tag{10.45}$$

Taking $\langle c\rangle = c_m$ and substituting the relation (10.44) for $\langle\tau_c\rangle$, we obtain the following relation for the mean free path:

$$\lambda_c = \frac{1}{\sqrt{2}\rho_N \sigma_c}. \tag{10.46}$$

If $\sigma_c = \pi D^2$, the relation for the mean free path becomes

$$\lambda_c = \frac{1}{\sqrt{2}\pi\rho_N D^2}. \tag{10.47}$$

Using the ideal gas equation of state to evaluate the density, the relation for λ_c from our simplistic analysis can be written

$$\lambda_c = \frac{k_B T}{\sqrt{2}\pi D^2 P}. \tag{10.48}$$

Example 10.2 For nitrogen gas at atmospheric pressure and 300 K, estimate the mean free path and the mean time between collisions if the effective diameter of the molecules is 4 Å.

Solution The mean free path is computed using Eq. (10.48). Substituting the specified values we get

$$\lambda_c = \frac{k_B T}{\sqrt{2}\pi D^2 P} = \frac{1.38 \times 10^{-23}\,(300)}{\sqrt{2}\pi(4.0 \times 10^{-10})^2\,(101,000)} = 5.76 \times 10^{-8}\,\text{m}.$$

The mean speed is given by Eq. (10.25):

$$c_m = \langle c \rangle = \left(\frac{8k_B T}{\pi m} \right)^{1/2} \left(\frac{(8)1.38 \times 10^{-23}(300)}{\pi(28/6.02 \times 10^{-26})} \right)^{1/2} = 476 \text{ m/s.}$$

The number density is computed from the ideal gas law as

$$\rho_N = \frac{P}{k_B T} = \frac{101,000}{1.38 \times 10^{-23}(300)} = 2.44 \times 10^{22} \text{ molecules/m}^{-3}.$$

Finally, we use these results to compute the mean time between collisions using Eq. (10.44):

$$\langle \tau_c \rangle = \frac{1}{\sqrt{2}\rho_N \sigma_c c_m} = \frac{1}{\sqrt{2}\rho_N \pi D^2 c_m}$$

$$= \frac{1}{\sqrt{2}(2.44 \times 10^{22})\pi(4.0 \times 10^{-10})^2 \, 476} = 1.21 \times 10^{-7} \text{ s.}$$

Transport Properties

We now wish to explore the relationship between the mean free path λ_c and transport in gases. To explore momentum transport, we consider the classic problem of flow between two parallel plates shown in Figure 10.8. For the circumstances shown in Figure 10.8, $\langle u \rangle$ varies from 0 to U between the plates. From our knowledge of macroscopic (laminar) fluid mechanics we expect $\langle u \rangle$ to vary linearly between the plates. It follows that the bulk momentum varies as

$$\langle P_x \rangle = \frac{mUy}{a}. \tag{10.49}$$

Although equally many molecules from above and below cross a plane at a given height y, the ones from above carry more momentum down than upward-moving ones carry upward. As a result, there exists a net downward flux of momentum.

Let j equal the flux of a transported quantity (here it is momentum). The number of molecules crossing a horizontal plane per unit area, per unit time, with velocity $\mathbf{c} = (u, v, w)$ is equal to the number of molecules in a parallelepiped with a base of unit area and a height equal to $|v|$. The volume of the parallelepiped is $|v|$ and the number of molecules in it with the specified velocity $\mathbf{c}$ is $|v|\rho_N f(u, v, w) \, du \, dv \, dw$.

Note that if v is positive, particles come from below, and if it is negative they come from above. Because these are equally probable, the net flux is zero. A molecule crossing the

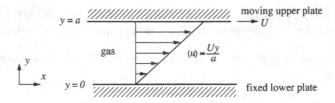

Figure 10.8

horizontal plane has, on the average, traveled a distance λ_c since its last collision. It follows that y', the average y location of its last collision, is

$$y' = y - v\lambda_c/|\mathbf{c}|, \tag{10.50}$$

where

$$|\mathbf{c}| = [u^2 + v^2 + w^2]^{1/2}. \tag{10.51}$$

We assume that during the last collision, the molecule came into local equilibrium with the environment and therefore carries the momentum associated with height y'. Expanding the x momentum in a Taylor series, we obtain

$$\langle p_x \rangle_{y=y'} = \langle p_x \rangle_{y=y} - \left(\frac{d\langle p_x \rangle}{dy} \right) \frac{v\lambda_c}{|\mathbf{c}|} + \cdots. \tag{10.52}$$

The net flow of momentum $\langle p_x \rangle$ is given by integrating the product of $\langle p_x \rangle_{y=y'}$ and the number of molecules crossing the surface from height y' over the entire range of velocities:

$$j(y) = \int_{-\infty}^{\infty} \int_{-\infty}^{\infty} \int_{-\infty}^{\infty} \langle p_x \rangle_{y=y'} v\rho_N f(u, v, w) \, du \, dv \, dw. \tag{10.53}$$

If we substitute the Taylor series expansion (10.52), the relation for $j(y)$ becomes

$$j(y) = \int_{-\infty}^{\infty} \int_{-\infty}^{\infty} \int_{-\infty}^{\infty} \left[\langle p_x \rangle_{y=y'} - \frac{d\langle p_x \rangle}{dy} \frac{v\lambda_c}{|\mathbf{c}|} \right] v\rho_N f(u, v, w) \, du \, dv \, dw. \tag{10.54}$$

Using the fact that $\langle p_x \rangle_{y=y'} = mUy/a$ and substituting Eq. (10.51) for $|\mathbf{c}|$, we obtain

$$j(y) = \int_{-\infty}^{\infty} \int_{-\infty}^{\infty} \int_{-\infty}^{\infty} \left[\frac{mUy}{a} \right] v\rho_N f(u, v, w) \, du \, dv \, dw.$$

$$- \rho_N \lambda_c \left(\frac{d\langle p_x \rangle}{dy} \right) \int_{-\infty}^{\infty} \int_{-\infty}^{\infty} \int_{-\infty}^{\infty} \left[\frac{v^2 f(u, v, w)}{[u^2 + v^2 + w^2]^{1/2}} \right] du \, dv \, dw. \tag{10.55}$$

For these conditions

$$f(u, v, w) = \left(\frac{m}{2\pi k_B T} \right)^{3/2} \exp \left\{ -\frac{m[(u - \langle u \rangle)^2 + v^2 + w^2]}{2k_B T} \right\}. \tag{10.56}$$

If $\langle u \rangle$ is small compared to the molecular velocities (since molecular speeds are comparable to the speed of sound this implies low Mach number), we can simply neglect it in the above expression for f. Evaluation of the triple integrals using this reduced relation for f is then straightforward. The first integral in Eq. (10.55) is zero and evaluation of the second yields

$$j(y) = -\left(\frac{1}{3} \right) \rho_N \lambda_c \langle c \rangle \left(\frac{d\langle p_x \rangle}{dy} \right), \tag{10.57}$$

where $\langle c \rangle$ is the mean speed defined by Eq. (10.25). Noting that $\langle p_x \rangle = m\langle u \rangle$ this can be written as

$$j(y) = -\left(\frac{1}{3} \right) \rho_N m \lambda_c \langle c \rangle \frac{d\langle u \rangle}{dy} \tag{10.58}$$

or more generally as

$$\mathbf{j}(y) = -\left(\tfrac{1}{3}\right)\rho_N \lambda_c \langle c \rangle \nabla \langle u \rangle. \tag{10.59}$$

We note the similarity between the above relation for j and Newton's law of viscous flow,

$$\tau = j(y) = -\mu_m \frac{d\langle u \rangle}{dy}, \tag{10.60}$$

which implies that for the gas

$$\mu_m = \left(\tfrac{1}{3}\right)\rho_N m \lambda_c \langle c \rangle. \tag{10.61}$$

Using previously derived relations for the mean free path and the mean speed and using the ideal gas relation to evaluate the density, we can convert this relation to

$$\mu_m = \left(\frac{1}{3}\right)\left(\frac{2}{\pi}\right)^{3/2} \frac{(m k_B T)^{1/2}}{D^2}. \tag{10.62}$$

It can be seen that μ_m is independent of density and pressure.

Note that in the above analysis we did not need to specify the form of $\langle p_x \rangle$ to derive the above relation for $j(y)$. When $\langle p_x \rangle$ is interpreted as momentum, j is the transport (flux) of momentum. The same analysis can be applied to heat transfer if $\langle p_x \rangle$ is replaced with $\langle \varepsilon \rangle$, the average energy per molecule. Setting $\langle p_x \rangle = \langle \varepsilon \rangle$ yields

$$\frac{d\langle p_x \rangle}{dy} = \frac{d\langle \varepsilon \rangle}{dy} = \frac{d\langle \varepsilon \rangle}{dT} \cdot \frac{dT}{dy} = \frac{\hat{c}_V}{N_A} \nabla T, \tag{10.63}$$

where $\hat{c}_V$ is the molar heat capacity. Substituting in the above analysis yields

$$\mathbf{j} = -\left(\frac{1}{3}\right)\left(\frac{\rho_N \lambda_c \hat{c}_V \langle c \rangle}{N_A}\right)\nabla T, \tag{10.64}$$

where $\mathbf{j}$ is now a heat flux vector. Comparing this with Fourier's law,

$$\mathbf{j}_Q = -k_t \nabla T, \tag{10.65}$$

we conclude that

$$k_t = \left(\frac{1}{3}\right)\left(\frac{\rho_N \lambda_c \hat{c}_V \langle c \rangle}{N_A}\right) = \frac{\mu_m \hat{c}_V}{m N_A}. \tag{10.66}$$

The above result implies that $\mu_m \hat{c}_V / m N_A k_t = 1$. A more refined analysis indicates that this ratio is constant but equals a value somewhat different from 1.

Although we will not pursue it here, this analysis can also be extended to mass diffusion in multicomponent gas mixtures (to predict the mass diffusion coefficient). More rigorous kinetic theory models can be constructed that treat the collision process in a more sophisticated manner. The interested reader can find more information on such treatments in References [1]–[3] listed at the end of this chapter.

10.2   The Boltzmann Transport Equation

System Behavior in Phase Space

In the previous section we have interpreted the distribution function as unchanging in time. This is consistent with our development of the statistical thermodynamics framework, which implicitly assumes that the system is in an equilibrium state characterized by an equilibrium distribution that is unchanging in time. In general, however, we expect that if the system is initially in a nonequilibrium state, the distribution will be different from the relation that applies at equilibrium. As the particles in the system move and collide, the distribution will change with time as the system configuration evolves, ultimately achieving the equilibrium distribution as time goes to infinity.

To begin our consideration of these issues we consider a classical system of N molecules. Each molecule has s degrees of freedom so that the number of coordinates needed to specify positions of all N molecules is $l = Ns$. Note that the coordinates may include angular coordinates and relative positions as well as translation coordinates. The l spatial coordinates $\hat{q}_i$ and l corresponding momenta $\hat{p}_j$ completely specify the classical mechanical state of the system. The initial $\hat{q}_i$ and $\hat{p}_j$ together with the equations of motion completely determine the future (and past) course of the classical system.

We now include in the analysis a conceptual Euclidean hyperspace of $2l$ dimensions, with a coordinate axis for each of the $2l$ coordinates and momenta. Gibbs termed this conceptual space a *phase space* for the system. The state of the classical N-body system at any time t is completely specified by the location of one point in phase space, referred to as a phase point. The evolution of the system state with time is completely described by the motion or trajectory of the phase point through phase space. The trajectory of the point is dictated by the equations of motion of the N bodies, as prescribed by Newton's laws and the force interactions among the molecules. These, in principle, can be integrated if the initial location of the point is specified. In practice, integration of such a large system of equations is not feasible.

We now consider a microcanonical ensemble of systems in phase space. There are A systems in the ensemble, each having the same N, V, and U. The classical state of each system in the ensemble is represented by a point in phase space. As each system evolves in time, each point will trace out its independent trajectory. It is important to note that the trajectories are independent because the systems in the ensemble are themselves isolated and independent. The postulate of equal a priori probabilities (see Chapter 2) requires that there is a representative phase point in phase space for each and every set of coordinates, consistent with the fixed macroscopic constraints (N, V, U). Associated with this ensemble we define a number density $f_{es}(\mathbf{p}, \mathbf{q}, t)$ such that

$$f_{es}(\mathbf{p}, \mathbf{q}, t)\, d\mathbf{p}\, d\mathbf{q} = \begin{pmatrix} \text{the number of systems in the ensemble that} \\ \text{have phase points in } d\mathbf{p}\, d\mathbf{q} \text{ about } \mathbf{p}, \mathbf{q} \text{ at time } t \end{pmatrix}.$$

$$(10.67)$$

In the above definition $\mathbf{p}$ and $\mathbf{q}$ are vectors $\hat{p}_1, \hat{p}_2, \ldots, \hat{p}_l$ and $\hat{q}_1, \hat{q}_2, \ldots, \hat{q}_l$, respectively, $d\mathbf{p} = d\hat{p}_1\, d\hat{p}_2 \ldots d\hat{p}_l$, and $d\mathbf{q} = d\hat{q}_1\, d\hat{q}_2 \ldots d\hat{q}_l$. From this definition, it follows directly that

$$\iiint_{\text{all }\mathbf{p},\mathbf{q}} \cdots \int f_{\text{es}}(\mathbf{p}, \mathbf{q}, t)\, d\mathbf{p}\, d\mathbf{q} = A. \tag{10.68}$$

The ensemble average of any function $\psi(\mathbf{p}, \mathbf{q})$ of the momenta and coordinates of the system is defined by

$$\langle \psi \rangle = \frac{1}{A} \iiint_{\text{all }\mathbf{p},\mathbf{q}} \cdots \int \psi(\mathbf{p}, \mathbf{q}) f_{\text{es}}(\mathbf{p}, \mathbf{q}, t)\, d\mathbf{p}\, d\mathbf{q}. \tag{10.69}$$

Gibbs postulated that this ensemble average is, in fact, the corresponding thermodynamic function.

Since the equations of motion determine the trajectory of each point, they must also determine the density of points at any time, if the dependence of f_{es} on $\mathbf{p}$ and $\mathbf{q}$ is known at some initial time t_0. Thus, for a classical system, the time dependence of f_{es} is dictated by the laws of classical mechanics.

We consider a small volume of phase space $\delta \hat{p}_1 \delta \hat{p}_2 \ldots \delta \hat{p}_l \delta \hat{q}_1 \delta \hat{q}_2 \ldots \delta \hat{q}_l$ about the point $\hat{p}_1, \hat{p}_2, \ldots, \hat{p}_l, \hat{q}_1, \hat{q}_2, \ldots, \hat{q}_l$. The number of system state points inside this volume of phase space at any instant is

$$\delta n = f_{\text{es}}\left(\hat{p}_1, \hat{p}_2, \ldots, \hat{p}_l, \hat{q}_1, \hat{q}_2, \ldots, \hat{q}_l, t\right) \delta \hat{p}_1 \delta \hat{p}_2 \ldots \delta \hat{p}_l \delta \hat{q}_1 \delta \hat{q}_2, \ldots \delta \hat{q}_l. \tag{10.70}$$

Note that δn will change with time as points enter and leave through faces of the differential volume element. Consider the two parallel surfaces of the element that are perpendicular to the $\hat{q}_1$ axis located at $\hat{q}_1$ and $\hat{q}_1 + d\hat{q}_1$. The number of phase points entering through the first face (at $\hat{q}_1$) per unit time is given by

$$\delta \dot{n}_{\hat{q}_1} = f_{\text{es}} \delta \hat{p}_1 \delta \hat{p}_2 \ldots \delta \hat{p}_l \delta \hat{q}_1 \delta \hat{q}_2 \ldots \delta \hat{q}_l \frac{\dot{\hat{q}}_1}{\delta \hat{q}_1}. \tag{10.71}$$

Since the volume element is of differential dimensions, the rate at which points pass out of the element through the other face can be expressed as

$$\delta \dot{n}_{\hat{q}_1 + d\hat{q}_1} = \left(f_{\text{es}} + \frac{\partial f_{\text{es}}}{\partial \hat{q}_1} \delta \hat{q}_1 \right) \delta \hat{p}_1 \delta \hat{p}_2 \ldots \delta \hat{p}_l \delta \hat{q}_1 \delta \hat{q}_2 \ldots \delta \hat{q}_l \left(\dot{\hat{q}}_1 + \frac{\partial \dot{\hat{q}}_1}{\partial \hat{q}_1} \delta \hat{q}_1 \right) \frac{1}{\delta \hat{q}_1}. \tag{10.72}$$

where f_{es} and the derivatives are evaluated at $\hat{q}_1$. Subtracting $\delta \dot{n}_{\hat{q}_1 + d\hat{q}_1}$ from $\delta \dot{n}_{\hat{q}_1}$ yields the net flow of phase points in the $\hat{q}_1$ direction into the volume element, which reduces to

$$\delta \dot{n}_{\hat{q}_1} - \delta \dot{n}_{\hat{q}_1 + d\hat{q}_1} = -\left(\frac{\partial f_{\text{es}}}{\partial \hat{q}_1} \dot{\hat{q}}_1 + f_{\text{es}} \frac{\partial \dot{\hat{q}}_1}{\partial \hat{q}_1} \right) \delta \hat{p}_1 \delta \hat{p}_2 \ldots \delta \hat{p}_l \delta \hat{q}_1 \delta \hat{q}_2 \ldots \delta \hat{q}_l. \tag{10.73}$$

In a similar manner, the following relation can be derived for the net flow into the volume in the $\hat{p}_1$ direction:

$$\delta \dot{n}_{\hat{p}_1} - \delta \dot{n}_{\hat{p}_1 + d\hat{p}_1} = -\left(\frac{\partial f_{\text{es}}}{\partial \hat{p}_1} \dot{\hat{p}}_1 + f_{\text{es}} \frac{\partial \dot{\hat{p}}_1}{\partial \hat{p}_1} \right) \delta \hat{p}_1 \delta \hat{p}_2 \ldots \delta \hat{p}_l \delta \hat{q}_1 \delta \hat{q}_2 \ldots \delta \hat{q}_l. \tag{10.74}$$

Similar relations can be derived for the net flow in all the other coordinate directions. The total net flow into the volume through all faces of the element is then found by summing contributions from all directions in phase space.

$$\delta\dot{n} = -\sum_{j=1}^{l} \left(\frac{\partial f_{es}}{\partial \hat{q}_j} \dot{\hat{q}}_j + f_{es} \frac{\partial \dot{\hat{q}}_j}{\partial \hat{q}_j} + \frac{\partial f_{es}}{\partial \hat{p}_j} \dot{\hat{p}}_j + f_{es} \frac{\partial \dot{\hat{p}}_j}{\partial \hat{p}_j} \right) \delta \hat{p}_1 \delta \hat{p}_2 \ldots \delta \hat{p}_l \delta \hat{q}_1 \delta \hat{q}_2 \ldots \delta \hat{q}_l.$$

(10.75)

This total net inflow must equal the change of δn with time, which implies that

$$\frac{\partial(\delta n)}{\partial t} = -\sum_{j=1}^{l} \left[f_{es} \left(\frac{\partial \dot{\hat{q}}_j}{\partial \hat{q}_j} + \frac{\partial \dot{\hat{p}}_j}{\partial \hat{p}_j} \right) + \frac{\partial f_{es}}{\partial \hat{q}_j} \dot{\hat{q}}_j + \frac{\partial f_{es}}{\partial \hat{p}_j} \dot{\hat{p}}_j \right] \delta \hat{p}_1 \delta \hat{p}_2 \ldots \delta \hat{p}_l \delta \hat{q}_1 \delta \hat{q}_2 \ldots \delta \hat{q}_l.$$

(10.76)

In Chapter 1 we noted that, for a system having many coordinates and momenta, the Hamiltonian $\hat{H}$ satisfies

$$\frac{\partial \hat{H}}{\partial \hat{q}_j} = -\dot{\hat{p}}_j, \qquad \frac{\partial \hat{H}}{\partial \hat{p}_j} = \dot{\hat{q}}_j.$$

(10.77)

Differentiating the first relation with respect to $\hat{p}_j$ and the second with respect to $\hat{q}_j$ and equating the expressions for the cross derivatives of $\hat{H}$ yields

$$\frac{\partial \dot{\hat{q}}_j}{\partial \hat{q}_j} = -\frac{\partial \dot{\hat{p}}_j}{\partial \hat{p}_j}.$$

(10.78)

This result implies that the factor multiplying f_{es} in Eq. (10.76) is zero. Using this fact and using Eq. (10.70) to eliminate δn transforms Eq. (10.76) to

$$\frac{\partial f_{es}}{\partial t} = -\sum_{j=1}^{l} \left[\frac{\partial f_{es}}{\partial \hat{q}_j} \dot{\hat{q}}_j + \frac{\partial f_{es}}{\partial \hat{p}_j} \dot{\hat{p}}_j \right].$$

(10.79)

Equation (10.79) is the *Liouville equation*, which is an important cornerstone of classical statistical mechanics.

Note that we can apply this analysis for an ensemble of systems to a distribution for a system of noninteracting particles in the following manner. We simply consider each particle in our system of interest to be a "system" in our ensemble. This is possible because a particle or molecule has a defined set of coordinates and momenta that describe its state just as the systems in the ensemble did. Each molecule in the system then can be viewed as moving through phase space as its microstate changes with time in the system due to its motion through physical space. Note that because we assumed that the systems in the ensemble did not interact, the Liouville equation can be applied only to particles that do not interact.

Designating the distribution function as f_{ep} for an ensemble of particles, we replace f_{es} in Eq. (10.79) with f_{ep} and rearrange slightly to obtain

$$\frac{\partial f_{ep}}{\partial t} + \sum_{j=1}^{l} \dot{\hat{q}}_j \frac{\partial f_{ep}}{\partial \hat{q}_j} + \sum_{j=1}^{l} \dot{\hat{p}}_j \frac{\partial f_{ep}}{\partial \hat{p}_j} = 0.$$

(10.80)

Since $f_{ep} = f_{ep}(\hat{q}_1, \hat{q}_2, \ldots, \hat{q}_l, \hat{p}_1, \hat{p}_2, \ldots, \hat{p}_l, t)$ Eq. (10.80) is equivalent to the total differential of f_{ep} with respect to t being equal to zero:

$$\frac{df_{ep}}{dt} = 0.$$

This is essentially a conservation equation for "convective" transport of particle state points through phase space. This result leads to several important conclusions. First, this equation implies that the density of particle phase points in the neighborhood of any selected moving phase point is constant as that phase point moves along its trajectory. Gibbs referred to this as the principle of conservation of density in phase. Also, if $(\mathbf{p}_0, \mathbf{q}_0, t_0)$ are the coordinates of a particle's phase point at time t_0, Liouville's equation implies that for all subsequent t

$$f_{es}(\mathbf{p}, \mathbf{q}, t) = f_{es}(\mathbf{p}_0, \mathbf{q}_0, t_0) \tag{10.81}$$

for that phase point as it moves along its trajectory through phase space. All subsequent points traversed by a particle in phase space are linked to a unique original point at time t_0. Since all the trajectories are dictated by the same equations of motion, two phase points could occupy the same location in phase space only if they started at the same point, which is not possible since no two particles can occupy the same physical location. It follows that, for noninteracting particles, trajectories of phase points can never cross, and the particle state points must move through phase space along nonintersecting trajectories.

The Boltzmann Transport Equation

By applying Eq. (10.80) to an ensemble of particles, the phase space corresponds to the combination of momenta and coordinates accessible to the particle. Since each momenta is always linearly related to corresponding velocity, we can define generalized coordinates $\hat{z}_j$ and generalized velocities $\hat{w}_j$ such that

$$\hat{q}_j = \hat{z}_j, \tag{10.82}$$

$$\hat{p}_j = \hat{m}_j \hat{w}_j, \tag{10.83}$$

where the $\hat{m}_j$ are generalized inertia factors. Using these relations to replace $\hat{p}_j$ with $\hat{w}_j$ and $\hat{q}_j$ with $\hat{z}_j$ and interpreting the distribution function in terms of velocities as

$$f_{pv}(\mathbf{w}, \mathbf{z}, t)\, d\mathbf{w}\, d\mathbf{z} = \begin{pmatrix} \textit{the number of particles in the ensemble that} \\ \textit{have phase points in } d\mathbf{w}\, d\mathbf{z} \textit{ about } \mathbf{w}, \mathbf{z} \textit{ at time } t \end{pmatrix} \tag{10.84}$$

we can transform Eq. (10.80) to

$$\frac{\partial f_{pv}}{\partial t} + \sum_{j=1}^{l} \dot{\hat{z}}_j \frac{\partial f_{pv}}{\partial \hat{z}_j} + \sum_{j=1}^{l} \dot{\hat{w}}_j \frac{\partial f_{pv}}{\partial \hat{w}_j} = 0. \tag{10.85}$$

In Eq. (10.84), $\mathbf{w}$ and $\mathbf{z}$ are vectors $\hat{w}_1, \hat{w}_2, \ldots, \hat{w}_l$ and $\hat{z}_1, \hat{z}_2, \ldots, \hat{z}_l$, respectively, $d\mathbf{w} = d\hat{w}_1 d\hat{w}_2 \ldots d\hat{w}_l$, and $d\mathbf{z} = d\hat{z}_1 d\hat{z}_2 \ldots d\hat{z}_l$. Since along particle flowlines in phase space $\dot{\hat{z}}_j = \hat{w}_j$, we can further modify Eq. (10.85) to

$$\frac{\partial f_{pv}}{\partial t} + \sum_{j=1}^{l} \hat{w}_j \frac{\partial f_{pv}}{\partial \hat{z}_j} + \sum_{j=1}^{l} \dot{\hat{w}}_j \frac{\partial f_{pv}}{\partial \hat{w}_j} = 0. \tag{10.86}$$

Thus, the Liouville equation indicates that if we follow the particles in a volume element along a flow line in phase space without collisions, the distribution is conserved. It follows that

$$f_{pv}(\mathbf{z} + d\mathbf{z}, \mathbf{w} + d\mathbf{w}, t + dt) = f_{pv}(\mathbf{z}, \mathbf{w}, t). \tag{10.87}$$

If collisions do take place, the distribution f_{pv} will change over a differential time interval dt by an amount $(\partial f_{pv}/\partial t)_{coll} dt$, and therefore

$$f_{pv}(\mathbf{z} + d\mathbf{z}, \mathbf{w} + d\mathbf{w}, t + dt) - f_{pv}(\mathbf{z}, \mathbf{w}, t) = (\partial f_{pv}/\partial t)_{coll} dt, \tag{10.88}$$

which rearranges to

$$\frac{f_{pv}(\mathbf{z} + d\mathbf{z}, \mathbf{w} + d\mathbf{w}, t + dt) - f_{pv}(\mathbf{z}, \mathbf{w}, t)}{dt} = \left(\frac{\partial f_{pv}}{\partial t}\right)_{coll}. \tag{10.89}$$

The left side of Eq. (10.89) is df_{pv}/dt, which from basic calculus is given by

$$\frac{df_{pv}}{dt} = \frac{\partial f_{pv}}{\partial t} + \sum_{j=1}^{l} \hat{w}_j \frac{\partial f_{pv}}{\partial \hat{z}_j} + \sum_{j=1}^{l} \dot{\hat{w}}_j \frac{\partial f_{pv}}{\partial \hat{w}_j}. \tag{10.90}$$

Replacing the left side of Eq. (10.89) with the right side of Eq. (10.90) we obtain

$$\frac{\partial f_{pv}}{\partial t} + \sum_{j=1}^{l} \hat{w}_j \frac{\partial f_{pv}}{\partial \hat{z}_j} + \sum_{j=1}^{l} \dot{\hat{w}}_j \frac{\partial f_{pv}}{\partial \hat{w}_j} = \left(\frac{\partial f_{pv}}{\partial t}\right)_{coll}. \tag{10.91}$$

Equation (10.91) is the *Boltzmann transport equation*. This equation plays a central role in the development of kinetic theory for nonequilibrium systems. Note that if we define a fractional distribution of the particles in the system as

$$f(\mathbf{w}, \mathbf{z}, t) \, d\mathbf{w} \, d\mathbf{z} = \begin{pmatrix} \textit{the fraction of system particles in the ensemble that} \\ \textit{have points in } d\mathbf{w} \ d\mathbf{z} \textit{ about } \mathbf{w}, \mathbf{z} \textit{ at time } t \end{pmatrix} \tag{10.92}$$

then

$$f(\mathbf{w}, \mathbf{z}, t) = \frac{f_{pv}(\mathbf{w}, \mathbf{z}, t)}{N}, \tag{10.93}$$

and by substitution into Eq. (10.91), it is clear that the fractional distribution also satisfies the Boltzmann transport equation,

$$\frac{\partial f}{\partial t} + \sum_{j=1}^{l} \hat{w}_j \frac{\partial f}{\partial \hat{z}_j} + \sum_{j=1}^{l} \dot{\hat{w}}_j \frac{\partial f}{\partial \hat{w}_j} = \left(\frac{\partial f}{\partial t}\right)_{coll}. \tag{10.94}$$

Thus, we are free to work with either a number density distribution for particles or a fractional distribution in our analysis of the time evolution of the system.

For many systems, the collision term in the Boltzmann transport equation $(df/dt)_{coll}$ may be treated with the introduction of a relation time τ_r defined such that

$$\left(\frac{\partial f}{\partial t}\right)_{coll} = -\frac{f - f_0}{\tau_r}, \tag{10.95}$$

where f_0 is the equilibrium distribution for the system. In general τ_r may be a function of velocity (or energy) and position in the system. Combining Eqs. (10.94) and (10.95) we obtain the Boltzmann transport equation with the relaxation time approximation:

$$\frac{\partial f}{\partial t} + \sum_{j=1}^{l} \hat{w}_j \frac{\partial f}{\partial \hat{z}_j} + \sum_{j=1}^{l} \dot{\hat{w}}_j \frac{\partial f}{\partial \hat{w}_j} = -\frac{f - f_0}{\tau_r}. \tag{10.96}$$

As an example, we consider a spatially uniform nonequilibrium distribution of velocities that is established in a system by external mechanisms that are suddenly removed. The system is expected to relax to its equilibrium distribution in the absence of external mechanisms and forces. If no forces act on the particles, the acceleration terms in Eq. (10.96) are zero. The spatial gradient terms are also zero since the distribution is spatially uniform. Using the relaxation time approximation, the Boltzmann transport equation reduces to

$$\frac{\partial f}{\partial t} = -\frac{f - f_0}{\tau_r}. \tag{10.97}$$

If the initial distribution is known to be f_i, the above equation can be solved by elementary methods to give

$$f = f_0 + (f_i - f_0)e^{-t/\tau_r}. \tag{10.98}$$

Example 10.3   A system contains atoms with mass m in a gaseous state. The state of the system may be assumed to be spatially uniform and effects of external force fields may be neglected. Initially the fractional distribution function f is given by

$$f = f_i = \begin{cases} \frac{1}{8v_c^3} & \text{for } -v_c < u < v_c, -v_c < v < v_c, -v_c < w < v_c, \\ 0 & \text{for } v_c < |u|, v_c < |v|, v_c < |w|. \end{cases}$$

Use the Boltzmann transport equation with the relaxation time approximation to explore the time evolution of the distribution function.

Solution   When the system reaches equilibrium, it will establish the Maxwell distribution

$$f = f_0 = \left(\frac{m}{2\pi k_B T}\right)^{3/2} e^{-m(u^2+v^2+w^2)/2k_B T}.$$

Multiplying f_i and f_0 by $(1/2)mc^2$, integrating over all u, v, and w values, and equating the results, we find that conservation of energy requires that the final temperature is related to v_c as follows:

$$v_c = \left(\frac{3k_B T}{m}\right)^{1/2}.$$

If we substitute into Eq. (10.98), the variation of f with time can be expressed as

$$f = (1 - e^{-t/\tau_r})\left(\frac{3}{2\pi}\right)^{1/2} e^{-3(u^2+v^2+w^2)/2v_c^2} + v_c^3 f_i e^{-t/\tau_r},$$

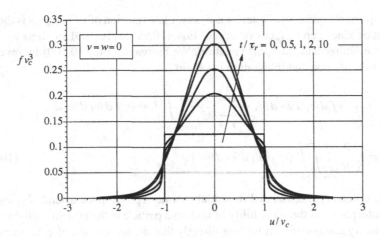

Figure 10.9

where f_i is the step function defined above. Note that f is a function of time and the three velocities that define phase space. To gain some insight into the temporal variation, the variation of f with u along the $v = 0$, $w = 0$ plane is plotted for different times in Figure 10.9. The plot indicates that the time required for the gas to establish its equilibrium distribution is about 5 to 10 times τ_r. If τ_r is on the order of the mean time between collisions, this implies that it takes an average of about 5 to 10 collisions per molecule to reach equilibrium.

Moments of the Boltzmann Equation

The link between the distribution of velocities and macroscopic transport phenomena is established by considering moments of the Boltzmann transport equation. Here we will limit our attention to a system with no external forces fields. We further limit our analysis to a system of particles having transportable quantities that are at most a function of particle translational velocity components. There are then only three translational velocity components to be considered. In such a system, the acceleration terms in the Boltzmann transport equation are zero, reducing it to the form

$$\frac{\partial f}{\partial t} + \sum_{j=1}^{3} \hat{w}_j \frac{\partial f}{\partial \hat{z}_j} = \left(\frac{\partial f}{\partial t}\right)_{\text{coll}}.$$ (10.99)

Let $\psi(\hat{w}_j)$ be any function of translational velocity components but independent of location and time. We multiply both sides of Eq. (10.99) by $\psi(\hat{w}_j)$ and integrate over all possible $\hat{w}_j$ values to obtain

$$\underset{-\infty \text{ to } \infty}{\iiint} \psi(\hat{w}_j)\left[\frac{\partial f}{\partial t} + \sum_{j=1}^{3} \hat{w}_j \frac{\partial f}{\partial \hat{z}_j}\right] d\hat{w}_1\, d\hat{w}_2\, d\hat{w}_3 = \underset{-\infty \text{ to } \infty}{\iiint} \psi(\hat{w}_j)\left(\frac{\partial f}{\partial t}\right)_{\text{coll}} d\hat{w}_1\, d\hat{w}_2\, d\hat{w}_3.$$

(10.100)

When the transportable quantity of interest is a power law function of velocity, as is the case for momentum or kinetic energy, conversion of Eqs. (10.99) to (10.100) amounts to taking a moment of the Boltzmann transport equation. We reorganize Eq. (10.100) by reversing the order of differentiation and integration to obtain

$$\frac{\partial}{\partial t} \underset{-\infty \text{ to } \infty}{\int\int\int} \psi f \, d\hat{w}_1 \, d\hat{w}_2 \, d\hat{w}_3 + \sum_{j=1}^{3} \frac{\partial}{\partial \hat{z}_j} \underset{-\infty \text{ to } \infty}{\int\int\int} \psi \hat{w}_j f \, d\hat{w}_1 \, d\hat{w}_2 \, d\hat{w}_3$$

$$= \left(\frac{\partial}{\partial t} \underset{-\infty \text{ to } \infty}{\int\int\int} \psi f \, d\hat{w}_1 \, d\hat{w}_2 \, d\hat{w}_3 \right)_{\text{coll}}. \tag{10.101}$$

Since f is the fraction of particles in a differential volume of phase space around $(\hat{w}_1, \hat{w}_2, \hat{w}_3)$, it can also be interpreted as the probability of finding a particle in the system with this combination of velocity components. It follows directly that the mean value of ψ is given by

$$\langle \psi \rangle = \underset{-\infty \text{ to } \infty}{\int\int\int} \psi f \, d\hat{w}_1 \, d\hat{w}_2 \, d\hat{w}_3 \tag{10.102}$$

and the mean value of the $\hat{w}_j \psi$ product is

$$\langle \hat{w}_j \psi \rangle = \underset{-\infty \text{ to } \infty}{\int\int\int} \psi f \hat{w}_j \, d\hat{w}_1 \, d\hat{w}_2 \, d\hat{w}_3. \tag{10.103}$$

Using Eqs. (10.102) and (10.103) to replace the first two integrals in Eq. (10.101), we obtain

$$\frac{\partial}{\partial t} \langle \psi \rangle + \sum_{j=1}^{3} \frac{\partial}{\partial \hat{z}_j} \langle \hat{w}_j \psi \rangle = \left(\frac{\partial}{\partial t} \langle \psi \rangle \right)_{\text{coll}}. \tag{10.104}$$

The relation (10.104) applies to any ψ that is a function of particle velocities. For concreteness, we will analyze this equation further for the specific case where ψ is the kinetic energy of a particle:

$$\psi = \tfrac{1}{2} m c^2, \tag{10.105}$$

where

$$c^2 = \hat{w}_1^2 + \hat{w}_2^2 + \hat{w}_3^2. \tag{10.106}$$

Substituting in Eq. (10.104) we get

$$\frac{1}{2} m \frac{\partial}{\partial t} \langle c^2 \rangle + \frac{1}{2} m \sum_{j=1}^{3} \frac{\partial}{\partial \hat{z}_j} \langle \hat{w}_j c^2 \rangle = \left(\frac{\partial}{\partial t} \langle m c^2 / 2 \rangle \right)_{\text{coll}}. \tag{10.107}$$

For elastic collisions, the total energy of the particles in the system does not change with time, and therefore the mean energy per particle in the system does not change with time. It follows that the right side of Eq. (10.107) is zero. It can therefore be written

$$\frac{\partial}{\partial t} \langle c^2 \rangle + \sum_{j=1}^{3} \frac{\partial}{\partial \hat{z}_j} \langle \hat{w}_j c^2 \rangle = 0. \tag{10.108}$$

The mean flux of kinetic energy in the $\hat{z}_j$ direction is given by

$$\langle j_{Q,j} \rangle = \iiint\limits_{-\infty \text{ to } \infty} \rho_N \hat{w}_j (1/2) m \left(\hat{w}_1^2 + \hat{w}_2^2 + \hat{w}_3^2 \right) f \, d\hat{w}_1 \, d\hat{w}_2 \, d\hat{w}_3, \qquad (10.109)$$

where the number density of particles in the system $\rho_N = N/V$ is taken to be constant. The right side of the above equation is therefore $(1/2)m\rho_N$ times $\langle \hat{w}_j c^2 \rangle$ and hence

$$\langle j_{Q,j} \rangle = \left(\tfrac{1}{2} \right) m \rho_N \langle \hat{w}_j c^2 \rangle. \qquad (10.110)$$

Combining Eq. (10.110) with (10.108), we find

$$\frac{\partial}{\partial t} \langle c^2 \rangle + \left(\frac{2}{m\rho_N} \right) \sum_{j=1}^{3} \frac{\partial}{\partial \hat{z}_j} \langle j_{Q,j} \rangle = 0. \qquad (10.111)$$

For a system of particles that stores energy by translation only, we demonstrated in Chapter 2 and in the previous section that

$$\langle c^2 \rangle = \frac{3k_B T}{m} = (2)\frac{3RT}{2\bar{M}} = \frac{2\hat{c}_V T}{\bar{M}}. \qquad (10.112)$$

If the system obeys the ideal gas law, $\hat{c}_V = \hat{c}_P - R = \hat{c}_P - P\bar{M}/m\rho_N T$ and

$$\langle c^2 \rangle = \frac{2\hat{c}_P T}{\bar{M}} - \frac{2P}{m\rho_N}. \qquad (10.113)$$

Substituting for $\langle c^2 \rangle$ in (10.111) and simplifying yields

$$\frac{m\rho_N \hat{c}_P}{\bar{M}} \frac{\partial T}{\partial t} - \frac{\partial P}{\partial t} + \sum_{j=1}^{3} \frac{\partial}{\partial \hat{z}_j} \langle j_{Q,j} \rangle = 0. \qquad (10.114)$$

If the pressure in the system is constant, this further reduces to

$$\frac{m\rho_N \hat{c}_P}{\bar{M}} \frac{\partial T}{\partial t} + \sum_{j=1}^{3} \frac{\partial}{\partial \hat{z}_j} \langle j_{Q,j} \rangle = 0. \qquad (10.115)$$

Equation (10.115) is one form of the well-known *energy transport equation* for a system of particles in the absence of bulk motion. If, as argued in the previous section, the energy flux is related to the temperature gradient as

$$\langle j_{Q,j} \rangle = J_{Q,\hat{z}_j} = -k_t \frac{\partial T}{\partial \hat{z}_j}, \qquad (10.116)$$

we obtain an energy transport equation that is equivalent to Fourier's Law, discussed in Chapter 9 from the point of view of macroscopic nonequilibrium thermodynamics,

$$\frac{m\rho_N \hat{c}_P}{\bar{M}} \frac{\partial T}{\partial t} - \sum_{j=1}^{3} \frac{\partial}{\partial \hat{z}_j} \left(k_t \frac{\partial T}{\partial \hat{z}_j} \right) = 0. \qquad (10.117)$$

Note that the kinetic theory analysis from which the transport equation was derived embodies that assumption that the distribution f is the Boltzmann distribution derived for an equilibrium system, Thus, although properties may vary spatially throughout the system, this analysis is based on the assumption that, locally, f is the Boltzmann distribution for the local temperature and density of the system. This is essentially equivalent to the macroscopic assumption of local thermodynamic equilibrium used widely in macroscopic analysis of nonequilibrium systems. If locally f deviates substantially from the Boltzmann distribution for the local conditions, the validity of the analysis is questionable.

With some additional effort, the energy transport equation with bulk motion effects can be derived in this manner, as well as relations for transport of momentum and transport of species in multicomponent systems. Each of these involve different moments of the Boltzmann transport equation. The interested reader can find the details of such derivations in References [1]–[3]. These derivations provide the conceptual link between the microscale analysis tools of kinetic theory and macroscopic transport theory.

10.3   Thermodynamics of Interfaces

Macroscopic thermodynamic and fluid mechanics treatments of the boundary between two phases often assume a sharp discontinuity in density and/or composition across the boundary. This sharp boundary is termed an *interface* between the phases. In considering the details of phase equilibria and physical processes at the interface, it is important to recognize that changes in properties, which can sometimes be quite large, actually occur over a very thin region centered about the nominal position of the interface. Figure 10.10 schematically shows the density distribution near a liquid–vapor interface. In the liquid, the density is lower in the interface region than in the bulk liquid phase. As described in Chapter 6, the molecules attract each other at moderate separation distances, and energy must be supplied to move them apart. Hence, the mean energy per molecule associated with this type of interaction is greater in the interfacial region than in the bulk liquid. The system thus has an additional free energy per unit area of

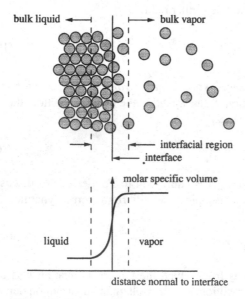

Figure 10.10

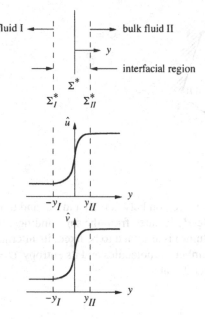

Figure 10.11

interface owing to the slightly larger mean spacing of the molecules in the interfacial region.

To explore these issues in more detail, we specifically consider the interfacial region between two pure fluids that is presumed to be bounded by two surfaces Σ_I and Σ_{II} that are parallel to the nominal interface surface Σ^* but are located in the corresponding bulk phases. As shown in Figure 10.11, the mean specific volume and internal energy actually vary continuously across the interfacial region. Because the continuous transition in mean property values between the bulk phases occurs over a region of finite thickness, the actual mass and internal energy of the region may differ from values that would be calculated by assuming that the bulk phase extended all the way to Σ^*. This difference is called the *surface excess* of the property. The *surface excess mass* $\Gamma_e^{\Sigma^*}$ is defined

$$\Gamma_e^{\Sigma^*} = \int_{-y_I}^{y_{II}} \frac{dy}{\hat{v}} - \frac{y_I}{\hat{v}_I} - \frac{y_{II}}{\hat{v}_{II}} \tag{10.118}$$

and the *surface excess internal energy* $U_e^{\Sigma^*}$ is given by

$$U_e^{\Sigma^*} = \int_{-y_I}^{y_{II}} \frac{\hat{u}\, dy}{\hat{v}} - \frac{\hat{u}_I y_I}{\hat{v}_I} - \frac{\hat{u}_{II} y_{II}}{\hat{v}_{II}}. \tag{10.119}$$

Note that both of these properties are defined per unit area of surface Σ^*. Surface excess values of other thermodynamic properties can be similarly defined. Note that these properties have been give a superscript Σ^* because they depend on the location of surface Σ^*. This concept facilitates a link between the idealized concept of a two-dimensional interface and the actual three-dimensional character of the interfacial region.

To examine the relationships among properties in the interfacial region and properties in the surrounding bulk phases at equilibrium, we consider the system in Figure 10.12. The

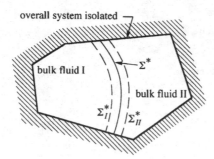

Figure 10.12

entire system, consisting of the interfacial region between Σ_I and Σ_{II} and bulk fluids I and II in Figure 10.12, is taken to be completely isolated from the surroundings. If the shape of the interfacial region (and hence its volume) is assumed to be fixed, its internal energy may be taken to be a function only of the number of molecules and its entropy, $U = U(N, S)$. It follows from results obtained in Chapter 2 that

$$dU = \mu\, dN + T\, dS, \tag{10.120}$$

where

$$T = \left(\frac{\partial U}{\partial S}\right)_{N,\text{shape}}, \qquad \mu = \left(\frac{\partial U}{\partial N}\right)_{S,\text{shape}}. \tag{10.121}$$

We consider possible perturbations that leave bulk fluid I unchanged while permitting exchange of energy and mass between the interfacial region and bulk fluid II. Because the overall system is isolated, we must have

$$dN_{\text{ir}} + dN_{II} = 0, \tag{10.122}$$

$$dS_{\text{ir}} + dS_{II} = 0, \tag{10.123}$$

$$\mu_{\text{ir}} dN_{\text{ir}} + T_{\text{ir}} dS_{\text{ir}} + \mu_{II} dN_{II} + T_{II} dS_{II} = 0. \tag{10.124}$$

Combining Eqs. (10.122)–(10.124) yields

$$(\mu_{\text{ir}} - \mu_{II})\, dN_{\text{ir}} + (T_{\text{ir}} - T_{II})\, dS_{\text{ir}} = 0. \tag{10.125}$$

Because Eq. (10.125) must be satisfied for all possible perturbations of N_{ir} and S_{ir}, necessary conditions for equilibrium are $\mu_{\text{ir}} = \mu_{II}$ and $T_{\text{ir}} = T_{II}$. An identical set of arguments can be presented for perturbation interactions between the interfacial region and bulk phase I. The net result is that, at equilibrium, the temperature and chemical potential must be the same in all three regions:

$$\mu_{\text{ir}} = \mu_I = \mu_{II}, \qquad T_{\text{ir}} = T_I = T_{II}. \tag{10.126}$$

Equation (10.120) applies to the interface region if the interface shape is fixed, regardless of the variation of thermodynamic properties within the region at equilibrium. Taking Eq. (10.120) as representing the actual property variations in the interfacial region and dropping the subscripts on μ and T since they are the same in all regions, we have

$$dU_{\text{ir}} = \mu\, dN_{\text{ir}} + T\, dS_{\text{ir}}. \tag{10.127}$$

An identical relation would apply if the properties equaled those of the respective bulk phases all the way to the interface surface Σ^*. Designating properties in that fictitious case with a prime we have

$$dU'_{ir} = \mu\, dN'_{ir} + T\, dS'_{ir}. \tag{10.128}$$

Subtracting Eq. (10.127) from Eq. (10.128) we obtain

$$d(U'_{ir} - U_{ir}) = \mu d(N'_{ir} - N_{ir}) + T d(S'_{ir} - S_{ir}). \tag{10.129}$$

But by definition, the terms in parentheses in (10.129) are excess properties,

$$U_e^{\Sigma^*} = U'_{ir} - U_{ir}, \qquad N_e^{\Sigma^*} = N'_{ir} - N_{ir}, \qquad S_e^{\Sigma^*} = S'_{ir} - S_{ir}, \tag{10.130}$$

and so Eq. (10.129) can be written

$$dU_e^{\Sigma^*} = \mu\, dN_e^{\Sigma^*} + T\, dS_e^{\Sigma^*} \quad \text{(fixed interface shape).} \tag{10.131}$$

If we now relax the fixed-shape restriction and allow deformation of the interface surface Σ^*, in general both its area and its curvature can change. If the principle radii of curvature for the surface are much greater than the interfacial region thickness, the effect of changing curvature on thermodynamic properties is expected to be very small. Since this is usually the case, we will neglect curvature effects on properties here. However, we do want to account for the effect of changes in interfacial area resulting from deformation of the interface. We therefore assume that in addition to being a function of $N_e^{\Sigma^*}$ and $S_e^{\Sigma^*}$, $U_e^{\Sigma^*}$ is also a function of the area of surface Σ^*: $U_e^{\Sigma^*} = U_e^{\Sigma^*}(N_e^{\Sigma^*}, S_e^{\Sigma^*}, A^{\Sigma^*})$. Thus, mathematically, it follows that

$$dU_e^{\Sigma^*} = \left(\frac{\partial U_e^{\Sigma^*}}{\partial N_e^{\Sigma^*}}\right)_{S_e^{\Sigma^*}, A^{\Sigma^*}} dN_e^{\Sigma^*} + \left(\frac{\partial U_e^{\Sigma^*}}{\partial S_e^{\Sigma^*}}\right)_{N_e^{\Sigma^*}, A^{\Sigma^*}} dS_e^{\Sigma^*} + \left(\frac{\partial U_e^{\Sigma^*}}{\partial A^{\Sigma^*}}\right)_{N_e^{\Sigma^*}, S_e^{\Sigma^*}} dA^{\Sigma^*}, \tag{10.132}$$

and since Eq. (10.132) must reduce to (10.131) if dA^{Σ^*} is zero, we conclude that

$$\left(\frac{\partial U_e^{\Sigma^*}}{\partial S_e^{\Sigma^*}}\right)_{N_e^{\Sigma^*}, A^{\Sigma^*}} = T, \qquad \left(\frac{\partial U_e^{\Sigma^*}}{\partial N_e^{\Sigma^*}}\right)_{S_e^{\Sigma^*}, A^{\Sigma^*}} = \mu. \tag{10.133}$$

We similarly define the *interfacial tension* σ_i as

$$\sigma_i = \left(\frac{\partial U_e^{\Sigma^*}}{\partial A^{\Sigma^*}}\right)_{N_e^{\Sigma^*}, S_e^{\Sigma^*}}, \tag{10.134}$$

which allows us to write Eq. (10.132) as

$$dU_e^{\Sigma^*} = \mu\, dN_e^{\Sigma^*} + T\, dS_e^{\Sigma^*} + \sigma_i\, dA^{\Sigma^*}. \tag{10.135}$$

For the interfacial region, the Helmholtz free energy F is given by

$$F = U - TS. \tag{10.136}$$

Subtracting from Eq. (10.136) the corresponding equation that would apply if the two parts of the interfacial region were occupied by bulk fluids I and II yields an analogous relation for the surface excess free energy:

$$F_e^{\Sigma^*} = U_e^{\Sigma^*} - T S_e^{\Sigma^*}. \tag{10.137}$$

Differentiating Eq. (10.137) and substituting the right side of Eq. (10.135) for $dU_e^{\Sigma*}$ yields

$$dF_e^{\Sigma*} = -S_e^{\Sigma*} dT + \mu \, dN_e^{\Sigma*} + \sigma_i \, dA^{\Sigma*}. \tag{10.138}$$

However, if we consider $F_e^{\Sigma*} = F_e^{\Sigma*}(T, N_e^{\Sigma*}, A^{\Sigma*})$ from a purely mathematical point of view, we can write

$$dF_e^{\Sigma*} = \left(\frac{\partial F_e^{\Sigma*}}{\partial T}\right)_{N_e^{\Sigma*}, A^{\Sigma*}} dT + \left(\frac{\partial F_e^{\Sigma*}}{\partial N_e^{\Sigma*}}\right)_{T, A^{\Sigma*}} dN_e^{\Sigma*} + \left(\frac{\partial F_e^{\Sigma*}}{\partial A^{\Sigma*}}\right)_{T, N_e^{\Sigma*}} dA^{\Sigma*}.$$

$$\tag{10.139}$$

Comparison of Eqs. (10.138) and (10.139) clearly indicates that the interfacial tension is related to the surface excess free energy as

$$\sigma_i = \left(\frac{\partial F_e^{\Sigma*}}{\partial A^{\Sigma*}}\right)_{T, N_e^{\Sigma*}}. \tag{10.140}$$

Thus σ_i is equal to the change in surface excess free energy produced by a unit increase in interface area $A^{\Sigma*}$.

An additional useful relation can be obtained by considering a shift in the nominal interface surface Σ^* uniformly toward region II by some amount dy. The position of Σ^* is somewhat arbitrary because of the continuous property variations in the interface region. But, if our analysis is to yield meaningful results, the values of properties in the three regions in Figure 10.12 should be independent of where Σ^* is located. In allowing Σ^* to shift by dy, we therefore expect that the overall free energy for the system as a whole is unchanged because no change occurs in the physical state. From Eq. (3.75) obtained in Chapter 3, we know that for the pure bulk fluids

$$dF = -S \, dT + \mu \, dN - P \, dV. \tag{10.141}$$

Using Eq. (10.141) for the bulk fluids and Eq. (10.138) for the interfacial region, we can write the total change in F for the three regions in Figure 10.12 as

$$dF = -S_e^{\Sigma*} dT + \mu \, dN_e^{\Sigma*} + \sigma_i \, dA^{\Sigma*}$$
$$- S_I \, dT + \mu \, dN_I - P_I \, dV_I - S_{II} \, dT + \mu \, dN_{II} - P_{II} \, dV_{II} = 0. \tag{10.142}$$

Because the total number of molecules and total volume do not change and there is no change of temperature associated with the position shift,

$$dN = dN_I + dN_{II} + dN_e^{\Sigma*} = 0, \tag{10.143}$$

$$dV = dV_I + dV_{II} = 0, \tag{10.144}$$

$$dT = 0. \tag{10.145}$$

Combining Eqs. (10.142)–(10.145), we can reduce the resulting relation to

$$P_I - P_{II} = \sigma_i \frac{dA^{\Sigma*}}{dV_I}. \tag{10.146}$$

Here we characterize the surface Σ^* by two radii of curvature measured in two perpendicular planes containing the local normal to Σ^*, as indicated in Figure 10.13. The changes

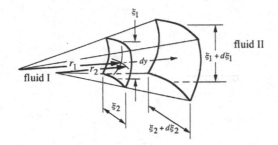

Figure 10.13

in A^{Σ^*} and V_I for a shift in Σ^* of dy are given by

$$dA^{\Sigma^*} = \xi_2\, d\xi_1 + \xi_1\, d\xi_2, \qquad\qquad (10.147)$$

$$dV_I = \xi_1\xi_2\, dy. \qquad\qquad (10.148)$$

Using simple geometric relations between sides of similar triangles, it can be argued that

$$d\xi_1 = \frac{\xi_1}{r_1}\, dy, \qquad d\xi_2 = \frac{\xi_2}{r_2}\, dy. \qquad\qquad (10.149)$$

Combining Eqs. (10.147)–(10.149) yields

$$\frac{dA^{\Sigma^*}}{dV_I} = \frac{1}{r_1} + \frac{1}{r_2}, \qquad\qquad (10.150)$$

which can be combined with Eq. (10.146) to obtain

$$P_I - P_{II} = \sigma_i\left(\frac{1}{r_1} + \frac{1}{r_2}\right). \qquad\qquad (10.151)$$

Equation (10.151) is usually called the *Young–Laplace equation*. It relates the pressure difference across the interface to the interfacial tension and the geometry of the interface at equilibrium. We have shown that the interfacial tension is related to the surface excess free energy as indicated by Eq. (10.140). This links the interfacial tension to the molecular properties that dictate the variation of the molecular spacing (density) and other properties across the interfacial region. The analysis in this section has considered two fluid phases of a pure substance. This type of analysis can also be extended to interfaces in multicomponent systems with slight modifications. Further discussion of interfacial phenomena in multicomponent systems can be found in References [4] and [5].

10.4   Molecular Transport at Interfaces

The mass and energy transfer during vaporization and condensation processes are important in a variety of technological applications. To fully understand the characteristics of such processes, we must consider the liquid–vapor interface at the molecular level. To do so, we will use some of the results from kinetic theory developed in Section 10.1 to interpret the motion of vapor molecules near a liquid–vapor interface. We consider a plane surface in the vapor phase but immediately adjacent to the interface as shown in Figure 10.14.

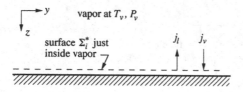

Figure 10.14

Even if no net vaporization or condensation occurs at the interface, a dynamic equilibrium is established in which the molecules from the vapor phase that enter the interfacial region and become part of the liquid phase are balanced, on the average, by an equal number of molecules that escape the liquid into the vapor region. When (net) condensation occurs, the flux of vapor molecules joining the liquid must exceed the flux of liquid molecules escaping into the vapor phase. When (net) vaporization occurs, the opposite must be true.

At least initially, we admit the possibility that the respective values of pressure and temperature in the two bulk phases shown in Figure 10.14 may be different. For the system shown in this figure, the net number flux of molecules across surface Σ_i^* immediately adjacent to the interface must be equal to the difference between the number fluxes j_l and j_v passing through surface Σ_i^* in opposite directions just inside the vapor region:

$$j_{\text{net}} = j_v - j_l. \tag{10.152}$$

To facilitate the evaluation of j_v and j_l in Eq. (10.152), we will assume that here the flux of molecules j_l is characterized by T_l and P_l whereas j_v is characterized by T_v and P_v. With this assumption, it would appear that j_v and j_l can be obtained by using Eq. (10.32), which, when combined with the ideal gas relation, can be written in the form

$$j = \left(\frac{\bar{M}}{2\pi N_A k_B T} \right)^{1/2} \frac{P}{m}, \tag{10.153}$$

where $\bar{M}$ is the molecular weight of the substance. However, use of Eq. (10.153) to evaluate j_v and j_l is inadequate in two respects. First, Eq. (10.153) predicts the flux of molecules in a stationary gas, whereas in the system in Figure 10.14, the vapor must have a net bulk velocity $w = w_0$ in the z direction as a result of the phase change at the interface. The bulk velocity is toward the interface ($w_0 > 0$) for condensation and away from the interface ($w_0 < 0$) for vaporization.

By extending the analysis presented in Section 10.1, it can be shown from kinetic theory (see Schrage [6]) that when the gas moves normal to a planar surface with a speed w_0, the flux of molecules through the plane in the direction of bulk motion is

$$j_{w+} = \Gamma_a \left(\frac{\bar{M}}{2\pi N_A k_B T} \right)^{1/2} \frac{P}{m} \tag{10.154}$$

and the flux of molecules in the direction opposite to that of the bulk motion is

$$j_{w-} = \Gamma_{-a} \left(\frac{\bar{M}}{2\pi N_A k_B T} \right)^{1/2} \frac{P}{m}, \tag{10.155}$$

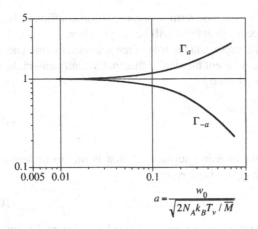

Figure 10.15

where

$$a = \frac{w_0}{(2N_A k_B T / \bar{M})^{1/2}}. \tag{10.156}$$

The factors Γ_a and Γ_{-a}, which correct for the effects of bulk gas motion, are given by the following relations:

$$\Gamma_a = \exp(a^2) + a\pi^{1/2}[1 + \mathrm{erf}(a)] \quad \text{(for } a > 0\text{)}, \tag{10.157}$$

$$\Gamma_{-a} = \exp(a^2) - a\pi^{1/2}[1 - \mathrm{erf}(a)] \quad \text{(for } a < 0\text{)}. \tag{10.158}$$

The variations of Γ_a and Γ_{-a} with a are shown in Figure 10.15.

Because the vapor in Figure 10.14 is expected to be in bulk motion relative to the surface Σ_i^*, it is more appropriate to compute j_v using Eq. (10.154) or (10.155). As suggested by Schrage [6], we assume that the surface Σ_i^* is an infinitesimally small distance from the interface so that there is no bulk motion effect on molecules emerging from the liquid and passing through the surface Σ_i^*. The net flux of molecules from the liquid j_l is therefore calculated using Eq. (10.153).

The second inadequacy in using Eq. (10.153) alone to compute fluxes at the interface is that doing so implicitly assumes that all molecules crossing the surface Σ_i^* in the negative z direction are actually from the liquid phase. Actually only a fraction $\hat{\sigma}_e$ is actually due to vaporization from the liquid. The remaining fraction $1 - \hat{\sigma}_e$ is due to "reflection" of vapor molecules that strike the interface but do not condense. Note that in this context "vaporization" is used to describe molecules escaping the liquid phase and "condensation" is used to describe molecules from the vapor phase being absorbed into the liquid phase. The fraction of molecules crossing the surface Σ_i^* in the positive z direction that condense and are not reflected is designated as $\hat{\sigma}_c$. If no phase change occurs at the interface, equilibrium requires that

$$\hat{\sigma}_e = \hat{\sigma}_c = \hat{\sigma} \quad \text{(at equilibrium)}.$$

Usually $\hat{\sigma}_e$ and $\hat{\sigma}_c$ are assumed to be equal even for the dynamic case when (net) phase change occurs at the interface, although the validity of this assumption is suspect. As a result, $\hat{\sigma}$ is variously referred to as a *vaporization coefficient, evaporation coefficient, condensation coefficient,* or *accommodation coefficient.* The term accommodation coefficient has been perhaps most widely used for $\hat{\sigma}$ in recent years and will be adopted here.

Allowing for the bulk motion of the vapor and the role of the accommodation coefficient as described above, the portion of the incident molecular flux that actually enters the liquid phase upon striking the liquid, j_{vc}, is given by

$$j_{vc} = \begin{cases} \hat{\sigma} \, j_{w+} & \text{for condensation,} \\ \hat{\sigma} \, j_{w-} & \text{for evaporation,} \end{cases} \tag{10.159}$$

and the portion of the molecular flux crossing surface Σ_i^* that is due to molecules that actually emerged from the liquid phase, j_{le}, is given by

$$j_{le} = \hat{\sigma} \, j. \tag{10.160}$$

The net molecular flux to or from the interface as a result of the phase change j_i is just equal to the difference between j_{vc} and j_{le}:

$$j_i = j_{vc} - j_{le}. \tag{10.161}$$

Substituting Eqs. (10.159) and (10.160) into (10.161), evaluating j_{w+} or j_{w-} at P_v and T_v and j at P_l and T_l, and rearranging yields

$$j_i = \frac{\hat{\sigma}}{m} \left(\frac{\bar{M}}{2\pi N_A k_B} \right)^{1/2} \left[\frac{\Gamma P_v}{T_v^{1/2}} - \frac{P_l}{T_l^{1/2}} \right]. \tag{10.162}$$

Macroscopically, the heat flux to the interface $J_{Q,i}$ must equal the net mole flux $j_i m / \bar{M}$ multiplied by the latent heat of vaporization $\Delta \hat{h}_{lv}$. Using Eq. (10.162) to evaluate j_i, we can write the resulting relation for the heat flux as

$$J_{Q,i} = \frac{\hat{\sigma} \Delta \hat{h}_{lv}}{(2\pi N_A k_B)^{1/2}} \left[\frac{\Gamma P_v}{T_v^{1/2}} - \frac{P_l}{T_l^{1/2}} \right]. \tag{10.163}$$

For constant ambient pressure, surrounding vapor must move to replace condensing vapor or it must move away to make room for vapor generated by evaporation. This implies that the bulk velocity of the vapor is related to the net molecular flux at the interface as

$$w_0 = \frac{j_i \hat{v}_v}{N_A} = \frac{j_i k_B T_v}{P_v}, \tag{10.164}$$

which implies that

$$a = \frac{j_i k_B T_v}{P_v} \left(\frac{\bar{M}}{2 N_A k_B T_v} \right)^{1/2}. \tag{10.165}$$

Since $J_{Q,i} = j_i m \Delta \hat{h}_{lv} / \bar{M}$, it also follows that

$$a = \frac{J_{Q,i} \bar{M} k_B T_v}{m \Delta \hat{h}_{lv} P_v} \left(\frac{\bar{M}}{2 N_A k_B T_v} \right)^{1/2}. \tag{10.166}$$

Because Γ is a function of a, Eq. (10.162) is actually an implicit relation for j_i, and Eq. (10.163) is actually an implicit relation for heat flux. If the liquid and vapor phases are in equilibrium (and the interface is planar), the pressure in each phase is expected to equal the saturation pressure at the temperature in that phase. If we assume that this is also true when net condensation or vaporization occurs, Eqs. (10.162) and (10.163) can be written

$$j_i = \frac{\hat{\sigma}}{m} \left(\frac{\bar{M}}{2\pi N_A k_B} \right)^{1/2} \left[\frac{\Gamma P_{sat}(T_v)}{T_v^{1/2}} - \frac{P_{sat}(T_l)}{T_l^{1/2}} \right], \tag{10.167}$$

$$J_{Q,i} = \frac{\hat{\sigma} \Delta \hat{h}_{lv}}{(2\pi \bar{M} N_A k_B)^{1/2}} \left[\frac{\Gamma P_{sat}(T_v)}{T_v^{1/2}} - \frac{P_{sat}(T_l)}{T_l^{1/2}} \right]. \tag{10.168}$$

With this assumption we have linked the molecular flux and the heat flux at the interface to the temperatures in the bulk phases T_v and T_l.

For vaporization and condensation processes at high temperatures, a is often small. For example, condensing or evaporating saturated pure water at 100°C at a heat flux of 100 kW/m^2 corresponds to an a value of 1.3×10^{-4}. In contrast, the values of a for vaporization or condensation processes at cryogenic temperatures (below 100 K) may be much higher. In the limit of small a, Γ_a and Γ_{-a} in Eqs. (10.157) and (10.158) are well approximated by

$$\Gamma = 1 + a\pi^{1/2}. \tag{10.169}$$

Substituting this relation into Eqs. (10.167) and (10.168) and using Eqs. (10.165) and (10.166), with a little manipulation, we obtain the following relations for the molecular flux and heat flux:

$$j_i = \frac{2\hat{\sigma}}{m(2 - \hat{\sigma})} \left(\frac{\bar{M}}{2\pi N_A k_B} \right)^{1/2} \left[\frac{P_{sat}(T_v)}{T_v^{1/2}} - \frac{P_{sat}(T_l)}{T_l^{1/2}} \right], \tag{10.170}$$

$$J_{Q,i} = \left[\frac{2\hat{\sigma}}{(2 - \hat{\sigma})} \right] \frac{\Delta \hat{h}_{lv}}{(2\pi \bar{M} N_A k_B)^{1/2}} \left[\frac{P_{sat}(T_v)}{T_v^{1/2}} - \frac{P_{sat}(T_l)}{T_l^{1/2}} \right]. \tag{10.171}$$

An alternate form of the heat flux relation can be obtained if the relations

$$\Delta P_{sat,vl} = P_{sat}(T_v) - P_{sat}(T_l), \tag{10.172}$$

$$\Delta T_{vl} = T_v - T_l \tag{10.173}$$

are substituted into Eq. (10.171) to eliminate P_l and T_l:

$$J_{Q,i} = \left[\frac{2\hat{\sigma}}{(2 - \hat{\sigma})} \right] \frac{\Delta \hat{h}_{lv}}{(2\pi \bar{M} N_A k_B)^{1/2}} \left[\frac{P_{sat}(T_v)}{T_v^{1/2}} - \frac{P_{sat}(T_v) - \Delta P_{sat,vl}}{(T_v - \Delta T_{vl})^{1/2}} \right]. \tag{10.174}$$

We next expand the second term in the square brackets in terms of $\Delta T_{vl}/T_v$ using the assumption that $\Delta P_{vl}/P_v \ll 1$ and $\Delta T_{vl}/T_v \ll 1$ so that terms of order $(\Delta P_{vl}/P_v)^2$ may be neglected. This leads to the relation

$$J_{Q,i} = \left[\frac{2\hat{\sigma}}{(2 - \hat{\sigma})} \right] \frac{\Delta \hat{h}_{lv}}{(2\pi \bar{M} N_A k_B T_v)^{1/2}} \left[\frac{\Delta P_{sat,vl}}{\Delta T_{vl}} - \frac{P_{sat}(T_v)}{2T_v} \right] \Delta T_{vl}. \tag{10.175}$$

Using the Clapeyron equation (8.81) to evaluate $\Delta P_{sat,vl}/\Delta T_{vl}$ as

$$\frac{\Delta P_{sat,vl}}{\Delta T_{vl}} \cong \left(\frac{dP}{dT} \right)_{sat} = \frac{\Delta \hat{h}_{lv}}{T_v \Delta \hat{v}_{lv}},$$

we can further simplify Eq. (10.175) to

$$
J_{Q,\mathrm{i}} = \left[\frac{2\hat{\sigma}}{(2-\hat{\sigma})} \right] \frac{(\Delta \hat{h}_{\mathrm{lv}})^2}{(2\pi \bar{M} N_A k_B T_v)^{1/2} T_v \Delta \hat{v}_{\mathrm{lv}}} \left[1 - \frac{P_{\mathrm{sat}}(T_v)\Delta \hat{v}_{\mathrm{lv}}}{2\Delta \hat{h}_{\mathrm{lv}}} \right] \Delta T_{\mathrm{vl}}.
$$

$$(10.176)$$

Equation (10.176) is an explicit relation between the heat flux associated with a condensation or vaporization process and the temperature difference across the interface that is required to drive the process. Note that this result applies for small a.

The above analysis of interfacial transport assumes that the fluxes of condensing and vaporizing molecules can be derived from kinetic theory for each flux separately and the results superimposed to obtain the net heat flux. The analysis does not consider nonequilibrium interactions between the molecules leaving the interface and those approaching the interface, which may have different mean energy levels because of the difference between the vapor and liquid temperatures. Inclusion of these effects makes analysis of the system considerably more difficult (see Wilhelm [7]).

Despite its approximate nature, the results of the above analysis provide a useful framework for assessing the effects of interfacial transport on vaporization and condensation processes. It should be noted that the equation relating $J_{Q,\mathrm{i}}$, and ΔT_{vl} derived above applies equally well to vaporization and condensation with the convention that $J_{Q,\mathrm{i}}$ is positive for condensation and negative for vaporization. In engineering analysis of vaporization and condensation processes, it is usually assumed that the vapor and liquid temperatures are both equal to the saturation temperature at the local pressure. The above results indicate that there is some deviation from this assumption of local thermodynamic equilibrium. Fortunately for most cases of practical interest, this deviation is very small.

Inspection of the relations derived above clearly indicates that the relation between heat flux and driving temperature difference depends directly on the value of the accommodation coefficient $\hat{\sigma}$. Quoted values of $\hat{\sigma}$ in the literature vary widely. The tabulation assembled by Paul [8] lists values ranging from 0.02 to 0.04 for some liquids to values very near 1.0 for others. Mills [9] has suggested that molecular accommodation should be imperfect only when the interface is impure. However, recent molecular dynamic simulation studies by Yasuoka et al. [10] imply that the accommodation coefficient for a pure water interface should be near 0.4. Paul [8] noted that in almost every case where $\hat{\sigma}$ was found to be appreciably less than one, the average condition of the vapor molecules was different from that for the liquid phase because of association, disassociation, or polymerization. This suggests that if a fluid is virtually pure and if association, disassociation, and polymerization do not occur to any significant degree, the accommodation coefficient should be close to unity. Because extreme purity is unlikely in most engineering systems, a value of $\hat{\sigma}$ less than one is expected in common applications. Values reported in the literature should therefore be regarded as typical values of $\hat{\sigma}$ rather than constants applicable to all systems since system contamination may alter values for a particular system.

Equation (10.176) indicates that a finite temperature difference is required to drive heat flow across an interface in the presence of vaporization or condensation. This is commonly interpreted as an additional resistance to heat flow. In many cases, this resistance is low and may justifiably be neglected in engineering analysis of such systems. However, Eq. (10.176) indicates that this resistance may be large at high heat flux levels and/or at low accommodation coefficient values. The importance of this resistance can be estimated from Eq. (10.176) with estimates of the system accommodation coefficient.

Example 10.4 In the latter stages of a vaporization process in an evaporator, a thin liquid film of water flows along the walls of the evaporator tube with mostly vapor flowing in the central core of the tube. The pressure in the tube at the location of interest is 100 kPa. A heat flux of 200 kW/m^2 is applied to outer surface of the tube and is conducted through the tube wall and the liquid film to the interface. For vaporization at this heat flux level, estimate the temperature difference across the interface if the accommodation coefficient is taken to be 0.5.

Solution For water at 100 kPa, $\Delta \hat{h}_{lv} = 40.6$ MJ/kmol, $\Delta \hat{v}_{lv} = 30.1$ m^3/kmol, and we take $P_{sat}(T_v) = 100$ kPa and $T_v = T_{sat}(100 \text{ kPa}) = 373$ K. Solving Eq. (10.176) for ΔT_{vl} and substituting for the properties, we obtain

$$\Delta T_{vl} = J_{Q,i} \left[\frac{2 - \hat{\sigma}}{2\hat{\sigma}} \right] \frac{2(2\pi \bar{M} N_A k_B T_v)^{1/2} T_v \Delta \hat{v}_{lv}}{(\Delta \hat{h}_{lv})[2\Delta \hat{h}_{lv} - P_{sat}(T_v) \Delta \hat{v}_{lv}]}$$

$$= 200{,}000 \left[\frac{2 - 0.5}{2(0.5)} \right]$$

$$\times \frac{2(2\pi (18)(6.02 \times 10^{26})(1.38 \times 10^{-23})373)^{1/2} \, 373(30.1)}{(40.6 \times 10^6)[2(40.6 \times 10^6) - (100{,}000)30.1]}$$

$$= 0.040 \text{ K}.$$

10.5 Phase Equilibria in Microscale Multiphase Systems

Small Bubbles in Pure Liquid

In Chapter 6, we discussed the fact that liquid may be superheated beyond its equilibrium saturation temperature at a specified pressure. We further noted that with increasing superheat, system density and energy fluctuations increase in magnitude. In superheated liquid, density fluctuations in the liquid may result in localized regions where the molecular density has been lowered to nearly that for saturated vapor. In real systems, extreme fluctuations of this type may result in the formation of an embryo vapor bubble within the bulk fluid. This process is known as *homogeneous nucleation*. Experimentally, homogeneous nucleation is observed to occur in superheated liquid, forming a very small bubble in the liquid. This type of nucleation process is depicted in Figures 10.16a & b. In the first portion of this section we will explore the necessary conditions for the bubble to be in equilibrium with the surrounding liquid and we will assess the stability of such an equilibrium.

To examine the stable equilibrium conditions for a spherical bubble, we consider the system shown in Figure 10.16b. Note that the liquid pressure is held constant in this system.

At equilibrium, the temperature in the vapor and the liquid must be the same, and the chemical potential in the two phases must be equal,

$$\hat{\mu}_l = \hat{\mu}_{ve}. \tag{10.177}$$

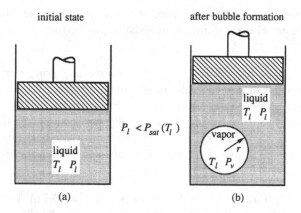

initial state   after bubble formation

Figure 10.16

Because of the curvature of the interface, however, the pressures in the vapor and in the liquid are not equal. For a spherical bubble or radius r_e, the Young–Laplace Eq. (10.151) requires that

$$P_{ve} = P_l + \frac{2\sigma_i}{r_e}. \tag{10.178}$$

Integrating the Gibbs–Duhem equation for a pure substance,

$$d\hat{\mu} = -\hat{s}\,dT + \hat{v}\,dP, \tag{3.50}$$

at constant temperature from $P = P_{sat}(T_l)$ to an arbitrary pressure P, we obtain

$$\hat{\mu} - \hat{\mu}_{sat} = \int_{P_{sat}(T_l)}^{P} \hat{v}\,dP. \tag{10.179}$$

We next use the ideal gas law ($\hat{v} = RT_v/P$) with $T_v = T_l$ to evaluate the integral on the right side of Eq. (10.179) for the vapor phase. This yields the following relation for the chemical potential of the vapor at its equilibrium pressure P_{ve}:

$$\hat{\mu}_{ve} = \hat{\mu}_{sat,v} + RT_l \ln\left\{\frac{P_{ve}}{P_{sat}(T_l)}\right\}. \tag{10.180}$$

For the liquid phase, we use Eq. (10.179) again, but this time, because the liquid is virtually incompressible, $\hat{v}$ is taken to be constant and equal to the value for saturated liquid at T_l, $\hat{v} = \hat{v}_l$. Evaluation of the integral in Eq. (10.179) for $P = P_l$ then yields

$$\hat{\mu}_l = \hat{\mu}_{sat,l} + \hat{v}_l[P_l - P_{sat}(T_l)]. \tag{10.181}$$

Note that for the "normal" flat interface saturation condition, equilibrium requires that $\hat{\mu}_{sat,v} = \hat{\mu}_{sat,l}$. Using this fact while combining Eqs. (10.179), (10.180), and (10.181) we obtain, after a little rearranging, the following relation for the vapor pressure inside the bubble at equilibrium:

$$P_{ve} = P_{sat}(T_l) \exp\left\{\frac{\hat{v}_l[P_l - P_{sat}(T_l)]}{RT_l}\right\}. \tag{10.182}$$

For superheated liquid, P_l must be less than $P_{sat}(T_l)$, as is evident for the isotherm shown in Figure 8.4. Consequently, the exponential term in Eq. (10.182) is less than one and P_{ve}

is less than $P_{sat}(T_1)$. Combining Eqs. (10.178) and (10.182) to eliminate P_{ve}, we obtain the following relation for r_e:

$$r_e = \frac{2\sigma_i}{P_{sat}(T_1)\exp\{\hat{v}_l[P_1 - P_{sat}(T_1)]/RT_1\} - P_1}. \tag{10.183}$$

Thus only bubbles having a radius equal to that given by the above relation will be in equilibrium with the surrounding superheated liquid at T_1 and P_1.

Alternatively, if Eqs. (10.178) and (10.182) are combined to eliminate P_1, the relation for P_{ve} becomes

$$P_{ve} = P_{sat}(T_1)\exp\left\{ \frac{\hat{v}_l[P_{ve} - P_{sat}(T_1) - 2\sigma_i/r_e]}{RT_1} \right\}. \tag{10.184}$$

In most cases, $P_{ve} - P_{sat}(T_1)$ is small compared to $2\sigma_i/r_e$. Hence, P_{ve} is well approximated for most systems by

$$P_{ve} = P_{sat}(T_1)\exp\left\{ \frac{-2\sigma_i\hat{v}_l}{r_e RT_1} \right\}. \tag{10.185}$$

Because the exponential term in the above equation is less than or equal to one, Eq. (10.185) implies, as did Eq. (10.182), that $P_{ve} \leq P_{sat}(T_1)$. The situation regarding the state points of vapor and liquid can be seen more clearly by considering Figure 10.17. If the liquid state point is on the superheated liquid curve at point 1, the vapor state point corresponding to an equal value of chemical potential $\hat{\mu}_{ve} = \hat{\mu}_1$ must lie on the superheated vapor curve at point 2. Hence, for a bubble of finite radius, equilibrium can be achieved only if the liquid is superheated and the vapor is superheated relative to the normal saturation state for a flat interface. Note also that the steep slope of the superheated vapor line results in values of P_{ve} that are much closer to $P_{sat}(T_1)$ than to P_1. Because $P_{ve} - P_1 = 2\sigma_i/r_e$, this provides justification for the assumption that $P_{sat}(T_1) - P_{ve} \ll 2\sigma_i/r_e$, which was used to obtain Eq. (10.185).

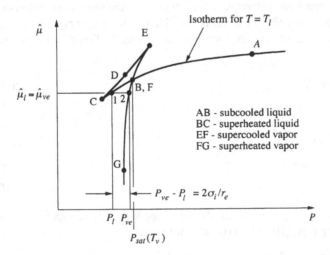

Figure 10.17

Example 10.5   Determine the size of a steam bubble in equilibrium with superheated liquid water at atmospheric pressure and 200°C. Also determine the vapor pressure inside the bubble.

Solution   For water at 200°C, $P_{sat}(T_l) = 1,554$ kPa, $\hat{v}_l = 0.02083$ m³/kmol, and $\sigma_i = 0.0304$ N/m. The size of the bubble at equilibrium is given by Eq. (10.183),

$$r_e = \frac{2\sigma_i}{P_{sat}(T_l)\exp\{\hat{v}_l[P_l - P_{sat}(T_l)]/RT_l\} - P_l}$$

$$= \frac{2(0.0304)}{1,554,000\exp\{.02083[101,000 - 1,554,000]/8308(200 + 273)\} - 101,000}$$

$$= 4.22 \times 10^{-8} \text{ m}.$$

The pressure in the vapor is given by Eq. (10.182),

$$P_{ve} = P_{sat}(T_l)\exp\left\{\frac{\hat{v}_l[P_l - P_{sat}(T_l)]}{RT_l}\right\}$$

$$= 1,554\exp\left\{\frac{.02083[101,000 - 1,554,000]}{8308(200 + 273)}\right\} = 1,542 \text{ kPa}.$$

The pressure in this case is only slightly different from the normal saturation pressure at 200°C.

We now wish to examine the stability of a bubble that satisfies the equilibrium requirement discussed above. We begin by considering the Helmholtz free energy of the system after a bubble of radius r has formed. By definition, $F = U - TS$. Since U and S are additive over subsystems in a composite system, and T is taken to be the same in all subsystems, it follows that F is additive over the subsystems. The Helmholtz free energy for the system as a whole is equal to the sum of contributions associated with the liquid, the vapor, and the interface:

$$F = F_l + F_v + F_i, \tag{10.186}$$

where

$$F_l = \hat{N}_l(\hat{u}_l - T_l\hat{s}_l), \tag{10.187}$$

$$F_v = \hat{N}_v(\hat{u}_v - T_l\hat{s}_v), \tag{10.188}$$

$$F_i = 4\pi r^2\sigma_i. \tag{10.189}$$

In the above relations, $\hat{N}_l$ and $\hat{N}_v$ are the number of moles of liquid and vapor in the system, respectively. Combining Eqs. (10.186)–(10.189), we obtain

$$F = \hat{N}_l(\hat{u}_l - T_l\hat{s}_l) + \hat{N}_v(\hat{u}_v - T_l\hat{s}_v) + 4\pi r^2\sigma_i. \tag{10.190}$$

By definition, the Gibbs free energy is given by

$$G = U - TS + PV. \tag{10.191}$$

Since $F = U - TS$, we can write

$$G = F + PV. \tag{10.192}$$

Since the volume of the system V must equal the sum of the vapor and liquid volumes,

$$V = \hat{N}_1 \hat{v}_1 + \hat{N}_v \hat{v}_v,$$

and we can combine Eqs. (10.190) and (10.191) to obtain

$$G = \hat{N}_1(\hat{u}_1 - T_1\hat{s}_1) + \hat{N}_v(\hat{u}_v - T_1\hat{s}_v) + 4\pi r^2 \sigma_i + P_1(\hat{N}_1\hat{v}_1 + \hat{N}_v\hat{v}_v). \tag{10.193}$$

By definition, the specific Gibbs free energies for the liquid and vapor are given by

$$\hat{g}_1 = \hat{u}_1 - T_1\hat{s}_1 + P_1\hat{v}_1, \tag{10.194}$$

$$\hat{g}_v = \hat{u}_v - T_1\hat{s}_v + P_v\hat{v}_v. \tag{10.195}$$

Reorganizing Eq. (10.193) and using Eqs. (10.194) and (10.195) we can write the relation for the system Gibbs free energy in the form

$$G = \hat{N}_v\hat{g}_v + \hat{N}_1\hat{g}_1 + 4\pi r^2 \sigma_i - \hat{N}_v\hat{v}_v(P_v - P_1). \tag{10.196}$$

The Gibbs free energy of the system before the bubble forms is given by

$$G_0 = (\hat{N}_v + \hat{N}_1)\hat{g}_1. \tag{10.197}$$

The change in Gibbs free energy associated with the bubble formation is therefore

$$\Delta G = G - G_0. \tag{10.198}$$

Combining Eqs. (10.196)–(10.198), we can write the relation for ΔG in the form

$$\Delta G = \hat{N}_v(\hat{g}_v - \hat{g}_1) + 4\pi r^2 \sigma_i - \hat{N}_v\hat{v}_v(P_v - P_1). \tag{10.199}$$

For a pure fluid, we demonstrated in Chapter 2 that $\hat{\mu} = \hat{g}$. Thus, for a bubble at equilibrium with the surrounding liquid, $\hat{\mu}_v = \hat{\mu}_1$, which, for a pure system, implies that $\hat{g}_v = \hat{g}_1$. In addition, at equilibrium, the Young–Laplace equation for the spherical bubble requires that

$$P_v - P_1 = 2\sigma_i/r_e.$$

The product $\hat{N}_v\hat{v}_v$ is the total volume of the bubble, which for a spherical bubble is related to the radius as

$$\hat{N}_v\hat{v}_v = (4/3)\pi r^3. \tag{10.200}$$

It follows that at equilibrium, Eq. (10.199) reduces to

$$\Delta G_e = (4/3)\pi r_e^2 \sigma_i, \tag{10.201}$$

where the radius of the bubble in equilibrium with the surrounding liquid is denoted as r_e. Substituting the right side of Eq. (10.200) for $\hat{N}_v\hat{v}_v$ allows us to write Eq. (10.199) as

$$\Delta G = \hat{N}_v(\hat{g}_v - \hat{g}_1) + 4\pi r^2 \sigma_i - (4/3)\pi r^3(P_v - P_1). \tag{10.202}$$

Because T_1 and P_1 are fixed, $\hat{g}_1$ is a function of T_1 and P_1, and $\hat{g}_v$ is a function of T_1 and P_v, Eq. (10.202) implies that ΔG is a function of P_v, $\hat{v}_v$, and r. Assuming that the vapor obeys the ideal gas law ($\hat{v}_v = RT_1/P_v$) and the bubble is in mechanical equilibrium and satisfies the Young–Laplace equation ($P_v - P_1 = 2\sigma_i/r$) regardless of size, the dependency of ΔG on P_v and $\hat{v}_v$ can be removed, with the result that ΔG is only a function of r. With these idealizations, we can explore the stability of the vapor bubble with a radius near the equilibrium radius r_e by expanding ΔG in a Taylor series about $r = r_e$:

$$\Delta G = \Delta G_e + \left(\frac{d\Delta G}{dr}\right)_{r=r_e} (r - r_e) + \frac{1}{2}\left(\frac{d^2\Delta G}{dr^2}\right)_{r=r_e} (r - r_e)^2 + \cdots .$$

$$(10.203)$$

Using the fact that the Gibbs–Duhem equation implies that

$$\left(\frac{\partial \hat{g}_v}{\partial P_v}\right)_T = \hat{v}_v \qquad (10.204)$$

together with the ideal gas law and the Young–Laplace equation, it can be shown directly by differentiating Eq. (10.202) that

$$\left(\frac{d\Delta G}{dr}\right)_{r=r_e} = 0, \qquad (10.205)$$

$$\left(\frac{d^2\Delta G}{dr^2}\right)_{r=r_e} = -\frac{8\pi\sigma_i}{3}\left[2 + \frac{1}{1 + 2\sigma_i r_e/P_1}\right]. \qquad (10.206)$$

Using these results to evaluate the derivatives in the expansion (10.203) for ΔG together with Eq. (10.201), we obtain

$$\Delta G = \frac{4}{3}\pi\sigma_i r_e^2 - \frac{4\pi\sigma_i}{3}\left[2 + \frac{1}{1 + 2\sigma_i r_e/P_1}\right](r - r_e)^2 + \cdots . \qquad (10.207)$$

Using Eq. (10.201) we can rearrange this equation to the form

$$\frac{\Delta G}{\Delta G_e} = \frac{\Delta G}{(4/3)\pi\sigma_i r_e^2} = 1 - \left[2 + \frac{1}{1 + 2\sigma_i r_e/P_1}\right]\left(\frac{r}{r_e} - 1\right)^2 + \cdots . \qquad (10.208)$$

Note that the term in square brackets in the expansion is approximately equal to 2 for most systems because usually $2\sigma_i r_e/P_1 \ll 1$.

The expansion (10.207) for ΔG indicates that ΔG has a local maximum at $r = r_e$, as indicated in Figure 10.18. It can also be argued that in the limit $r \to \infty$, all the metastable superheated liquid would undergo transition to superheated vapor. Note in Figure 10.17 that a transition from superheated liquid to superheated vapor results in a decrease in $\hat{\mu}$, which implies a decrease in $\hat{g}$ and in the system Gibbs free energy. Thus ΔG must be negative at large r. Because we also know that ΔG must approach zero as $r \to 0$, the overall variation of ΔG with r must look like the solid-line representation of Eq. (10.208) plus the broken-line extensions shown in Figure 10.18.

The variation of ΔG with r shown in Figure 10.18 has several noteworthy features. We noted in Chapter 3 that for a system held at constant temperature and pressure, the Gibbs free energy must be at a minimum for a stable equilibrium. It follows that ΔG must be at a minimum for stable equilibrium in the system considered here, and it is clear from

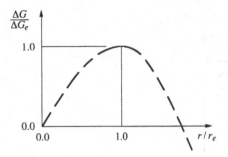

Figure 10.18

Figure 10.18 that a bubble of radius $r = r_e$ is in an unstable equilibrium. We know from our analysis in Section 10.4 that at the interface there is a continual exchange of molecules between the vapor and liquid phases. If a bubble is in equilibrium with $r = r_e$, the loss of one molecule from the bubble decreases its radius into the range $0 < r < r_e$ where $d\Delta G/dr$ is positive. Further reduction of r is expected in this range because spontaneous processes that decrease the Gibbs free energy are most probable. The end result is that once entering this range of r, the bubble is likely to collapse completely. If a bubble of radius $r = r_e$ gains one molecule and enters the range $r_e < r$, decreases as r increases and the bubble is most likely to spontaneously grow.

The above observations provide insight into the nature of homogeneous nucleation of vapor in a superheated liquid. If density fluctuations in a metastable superheated liquid produce an embryo bubble of radius $r < r_e$, the bubble is likely to collapse. Conversely, if the embryo bubble has a radius greater than r_e, it is expected to spontaneously grow, resulting in homogeneous nucleation of the vapor phase in the system.

In the above analysis, the assumption of mechanical equilibrium embodied in the use of the Young–Laplace equation is reasonable if the bubble radius readjusts rapidly to mechanical force variations. However, if the mechanical forces change too rapidly for the liquid to readjust, then mechanical equilibrium may not be maintained. For such conditions, a better assumption would be that the values of chemical potential in the two phases are equal. Use of this idealization, along with the others noted above, results in an expansion for ΔG that is identical to Eq. (10.207) except that the term in square brackets is simply equal to 3. Zeldovich [11] and Kagen [12] have shown that this modified form of the expansion for ΔG is more appropriate for *cavitation* processes in which a superheated liquid condition is created by a sudden (isothermal) decrease in pressure. Regardless of whether chemical or mechanical equilibrium is assumed, the stability characteristics of the vapor bubble are qualitatively the same and quantitatively only slightly different.

The thermodynamic analysis presented in this section forms the foundation for analysis of the kinetics of bubble embryo formation in superheated liquids. The interested reader can find more information on analysis of the kinetic limit of superheat in the references by Carey [13] and Debenedetti [14].

Small Droplets in Pure Vapor

The vapor phase of a pure substance may be brought to a supersaturated state by either cooling it by transferring heat through the walls of a containing structure or by

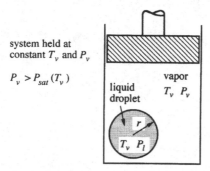

Figure 10.19

rapidly dropping the pressure and temperature of the gas in an adiabatic process. For either
type of process, once a saturated state is reached, further reduction in the system pressure
and temperature or further removal of heat may initiate condensation of some of the vapor
to liquid if nuclei of the liquid phase are present in the system. Often there are adsorbed
molecules in a near-liquid state on the containment walls or on dust particles suspended in
the vapor that serve as nuclei. However, in the absence of these types of nuclei, the processes
mentioned above may reduce the pressure or cool the vapor to a state point in the metastable
vapor region of Figure 8.4.

Heterophase fluctuations of density in a metastable supersaturated vapor may result
in the formation of an embryo droplet that may initiate the condensation process. Such
a homogeneous nucleation process will, as in the case of vaporization, depend on the
thermodynamic stability of droplet embryos in the vapor. To explore these matters further,
we will specifically consider the system shown in Figure 10.19 containing a liquid droplet of
radius r in equilibrium with a surrounding vapor held at fixed temperature T_v and pressure P_v.

At equilibrium, the chemical potential in the vapor and the droplet must be equal:

$$\hat{\mu}_v = \hat{\mu}_l. \tag{10.209}$$

The temperature in both phases must be the same and the pressures in the vapor and liquid
must be related through the Young–Laplace equation,

$$P_{le} = P_v + \frac{2\sigma_i}{r_e}. \tag{10.210}$$

Using the constant-temperature integrated form of the Gibbs–Duhem equation derived
above,

$$\hat{\mu} - \hat{\mu}_{sat} = \int_{P_{sat}(T_l)}^{P} \hat{v}\, dP, \tag{10.179}$$

we evaluate the integral on the right side using the ideal gas law ($\hat{v} = RT_v/P$) for the vapor
to obtain

$$\hat{\mu}_{ve} = \hat{\mu}_{sat,v} + RT_v \ln\left\{ \frac{P_v}{P_{sat}(T_l)} \right\}. \tag{10.211}$$

For the liquid phase inside the droplet, the chemical potential can again be evaluated
using Eq. (10.179). The liquid is taken to be incompressible, with $\hat{v}$ equal to the value for

saturated liquid at T_v, which we designate as $\hat{v}_l$. With this assumption, evaluation of the integral in Eq. (10.179) for $P = P_{le}$ yields

$$\hat{\mu}_{le} = \hat{\mu}_{sat,l} + \hat{v}_l[P_{le} - P_{sat}(T_l)].$$ (10.212)

Equating the values of $\hat{\mu}_{ve}$ and $\hat{\mu}_{le}$ given by Eqs. (10.211) and (10.212) to satisfy Eq. (10.209), and using the fact that $\hat{\mu}_{sat,v} = \hat{\mu}_{sat,l}$, we obtain the following relation:

$$P_v = P_{sat}(T_v)\exp\left\{\frac{\hat{v}_l[P_{le} - P_{sat}(T_v)]}{RT_v}\right\}.$$ (10.213)

Comparison of the above relation with Eq. (10.182) reveals that the relation (10.213) for the equilibrium vapor pressure surrounding a small liquid droplet is similar in form to that for the pressure inside a small vapor bubble in equilibrium with a surrounding liquid pool. As seen in Figure 10.20, if the vapor state point is on the metastable supercooled vapor curve at point 1, the liquid state corresponding to equal $\hat{\mu}$ must lie on the subcooled liquid line at point 2. For a liquid droplet with finite radius, equilibrium can therefore be achieved only if the liquid is subcooled and the vapor is supersaturated relative to the normal saturation state for a flat interface. Equation (10.213) also indicates that if $P_v > P_{sat}(T_v)$, then P_{le} must also be greater than $P_{sat}(T_v)$, which is consistent with the state points indicated in Figure 10.20.

If we substitute Eq. (10.210) to eliminate P_{le}, Eq. (10.213) becomes

$$P_v = P_{sat}(T_v)\exp\left\{\frac{\hat{v}_l[P_v - P_{sat}(T_v) + 2\sigma_i/r_e]}{RT_v}\right\}.$$ (10.214)

In most instances, the steep slope of the supercooled vapor line in Figure 10.20 requires that the value of P_v be much closer to $P_{sat}(T_v)$ than to P_{le}. When this is the case, $P_v - P_{sat}(T_v)$ is small compared to $2\sigma_i/r_e = P_{le} - P_v$, and the relation (10.214) for P_v is well approximated by

$$P_v = P_{sat}(T_v)\exp\left\{\frac{2\hat{v}_l\sigma_i}{r_e RT_v}\right\}.$$ (10.215)

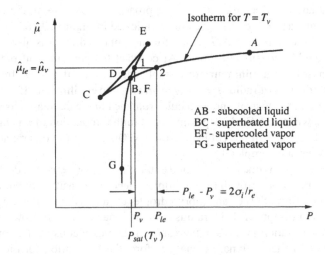

Figure 10.20

Equations (10.214) and (10.215) can also be inverted to solve for the equilibrium droplet radius for a given set of vapor pressure and temperature conditions:

$$r_e = \frac{2\sigma_i}{(RT_v/\hat{v}_l)\ln\{P_v/P_{sat}(T_v)\} - P_v + P_{sat}(T_v)},$$ (10.216)

$$r_e = \frac{2\sigma_i\hat{v}_l}{RT_v\ln\{P_v/P_{sat}(T_v)\}} \quad \text{for } P_v - P_{sat}(T_v) \ll 2\sigma_i/r_e.$$ (10.217)

The stability of an embryo droplet that forms in a system initially containing supercooled vapor can be analyzed in a manner almost identical to that used above to analyze the stability of a bubble in superheated liquid. Such an analysis again considers the change in the Gibbs free energy ΔG associated with the formation of a droplet in a system held at constant temperature T_v and constant pressure P_v, as indicated in Figure 10.19. Proceeding in the manner described earlier in this section for a bubble in superheated liquid, it can be shown that the change in the Gibbs free energy of the system for formation of a droplet with just the right radius r_e to be in equilibrium with the surrounding vapor is

$$\Delta G_e = \tfrac{4}{3}\pi\sigma_i r_e^2.$$ (10.218)

The above result is in fact, identical to the corresponding result obtained in the analysis of the bubble in superheated liquid. Following the same line of analysis used above in analysis of the bubble, it can be shown that for the droplet formation process depicted in Figure 10.19, the leading terms in a Taylor series expansion for ΔG about $r = r_e$ are

$$\Delta G = \tfrac{4}{3}\pi\sigma_i r_e^2 - 4\pi\sigma_i(r - r_e)^2 + \cdots.$$ (10.219)

These results imply that the variation of ΔG with r is qualitatively similar to that shown in Figure 10.18. The expansion indicates that ΔG has a local maximum at $r = r_e$. In the limit $r \to \infty$, all the metastable vapor would undergo a transition to subcooled liquid. Note in Figure 10.20 that a transition from supercooled vapor to subcooled liquid results in a decrease in $\hat{\mu}$, which implies a decrease in $\hat{g}$ and in the system Gibbs free energy. Thus, ΔG must be negative at large r. Because ΔG must approach zero as $r \to 0$, the overall variation of ΔG with r must again look like the curve indicated in Figure 10.18.

The interpretation of the variation of ΔG with r for the embryo droplet is also similar to that for the embryo bubble considered above. At constant temperature and pressure, the Gibbs free energy must be at a minimum for a stable equilibrium. It follows from the expansion (10.219) that a droplet of radius $r = r_e$ is in an unstable equilibrium. If a droplet is in equilibrium with $r = r_e$, the loss of one molecule from the bubble decreases its radius into the range $0 < r < r_e$ where $d\Delta G/dr$ is positive, and further reduction of r is favored because of the associated decrease in the Gibbs free energy. Once entering this range of r, the droplet is likely to evaporate completely.

If a droplet of radius $r = r_e$ gains one molecule and enters the range $r_e < r$, ΔG decreases as r increases, which favors further growth of the droplet. Hence, if density fluctuations in a metastable vapor result in formation of an embryo droplet of radius $r < r_e$, the droplet is likely to lose molecules and disappear. If its radius is greater than r_e, it will grow spontaneously, initiating the condensation process in the vapor via homogeneous nucleation.

The results of this type of thermodynamic analysis form the foundation for modeling of the kinetics of droplet embryo formation in supersaturated vapor. The interested reader

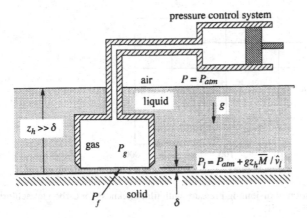

Figure 10.21

can find more information on analysis of the kinetic limit of supersaturation for vapors in the reference by Carey [13].

Ultrathin Liquid Films

We now will turn our attention to very thin liquid films on solid surfaces. If a solid surface is initially in contact with a gas or vapor phase and a small quantity of liquid is brought into contact with the surface, the liquid may bead up into a droplet if it poorly wets the surface, or it may spread over the surface to form a thin liquid film. The latter behavior generally results when there is a strong attractive force interaction between the molecules in the liquid and the molecules of the solid near the interface. To analyze more fully the thin film that occurs for such circumstances, we consider the system shown in Figure 10.21 in which liquid is in contact with and fully wets the horizontal upward-facing solid surface.

A housing with a trapped volume of gas inside is brought into close proximity to the solid surface so that a thin film of liquid of thickness δ exists between the vapor–liquid and liquid–solid interfaces. The liquid does not wet the inside of the housing and a tube connects the vapor space inside the housing to a system that holds the pressure constant.

If the pressure in the gas inside the housing is P_g and the liquid–gas interface is flat, then at equilibrium, P_g must balance the liquid pressure across the interface in the liquid. If δ is large, the pressure across the interface in the liquid film P_f will just equal the local ambient pressure in the liquid P_l given by

$$P_l = \frac{g_e z_h \bar{M}}{\hat{v}_l} + P_{atm}.$$ (10.220)

However, if the housing is brought very close to the solid surface, the pressure inside the housing must not only balance the local liquid pressure P_l, it must also counteract the attractive forces between the liquid molecules and the solid surface, which otherwise would draw liquid into the film to establish a thicker film of liquid on the surface. When the film is very thin, these attractive forces act to pull liquid into the layer as if the pressure in the layer were reduced below the ambient pressure P_l by an amount P_d, which is known as the

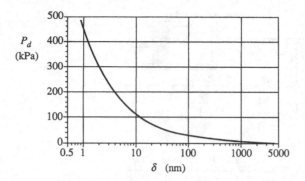

Figure 10.22   Variation of disjoining pressure with film thickness for carbon tetrachloride on glass at 77°C (from Ref. [15]).

disjoining pressure. By convention, if the affinity of the liquid for the solid draws liquid into the film, P_d is taken to be negative.

For the circumstances shown in Figure 10.21, the local pressure P_l and the disjoining pressure effect (due to solid–liquid attraction) act in tandem to thicken the film. To maintain a thin liquid film, the gas pressure force in the housing must balance both effects, which implies that at equilibrium

$$P_g = P_l + |P_d| = P_l - P_d, \tag{10.221}$$

where P_l is given by Eq. (10.220). Note that in Eq. (10.221) because P_d is negative for a liquid that is attracted to the solid surface, $-P_d$ is positive for this system. Because the attractive forces between liquid molecules and those of the solid are expected to be stronger for liquid molecules that are closer to the interface, the disjoining pressure difference $-P_d$ required to counteract them is expected to increase as δ gets smaller. This is reflected in the variation of disjoining pressure with δ indicated in Figure 10.22 for carbon tetrachloride on glass.

If we consider the process of slowly bringing the housing progressively closer to the solid wall while adjusting P_g to maintain equilibrium, the pressure required to thin the layer becomes continually larger until finally the last monolayer of liquid molecules is removed from the surface. The molecules of gas are then in contact with those of the solid to within an interfacial separation δ_0, which is less than the thickness of the liquid monolayer just removed. The work done per unit area of the solid surface, W_f, is given by

$$W_f = \int_{\delta_0}^{\infty} -P_d(\delta)\, d\delta. \tag{10.222}$$

Note that in removing the liquid film on a unit area of solid surface we have created a unit area of solid–gas interface and destroyed a liquid–solid interface and a liquid–gas interface. In Section 10.3, we identified the excess free energy associated with a gas–liquid interface as

$$\sigma_{lg} = \left(\frac{\partial F_e^{\Sigma_{lg}}}{\partial T} \right)_{T, N_e^{\Sigma_{lg}}}. \tag{10.141}$$

We can extend this concept to define the surface excess free energy associated with a unit area of the solid–liquid and solid–gas interfaces:

$$\sigma_{sg} = \left(\frac{\partial F_e^{\Sigma_{sg}}}{\partial T} \right)_{T, N_e^{\Sigma_{sg}}}, \tag{10.223}$$

$$\sigma_{sl} = \left(\frac{\partial F_e^{\Sigma_{sl}}}{\partial T} \right)_{T, N_e^{\Sigma_{sl}}}. \tag{10.224}$$

It follows that if our process of removing the liquid film from the surface is reversible, the work done must equal the free energy gain associated with creation of the solid–vapor interface minus the free energy losses associated with destruction of the liquid–vapor and liquid–solid interfaces. We can state this mathematically as

$$W_f = \sigma_{sg} - \sigma_{gl} - \sigma_{sl}. \tag{10.225}$$

This provides a theoretical link between interfacial tensions and the work required to remove a liquid film from a surface, which we also have related to the concept of disjoining pressure. The right side of Eq. (10.225) is also referred to as the *spreading coefficient* for the liquid since it equals the net free energy gain associated with the spreading of liquid over a solid surface. Consistent with the second law, such spreading will be spontaneous (in a system at constant T, N, and V) if the net free energy change is positive. This type of spontaneous spreading is exhibited by some hydrocarbon liquids on metal surfaces.

For a thin liquid film on a solid surface, like that shown in Figure 10.23, the affinity of the liquid for the solid may affect the equilibrium vapor pressure near the interface. More information about disjoining pressure effects in such circumstances can be obtained by examining the features of a liquid film on a solid surface at a molecular level. We consider first the interaction between a single fluid molecule and the molecules of the solid. This circumstance is shown schematically in Figure 10.24.

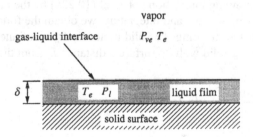

Figure 10.23

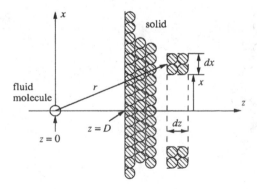

Figure 10.24

We assume here that the separation distance between the fluid molecule and the solid molecules is sufficiently large that only the long-range attractive portion of their pair potential is important. We therefore take the form of the potential to be

$$\phi(r) = -\frac{C_\phi}{r^n}. \tag{10.226}$$

Note that if we take this to be equivalent to the outer portion of the Lennard–Jones 6-12 potential,

$$\phi(r) = 4\varepsilon_0 \left[\left(\frac{\sigma_{LJ}}{r}\right)^{12} - \left(\frac{\sigma_{LJ}}{r}\right)^6 \right], \tag{10.227}$$

then

$$C_\phi = 4\varepsilon_0 \sigma_{LJ}^6, \qquad n = 6. \tag{10.228}$$

We further assume that the force interactions between the single fluid molecule and the molecules in the solid are additive. It follows that the potential energy of the system associated with attractive forces is the sum of the energy increase for each solid molecule as it is brought from infinitely far away ($z \to \infty$) into its position in the solid such that the fluid molecule is a distance $z = D$ from the surface. Consistent with our sign convention (work is positive if done by the system) this must equal the negative of the work done by the system to bring the solid molecules from infinitely far away into place in the solid wall. The work done by the system can be computed by summing the contributions for each molecule in the solid. The number of molecules in a differential ring element of the solid in Figure 10.25 is $2\pi \rho_{N,s} x \, dx \, dz$. Each of these is at a distance $r = \sqrt{x^2 + z^2}$ from the fluid molecule. Weighting the corresponding interaction potential (10.226) by the number of solid molecules in the differential ring volume and integrating, we obtain the following relation for the work done by the system to bring the solid molecule from infinitely far away from the surface to its place in the solid with its surface a distance D from the fluid molecule:

$$
\begin{aligned}
W_D &= \int_{z=D}^{z=\infty} \int_{x=0}^{x=\infty} -\frac{2\pi \rho_{N,s} x C_\phi}{(x^2 + z^2)^{n/2}} \, dx \, dz \\
&= -2\pi \rho_{N,s} C_\phi \int_{z=D}^{z=\infty} \int_{x=0}^{x=\infty} \frac{x}{(x^2 + z^2)^{n/2}} \, dx \, dz.
\end{aligned}
\tag{10.229}
$$

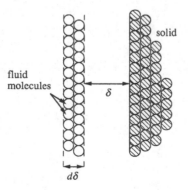

Figure 10.25

The integral over x can be evaluated from standard integral tables. Doing so and evaluating the results at the indicated limits yields

$$W_D = -2\pi \rho_{N,s} C_\phi \int_{z=D}^{z=\infty} \left[\frac{1}{(n-2)z^{n-2}} \right] dz. \tag{10.230}$$

Executing the integration over z, gives

$$W_D = \frac{-2\pi \rho_{N,s} C_\phi D^{3-n}}{(n-2)(n-3)}. \tag{10.231}$$

Note that for $n > 3$, W_D is negative. The work done by the system is therefore negative, which implies that work must be done on the system by some outside agent to move the solid molecules away from the fluid molecule, or conversely, to move the fluid molecule away from the surface.

We now consider a thin slab of fluid molecules a distance delta from the solid surface and having a differential thickness $d\delta$, as indicated in Figure 10.25.

Assuming that we can obtain the net effect by summing the contributions of the fluid molecules in the layer, the work that must be done on the system per unit area to move the thin layer of liquid molecules from distance δ from the surface to infinity is

$$dW_f = \frac{2\pi \rho_{N,s} C_\phi \delta^{3-n}}{(n-2)(n-3)} \rho_{N,l} \, d\delta. \tag{10.232}$$

If the solid was initially in contact with a semi-infinite region of liquid, the work required to remove liquid to a thickness δ_0 is obtained by integrating the above relation:

$$W_f = \int_{\delta_0}^{\infty} \frac{2\pi \rho_{N,s} C_\phi \delta^{3-n}}{(n-2)(n-3)} \rho_{N,l} \, d\delta. \tag{10.233}$$

Note, however, that we previously obtained Eq. (10.222) for W_f from our macroscopic analysis,

$$W_f = \int_{\delta_0}^{\infty} -P_d(\delta) \, d\delta. \tag{10.222}$$

Clearly, for both of the above equations for W_f to be valid, the disjoining pressure must be given by

$$-P_d(\delta) = \frac{2\pi \rho_{N,s} \rho_{N,l} C_\phi \delta^{3-n}}{(n-2)(n-3)}. \tag{10.234}$$

It is conventional to define the Hamaker constant A_H as

$$A_H = \pi^2 \rho_{N,s} \rho_{N,l} C_\phi. \tag{10.235}$$

With this definition, the disjoining pressure relation can be written

$$-P_d(\delta) = \frac{2A_H}{\pi(n-2)(n-3)} \delta^{3-n}. \tag{10.236}$$

If the long-range attractive force potential is equivalent to the long-range portion of the Lennard–Jones 12-6 potential, then $n = 6$ and the above relation reduces to

$$-P_d(\delta) = \frac{A_H}{6\pi} \delta^{-3}. \tag{10.237}$$

Equations (10.235) and (10.237) link disjoining pressure effects to the intermolecular long-range attractive forces between molecules in the liquid near the solid surface and molecules within the solid.

Example 10.6 For carbon tetrachloride (CCl_4), the constants $\varepsilon_0 = 4.51 \times 10^{-21}$ J and $\sigma_{LJ} = 5.9 \times 10^{-10}$ m are suggested for use in the Lennard–Jones 6–12 pair potential. Use these data to estimate the dependence of disjoining pressure on liquid film thickness for carbon tetrachloride on glass (SiO_2).

Solution As an idealized model, we will assume here that the long-range force interaction between a CCl_4 molecule and a molecule in the solid is the same as that between two CCl_4 molecules. It follows that the long-range portion of the pair potential for CCl_4–solid molecule interactions,

$$\phi(r) = -\frac{C_\phi}{r^n}, \tag{10.226}$$

must be equivalent to the outer portion of the Lennard–Jones 6–12 potential (10.227) for CCl_4 pairs interactions. This requires that $n = 6$ and from Eq. (10.228)

$$C_\phi = 4\varepsilon_0\sigma_{LJ}^n = 4(4.51 \times 10^{-21})(5.9 \times 10^{-10})^6 = 7.61 \times 10^{-76} \ Jm^6.$$

The density of the fluid, CCl_4, and the solid, SiO_2, are 1,590 kg/m^3 and 2,210 kg/m^3, respectively. These are converted to number densities by dividing by the corresponding molecular mass and multiplying by Avogadro's number. This gives

$$\rho_{N,l} = \frac{\rho_l}{M_l}N_A = \frac{1590}{154.0}(6.02 \times 10^{26}) = 6.22 \times 10^{27} \ \text{molecules/m}^3,$$

$$\rho_{N,s} = \frac{\rho_s}{M_s}N_A = \frac{2210}{60.1}(6.02 \times 10^{26}) = 2.21 \times 10^{28} \ \text{molecules/m}^3.$$

For $n = 3$, Eq. (10.236) becomes

$$-P_d = A\delta^B,$$

where $A = A_H/6\pi$ and $B = -3$. Using Eq. (10.235), we then compute A as

$$A = \frac{A_H}{6\pi} = \frac{\pi}{6}\rho_{N,l}\rho_{N,s}C_\phi = \frac{\pi}{6}(6.22 \times 10^{27})(2.21 \times 10^{28})7.61 \times 10^{-76}$$

$$= 5.48 \times 10^{-20} \ Pa\,m^{-3}.$$

This idealized analysis indicates that $A = 5.48 \times 10^{-20}$ Pa m^{-3} and $B = -3.0$ in the disjoining pressure relation $-P_d = A\delta^B$. This agrees well with data for thick films ($\delta > 10^{-8}$ m) examined by Potash and Wayner [15]. However, for film thicknesses less than 10^{-8} m, these investigators indicated that $A = 1.782$ Pa m$^{-0.6}$ and $B = -0.6$ provides a better fit to data. This implies that the Lennard–Jones-type pair potential considered in the simple analysis in this section is somewhat inadequate for very thin films.

The equilibrium conditions for a very thin film in contact with a solid surface and its vapor can be examined by considering the system shown in Figure 10.23. At a specified vapor ambient pressure P_{ve}, the system is at equilibrium at temperature T_e. Two necessary conditions for equilibrium are that the temperature and chemical potential in the liquid and vapor phases must be equal. As shown earlier in this section, integration of the Gibbs–Duhem equation along a line of constant temperature from saturation conditions to an arbitrary point on the isotherm in the vapor and liquid regions yields the following two relations:

$$\hat{\mu}_{ve} = \hat{\mu}_{sat,v} + RT_e \ln \left\{ \frac{P_{ve}}{P_{sat}(T_e)} \right\}, \tag{10.238}$$

$$\hat{\mu}_l = \hat{\mu}_{sat,l} + \hat{v}_l[P_l - P_{sat}(T_e)]. \tag{10.239}$$

Note that we have taken the isotherm temperature to be T_e. Setting $\hat{\mu}_{ve} = \hat{\mu}_l$ and solving for P_{ve}, we obtain

$$P_{ve} = P_{sat}(T_e) \exp \left\{ \frac{\hat{v}_l[P_l - P_{sat}(T_e)]}{RT_e} \right\}. \tag{10.240}$$

The analysis above links the disjoining pressure to the film thickness δ by a relation of the form

$$P_d = -A\delta^{-B}. \tag{10.241}$$

(See also the discussion of disjoining pressure in References [15] and [16].) It follows from the above relation that for the system in Figure 10.23

$$P_l = P_{ve} - A\delta^{-B}. \tag{10.242}$$

Substituting this expression into Eq. (10.240) yields

$$P_{ve} = P_{sat}(T_e) \exp \left\{ \frac{\hat{v}_l[P_{ve} - P_{sat}(T_e) - A\delta^{-B}]}{RT_e} \right\}. \tag{10.243}$$

For a specified film thickness, this relation can be used to determine the equilibrium vapor pressure for a given temperature. Alternatively, for a specified system pressure and film thickness, the equilibrium temperature can be determined. By rearranging this relation, we can also obtain an expression for the equilibrium film thickness for a specified set of vapor pressure and temperature conditions:

$$\delta = \left[\frac{A}{P_{sat}(T_e)} \right]^{1/B} \left\{ \frac{P_{ve}}{P_{sat}(T_e)} - 1 - \frac{RT_e}{\hat{v}_l P_{sat}(T_e)} \ln \left\{ \frac{P_{ve}}{P_{sat}(T_e)} \right\} \right\}^{-1/B}. \tag{10.244}$$

At a given vapor pressure, this equation indicates that if the film is thin enough, it can be superheated ($T_e > T_{sat}(P_v)$) and be in equilibrium with the vapor. This means that the liquid is in a superheated metastable state and the vapor is also superheated, as indicated in Figure 10.17. It also implies that heating the solid surface, and hence the liquid film, to a temperature slightly above its normal saturation temperature will not necessarily evaporate the liquid film completely. Instead, an equilibrium thin film may remain on the surface with a thickness that satisfies Eq. (10.244).

Equation (10.244) can be interpreted as indicating that a thin film can exist on a solid surface even when its vapor pressure in the surrounding gas is below the normal saturation pressure for the system temperature. This is consistent with the observed adsorption of thin liquid films onto surfaces such as metal, even when the vapor pressure of the adsorbed species is below the normal saturation value at the system temperature.

The liquid film thickness generally must be exceedingly small for the equilibrium saturation conditions to differ significantly from normal flat-interface saturation conditions. For carbon tetrachloride on glass, Potash and Wayner [15] recommend the values $A = 1.782$ Pa m^B and $B = 0.6$. Using these values with thermodynamic data for CCl$_4$, Eq. (10.231) was used to predict the fractional change in the equilibrium saturation pressure with film thickness at a system temperature of 77°C. The resulting variation is shown in Figure 10.26. For a liquid film thickness of 1 mm (1,000 μm) the fractional change in the vapor pressure is less than 10^{-4}, or less than 0.01%. A 1% change in the vapor pressure would require a film thickness of about 0.1 μm. For films thicker than 1 μm, the deviation from normal equilibrium conditions is insignificant.

For highly wetting liquids (those that tend to spontaneously spread over a surface), the above considerations also affect the thermophysics at the location where a pool of liquid comes into contact with the wall of a container. Such circumstances are shown in Figure 10.27. A highly wetting fluid establishes an extended meniscus at the wall, which

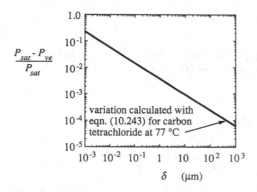

Figure 10.26

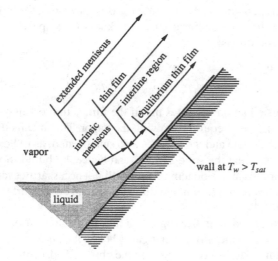

Figure 10.27

includes an intrinsic meniscus region due to interfacial tension effects as well as a thin film that results from the liquid's tendency to spontaneously spread over the surface. If the wall temperature is above the normal saturation temperature, heat may be conducted across the liquid film in this region, resulting in evaporation at the interface. However, because the interface is curved in the intrinsic meniscus region, the wall superheat will have to exceed that required for equilibrium with the interface curvature present. In addition, because of the disjoining pressure effects described above, beyond the intrinsic meniscus a thin film can exist on the wall at equilibrium, even when the wall is superheated above the normal saturation temperature for the local vapor pressure. This thin film can exist without evaporating if its thickness is equal to that specified by Eq. (10.244).

Closing Remarks

For all of the microscale multiphase systems considered in this section, the thermophysics involves dynamic equilibrium with molecular exchange at the interface, and one of the phases exists in a metastable state. Our statistical thermodynamics results in earlier chapters indicates that the fluctuations associated with the metastable state will be low for weak departures from normal saturation conditions. Note that this permits coexistence of a metastable phase with another stable phase in these microscale multiphase systems. Our analysis also indicates, however, that there are limits as to how far the metastable phase can depart from normal saturation conditions. If we consider a bubble in equilibrium with superheated liquid at a fixed temperature, for example, Figure 10.17 indicates that if the radius of the bubble becomes progressively smaller, eventually the equilibrium liquid pressure becomes so low that it goes beyond the limit of intrinsic stability for the liquid. When this limit is exceeded, the system cannot achieve an equilibrium condition. Similar limitations apply to the cases of a droplet in equilibrium with supersaturated vapor and equilibrium for a thin liquid film on a solid surface.

10.6   Microscale Aspects of Electron Transport in Conducting Solids

We now will examine the relationship between microscopic thermophysics of electron behavior in solids and the macroscopic transport of heat and electric current. To do so, it is necessary to determine the heat flux and current using a model that accounts for the velocity and energy distributions of the electrons in the solid. Such an analysis is based on the Boltzmann transport equation developed in Section 10.2.

The system of interest is shown schematically in Figure 10.28. We consider first a metal in which the electron transport can be modeled as transport in an electron gas. In Section 10.2 we demonstrated that the Boltzmann transport equation applies to the distribution f_{pv} defined so that $f_{pv} d\mathbf{u}\, d\mathbf{z}$ is the number of particles in the ensemble (system) that have points at a specific location of phase space (which defines a microstate). Here, we consider each combination of u, v, and w to be a microstate. For an electron gas, we noted in Chapter 6 that, at equilibrium, the mean number of electrons in any microstate with energy ε is given by the Fermi distribution

$$f_F = \frac{1}{1 + e^{(\varepsilon - \varepsilon_F)/k_B T}}. \qquad (10.245)$$

For a specified set of u, v, and w values, the mean occupancy of the velocity microstate is taken to equal the value predicted by the Fermi distribution at $\varepsilon = (1/2)m_e(u^2 + v^2 + w^2)$.

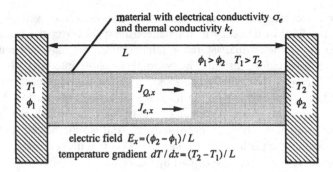

material with electrical conductivity σ_e and thermal conductivity k_t

L

$\phi_1 > \phi_2 \quad T_1 > T_2$

T_1
ϕ_1

$J_{Q,x} \rightarrow$

$J_{e,x} \rightarrow$

T_2
ϕ_2

electric field $E_x = (\phi_2 - \phi_1)/L$

temperature gradient $dT/dx = (T_2 - T_1)/L$

Figure 10.28

To properly convert between the velocity component microstates and the energy microstates on which the Fermi distribution is based, we must properly normalize the distribution function. We therefore take the equilibrium distribution for the electrons to be

$$f_{n0} = \frac{A_f}{1 + e^{(\varepsilon - \varepsilon_F)/k_B T}}. \tag{10.246}$$

If we integrate over all possible values of the velocity components, we account for the contributions of all microstates. If we consider a unit volume of the conducting solid, integrating f_{n0} over all u, v, and w must equal the number of electrons in the unit volume, which is the number density of electrons ρ_N:

$$\rho_N = \int_{-\infty}^{\infty} \int_{-\infty}^{\infty} \int_{-\infty}^{\infty} f_{n0} \, du \, dv \, dw. \tag{10.247}$$

To convert the triple velocity integral to an integral over energy, we first convert the Cartesian velocities to spherical coordinate representation using

$$u = c \sin \theta \cos \phi, \tag{10.248a}$$

$$v = c \sin \theta \sin \phi, \tag{10.248b}$$

$$w = c \cos \theta, \tag{10.248c}$$

where the angles are defined in Figure 10.1. Equation (10.247) then becomes

$$\rho_N = \int_0^{\infty} \int_0^{2\pi} \int_0^{\pi} f_{n0} c^2 \sin \theta c \, d\theta \, d\phi \, dc. \tag{10.249}$$

Aside from the sine term, the integrand is independent of θ and ϕ. We therefore execute the integrations over these variables to obtain

$$\rho_N = 4\pi \int_0^{\infty} f_{n0} \, c^2 dc. \tag{10.250}$$

We next use the fact that

$$c = \left(\frac{2}{m_e}\right)^{1/2} \varepsilon^{1/2} \tag{10.251}$$

to convert the variable of integration from c to ε. Doing so yields

$$\rho_N = \frac{4\pi\sqrt{2}}{m_e^{3/2}} \int_0^\infty \frac{A_f}{1 + e^{(\varepsilon-\varepsilon_F)/k_B T}} \varepsilon^{1/2} \, d\varepsilon. \tag{10.252}$$

Since the Fermi distribution is virtually a step function except at very high temperatures, we evaluate the integral as

$$\rho_N = \frac{4\pi\sqrt{2}}{m_e^{3/2}} \int_0^{\varepsilon_F} A_f \varepsilon^{1/2} \, d\varepsilon = \frac{8\pi\sqrt{2}}{3m_e^{3/2}} A_f \varepsilon_F^{3/2}. \tag{10.253}$$

Solving the above equation for A_f, we obtain

$$A_f = \frac{3m_e^{3/2}\rho_N}{8\pi\sqrt{2}\varepsilon_F^{3/2}}, \tag{10.254}$$

from which it follows that the equilibrium distribution is

$$f_{n0} = \left(\frac{3m_e^{3/2}\rho_N}{8\pi\sqrt{2}\varepsilon_F^{3/2}} \right) \frac{1}{1 + e^{(\varepsilon-\varepsilon_F)/k_B T}}. \tag{10.255}$$

When the system departs from equilibrium, we expect that the distribution obeys the Boltzmann transport equation:

$$\frac{\partial f_n}{\partial t} + \sum_{j=1}^{l} \hat{w}_j \frac{\partial f_n}{\partial \hat{z}_j} + \sum_{j=1}^{l} \dot{\hat{w}}_j \frac{\partial f_n}{\partial \hat{w}_j} = \left(\frac{\partial f_n}{\partial t} \right)_{\text{coll}}. \tag{10.256}$$

In our model analysis here, we will make the assumption that the distribution function with temperature gradients and electric fields present is not very much different from the equilibrium distribution, and the collision term can be represented by a relaxation time τ_r:

$$\left(\frac{\partial f_n}{\partial t} \right)_{\text{coll}} = -\frac{f_n - f_{n0}}{\tau_r}, \tag{10.257}$$

where f_{n0} is the equilibrium distribution for the system. Combining Eqs. (10.256) and (10.257), the Boltzmann transport equation becomes.

$$\frac{\partial f_n}{\partial t} + \sum_{j=1}^{l} \hat{w}_j \frac{\partial f_n}{\partial \hat{z}_j} + \sum_{j=1}^{l} \dot{\hat{w}}_j \frac{\partial f_n}{\partial \hat{w}_j} = -\frac{f_n - f_{n0}}{\tau_r}. \tag{10.258}$$

When only an electric field in the x direction of magnitude E_x is present, the Boltzmann equation reduces to

$$u \frac{\partial f_n}{\partial x} + \left(\frac{-e_e E_x}{m_e} \right) \frac{\partial f_n}{\partial u} = -\frac{f_n - f_{n0}}{\tau_r}, \tag{10.259}$$

where we have replaced the acceleration in the x direction $\dot{u}$ with the electric field force $-e_e E_x$ divided by the mass of the charge carrier m_e.

Based on our assumption that the distribution function is close to the equilibrium distribution for the system f_{n0}, we let $f_n = f_{n0}$ on the left side of Eq. (10.259) and solve for f_n to obtain

$$f_n = f_{n0} + \left(\frac{e_e E_x \tau_r}{m_e}\right)\frac{\partial f_{n0}}{\partial u} - \tau_r u \frac{\partial f_{n0}}{\partial x}. \tag{10.260}$$

Differentiating Eq. (10.255) to obtain $\partial f_{n0}/\partial T$ and $\partial f_{n0}/\partial \varepsilon$ yields

$$\frac{\partial f_{n0}}{\partial T} = -\left(\frac{3m_e^{3/2}\rho_N}{8\pi\sqrt{2}\varepsilon_F^{3/2}}\right)\frac{d}{dT}\left[\frac{\varepsilon - \varepsilon_F}{k_B T}\right]\frac{e^{(\varepsilon-\varepsilon_F)/k_B T}}{(1 + e^{(\varepsilon-\varepsilon_F)/k_B T})^2}, \tag{10.261}$$

$$\frac{\partial f_{n0}}{\partial \varepsilon} = -\left(\frac{3m_e^{3/2}\rho_N}{8\pi\sqrt{2}\varepsilon_F^{3/2}}\right)\left(\frac{1}{k_B T}\right)\frac{e^{(\varepsilon-\varepsilon_F)/k_B T}}{(1 + e^{(\varepsilon-\varepsilon_F)/k_B T})^2}. \tag{10.262}$$

Combining the above two equations to eliminate the exponential terms, we obtain

$$\frac{\partial f_{n0}}{\partial T} = k_B T \frac{d}{dT}\left[\frac{\varepsilon - \varepsilon_F}{k_B T}\right]\frac{\partial f_{n0}}{\partial \varepsilon}. \tag{10.263}$$

Using the chain rule we can write

$$\frac{\partial f_{n0}}{\partial x} = \frac{\partial f_{n0}}{\partial T}\frac{dT}{dx}. \tag{10.264}$$

Substituting the right side of Eq. (10.264) for $\partial f_{n0}/\partial T$ in Eq. (10.263) yields

$$\frac{\partial f_{n0}}{\partial x} = k_B T \frac{d}{dT}\left[\frac{\varepsilon - \varepsilon_F}{k_B T}\right]\frac{\partial f_{n0}}{\partial \varepsilon}\frac{dT}{dx}. \tag{10.265}$$

We can similarly use the chain rule to obtain

$$\frac{\partial f_{n0}}{\partial u} = \frac{\partial f_{n0}}{\partial \varepsilon}\frac{\partial \varepsilon}{\partial u}, \tag{10.266}$$

and since

$$\varepsilon = (1/2)m_e(u^2 + v^2 + w^2) \tag{10.267}$$

it follows that

$$\frac{\partial \varepsilon}{\partial u} = m_e u \tag{10.268}$$

and Eq. (10.266) can be written as

$$\frac{\partial f_{n0}}{\partial u} = m_e u \frac{\partial f_{n0}}{\partial \varepsilon}. \tag{10.269}$$

Using Eqs. (10.265) and (10.267), we can reorganize Eq. (10.260) to the form

$$f_n = f_{n0} + \left(\frac{e_e E_x \tau_r}{m_e}\right)m_e u \frac{\partial f_{n0}}{\partial \varepsilon} - \tau_r u k_B T \frac{d}{dT}\left[\frac{\varepsilon - \varepsilon_F}{k_B T}\right]\frac{\partial f_{n0}}{\partial \varepsilon}\frac{dT}{dx}. \tag{10.270}$$

Executing the differentiation on the right side then gives

$$f_n = f_{n0} + \tau_r u \frac{\partial f_{n0}}{\partial \varepsilon}\left[e_e E_x + \left(\frac{\varepsilon}{T}\right)\frac{dT}{dx} + T\frac{d}{dT}\left(\frac{\varepsilon_F}{T}\right)\frac{dT}{dx}\right]. \tag{10.271}$$

For the system considered here, the electron flux in the x direction is given by

$$J_{e,x} = -\int_{-\infty}^{\infty}\int_{-\infty}^{\infty}\int_{-\infty}^{\infty} e_e u f_n du\, dv\, dw. \tag{10.272}$$

Substituting the right side of Eq. (10.271) for f_n in Eq. (10.272) yields

$$J_{e,x} = -\int_{-\infty}^{\infty}\int_{-\infty}^{\infty}\int_{-\infty}^{\infty} e_e u f_{n0}\, du\, dv\, dw - \int_{-\infty}^{\infty}\int_{-\infty}^{\infty}\int_{-\infty}^{\infty} e_e \tau_r u^2$$
$$\times \frac{\partial f_{n0}}{\partial \varepsilon}\left[e_e E_x + \left(\frac{\varepsilon}{T}\right)\frac{dT}{dx} + T\frac{d}{dT}\left(\frac{\varepsilon_F}{T}\right)\frac{dT}{dx}\right] du\, dv\, dw. \tag{10.273}$$

Because f_{n0} is an even function of u, v, and w, and u is odd, the first triple integral in the above relation is zero, whereupon the relation reduces to

$$J_{e,x} = -\int_{-\infty}^{\infty}\int_{-\infty}^{\infty}\int_{-\infty}^{\infty} e_e \tau_r u^2 \frac{\partial f_{n0}}{\partial \varepsilon}\left[e_e E_x + \left(\frac{\varepsilon}{T}\right)\frac{dT}{dx} + T\frac{d}{dT}\left(\frac{\varepsilon_F}{T}\right)\frac{dT}{dx}\right] du\, dv\, dw. \tag{10.274}$$

We first convert the Cartesian velocities to spherical coordinate representation using Eqs. (10.248a)–(10.248c). Equation (10.274) then becomes

$$J_{e,x} = -\int_0^{\infty}\int_0^{2\pi}\int_0^{\pi} e_e \tau_r \frac{\partial f_{n0}}{\partial \varepsilon}\left[e_e E_x + \left(\frac{\varepsilon}{T}\right)\frac{dT}{dx} + T\frac{d}{dT}\left(\frac{\varepsilon_F}{T}\right)\frac{dT}{dx}\right] c^4$$
$$\times \sin^3\theta \cos^2\phi\, d\theta\, d\phi\, dc. \tag{10.275}$$

Executing the integrations over θ and ϕ, we obtain

$$J_{e,x} = -\frac{4\pi}{3} e_e \int_0^{\infty} \tau_r \frac{\partial f_{n0}}{\partial \varepsilon}\left[e_e E_x + \left(\frac{\varepsilon}{T}\right)\frac{dT}{dx} + T\frac{d}{dT}\left(\frac{\varepsilon_F}{T}\right)\frac{dT}{dx}\right] c^4\, dc. \tag{10.276}$$

Using the relation (10.251), we next convert the variable of integration from c to ε. Doing so yields

$$J_{e,x} = -\frac{\rho_N e_e}{A_f m_e \varepsilon_F^{3/2}} \int_0^{\infty} \tau_r \frac{\partial f_{n0}}{\partial \varepsilon}\left[e_e E_x + \left(\frac{\varepsilon}{T}\right)\frac{dT}{dx} + T\frac{d}{dT}\left(\frac{\varepsilon_F}{T}\right)\frac{dT}{dx}\right] \varepsilon^{3/2}\, d\varepsilon. \tag{10.277}$$

This relation can be rearranged to the form

$$J_{e,x} = -\frac{\rho_N e_e}{A_f m_e \varepsilon_F^{3/2}}\left[e_e E_x + T\frac{d}{dT}\left(\frac{\varepsilon_F}{T}\right)\frac{dT}{dx}\right] \int_0^{\infty} \tau_r \frac{\partial f_{n0}}{\partial \varepsilon} \varepsilon^{3/2}\, d\varepsilon$$
$$- \frac{\rho_N e_e}{A_f m_e \varepsilon_F^{3/2}}\left(\frac{1}{T}\right)\frac{dT}{dx}\int_0^{\infty} \tau_r \frac{\partial f_{n0}}{\partial \varepsilon} \varepsilon^{5/2}\, d\varepsilon. \tag{10.278}$$

The heat flux in the x direction is similarly given by

$$J_{Q,x} = \int_{-\infty}^{\infty} \int_{-\infty}^{\infty} \int_{-\infty}^{\infty} u\varepsilon f_n \, du \, dv \, dw. \tag{10.279}$$

Converting the integrals to polar coordinate form, integrating over the angles, and converting the integral over c to an integral over energy gives

$$J_{Q,x} = \frac{\rho_N}{A_f m_e \varepsilon_F^{3/2}} \left[e_e E_x + T \frac{d}{dT}\left(\frac{\varepsilon_F}{T}\right)\frac{dT}{dx} \right] \int_0^{\infty} \tau_r \frac{\partial f_{n0}}{\partial \varepsilon} \varepsilon^{5/2} \, d\varepsilon$$

$$+ \frac{\rho_N}{A_f m_e \varepsilon_F^{3/2}} \left(\frac{1}{T}\right)\frac{dT}{dx} \int_0^{\infty} \tau_r \frac{\partial f_{n0}}{\partial \varepsilon} \varepsilon^{7/2} \, d\varepsilon. \tag{10.280}$$

For convenience, we define the following integrals:

$$I_{3/2} = -\frac{1}{A_f} \int_0^{\infty} \tau_r \frac{\partial f_{n0}}{\partial \varepsilon} \varepsilon^{3/2} \, d\varepsilon, \tag{10.281}$$

$$I_{5/2} = -\frac{1}{A_f} \int_0^{\infty} \tau_r \frac{\partial f_{n0}}{\partial \varepsilon} \varepsilon^{5/2} \, d\varepsilon, \tag{10.282}$$

$$I_{7/2} = -\frac{1}{A_f} \int_0^{\infty} \tau_r \frac{\partial f_{n0}}{\partial \varepsilon} \varepsilon^{7/2} \, d\varepsilon. \tag{10.283}$$

With these definitions, the electron flux and heat flux relations can be written

$$J_{e,x} = \frac{\rho_N e_e}{m_e \varepsilon_F^{3/2}} \left(\left[e_e E_x + T \frac{d}{dT}\left(\frac{\varepsilon_F}{T}\right)\frac{dT}{dx} \right] I_{3/2} + \left(\frac{1}{T}\right)\frac{dT}{dx} I_{5/2} \right), \tag{10.284}$$

$$J_{Q,x} = -\frac{\rho_N}{m_e \varepsilon_F^{3/2}} \left(\left[e_e E_x + T \frac{d}{dT}\left(\frac{\varepsilon_F}{T}\right)\frac{dT}{dx} \right] I_{5/2} + \left(\frac{1}{T}\right)\frac{dT}{dx} I_{7/2} \right). \tag{10.285}$$

The electrical conductivity for an isothermal conductor, defined as

$$\sigma_e = (J_{e,x}/E_x)_T \tag{10.286}$$

is determined by manipulating Eq. (10.284) to give

$$\sigma_e = \frac{\rho_N e_e^2 I_{3/2}}{m_e \varepsilon_F^{3/2}}. \tag{10.287}$$

Note that in the above equations, the electron number density ρ_N is computed as the atom number density in the metal multiplied by the number of valence electrons per atom. If we set $J_{e,x}$ to zero and combine Eqs. (10.284) and (10.285), we obtain the following relation for the zero-current heat flux:

$$J_{Q,x} = -\frac{\rho_N}{m_e \varepsilon_F^{3/2} T} \left(\frac{I_{7/2} I_{3/2} - I_{5/2}^2}{I_{3/2}} \right) \frac{dT}{dx}. \tag{10.288}$$

The definition of the zero-current thermal conductivity is

$$k_t = -\frac{J_{Q,x}}{dT/dx}. \tag{10.289}$$

Using Eq. (10.288), the thermal conductivity is determined to be

$$k_t = \frac{\rho_N}{m_e \varepsilon_F^{3/2} T} \left(\frac{I_{7/2} I_{3/2} - I_{5/2}^2}{I_{3/2}} \right). \tag{10.290}$$

Computational evaluation of the fluxes and conductivities requires that we evaluate the integrals defined in Eqs. (10.281)–(10.283). Since f_{n0} is known, the main difficulty is accounting for the relaxation time τ_r. In general, τ_r is proportional to the time between collisions and is therefore a function of the energy of the particle. All the integrals are of the form

$$I_n = -\frac{1}{A_f} \int_0^\infty \tau_r \frac{\partial f_{n0}}{\partial \varepsilon} \varepsilon^n d\varepsilon. \tag{10.291}$$

We first expand the relation (10.262) in a Taylor series about ε_F. Retaining terms to second order, we find

$$\frac{\partial f_{n0}}{\partial \varepsilon} = -\left(\frac{3m_e^{3/2} \rho_N}{32\pi \sqrt{2} \varepsilon_F^{3/2} k_B T} \right) \left[1 - \frac{\xi^2}{4} + \cdots \right], \tag{10.292}$$

where

$$\xi = (\varepsilon - \varepsilon_F)/k_B T. \tag{10.293}$$

Since to second order

$$e^{-x^2} = 1 - x^2 + \cdots, \tag{10.294}$$

we choose to modify Eq. (10.292) to the form

$$\frac{\partial f_{n0}}{\partial \varepsilon} = -\left(\frac{3m_e^{3/2} \rho_N}{32\pi \sqrt{2} \varepsilon_F^{3/2} k_B T} \right) e^{-\xi^2/4}, \tag{10.295}$$

where we expect this approximation to be accurate to second order in ξ. Also, by definition

$$\varepsilon^n = (\varepsilon_F + k_B T \xi)^n = \varepsilon_F^n (1 + k_B T \xi / \varepsilon_F)^n. \tag{10.296}$$

We expand the second factor on the right of the above relation for small $k_B T \xi / \varepsilon_F$ to obtain

$$\varepsilon^n = \varepsilon_F^n \left[1 + n \frac{k_B T \xi}{\varepsilon_F} + \frac{n(n-1)}{2} \left(\frac{k_B T \xi}{\varepsilon_F} \right)^2 + \cdots \right]. \tag{10.297}$$

To account for the variation of τ_r with ε, we also expand τ_r in a Taylor series about ε_F:

$$\tau_r = \tau_{r,F} + \left(\frac{d\tau_r}{d\varepsilon} \right)_{\varepsilon_F} (\varepsilon - \varepsilon_F) + \frac{1}{2} \left(\frac{d^2\tau_r}{d\varepsilon^2} \right)_{\varepsilon_F} (\varepsilon - \varepsilon_F)^2 + \cdots. \tag{10.298}$$

Assuming a power law variation of τ_r with ε,

$$\tau_r = A_r \varepsilon^{-\gamma}, \tag{10.299}$$

we can evaluate the derivatives in the above equation and the expansion can be manipulated to the form

$$\tau_r = \tau_{r,F} \left[1 - \gamma \frac{k_B T \xi}{\varepsilon_F} + \frac{\gamma(\gamma+1)}{2} \left(\frac{k_B T \xi}{\varepsilon_F} \right)^2 + \cdots \right]. \tag{10.300}$$

Inserting the right sides of Eqs. (10.295), (10.297), and (10.300) into the appropriate locations in (10.291), multiplying the series, and keeping terms to second order, we obtain

$$I_n = -\frac{\tau_{r,F}\varepsilon_F^n}{4} \int_{-\varepsilon_F/k_BT=-\infty}^{\infty} \left[1 + (n-\gamma)\frac{k_BT\xi}{\varepsilon_F} + \left(\frac{\gamma^2+\gamma}{2} - \gamma n + \frac{n(n-1)}{2}\right)\right.$$
$$\left. \times \left(\frac{k_BT\xi}{\varepsilon_F}\right)^2 + \cdots\right]e^{-\xi^2/4}d\xi. \tag{10.301}$$

The integral can be evaluated term by term using standard integral tables, which yields the following relation:

$$I_n = -\frac{\tau_{r,F}\varepsilon_F^n\sqrt{\pi}}{2}\left[1 + (\gamma^2 + (1-2n)\gamma + n(n-1))\left(\frac{k_BT}{\varepsilon_F}\right)^2 + \cdots\right]. \tag{10.302}$$

We can use this relation to evaluate the integrals in the flux and conductivity relations.

Another interesting result is obtained by considering the ratio known as the Lorenz number Lz:

$$Lz = \frac{k_t}{\sigma_e T}. \tag{10.303}$$

Substituting the relations (10.287) and (10.290) for the electrical and thermal conductivities yields

$$Lz = \left(\frac{I_{7/2}I_{3/2} + I_{5/2}^2}{k_B^2 T^2 I_{3/2}^2}\right)\frac{k_B^2}{e_e^2}. \tag{10.304}$$

Equation (10.302) can be used to evaluate $I_{3/2}$, $I_{5/2}$, and $I_{7/2}$ in Eq. (10.296). Doing so indicates that to second order in k_BT/ε_F, the factor in parentheses in Eq. (10.304) is 4, regardless of the value of γ. This implies that

$$Lz = 4\frac{k_B^2}{e_e^2}. \tag{10.305}$$

A more rigorous treatment indicates that the factor in parentheses in Eq. (10.304) is about 3.3. The value of $Lz(e_e^2/k_B^2)$ for most metals is, in fact, close to 3.3, as indicated in Figure 10.29.

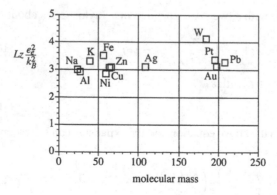

Figure 10.29

Although approximate, our analysis here indicates a value that is within about 25% of that for most metals. This relation between the thermal and electrical conductivity for metals is known as the *Weidermann–Franz law*.

This line of analysis can be applied to semiconductors, but the analysis becomes much more complicated if the accessible electron energies have a band structure. If the energy distribution is well approximated by a continuous relation, the analysis is somewhat easier. The Fermi distribution used above applies to metals where the number density of free electrons in the gas is high. In semiconductors, the number density of electrons becomes low enough that the allowable energy microstates are not densely populated. In these circumstances the system exhibits dilute occupancy and the distribution deviates from that for a fully degenerate fermion gas, approaching the classical form of a Maxwellian distribution with a mean energy approximately equal to the Fermi energy. As a first approximation, semiconductors have been modeled using a Boltzmann distribution centered on the Fermi energy. Such a distribution would have the form

$$f_{n0} = C_{\mathrm{f}} \left(\frac{h^2}{2\pi m_e k_{\mathrm{B}} T} \right)^{3/2} e^{-(\varepsilon - \varepsilon_{\mathrm{F}})/k_{\mathrm{B}} T}, \tag{10.306}$$

where C_{f} is an appropriate normalization constant. Note that this is consistent with our observation in Chapter 6 that fermion gas behavior approaches that of a classical gas in the limit of dilute occupancy. Constructing a model of this type with a power-law dependence of τ_{c} on ε yields results similar to those above for metals, except that $Lz(e_e^2/k_{\mathrm{B}}^2)$ depends on the power-law exponent γ.

The analysis of electron transport presented in this section can also be extended to derive information about thermoelectric effects. For example, setting $J_{e,x} = 0$ in Eq. (10.284), we obtain the relation for the electric field induced by the presence of a temperature gradient:

$$E_x = \left(\frac{1}{e_e T} \right) \frac{dT}{dx} \left(\frac{\varepsilon_{\mathrm{F}} I_{3/2} - I_{5/2}}{I_{3/2}} \right). \tag{10.307}$$

In Chapter 9, we defined the absolute thermoelectric power as the change in electric potential per unit change in temperature along a conductor. Thus, for material A,

$$\hat{\varepsilon}_A = \frac{\Delta \phi}{\Delta T} = \frac{(d\phi/dx)\Delta x}{(dT/dx)\Delta x} = \frac{(d\phi/dx)}{(dT/dx)}, \tag{10.308}$$

and since basic electrostatics dictates that $E_x = d\phi/dx$, we can combine Eqs. (10.307) and (10.308) to obtain

$$\hat{\varepsilon}_A = \left(\frac{1}{e_e T} \right) \left(\frac{\varepsilon_{\mathrm{F}} I_{3/2} - I_{5/2}}{I_{3/2}} \right). \tag{10.309}$$

Using Eq. (10.302) to evaluate and $I_{3/2}$ and $I_{5/2}$ leads to

$$\hat{\varepsilon}_A = (2\gamma - 6) \left(\frac{k_{\mathrm{B}} T}{\varepsilon_{\mathrm{F}}} \right) \left(\frac{k_{\mathrm{B}}}{e_e} \right). \tag{10.310}$$

Note that since $k_{\mathrm{B}} T/\varepsilon_{\mathrm{F}}$ is the Fermi temperature θ_{F}, this can be written in the form

$$\hat{\varepsilon}_A = (2\gamma - 6) \left(\frac{T}{\theta_{\mathrm{F}}} \right) \left(\frac{k_{\mathrm{B}}}{e_e} \right). \tag{10.311}$$

The ratio k_B/e_e is equal to 86.1 μV/K. Because the Fermi temperature for most metals is typically greater than 10,000 K (see Table 7.1), the absolute thermoelectric power for most metals is very low, being on the order of 1–10 μV/K. With some additional effort, relations for other thermoelectric properties can also be obtained.

Example 10.7   Use the results of the analysis in this section to estimate the absolute thermoelectric power for aluminum at 300 K. Assume that the relaxation time τ_r is independent of energy.

   Solution   For copper, the Fermi temperature is approximately 134,900 K (see Table 7.1). Since the relaxation time is assumed to be constant, we use Eq. (10.311) with $\gamma = 0$:

$$\hat{\varepsilon}_{Al} = -6\left(\frac{T}{\theta_F}\right)\left(\frac{k_B}{e_e}\right).$$

Substituting the parameter values, we obtain

$$\hat{\varepsilon}_{Al} = -6\left(\frac{300}{81,200}\right)(86.1) = -1.2\,\mu\text{V/K}.$$

The value predicted by this relation is only slightly different from the value of about $-1.7\,\mu$V/K indicated by measured data.

   This line of analysis using the Boltzmann transport equation provides a linkage between the microphysics of the electron behavior in the conductor and the macroscopic properties of the material. It also provides a kinetic theory foundation for the linear transport models discussed in Chapter 9. It should be noted that in this model we have neglected interactions between electrons and the ions in the metal lattice as well as the scattering of electrons off of contaminants and imperfections in the lattice. Because energy storage in the lattice is associated with phonons (vibrations of the lattice; see Chapter 7), the interaction of electrons with lattice waves is modeled as electron–phonon collisions. Scattering of electrons due to imperfections in the lattice and collisions with phonons inhibit electron transport and neglecting these effects generally overestimates the electrical conductivity. Phonons as well as electrons transport heat in the metal. In metals, the electron concentration is large and the electron contribution dominates the thermal conductivity of the material. Although the kinetic theory model developed here is highly idealized, its predictions are comparable to the behavior of real metals, and the model provides useful insight into how free electron behavior affects the thermoelectric properties of the metal. The interested reader can find further information about the microphysics of transport phenomena in conducting materials in References [17] and [18] at the end of this chapter.

10.7   The Breakdown of Classical and Continuum Theories at Small Length and Time Scales

   The purpose of this final section is to summarize conclusions regarding the breakdown of classical and continuum theories in systems characterized by small length or times scales. A primary objective throughout this text has been to construct theoretical machinery that can be applied to macroscopic system processes from information about the microstructure of matter. In doing so we have taken advantage of two features. The first is that quantum effects vanish at high temperatures and/or large system sizes. The second is that the statistics of a system of particles becomes increasingly deterministic as the number of particles in the system becomes large.

Length Scales

We have identified two length scales that play a central role in assessing the appropriateness of classical, continuum models of thermal phenomena. The first is the thermal de Broglie wavelength Λ defined as

$$\Lambda = \left(\frac{h^2}{2\pi m k_B T} \right)^{1/2}.$$ (10.312)

As discussed in Chapter 4, if the characteristic dimension of the system, L, is comparable to or smaller than Λ, quantum effects are important and must be included in any model analysis. Thus

$$L \cong \Lambda$$ (10.313)

defines one lower bound on the system size for applicability of classical thermodynamics theory. Our analysis in Chapter 4 indicated that $L \leq \Lambda$ is likely only for very low temperatures and/or systems containing particles of very low mass.

In gaseous systems, the mean free path of the particles between collisions λ_c is another limiting length scale. If the length scale of the system is comparable to or smaller than λ_c, the particles interact with a boundary of the system and then, on the average, travel a distance more that L before interacting with another particle in the system. At mean free path values comparable to L, the system cannot be accurately modeled as a continuum. Depending on the type of boundary conditions on the system, the energy distribution within the system may depart substantially from the Boltzmann distribution and consequently the behavior of the system may depart from predictions of equilibrium theory.

In condensed matter systems such as solid crystals or liquids, the particles continuously interact and so the concept of mean free path for the atoms or molecules is inapplicable. However, it reappears when the electron gas model is applied or energy storage or transport in the lattice is modeled in terms of phonons. When the system characteristic length becomes comparable to the mean free path of the electron or phonon, local thermodynamic equilibrium may not be achieved and the behavior of the system will deviate from the mean behavior for a continuum.

Another intrinsic length scale for a condensed matter systems is the mean spacing between atoms or molecules L_{ms}. This can be computed approximately from the molar specific volume as

$$L_{ms} = (\hat{v}_l/N_A)^{1/3}.$$ (10.314)

When the dimensions of the system approach L_{ms}, the number of molecules becomes too small to effectively average in statistical theory. Also, as noted in our consideration of fluctuations in systems of molecules, in general, the relative magnitude of fluctuations in density and energy generally increases as the number of molecules becomes small. The system therefore will not have well-defined average properties in the usual thermodynamic sense.

Interfacial regions at a liquid–gas interface and thin adsorbed liquid films are examples of cases where the system dimension is beginning to approach the mean separation distance between molecules. We have extended our thermodynamic analysis to such systems by introducing a macroscopic treatment of the special microscale features of the system. Clearly, as this length scale is approached, classical continuum theory becomes suspect and its accuracy must be reassessed.

In two-phase gas–liquid systems we have seen that if the dimensions of the dispersed phase become small, the conditions for equilibrium become much different from those in which length scales of both phases are large. The departures from normal (flat interface) equilibrium conditions become large as the limit of intrinsic stability is approached. In superheated liquids, this limit corresponds approximately to about 0.9 times the critical temperature. Applying Eq. (10.183) at this temperature, we conclude that the departure from normal equilibrium conditions becomes substantial when the size of a bubble approaches r_{il}, given by

$$r_{il} = \frac{2\sigma_i}{P_{sat}(0.9T_c) \exp\{\hat{v}_l[P_l - P_{sat}(0.9T_c)]/R(0.9T_c)\} - P_l}. \tag{10.315}$$

Similar considerations apply to the limiting dimensions of liquid droplets in supersaturated vapor and thin liquid films in superheated vapor.

In addition to the physical length scale for the system, the length scale over which temperature changes significantly is also important in assessing the viability of classical continuum theory. This length scale $L_{\Delta T}$ is defined as

$$L_{\Delta T} = \frac{T}{\nabla T}. \tag{10.316}$$

If boundary conditions are established in a system that result in $L_{\Delta T}$ being comparable to the mean free path of the particles, the continuum model will not accurately represent the system behavior. From the definition, this clearly is likely to occur only if the system temperature is low and/or the temperature gradient is very large. When such conditions exist, use of classical continuum models must be reconsidered.

Time Scales

In gaseous systems, collisions redistribute energy in the system, which facilitates its evolution toward equilibrium. Equilibrium statistical thermodynamics provides useful predictions of physical properties in gaseous systems if the time scale over which the system is permitted to reach equilibrium is much longer than the mean time between collisions $\langle \tau_c \rangle$. Fortunately, in most systems $\langle \tau_c \rangle$ is very small and this requirement is satisfied. However, in some systems this is not satisfied and the system may exist in a nonequilibrium state for long periods on time, making it impossible to analyze the system behavior with classical thermodynamics tools.

Within gaseous systems, the number of collisions per molecule required to reach an equilibrium distribution may be different for different modes of energy storage. In some systems, for example, the translational energy may achieve a Boltzmann distribution in about 5 collisions per molecule, while it may take an average of 20 collisions for rotational energy and an average of 50 collisions per molecule for the vibrational energy to achieve equilibrium distributions. This difference occurs because rotational and vibrational energy are not exchanged in every collision. In high speed gas flow across a shock wave, the residence time of the gas in the system may be small compared to the time required for the gas to reach full equilibrium, resulting in a gas flow downstream of the shock that is far from equilibrium. If the rotation and/or vibration modes are very slow to equilibrate, they are sometimes said to be "frozen" in the distributions that existed before the gas passed through the shock.

In solids, the time constants of interest are the mean time between collisions for phonons and electrons. In an analogous manner to gases, these time constants dictate the rate at which energy is transported and distributed in the material. In a solid crystal, the characteristic time associated with lattice vibrations is on the order of the mean of the inverse of the frequency of lattice vibrations:

$$\langle \tau_c \rangle \sim \left\langle \frac{1}{\nu} \right\rangle. \tag{10.317}$$

At very low temperatures the lower energy, lower frequency modes contain most of the energy and the time between phonon collisions is longer than at higher temperatures where higher frequencies contain more energy resulting in a higher mean time between phonon collisions. In metals where a substantial portion of energy transport is by free electrons, the mean time between electron collisions with electrons and phonons is a limiting time scale for equilibration of the system and transport of heat.

Very short time scale pulsed laser heating of solids is now possible in which energy is delivered to a localized region of the solid over a time interval that is comparable to the mean time between collisions of phonons and electrons. Processes of this type cannot be modeled using classical tools based on local thermodynamic equilibrium because the system cannot react fast enough to reach local equilibrium. These time scales thus define limits to the applicability of classical continuum tools that are commonly used in larger-scale slower processes.

Other Considerations

Another feature of systems that can cause them to deviate from conventional classical thermodynamic behavior is high levels of property fluctuations. We have shown that pure liquids and gases may exist in metastable states in which $-(\partial P / \partial \hat{v})_T$ may be small, resulting in very large density and energy fluctuations. When this is so, the properties vary significantly with time. This is quite unlike systems in which $-(\partial P / \partial \hat{v})_T$ is large and the system properties are virtually invariant with time. The classical definition of an equilibrium state does not exist for systems that exhibit severe property fluctuations. As we have seen in previous chapters, this type of nonclassical behavior is exhibited near the limits of intrinsic stability and near the critical point. Small $-(\partial P / \partial \hat{v})_T$ thus represents another limit of the range of applicability of classical thermodynamic theory. We have also shown in Section 10.4 that at high vaporization or condensation rates, the conditions at liquid–vapor interfaces may deviate significantly from those for local thermodynamic equilibrium.

Other Analysis Tools

The above discussion describes several instances in which classical thermodynamics and continuum transport theory based on local thermodynamic equilibrium no longer provide accurate models of system behavior. The obvious question at this point is: What can be done to model systems that exhibit nonclassical or noncontinuum behavior? There are at least three avenues to pursue. The first is to adapt the statistical thermodynamics framework described here to handle nonequilibrium systems. Nonequilibrium statistical mechanics continues to be an area of ongoing research. Much progress has been made, and work in this area continues to develop the analytical machinery that can be applied to nonequilibrium systems.

The second avenue of analysis for nonequilibrium systems involves solving the Boltzmann transport equation. In this chapter we have developed only an approximate analysis for systems that deviate only slightly from equilibrium. Solving the Boltzmann transport equation for a system that departs strongly from equilibrium is a challenging problem, as was recognized by Boltzmann himself. Although Boltzmann derived the equation in 1872, fruitful attempts to solve the equation did not come until much later. In 1905 Lorenz applied Boltzmann's transport theory to the electron gas model of electrical conduction. In 1917 David Enskog and Sydney Chapmann independently developed the mathematical theory of nonequilibrium gases based on Boltzmann's transport theory. Harold Grad solved the Boltzmann equation by expanding in a series of orthogonal polynomials in 1949. Since the 1950s, computer-based methods to solve the Boltzmann transport equation have also been developed and have been used successfully to determine transport in nonequilibrium systems. Computer-based methods for solving the Boltzmann transport equation show promise as a means of analyzing nonequilibrium systems that are far from equilibrium.

The third alternative is to use computer molecular dynamic simulations to predict system thermophysical properties and/or transport in nonequilibrium systems. This category of techniques includes both deterministic molecular dynamic simulations and stochastic Monte Carlo simulation schemes. Deterministic schemes have the advantage that they account for complex molecular interactions in the system in a relatively straightforward way. Stochastic methods use probability distributions derived from kinetic theory to represent the interactions among molecules. The use of such distributions tends to average the characteristics of such interactions but offer improved computational efficiency. With the projected continued increase in accessible computing power, molecular simulation schemes are likely to be used with increasing frequency for analyzing systems that depart strongly from equilibrium conditions. More information on such methods can be obtained from References [19] and [20] at the end of this chapter.

Exercises

10.1 Estimate the mean free path and mean time between collisions for nitrogen gas at atmospheric pressure and 80 K. You may assume that nitrogen molecules have an effective diameter of 3.7 Å.

10.2 The thermal conductivity of nitrogen gas at atmospheric pressure and 300 K is reported to be 0.0259 W/mK. Use the results of the simple kinetic theory developed in Section 10.1 together with the known behavior of ideal gases to estimate the mean effective diameter of a nitrogen molecule.

10.3 If the gas in Example 10.3 is argon at atmospheric pressure and 300 K, estimate the time required for the gas to reach equilibrium if the relaxation time is taken to be that required for each atom to undergo three collisions.

10.4 Derive Eq. (10.150) from Eq. (10.32).

10.5 Film condensation of pure oxygen vapor on a cooled, flat surface occurs at atmospheric pressure. The heat flux removed at the solid wall is a constant and uniform value of $100 \, \text{kW/m}^2$. Assuming an accommodation coefficient of 1.0, estimate the temperature difference across the interface during this process.

10.6 A thin film of liquid nitrogen flows along the walls of the evaporator tube with mostly vapor flowing in the central core of the tube. The pressure in the tube at the location of interest is 208 kPa. A heat flux of $50 \, \text{kW/m}^2$ is applied to the outer surface of the tube and is conducted through the tube wall and the liquid film to the interface. For vaporization at this heat flux level, estimate the temperature difference across the interface if the accommodation coefficient is taken to be 0.9.

10.7 Determine the equilibrium vapor pressure for nitrogen vapor in equilibrium with droplets having a diameter of $0.5 \, \mu\text{m}$ at 80 K.

10.8 From Eq. (10.202), the ideal gas law, and the Young–Laplace equation, derive Eqs. (10.205) and (10.206).

10.9 In superheated water at atmospheric pressure and 460 K, spontaneous fluctuations create a bubble with a radius of 4.0×10^{-8} m. What do you expect to happen to this bubble? Briefly explain your answer. Properties of water at 460 K are: $\hat{v}_l = 0.0205 \, \text{m}^3/\text{kmol}$, $P_{\text{sat}} = 1{,}172$ kPa, and $\sigma_i = 0.0407$ N/m.

10.10 Estimate the number of molecules needed to make up a bubble in equilibrium with superheated liquid nitrogen at atmospheric pressure and 100 K.

10.11 A water droplet with a radius of 2.0 nm (nanometers) spontaneously forms in supersaturated steam at atmospheric pressure and 80°C. What do you expect to happen to this bubble? Briefly explain your answer. Properties of water at 80°C are: $\hat{v}_l = 0.0185 \, \text{m}^3/\text{kmol}$, $P_{\text{sat}} = 47.4$ kPa, and $\sigma_i = 0.0629$ N/m.

10.12 In the last stages of the vaporization process inside the tubes of an evaporator, the two-phase flow consists of $5 \, \mu\text{m}$ diameter droplets of R-134a liquid entrained in the flowing vapor. If the vapor were in equilibrium with the droplets at a vapor pressure of 315 kPa, by how much would the equilibrium temperature differ from the standard (flat interface) saturation temperature?

10.13 Droplets of liquid water in atmospheric fog have a mean diameter of $10 \, \mu\text{m}$. If the air temperature is 14°C, estimate the partial pressure of water vapor in the air. (Hint: Treat the air and water vapor as a mixture of ideal gases.)

10.14 Derive the equation for $J_{Q,x}$ at zero current (10.288) from Eqs. (10.284) and (10.285).

10.15 Derive Eq. (10.302) for I_n from Eq. (10.291) using Eqs. (10.295), (10.297), and (10.300).

10.16 Use Eqs. (10.304) and (10.302) to show that Eq. (10.305) is valid to second order in $k_B T/\varepsilon_F$.

10.17 Use the results of the analysis in Section 10.6 to estimate the absolute thermoelectric power for gold at 300 K. Assume that the relaxation time τ_r is independent of energy.

10.18 Equation (10.311) predicts that the absolute thermoelectric power varies linearly with temperature. Assuming that the relaxation time τ_r is independent of energy, compute and plot the variation for copper between 250 and 350 K and compare the results to the following curvefit to experimentally determined values:

$$\hat{\varepsilon}_{\text{Cu}} = 1.83 - 0.0032(300 - T), \tag{10.318}$$

where T is in K and the absolute thermoelectric power is in μV/K.

10.19 Use the results of the analysis in Section 10.6 to estimate the absolute thermoelectric power for aluminum at 300 K. Determine the value of γ that matches the model prediction to the reported value of $-1.66\ \mu V/K$.

10.20 For liquid methane at atmospheric pressure, estimate the size of a bubble in equilibrium with surrounding superheated liquid at which classical theory is expected to fail because large density and energy fluctuations occur in the liquid.

References

[1] Chapman, S. and T. G. Cowling, *The Mathematical Theory of Non-Uniform Gases*, 3rd ed., Cambridge University Press, Cambridge, UK, 1970.

[2] Gombosi, T. I., *Gaskinetic Theory*, Cambridge University Press, Cambridge, UK, 1994.

[3] Hirschfelder, J. O., C. F. Curtiss, and R. B. Bird, *Molecular Theory of Gases and Liquids*, J. Wiley & Sons, New York, 1954.

[4] Adamson, A. W., *Physical Chemistry of Surfaces*, 4th ed., J. Wiley & Sons, New York, 1982.

[5] Miller, C. A. and P. Neogi, *Interfacial Phenomena*, Marcel Dekker, New York, 1985.

[6] Schrage, R. W., *A Theoretical Study of Interphase Mass Transfer*, Columbia University Press, New York, 1953.

[7] Wilhelm, D. J., "Condensation of Metal Vapors and the Kinetic Theory of Condensation," Argonne National Laboratory Report ANL-6948, 1964.

[8] Paul, B., "Compilation of Evaporation Coefficients," *ARS Journal*, 32:1321, 1962.

[9] Mills, A. F., "The Condensation of Steam at Low Pressures," Technical Report NSF GP-2520, series no. 6, issue no. 39, Space Sciences Laboratory, University of California at Berkeley, 1965.

[10] Yasuoka, K., M. Matsumoto, and Y. Kataoka, "Molecular Simulation of Evaporation and Condensation I. Self Condensation and Molecular Exchange," *Proc. ASME/JSME Thermal Enginerring Joint Conf.*, 2:459, 1995.

[11] Zeldovich, Y. B., "On the Theory of New Phase Formation: Cavitation," *Acta Physiochem. URSS*, 18:1, 1943.

[12] Kagen, Y., "The Kinetics of Boiling of a Pure Liquid," *Russian J. Phys. Chem.*, 34:42, 1960.

[13] Carey, V. P., *Liquid–Vapor Phase-Change Phenomena*, Taylor and Francis, New York, 1992.

[14] Debenedetti, P. G., *Metastable Liquids, Concepts and Principles*, Princeton University Press, Princeton, NJ, 1996.

[15] Potash, M., Jr. and P. C. Jr. Wayner, "Evaporation from a Two-Dimensional Extended Meniscus," *Int. Journal of Heat and Mass Transfer*, 15:1851, 1972.

[16] Deryagin, B. V. and A. M. Zorin, "Optical Study of the Adsorption and Surface Condensation of Vapors in The Vacinity of Saturation on a Smooth Surface," *Proc. 2nd Int. Congress on Surface Activity* (London), 2:145, 1957.

[17] Kittel, C., *Introduction to Solid State Physics*, 6th ed., John Wiley & Sons, New York, 1986.

[18] Ziman, J. M., *Principles of the Theory of Solids*, 2nd ed., Cambridge University Press, 1972.

[19] Haile, J. M., *Molecular Dynamics Simulation, Elementary Methods*, John Wiley & Sons, New York, 1992.

[20] Frenkel, D. and B. Smit, *Understanding Molecular Simulation, From Algorithms to Applications*, Academic Press, San Diego, CA, 1996.

Some Mathematical Fundamentals

I.1   Probability Distributions

Let u be a variable that can assume M discrete values $u_1, u_2, \ldots, u_M$ with corresponding probabilities $\tilde{P}(u_1), \tilde{P}(u_2), \ldots, \tilde{P}(u_M)$. The variable u is said to be a *discrete variable* and $\tilde{P}(u)$ is said to be a *discrete distribution*. The mean value of u is computed as

$$\langle u \rangle = \frac{\sum_{j=1}^{M} u_j \tilde{P}(u_j)}{\sum_{j=1}^{M} \tilde{P}(u_j)}. \tag{I.1}$$

Since the probability distribution must be normalized,

$$\sum_{j-1}^{M} \tilde{P}(u_j) = 1, \tag{I.2}$$

the relation for the mean value of u reduces to

$$\langle u \rangle = \sum_{j=1}^{M} u_j \tilde{P}(u_j). \tag{I.3}$$

The mean of any function $f(u)$ is given by

$$\langle f(u) \rangle = \sum_{j=1}^{M} f(u_j) \tilde{P}(u_j). \tag{I.4}$$

If the random variable u is continuous rather than discrete, we define $\tilde{P}(u)du$ as the probability that the random variable u lies between u and $u + du$. If the probability density function $\tilde{P}(u)$ is normalized so that

$$\int_{u_{\min}}^{u_{\max}} \tilde{P}(u)\,du = 1 \tag{I.5}$$

the mean value of u is given by

$$\langle u \rangle = \int_{u_{\min}}^{u_{\max}} u \tilde{P}(u)\,du, \tag{I.6}$$

where the integration limits in Eqs. (I.5) and (I.6) imply integration over the entire accessible range of u values. In a similar fashion, the mean value of a function of u is computed as

$$\langle f(u) \rangle = \int_{u_{\min}}^{u_{\max}} f(u) \tilde{P}(u)\,du, \tag{I.7}$$

where again integration is executed over the entire range of u.

391

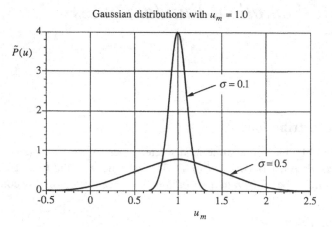

Figure I.1

An important and frequently encountered continuous probability distribution is the *Gaussian distribution* defined as

$$\tilde{P}(u) = \frac{1}{\sqrt{2\pi\sigma^2}} \exp\left\{-\frac{(u - u_{\mathrm{m}})^2}{2\sigma^2}\right\}, \tag{I.8}$$

where u_{m} is the mean value of u computed using Eq. (I.6). The quantity σ^2, termed the *variance*, dictates the width of the Gaussian distribution. As indicated in Figure I.1, the smaller the value of σ, the narrower the peak of the distribution. In the limit $\sigma \to 0$, the Gaussian distribution becomes a delta function.

Further discussion of probability distributions and their use can be found in References [1] and [2].

I.2   Some Results of Combinatorial Analysis

The development of statistical thermodynamics presented in this text makes heavy use of results from a branch of mathematics known as combinatorial analysis. Derivation of these results is beyond the scope of this text. However, in this section some useful results are summarized and examples are given to illustrate their use.

For N distinguishable objects

$$\left\{\begin{array}{l} \textit{the number of ways of arranging} \\ \textit{N distinguishable objects} \end{array}\right\} = N! = N(N-1)(N-2)\cdots(2)(1) \tag{I.9}$$

and

$$\left\{\begin{array}{l} \textit{the number of ways of putting N distinguishable} \\ \textit{objects into r distinguishable boxes so that there} \\ \textit{are } N_1 \textit{ in box 1, } N_2 \textit{ in box 2, } \ldots, N_r \textit{ in box r, and} \\ \textit{order in any box does not matter} \end{array}\right\} = \frac{N!}{(N_1!)(N_2!)\cdots(N_r!)}. \tag{I.10}$$

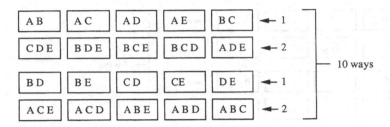

Figure I.2

To illustrate use of Eq. (I.10), suppose we have objects A, B, C, D, and E and we are to put them in two boxes (1 and 2) so that there are two in box one and three in box two. Thus $N = 5$, $N_1 = 2$, and $N_2 = 3$ and

$$\{the\ number\ of\ ways\} = \frac{5 \cdot 4 \cdot 3 \cdot 2 \cdot 1}{(2 \cdot 1)(3 \cdot 2 \cdot 1)} = 10.$$

The ten different ways are illustrated in Figure I.2.

Additional useful results for distinguishable objects are

$$\left\{ \begin{array}{l} the\ number\ of\ ways\ of\ selecting\ N\ distinguishable \\ objects\ from\ a\ set\ of\ G\ distinguishable\ objects \end{array} \right\} = \frac{G!}{(N!)(G - N)!} \quad \text{(I.11)}$$

and

$$\left\{ \begin{array}{l} the\ number\ of\ ways\ of\ N\ distinguishable\ objects \\ can\ be\ put\ in\ G\ distinguishable\ boxes \end{array} \right\} = G^N. \quad \text{(I.12)}$$

To illustrate use of Eq. (I.11), suppose we have objects A, B, C, D, and E and we are to select three objects. Thus $G = 5$, $N = 3$, and

$$\{the\ number\ of\ ways\} = \frac{5 \cdot 4 \cdot 3 \cdot 2 \cdot 1}{(2 \cdot 1)(3 \cdot 2 \cdot 1)} = 10.$$

The ten different combinations are

$$ABC, ABD, ABE, ACD, ACE, ADE, BCD, BCE, BDE, CDE.$$

To illustrate use of Eq. (I.12), we consider objects A, B, and C, which we are to put in box number 1 and box number 2. Thus $G = 2$, $N = 3$, and

$$\{the\ number\ of\ ways\} = (2)^3 = 8.$$

The eight different ways are illustrated in Figure I.3.

A useful result for indistinguishable objects is

$$\left\{ \begin{array}{l} the\ number\ of\ ways\ of\ N\ indistinguishable\ objects \\ can\ be\ put\ in\ G\ distinguishable\ boxes \end{array} \right\} = \frac{(N + G - 1)!}{N!(G - 1)!}. \quad \text{(I.13)}$$

To illustrate use of this relation we consider four indistinguishable objects A, A, A, A, which we are to put in two boxes. It follows that $N = 4$, $G = 2$, and

$$\{the\ number\ of\ ways\} = \frac{5 \cdot 4 \cdot 3 \cdot 2 \cdot 1}{(4 \cdot 3 \cdot 2 \cdot 1)(1)} = 5.$$

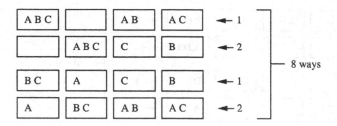

Figure I.3

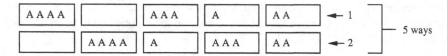

Figure I.4

The five different ways of putting the four objects into the two boxes are illustrated in Figure I.4.

More detailed discussion of the basic elements and application of combinatorial analysis can be found in References [3] and [4].

I.3   The Method of Lagrange Multipliers

We wish to address the problem of maximizing a function of several variables $f(x_1, x_2, \ldots, x_r)$ when other constraints, expressible as $g_1(x_1, x_2, \ldots, x_r) = 0$, $g_2(x_1, x_2, \ldots, x_r) = 0$, etc., are imposed. If it were not for the constraints, the maximum of $f(x_1, x_2, \ldots, x_r)$ is dictated by the requirement that for arbitrary changes in the independent variables δx_j

$$\delta f = \sum_{j=1}^{r} \left(\frac{\partial f}{\partial x_j} \right)_0 \delta x_j = 0. \tag{I.14}$$

In the above relation, the zero subscript indicates that the equation equals zero only when all partial derivatives are evaluated at the maximum (or minimum) of f.

If we have an additional constraint $g_1(x_1, x_2, \ldots, x_r) = 0$, the constraint relation can be differentiated to obtain

$$\delta g_1 = \sum_{j=1}^{r} \left(\frac{\partial g_1}{\partial x_j} \right)_0 \delta x_j = 0. \tag{I.15}$$

We next multiply both sides by a constant λ_1 and subtract the resulting relation from Eq. (I.14) to obtain

$$\sum_{j=1}^{r} \left(\frac{\partial f}{\partial x_j} - \lambda_1 \frac{\partial g_1}{\partial x_j} \right)_0 \delta x_j = 0. \tag{I.16}$$

Since this must hold at any combination of δx_j values, it follows that

$$\left(\frac{\partial f}{\partial x_j} \right)_0 - \lambda_1 \left(\frac{\partial g_1}{\partial x_j} \right)_0 = 0, \qquad 1 \leq j \leq r. \tag{I.17}$$

We thus have a set of r equations that, in principle, can be solved for r values of the x_j at the extremum: $x_j = x_{j,0}$. The values of $x_{j,0}$ thus obtained will depend on λ_1. In practice, the multiplier value is usually dictated by physical requirements of some sort.

Lagrange's method is easily extended to multiple constraints. Expressing the constraints as $g_i(x_1, x_2, \ldots, x_r) = 0$, we differentiate to yield for each

$$\delta g_i = \sum_{j=1}^{r} \left(\frac{\partial g_i}{\partial x_j} \right)_0 \delta x_j = 0. \tag{I.18}$$

Multiplying each relation of this type by λ_i and subtracting it from the δf relation yields

$$\sum_{j=1}^{r} \left[\frac{\partial f}{\partial x_j} - \sum_i \lambda_i \left(\frac{\partial g_i}{\partial x_j} \right) \right]_0 \delta x_j = 0. \tag{I.19}$$

It follows that the $x_{j,0}$ values at the extremum are determined by simultaneous solution of the equations

$$\frac{\partial f}{\partial x_j} - \sum_i \lambda_i \frac{\partial g_i}{\partial x_j} = 0. \tag{I.20}$$

As in the single-constraint case, the multipliers λ_i will generally be dictated by physical considerations.

I.4 Stirling's Approximation

Because $N! = N(N-1)(N-2)\cdots(2)(1)$, $\ln N!$ is given by

$$\ln N! = \sum_{m=1}^{N} \ln m. \tag{I.21}$$

In the plot in Figure I.5, the summation in Eq. (I.21) is equal to the sum of the areas in the rectangles. The function $\ln x$ forms an envelope to the rectangles such that the area under the curve approximates the area in the rectangles. This approximation gets more accurate as N and x increase. This implies that

$$\ln N! = \sum_{m=1}^{N} \ln m \cong \int_1^N \ln x \, dx = [x \ln x - x]_1^N. \tag{I.22}$$

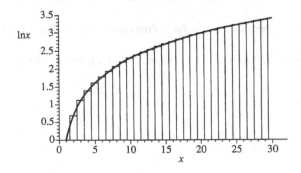

Figure I.5

Neglecting 1 compared to N for large N, we find Stirling's approximation for large N:

$$\ln N! \cong N \ln N - N. \tag{I.23}$$

I.5   Relations among Partial Derivatives

In our thermodynamics analysis, we are frequently interested in functional relations among three variables such as $x = x(y, z)$. For differential changes in the variable, it follows from basic calculus that

$$dx = \left(\frac{\partial x}{\partial y}\right)_z dy + \left(\frac{\partial x}{\partial z}\right)_y dz. \tag{I.24}$$

If the relation among the three variables can be inverted to the form $y = y(x, z)$, it follows from the same reasoning that

$$dy = \left(\frac{\partial y}{\partial x}\right)_z dx + \left(\frac{\partial y}{\partial z}\right)_x dz. \tag{I.25}$$

Substituting the right side of (I.25) into (I.24) for dy and rearranging yields

$$\left[\left(\frac{\partial x}{\partial y}\right)_z \left(\frac{\partial y}{\partial x}\right)_z - 1\right] dx + \left[\left(\frac{\partial x}{\partial y}\right)_z \left(\frac{\partial y}{\partial z}\right)_x + \left(\frac{\partial x}{\partial z}\right)_y\right] dz = 0. \tag{I.26}$$

Equation (I.26) must hold for arbitrary values of two of the three variables x, y, and z. If we choose x and z as the independent variables, Eq. (I.26) must hold for arbitrary dx and dz. This is true only if the terms in square brackets in Eq. (I.26) are identically zero. Setting the expressions in square brackets equal to zero, we obtain the following two relations among partial derivatives that must be satisfied:

$$\left(\frac{\partial x}{\partial y}\right)_z \left(\frac{\partial y}{\partial x}\right)_z = 1, \tag{I.27}$$

$$\left(\frac{\partial x}{\partial y}\right)_z \left(\frac{\partial y}{\partial z}\right)_x \left(\frac{\partial z}{\partial x}\right)_y = -1. \tag{I.28}$$

References

[1] Kreysig, E., *Advanced Engineering Mathematics*, 5th ed., John Wiley & Sons, New York, 1983.

[2] J. L. Devore, *Probability and Statistics for Engineering and the Sciences*, 3rd ed., Duxbury Press, Belmont, CA, 1991.

[3] Riordan, J., *An Introduction to Combinatorial Analysis*, Princeton University Press, Princeton, NJ, 1980.

[4] Slomson, A. B., *An Introduction to Combinatorics*, Chapman and Hall, London, 1991.

Physical Constants and Prefix Designations

Table II.1 *Physical Constants*[a]

Avogadro's number, $N_A = 6.02252 \times 10^{26}$ kmol^{-1}
Boltzmann's constant, $k_B = 1.38054 \times 10^{-23}$ J/K
Electron rest mass, $m_e = 9.1091 \times 10^{-31}$ kg
Electron charge, $e_e = 1.6021 \times 10^{-19}$ coulomb
Earth normal gravitational acceleration, $g_e = 9.80665$ m/s^2
Planck's constant, $h = 6.6256 \times 10^{-34}$ J·s
Stephan–Boltzmann constant, $\sigma_{SB} = 5.6697 \times 10^{-8}$ W/m^2·K^4
Universal gas constant, $R(= N_A/k_B) = 8.3143$ kJ/kmol·K
Atomic mass unit, $u_{am} = 1.66043 \times 10^{-27}$ kg
Speed of light in vacuum, $c_l = 2.997925 \times 10^8$ m/s

[a] *Source:* "New Values for the Physical Constants," *Phys. Today*, 17:48–9, Feb. 1964.

Table II.2 *Prefix Designations*

Fraction	Prefix	Abbreviation
10^{-1}	deci	d
10^{-2}	centi	c
10^{-3}	milli	m
10^{-6}	micro	μ
10^{-9}	nano	n
10^{-12}	pico	p
10^{-15}	femto	f
10^{-18}	atto	a

Multiple	Prefix	Abbreviation
10	deka	da
10^2	hecto	h
10^3	kilo	k
10^6	mega	M
10^9	giga	G
10^{12}	tera	T
10^{15}	peta	P
10^{18}	exa	E

Thermodynamic Properties of Selected Materials

Figures and tables in this appendix are reprinted by permisson of the American Society of Heating, Refrigerating and Air-Conditioning Engineers, Atlanta, Georgia, from the 1997 *ASHRAE Handbook – Fundamentals* (SI edition).

Table III.1 *Argon – Properties of Saturated Liquid and Vapor*

Temp. K	Press. MPa	Density kg/m³ Liquid	Volume m³/kg Vapor	Enthalpy kJ/kg		Entropy kJ/(kg K)	
				Liquid	Vapor	Liquid	Vapor
**83.80	0.068950	0.24746	1415.0	−121.82	42.417	1.3250	3.2848
85	0.078975	0.21846	1407.8	−120.46	42.892	1.3410	3.2628
90	0.13362	0.13476	1377.0	−114.74	44.735	1.4060	3.1779
95	0.21321	0.08758	1345.2	−108.98	46.317	1.4676	3.1023
100	0.32401	0.05938	1312.0	−103.15	47.594	1.5266	3.0340
105	0.47259	0.04167	1277.3	−97.202	48.520	1.5835	2.9713
110	0.66575	0.03006	1240.8	−91.099	49.943	1.6388	2.9128
115	0.91049	0.02217	1202.0	−84.795	49.100	1.6931	2.8574
120	1.2139	0.01663	1160.4	−78.234	48.610	1.7467	2.8038
125	1.5835	0.01263	1115.4	−71.340	47.464	1.8004	2.7508
130	2.0269	0.009661	1065.6	−63.998	45.495	1.8548	2.6970
135	2.5527	0.007393	1009.2	−56.016	42.431	1.9112	2.6404
140	3.1704	0.005600	941.94	−47.017	37.749	1.9720	2.5775
145	3.8921	0.004103	853.13	−36.042	30.180	2.0433	2.5000
*150.73	4.865	0.00187	535.	1.87	1.87	2.27	2.27

** Triple point.
* Critical point.

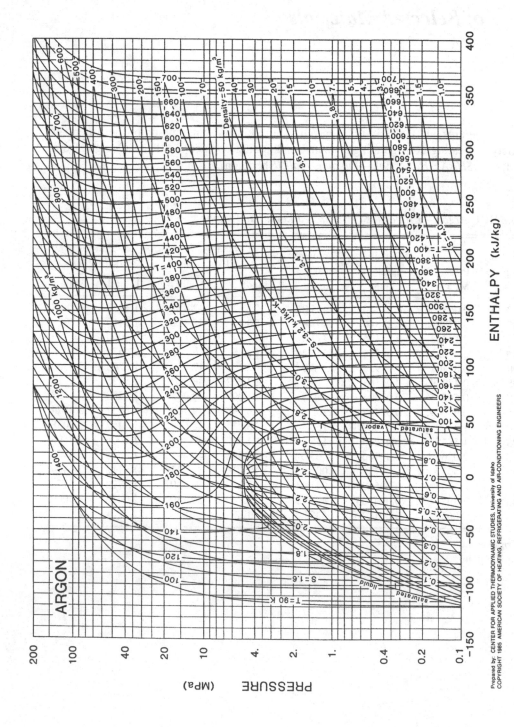

Figure III.1

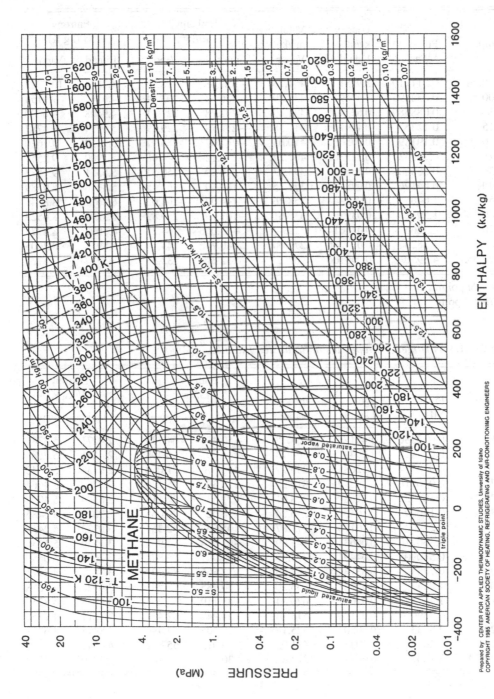

ENTHALPY (kJ/kg)

PRESSURE (MPa)

Figure III.2

Table III.2 *Methane – Properties of Saturated Liquid and Saturated Vapor*

Temp. K	Press. MPa	Density kg/m³ Liquid	Volume m³/kg Vapor	Enthalpy kJ/kg Liquid	Vapor	Entropy kJ/(kg K) Liquid	Vapor
**90.68	0.011719	3.9781	451.23	−357.68	185.75	4.2894	10.2823
100	0.034495	1.4769	438.89	−326.63	203.44	4.6147	9.9154
110	0.088389	0.6234	424.89	−292.28	221.11	4.9408	9.6080
115	0.12357	0.42971	417.58	−274.79	229.29	5.0954	9.4787
120	0.19189	0.30565	410.05	−247.07	236.97	5.2450	9.3620
125	0.26933	0.22330	402.27	−239.12	244.06	5.3900	9.2555
130	0.36800	0.16688	394.23	−220.89	250.51	5.5311	9.1572
140	0.64196	0.098424	377.15	−183.40	261.15	5.8036	8.9789
150	1.0405	0.061244	358.26	−144.02	268.02	6.0677	8.8146
160	1.5918	0.039387	336.61	−101.79	269.82	6.3299	8.6525
170	2.3266	0.025614	310.47	−55.066	264.21	6.5992	8.4773
180	3.2820	0.016266	276.00	−0.242	245.79	6.8937	8.2605
*190.555	4.5950	0.006166	262.2	132.3	132.3	7.572	7.572

** Triple point.
* Critical point.

Table III.3 *Nitrogen – Properties of Saturated Liquid and Vapor*

Temp. K	Press. MPa	Density kg/m³ Liquid	Volume m³/kg Vapor	Enthalpy kJ/kg Liquid	Vapor	Entropy kJ/(kg K) Liquid	Vapor
**63.15	0.012530	1.4817	867.78	−150.45	64.739	2.4271	5.8381
65	0.017418	1.0942	860.78	−146.79	66.498	2.4841	5.7688
70	0.038584	0.52685	840.77	−136.67	71.058	2.6338	5.6042
75	0.076116	0.28217	819.22	−126.39	75.275	2.7750	5.4664
77.35	0.101325	0.21680	808.61	−121.53	77.113	2.8384	5.4090
80	0.13699	0.16409	796.24	−116.02	79.065	2.9078	5.3486
85	0.22903	0.10174	771.87	−105.56	82.334	3.0333	5.2455
90	0.36071	0.06631	745.99	−94.914	84.982	3.1530	5.1531
95	0.54090	0.04491	718.38	−83.991	86.890	3.2684	5.0680
100	0.77886	0.03132	688.65	−72.666	87.901	3.3811	4.8973
105	1.0842	0.02228	656.20	−60.785	87.791	3.4926	4.9078
110	1.4671	0.01602	620.04	−48.119	86.203	3.6048	4.8258
115	1.9390	0.01150	578.14	−34.247	82.471	3.7211	4.7358
120	2.5133	0.008035	525.12	−18.105	74.996	3.8495	4.6251
125	3.2099	0.004863	431.03	6.015	55.882	4.0342	4.4331
*126.20	3.400	0.003184	314.0	30.70	30.70	4.2270	4.2270

** Triple point.
* Critical point.

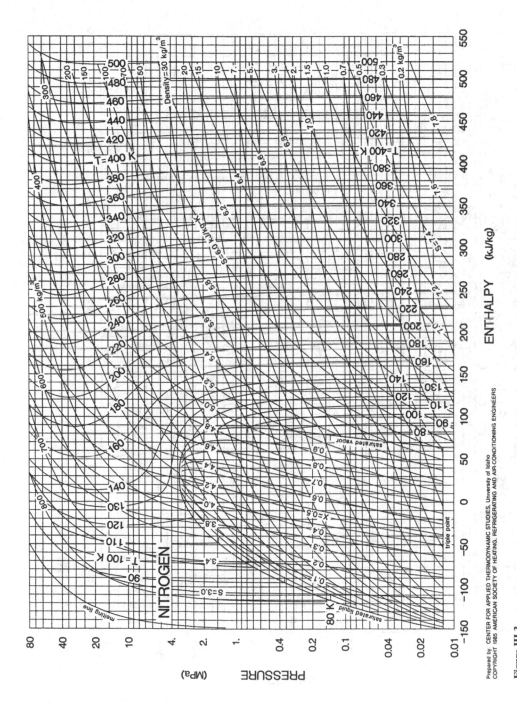

Prepared by CENTER FOR APPLIED THERMODYNAMIC STUDIES, University of Idaho
COPYRIGHT 1985 AMERICAN SOCIETY OF HEATING, REFRIGERATING AND AIR-CONDITIONING ENGINEERS

Figure III.3

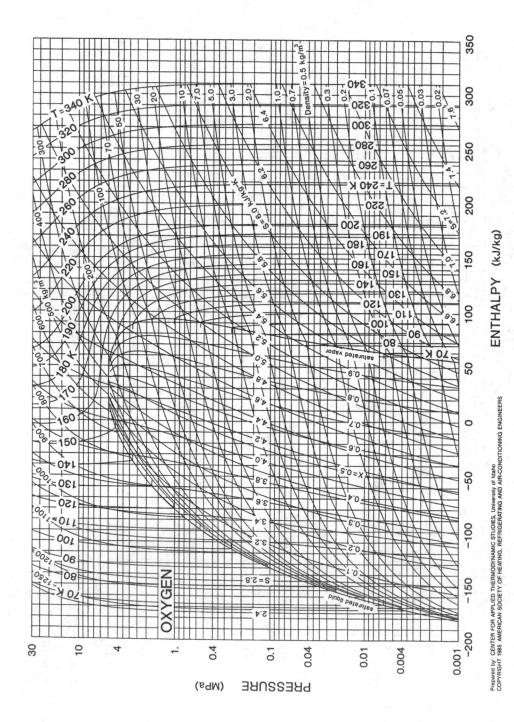

Figure III.4

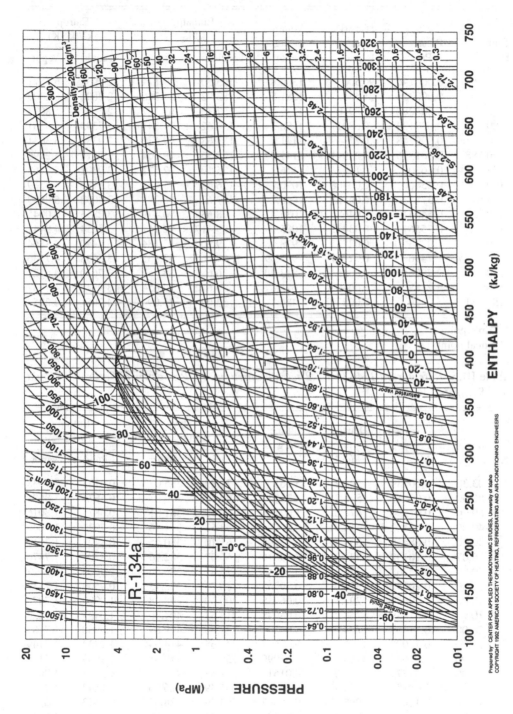

Prepared by: CENTER FOR APPLIED THERMODYNAMIC STUDIES, University of Idaho
COPYRIGHT 1992 AMERICAN SOCIETY OF HEATING, REFRIGERATING AND AIR-CONDITIONING ENGINEERS

Figure III.5

Table III.4 *Oxygen – Properties of Saturated Liquid and Vapor*

Temp. K	Press. MPa	Density kg/m³ Liquid	Volume m³/kg Vapor	Enthalpy kJ/kg Liquid	Enthalpy kJ/kg Vapor	Entropy kJ/(kg K) Liquid	Entropy kJ/(kg K) Vapor
**54.36	0.000146	96.500	1306.8	−193.50	49.174	2.0931	6.5571
60	0.000726	21.460	1281.5	−183.92	54.266	2.2608	6.2303
70	0.006262	2.8910	1236.6	−167.35	63.127	2.5161	5.8083
80	0.030124	0.68011	1190.2	−150.62	71.541	2.7393	5.5161
90	0.099348	0.22749	1141.9	−133.68	79.170	2.9381	5.3031
90.188	0.101325	0.22341	1141.0	−133.36	79.304	2.9417	5.2996
95	0.16308	0.14418	1116.7	−125.10	82.572	3.0302	5.2163
100	0.25400	0.09568	1090.7	−116.43	85.629	3.1183	5.1389
105	0.37853	0.06593	1063.7	−107.62	88.285	3.3032	5.0689
110	0.54341	0.04685	1035.4	−98.634	90.479	3.2853	5.0045
115	0.75561	0.03412	1005.5	−89.417	92.141	3.3654	4.9441
120	1.0223	0.02535	973.78	−79.909	93.184	3.4440	4.8864
130	1.7490	0.01457	902.39	−59.663	92.932	3.5998	4.7736
140	2.7876	0.008541	813.20	−36.684	88.067	3.7610	4.6521
150	4.2188	0.004660	675.87	−6.672	72.654	3.9543	4.4832
*154.581	5.0429	0.002293	436.1	33.09	33.09	4.205	4.205

** Triple point.
* Critical point.

Table III.5 *Refrigerant 134a (1,1,1,2-tetrafluoroethane) – Properties of Saturated Liquid and Saturated Vapor*

Temp. K	Press. MPa	Density kg/m³ Liquid	Volume m³/kg Vapor	Enthalpy kJ/kg Liquid	Enthalpy kJ/kg Vapor	Entropy kJ/(kg K) Liquid	Entropy kJ/(kg K) Vapor
**−103.30	0.00039	1591.2	35.263	71.89	335.07	0.4143	1.9638
−80.00	0.00369	1526.2	4.2504	99.65	349.03	0.5674	1.8585
−60.00	0.01594	1471.0	1.0770	123.96	361.51	0.6871	1.8016
−40.00	0.05122	1414.8	0.36095	148.57	374.16	0.7973	1.7649
−30.00	0.08436	1385.9	0.22596	161.10	380.45	0.8498	1.7519
−26.07	0.10132	1374.3	0.19016	166.07	382.90	0.8701	1.7476
−20.00	0.13268	1356.2	0.14744	173.82	386.66	0.9009	1.7417
−10.00	0.20052	1325.6	0.09963	186.78	392.75	0.9509	1.7337
0.00	0.29269	1293.7	0.06935	200.00	398.68	1.0000	1.7274
10.00	0.41449	1260.2	0.04948	213.53	404.40	1.0483	1.7224
20.00	0.57159	1224.9	0.03603	227.40	409.84	1.0960	1.7183
30.00	0.77008	1187.2	0.02667	241.65	414.94	1.1432	1.7149
40.00	1.0165	1146.5	0.01999	256.34	419.58	1.1903	1.7115
50.00	1.3177	1102.0	0.01511	271.59	423.63	1.2373	1.7078
60.00	1.6815	1052.4	0.01146	287.49	426.86	1.2847	1.7031
70.00	2.1165	995.6	0.00867	304.29	428.89	1.3332	1.6963
80.00	2.6331	927.4	0.00646	322.41	429.02	1.3837	1.6855
90.00	3.2445	836.9	0.00461	343.01	425.48	1.4392	1.6663
100.00	3.9721	646.7	0.00265	374.02	407.08	1.5207	1.6093
*101.03	4.0560	513.3	0.00195	389.79	389.79	1.5593	1.5593

Table III.5 *Refrigerant 134a Properties (cont'd)*

Temp. °C	Press. MPa	Spec. Heat, c_p kJ/(kg K)		Vapor c_p/c_v	Sound Speed m/s		Viscosity μPa s		Thermal Cond. mW/(m K)		Surface Tension
		Liquid	Vapor	Vapor	Liquid	Vapor	Liquid	Vapor	Liquid	Vapor	mN/m
**−103.30	0.00039	1.147	0.585	1.163	1135.	127.	2186.6	6.63	—	—	28.15
−80.00	0.00369	1.211	0.637	1.151	999.	134.	1109.9	7.57	—	—	24.11
−60.00	0.01594	1.220	0.685	1.146	904.	139.	715.4	8.38	121.1	—	20.81
−40.00	0.05122	1.243	0.740	1.148	812.	144.	502.2	9.20	111.9	8.19	17.66
−30.00	0.08436	1.260	0.771	1.152	765.	145.	430.4	9.62	107.3	9.16	16.13
−26.07	0.10132	1.268	0.784	1.154	747.	146.	406.4	9.79	105.4	9.52	15.54
−20.00	0.13268	1.282	0.805	1.157	719.	146.	373.1	10.05	102.6	10.07	14.63
−10.00	0.20052	1.306	0.842	1.166	672.	147.	326.3	10.49	98.0	10.93	13.16
0.00	0.29269	1.335	0.883	1.178	626.	147.	287.4	10.94	93.4	11.79	11.71
10.00	0.41449	1.367	0.930	1.193	579.	146.	254.3	11.42	88.8	12.66	10.30
20.00	0.57159	1.404	0.982	1.215	532.	145.	225.8	11.92	84.2	13.57	8.92
30.00	0.77008	1.447	1.044	1.244	484.	143.	200.7	12.48	79.6	14.56	7.5
40.00	1.0165	1.500	1.120	1.285	436.	140.	178.2	13.10	75.0	15.64	6.27
50.00	1.3177	1.569	1.218	1.345	387.	137	157.7	13.83	70.4	16.84	5.01
60.00	1.6815	1.663	1.354	1.438	338.	132.	138.6	14.71	64.8	18.19	3.81
70.00	2.1165	1.806	1.567	1.597	287.	126.	120.3	15.85	61.2	19.72	2.67
80.00	2.6331	2.069	1.967	1.917	235.	118.	102.1	17.46	56.6	21.46	1.63
90.00	3.2445	2.766	3.064	2.832	178.	108.	82.6	20.15	—	—	0.72
100.00	3.9721	—	—	—	105.	94.	53.0	28.86	—	—	0.03
*101.03	4.0560	∞	∞	∞	0	0	—	—	∞	∞	0

**Triple point.
*Critical point.

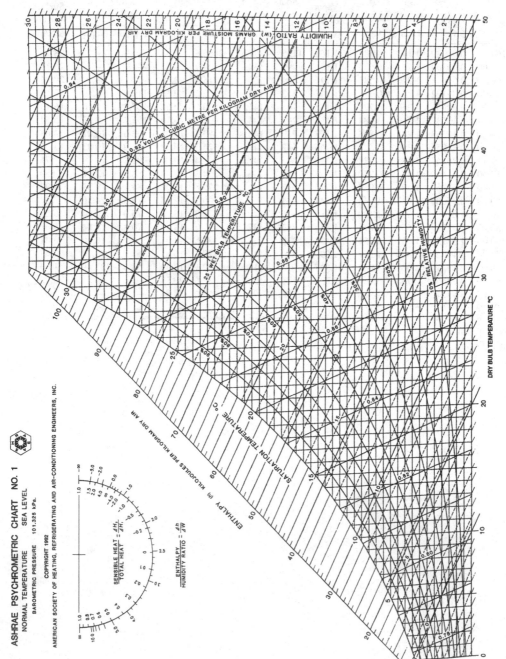

Figure III.6

Typical Force Constants for the Lennard–Jones 6-12 Potential[†]

Constants listed are inferred from second virial coefficients.

Atom or Molecule	ε_0/k_B (K)	σ_{LJ} (Å)[a]
Ar	120	3.41
Ne	35.6	2.75
N_2	95	3.70
O_2	118	3.58
CO	100	3.76
NO	131	3.17
CO_2	198	4.33
CH_4	149	3.78
N_2O	189	4.59
CF_4	153	4.70
$CH_2{=}CH_2$	199	4.52
C_2H_6	243	3.95
C_3H_8	242	5.64
$n\text{-}C_7H_{16}$	282	8.88

[a] $1 Å = 10^{-10}$ m.

[†]Source: Hirschfelder, J. O., Curtiss, C. F., and Bird, R. B., *Molecular Theory of Gases and Liquids*, John Wiley, New York, 1954.

409

Index

411

Printed in the United States
By Bookmasters